physica status solidi

AKADEMIE-VERLAG · BERLIN

VOLUME 6 · NUMBER 2 · 1964

Contents

physica status solidi

Volume 6 · Number 2 · Pages 275 to 568 and K 69 to K 128

August 1, 1964

AKADEMIE-VERLAG · BERLIN

Subscriptions and orders for single copies should be addressed to

AKADEMIE-VERLAG GmbH, Berlin W 8, Leipziger Straße 3 – 4 or to Buchhandlung KUNST UND WISSEN, Erich Bieber, 7 Stuttgart S, Wilhelmstr. 4 – 6 or to Deutsche Buch-Export und -Import GmbH, Leipzig C 1, Postschließfach 276

Editorial Note:

"physica status solidi" undertakes that an original paper *accepted for publication* before the 8th of any month will be published within 50 days of this date unless the author requests a postponement. In special cases there may be some delay between *receipt* and *acceptance* of a paper due to the review and, if necessary, revision of the paper.

Schriftleiter und verantwortlich für den Inhalt: Professor Dr. Dr. h. c. P. Görlich, Berlin C 2, Neue Schönhauser Str. 20 bzw. Jena, Humboldtstr. 26. Redaktionskollegium: Dr. S. Oberländer, Dr. E. Gutsche, Dr. W. Borchardt. Anschrift der Schriftleitung: Berlin C 2, Neue Schönhauser Str. 20, Fernruf: 426788. Verlag: Akademie-Verlag GmbH, Berlin W 8, Leipziger Str. 3–4, Fernruf: 220441, Telex-Nr. 011773, Postscheckkonto: Berlin 35021. — Die Zeitschrift „physica status solidi" erscheint jeweils am 1. des Monats. Bezugspreis eines Bandes DM 60,—. Bestellnummer dieses Bandes: 1068/6. Für 1964 sind 4 Bände vorgesehen. Gesamtherstellung: VEB Druckerei „Thomas Müntzer" Bad Langensalza. — Veröffentlicht unter der Lizenznummer 1310 des Presseamtes beim Vorsitzenden des Ministerrates der Deutschen Demokratischen Republik.

Review Article

phys. stat. sol. 6, 277 (1964)

Jena, Carl-Zeiß-Str. 1

Spectroscopic Properties of Activated Laser Crystals (II)

By

P. GÖRLICH, H. KARRAS, G. KÖTITZ, and R. LEHMANN

4.1.2 Experimental data

4.1.2.1 Cerium, Z = 58

	Ce^{3+}	Ce^{2+}
Electron configuration	$4f^1$	—
Ground state	$^2F_{5/2}$	—

The Ce^{3+} ion is distinguished by its $4f$ shell having only one electron, which theoretically can only be in the ground state $^2F_{5/2}$ or in the $^2F_{7/2}$ state. For free ions the separation of these levels amounts to 2253 cm^{-1}, which corresponds to a wavelength of about 4.44 μm. FEOFILOV [48, 55][1]) succeeded in detecting this weak forbidden transition on CaF_2:Ce^{3+} crystals (10^{-4} to 10^{-2} per cent by weight Ce^{3+}) of type I and type II (Fig. 6); this was verified by MANDEL and co-authors

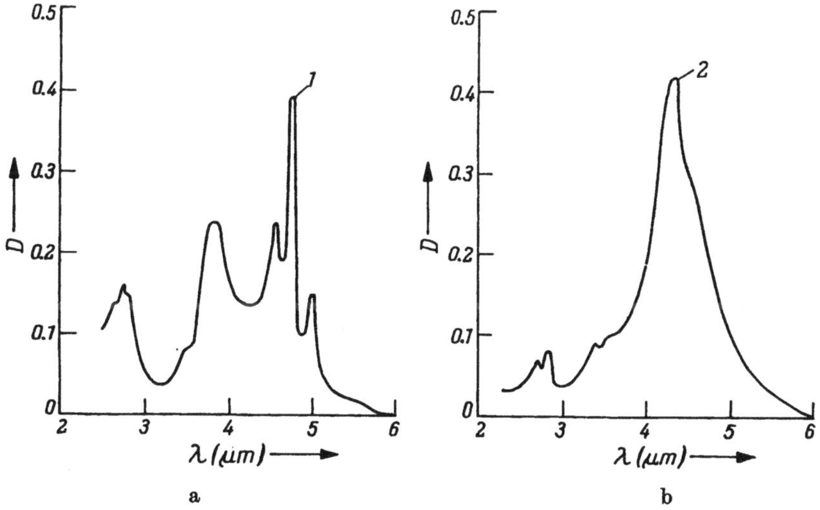

Fig. 6. Infrared absorption spectra of CaF_2:Ce single crystals. a) type I, b) type II; $T = 20\,°C$ (after FEOFILOV [48])

[1]) The references [1] to [75] have been published in Part I of this article (phys. stat. sol. 5, 437 (1964)).

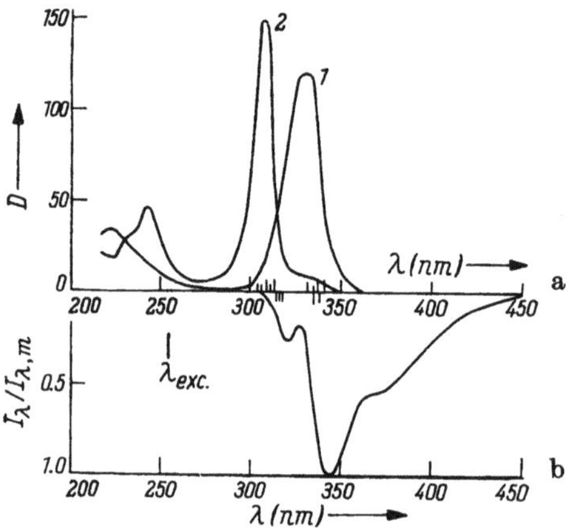

Fig. 7. Absorption (a) and luminescent (b) spectra of CaF_2: Ce single crystals in the ultraviolet region. 1: type I, 2: type II; $T = 20\,°C$. The line segments around the horizontal axis mark the appearance of narrow lines in the cooling of the samples; $\lambda_{exc} = 254$ nm (after FEOFILOV [48])

[76] on CdF_2: Ce^{3+} (CdF_2 having likewise fluorite structure). The oscillator strength was determined by them to be 2×10^{-6}. A fluorescence ($\lambda \sim 5$ to 6 μm) resulting from this $4f - 4f$ transition has to date not been detected. Thus, apart from Pm and Lu, Ce^{3+} is the only ion of the rare earths, on which no $4f - 4f$ fluorescence has been observed. In absorption [77, 78] as well as in fluorescence [43, 55] the occurring wavelengths of the $4f - 5d$ transitions (their intensities being higher by 2 to 3 orders of magnitude) lie in the ultraviolet region (Fig. 7). On cooling down to 77 or 4.2 °K these bands are splitting up into a multiplicity of lines. More detailed data on this as well as on Ce^{3+} in SrF_2 and BaF_2 were recently presented by KAPLYANSKII and co-authors [79]. The occuring lines are assigned to transitions between the energetically lowest d states and the $^2F_{7/2}$ or $^2F_{5/2}$ level, respectively.

4.1.2.2, Praseodymium Z = 59

	Pr^{3+}	Pr^{2+}
Electron configuration	$4f^2$	$4f^3$
Ground state	$^3H_4$	$^4I_{9/2}$

According to STEPANOV and FEOFILOV Pr^{3+} in CaF_2 fluoresces in the form of lines at about 450 nm as well as in the region between 600 and 650 nm. YARIV and co-workers [14] were able to detect a strong fluorescence of $CaWO_4$: Pr^{3+} at about 1.05 μm and achieved laser emission at 1.0468 μm. After Sm^{2+} and Nd^{3+}, Pr^{3+} was the third element of the rare earths showing stimulated emission.

$CaWO_4$: Pr^{3+} crystals (0.05%) develop a light green colour caused by the three absorption bands in the region between 400 nm and 490 nm (Fig. 8). The nomenclature of the absorption transitions and the energy level scheme for Pr^{3+}, which was adopted by YARIV and co-authors for Pr^{3+} in $CaWO_4$ (Fig. 9), were suggested by DIEKE and SARUP [80]. That part which is of importance for the explanation of the laser emission is shown in Fig. 10. The transitions between the ground state $^3H_4$ and the excited states $^3P_0$, $^3P_1$, and $^3P_2$ serve for pumping (Fig. 8). The width of these multi-line multiplet transitions is 50 to 70 cm^{-1} and oscillator strengths

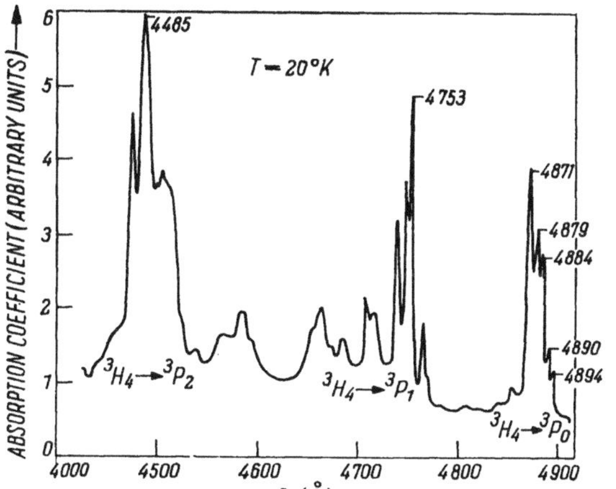

Fig. 8. Absorption curve for the $^3H_4 \rightarrow {^3P_{0,1,2}}$ transitions in $CaWO_4$: Pr^{3+} at 20 °K (after YARIV and co-authors [14])

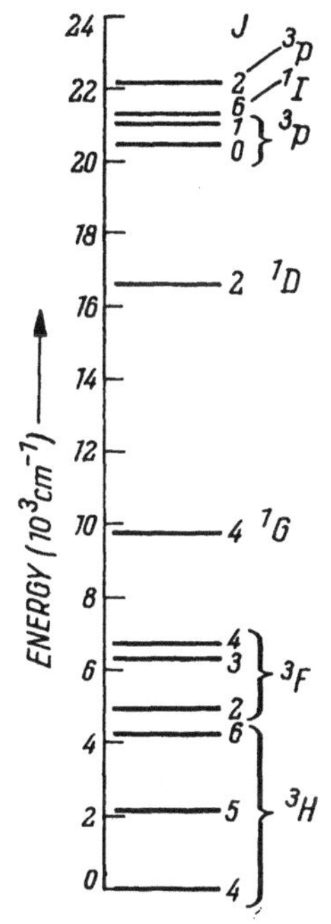

Fig. 9. The energy levels of Pr^{3+} in $PrCl_3$. No Stark splitting is shown (after DIEKE and SARUP [80]) ▶

Fig. 10. Energy level diagram for levels partaking directly in maser action in $CaWO_4$:Pr^{3+} (after YARIV and co-authors [14]) ▼

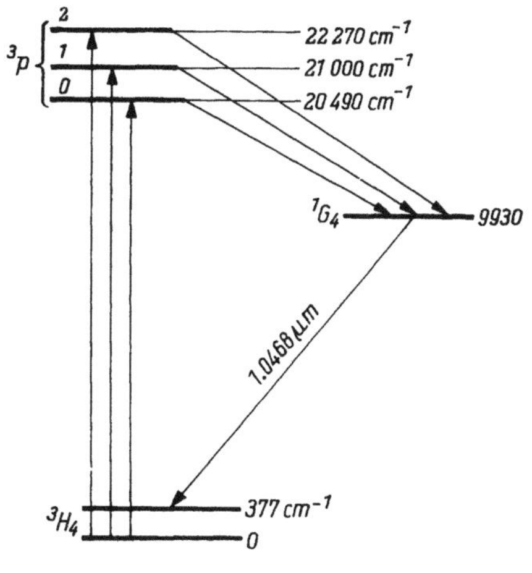

of 1 to 2×10^{-5} are reported. In the $^3H_4$ — $^3P_0$ transition lines occur spaced 0.4 to 0.7 nm ($\sim$ 8 to 14 cm^{-1}) apart. The initial level of the stimulated emission is the metastable Stark splitting level $^1G_4$ having a lifetime of $50 \times 10^{-6} \pm 5 \times 10^{-6}$ s. The terminal state is a level 377 cm^{-1} above the ground state $^3H_4$ (four-level laser). In order to depopulate this level temperatures below 100 °K are required. Under the operating conditions given [14] a $CaWO_4$:Pr^{3+} (0.5 mole %) laser resonator could be operated at 90, 70, 20 and 4.2 °K. For all these temperatures a relatively low threshold value of about 20 J is reported.

PORTO, SODEN, and YARIV [15] announced the publication of data on the laser emission of $SrMoO_4$:Pr^{3+} at 1.04 μm.

The incorporation of Pr^{3+} in Ca $(NbO_3)_2$ together with Ti^{4+} for charge compensation has recently been described by BALLMANN, PORTO, and YARIV [16], who achieved the stimulated emission with a wavelength of 1.04 μm at 77 °K and threshold values between 20 and 25 J.

Apart from a short reference to a strong infrared fluorescence (without wavelength data) of Pr^{2+} in CaF_2 by JOHNSON [6] nothing has been reported on Pr^{2+} as an activator.

4.1.2.3 Neodymium, Z = 60

	Nd^{3+}	Nd^{2+}
Electron configuration	$4f^3$	—
Ground state	$^4I_{9/2}$	—

To date 11 substances have become known, which when activated with Nd^{3+} exhibit laser properties. Johnson and co-workers detected laser action of Nd^{3+} in $CaWO_4$ [17, 18, 19, 20], $SrWO_4$ [6], $SrMoO_4$ [21], $CaMoO_4$ [6], $PbMoO_4$ [6], CaF_2 [22], SrF_2 [6], BaF_2 [6], and LaF_3 [6] (Table 1 — see Part I of this article published in 5, 437 (1964)), but not all results have as yet been published. Recently Ballman, Porto, and Yariv [16] described the stimulated emission of Nd^{3+} in Ca $(NbO_3)_2$, while Snitzer [81] was the first to report laser action of Nd^{3+} in glass. The first communication on $SrF_2:Nd^{3+}$ lasers has been published by Kariss and Feofilov [23]. In all cases the laser wavelengths are between 1.04 and 1.07 μm.

With the exception of CaF_2 and BaF_2 as host lattices it is possible even at room temperature to excite Nd^{3+} to stimulated emission; in the case of $CaWO_4:Nd^{3+}$ even continuous laser operation can be achieved. With 1500 W excitation Johnson obtained the relatively high emission of 0.5 W [6].

Ignoring for the present the fine structure within a single multiplet transition in the absorption and fluorescence spectra as well as the different threshold values of Nd^{3+} in host crystals of different lattice types, or the same lattice types but different lattice constants, as well as in one and the same substance, one will find some striking similarities.

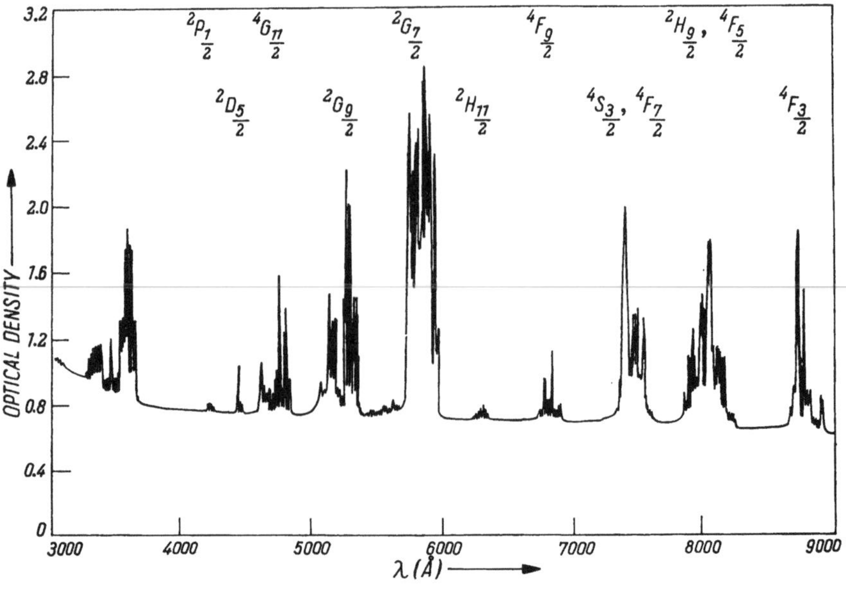

Fig. 11. Absorption spectrum of 1% Nd^{3+} in $CaWO_4$ at 77 °K. Sample thickness = 0.4 cm (after Johnson [6])

The absorption spectrum (Fig. 11) consists of a series of groups of closely spaced sharp absorption lines, of which the conspicuous group at 570 to 590 nm ($f \sim 3 \times 10^{-6}$ for many of these lines) is preferably used for optical pumping. For different host lattices the multiplet transitions corresponding to the groups merely exhibit differences in the fine structure splitting as a result of the different crystal field. The notation of the terms was suggested by CARLSON and DIEKE [82, 83], who derived it from absorption and fluorescence measurements of Nd^{3+} in $LaCl_3$. Fig. 12 shows the energy level scheme for Nd^{3+} in $CaWO_4$.

Excitation with white light entails a strong infrared fluorescence (Fig. 13). The following transitions can be observed:

$^4F_{3/2}$ — $^4I_{9/2}$ transition at about 0.9 μm,

$^4F_{3/2}$ — $^4I_{11/2}$ transition at about 1.06 μm,

$^4F_{3/2}$ — $^4I_{13/2}$ transition at about 1.35 μm.

As the fine structures of the individual transitions are not apparent from Fig. 13, we shall deal with each of these multiplet transitions separately.

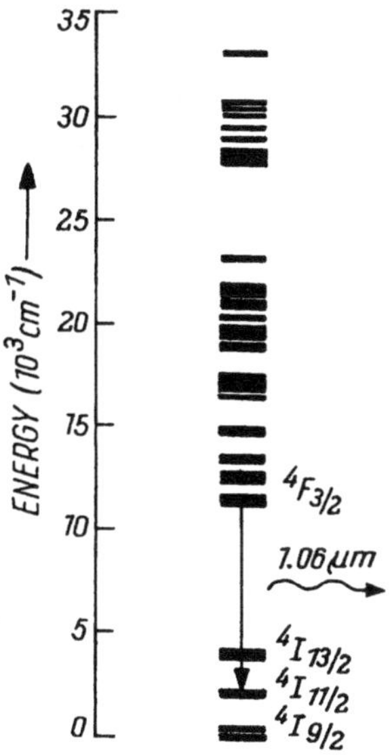

Fig. 12. Energy levels of Nd^{3+} in $CaWO_4$. Maser transitions are indicated by the arrow (after JOHNSON [6]) ▶

Fig. 13. Fluorescence spectrum of Nd^{3+} in $CaWO_4$ at 77 °K, with transitions designated (after JOHNSON and NASSAU [17]) ▼

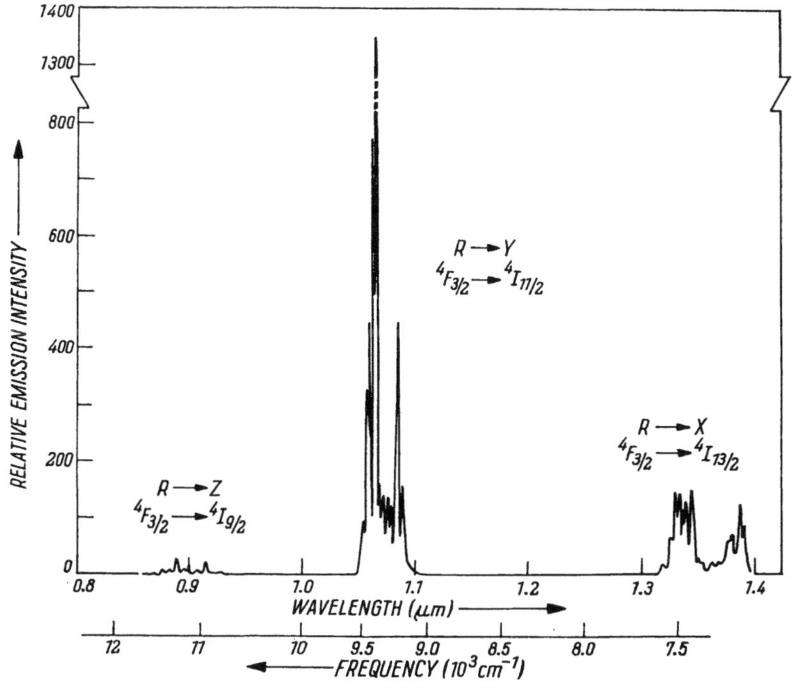

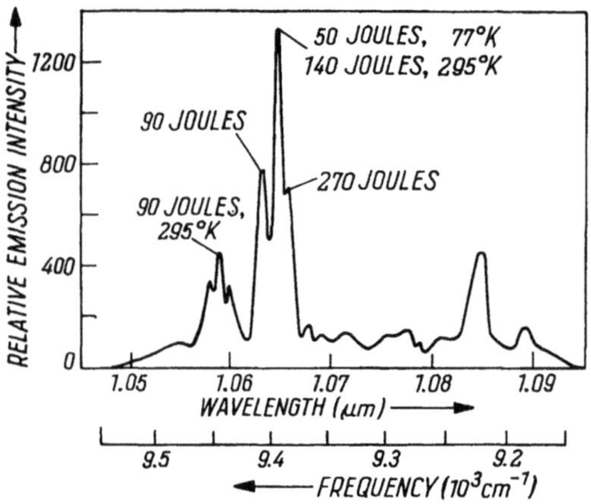

Fig. 14. $^4F_{3/2} \rightarrow {}^4I_{11/2}$ emission from $CaWO_4$:Nd^{3+} (0.3%) at 77 °K. Laser lines and threshold energies measured in a FT 524 helical xenon lamp are designated (after JOHNSON [6])

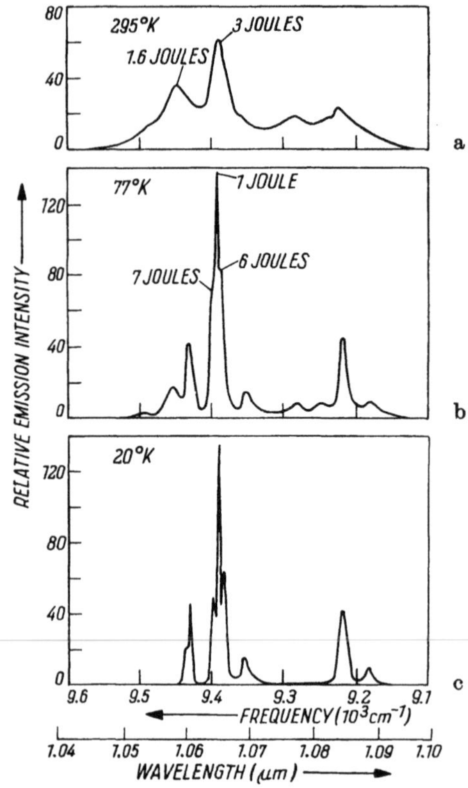

Fig. 15. $^4F_{3/2} \rightarrow {}^4I_{11/2}$ emission in $CaWO_4$:Nd^{3+} with Na^+ compensation: a) 295 °K, b) 77 °K, c) 20° K. Laser lines and thresholds are incidated (after JOHNSON [6])

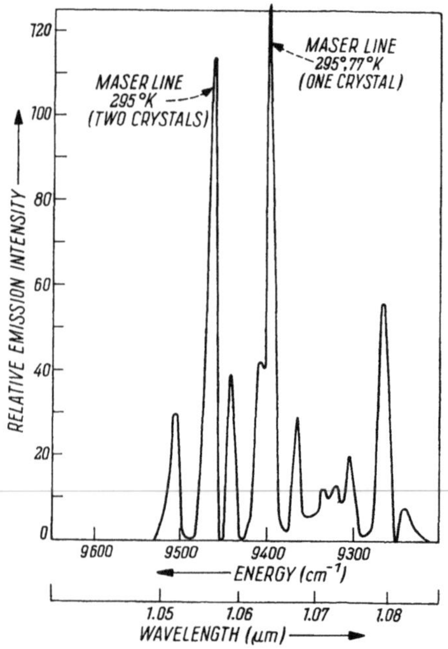

Fig. 16. $^4F_{3/2} \rightarrow {}^4F_{11/2}$ emission from $SrMoO_4$:Nd^{3+} at 77 °K, without charge compensation. The relative strength of several lines varies from crystal to crystal (after JOHNSON [6])

$^4F_{3/2} - {}^4I_{11/2}$:

This transition is the most intensive (~ 80%) and has been used in all host substances for stimulated emission. Owing to the large distance of about 2000 cm^{-1} between the terminal state and the ground state the former may be considered as empty even at room temperature, so that stimulated emission becomes possible at ambient temperature. The detailed spectrum of the spontaneous fluorescence of this transition is shown in Fig. 14.

Fig. 17. $^4F_{3/2} \to {}^4I_{11/2}$ emission from $SrMoO_4:Nd^{3+}$ at 77 °K, with Li^+ compensation. Laser lines and thresholds are designated (after JOHNSON [6])

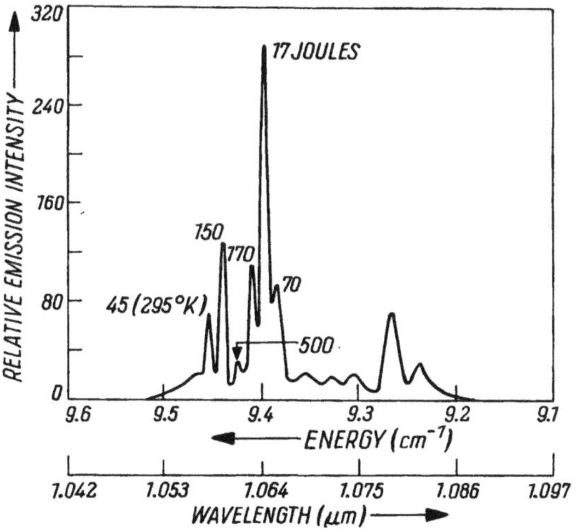

The absorption and fluorescence spectra of Nd^{3+} in $CaWO_4$ are in reality much more complex than hitherto described. This is mainly due to the fact that the trivalent activator ion in the host lattice isomorphously occupies the position of a divalent calcium ion. There are several possibilities for the necessary local charge compensation, such as Ca^{2+} vacancies, oxygen in interstitial sites, paired neodymium ions, etc. Since this charge compensation does generally not take place uniformly, each Nd^{3+} ion due to its special position in the crystal field gives rise to an individual spectrum, which leads to line broadening and an increase of the threshold value. Furthermore, according to JOHNSON [6] changes can be observed, from crystal to crystal, in the spectral position of the lines showing stimulated emission, as well as in their relative fluorescence intensities. A uniform charge compensation is achieved by incorporating in the host lattice a quantity of monovalent alkaline metal ions (Na^+, K^+, Li^+) that is equivalent to the activator concentration. The changes in the fluorescence spectrum can be seen from Fig. 15. The spectrum reduces to a few lines, of which one with a smaller half width becomes dominant. At equal Nd^{3+} concentration the threshold value reduces by a factor of 3. This effect had been discovered by JOHNSON [6] when investigating the stimulated emission on $SrMoO_4$: Nd^{3+} (Fig. 16, 17).

Fig. 18 shows the energy level scheme for $CaWO_4:Nd^{3+}$ with Na^+ compensation, which was derived from the absorption spectrum of Fig. 11 and the fluorescence spectrum of Fig. 15. Strictly speaking, it is only valid for the $CaWO_4:Nd^{3+}$ crystal, on which the spectroscopic data used were determined.

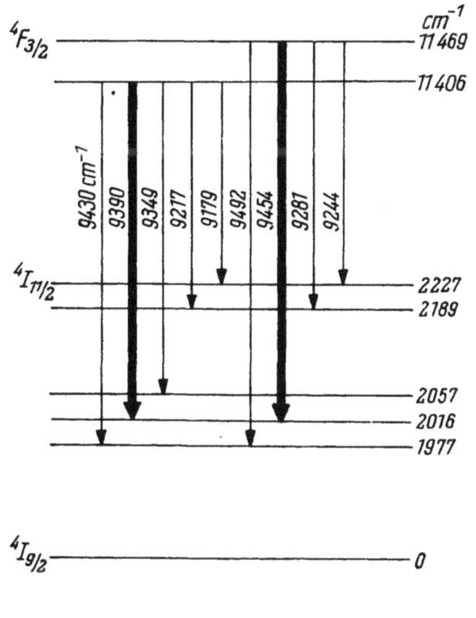

Fig. 18. Laser transitions and levels of $^4F_{3/2}$ and $^4I_{11/2}$ multiplets in $CaWO_4:Nd^{3+}$ with Na^+ compensation. At 77 °K stimulated emission is observed only in the line at 9390 cm^{-1} (after JOHNSON [6])

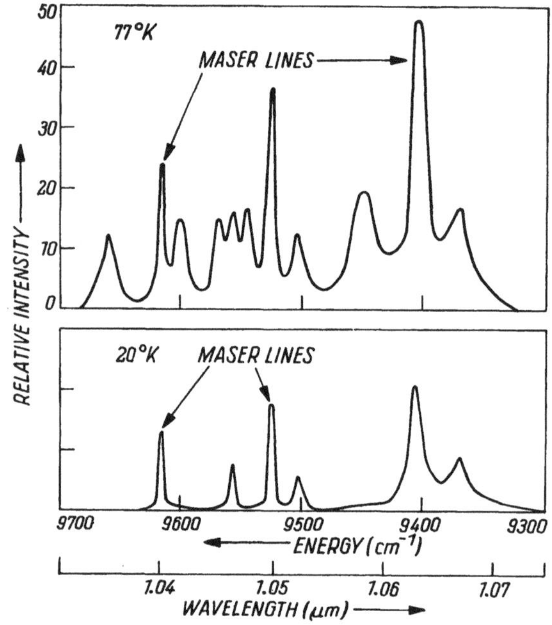

Fig. 19. $^4F_{3/2} \rightarrow {^4I_{11/2}}$ emission in LaF_3: Nd^{3+} at 77 °K and 20 °K (after Johnson [6])

LaF_3 is the first and so far the only host crystal, in which the Nd^{3+} ions occupy sites of trivalent lattice ions. The problem of charge compensation does not exist in this case. The 6 lines measured in the fluorescence spectrum at 20 °K (Fig. 19) allow the 6 levels of the $^4I_{11/2}$ multiplet to be determined (Fig. 20). From the measurements at 77 °K the upper state of the $^4F_{3/2}$ term could be identified.

Of the alkaline earth fluorides as host crystals, CaF_2, SrF_2, and BaF_2 have successfully been used as lasers at about 1.06 μm. According to Johnson they are inferior to the tungstates and molybdates with regard to line width and fluorescence intensity. The threshold of CaF_2:Nd^{3+} lies at 77 °K, but not higher than for $CaMoO_4$:Nd^{3+}, $PbMoO_4$:Nd^{3+}, or LaF_3:Nd^{3+}. Just as CaF_2:Nd^{3+}, also BaF_2:Nd^{3+} may be excited to stimulated emission only at low temperatures (77 °K). Things are different for SrF_2:Nd^{3+} which according to Johnson [6] shows laser emission even at room temperature. Prior to Johnson, Kariss, and Feofilov [23] already reported the detection of the stimulated emission on SrF_2:Nd^{3+} (0.2 mole %). The absorption spectrum (Fig. 21) of Nd^{3+} in a crystal with fluorite structure given by them permits a comparison with that of Nd^{3+} in a crystal with scheelite

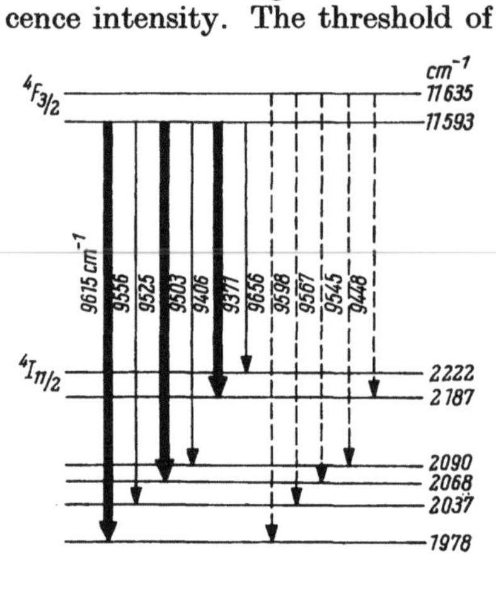

Fig. 20. Laser transitions and levels of $^4F_{3/2}$ and $^4I_{11/2}$, LaF_3:Nd^{3+}. Fluorescence lines observed at 20 °K are designated by solid arrows. Dotted arrows indicate additional lines at 77° K (after Johnson [6])

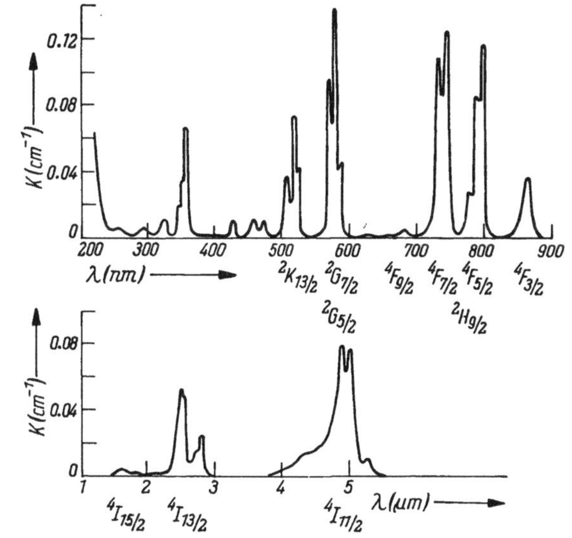

Fig. 21. Absorption spectrum of $SrF_2:Nd^{3+}$ (0.2 mole %) at 300° K (d = 10 mm) (after Kariss and Feofilov [23])

Fig. 22. Luminescence spectrum of $SrF_2:Nd^{3+}$ at 77 °K (after Kariss and Feofilov [23]) (Read 8696 Å instead of 8896 Å) ▼

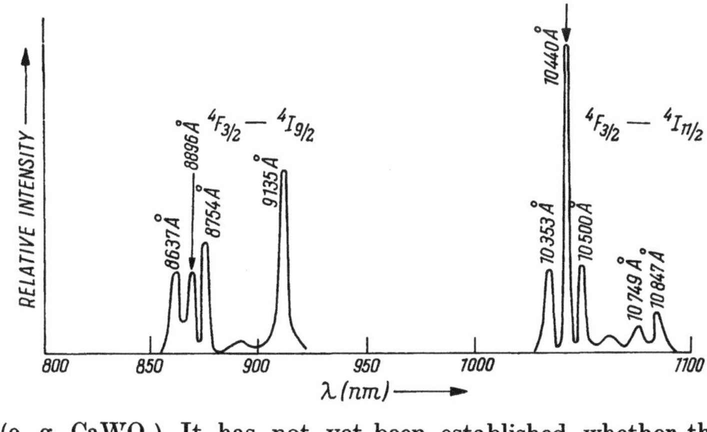

structure (e. g. $CaWO_4$). It has not yet been established, whether the slighter splitting of the line groups is primarily due to the different crystal fields or the different measuring temperatures. Extending the range of Johnson's measurements they determined further absorption peaks at about 1.6, 2.5, and 5 µm, which in the opinion of the authors can possibly be explained by transitions in the ^4I-multiplet between the ground state $^4I_{9/2}$ and the $^4I_{11/2}$, $^4I_{13/2}$, or $^4I_{15/2}$ terms, respectively.

In the spectrum of the spontaneous fluorescence (Fig. 22) lines occurred at 1.0353, 1.0440, 1.0500, 1.0749, 1.0847 µm. The comparable lines in the case of $CaF_2:Nd^{3+}$ appeared at 1.0335, 1.0465, 1.0544, 1.082, 1.095 µm. The values published by them for the luminescence duration were 1.3 to 1.4 ms at 300 °K and 1.6 ms at 77 °K. No particulars were given relative to the threshold value.

$^4F_{3/2} - {}^4I_{9/2}$:

Using dielectric multiple layers Johnson and Thomas [20] recently succeeded in detecting the stimulated emission with a wavelength of 0.9145 µm and ~ 1.34µm on $CaWO_4:Nd^{3+}$ for the $^4F_{3/2} - {}^4I_{9/2}$ and $^4F_{3/2} - {}^4I_{13/2}$ transitions as well. From

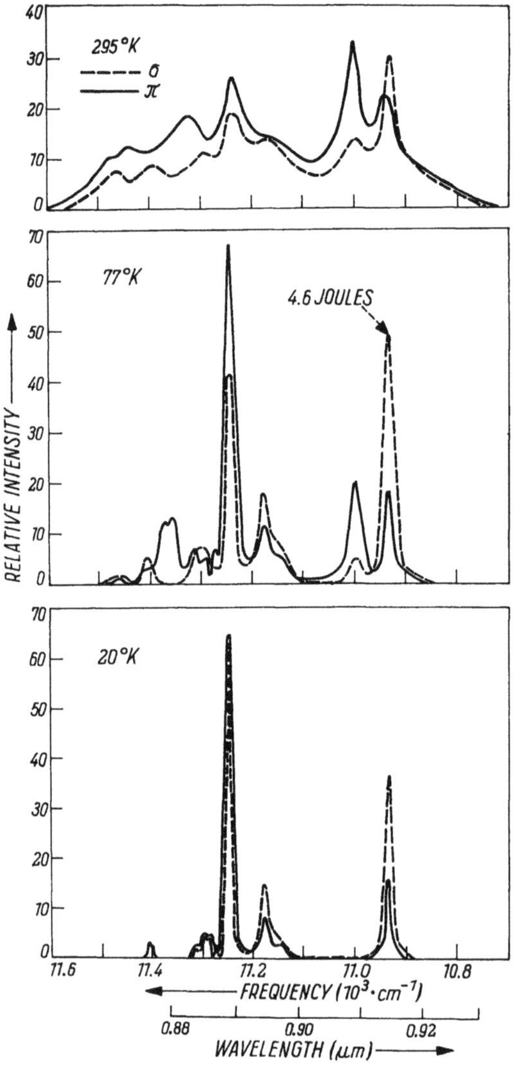

Fig. 23. Polarization of the $^4F_{3/2} \rightarrow {}^4I_{9/2}$ emission in $CaWO_4:Nd^{3+}$ (Na^+ compensation) at 295, 77, and 20 °K. Laser oscillation is observed at 77 °K in the line at 0.9145 μm (10935 cm^{-1}) (after Johnson and Thomas [20])

Fig. 24. Energy levels of $^4F_{3/2}$ and $^4I_{9/2}$ multiplets. Dashed lines indicate additional transitions seen at 77 °K (after Johnson and Thomas [20])

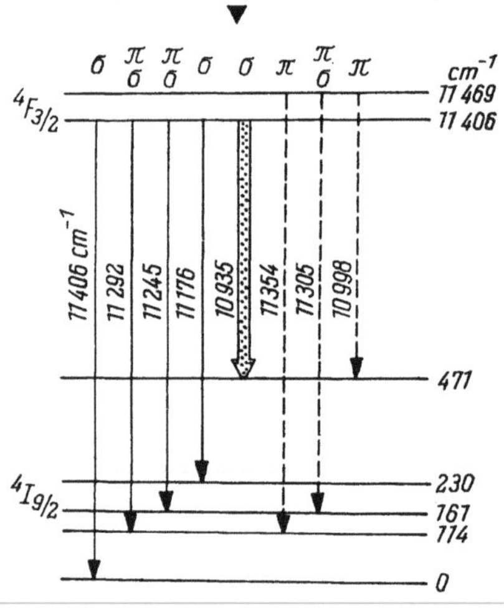

the fluorescence spectrum of the $^4F_{3/2} - {}^4I_{9/2}$ transition (Fig. 23) the authors inferred the energy level scheme shown in Fig. 24.

As the terminal state is seen to be lying merely 471 cm^{-1} above the ground state, the inverted population could only be achieved at 77 °K. At this temperature the half width of the spontaneous fluorescence of the laser line was 15 cm^{-1}, but it was as small as 10 cm^{-1} at 20 °K. Compared with the threshold value of 0.8 J for the strongest stimulated emission at 1.065 μm, it was about 5 to 6 times that value (4.6 J) for the 0.9145 μm emission. The changes of a continuous operation are less than those of the emissions at 1.06 and 1.34 μm.

For Nd^{3+} in SrF_2 (Fig. 22) and CaF_2 Kariss and Feofilov [23] recorded the following spontaneous fluorescence lines for the $^4F_{3/2} - {}^4I_{9/2}$ transition:

SrF_2:	863.7	869.6	875.4	913.5 nm
CaF_2:	859.0	866.0	874.5	917.0 nm

Nothing has so far been published about a corresponding stimulated emission in SrF_2 and CaF_2.

$^4F_{3/2} - {^4I_{13/2}}$:

The spectrum of the polarized luminescence of this transition may be seen from Fig. 25. While at room temperature only the line at 1.339 μm was ascertained for laser emission, there are 3 lines at 77 °K (see Table 1), of which the one at 1.3372 μm has the relatively lowest threshold value of 2.1 J and a half width of the spontaneous fluorescence of 6.4 cm^{-1} (at 20 °K 2.4 cm^{-1}). Lines having lowest threshold values exhibit π-polarization both at 295 °K and 77 °K in contrast to the line at 1.06 μm which shows σ-polarization at 77 °K and π-polarization at

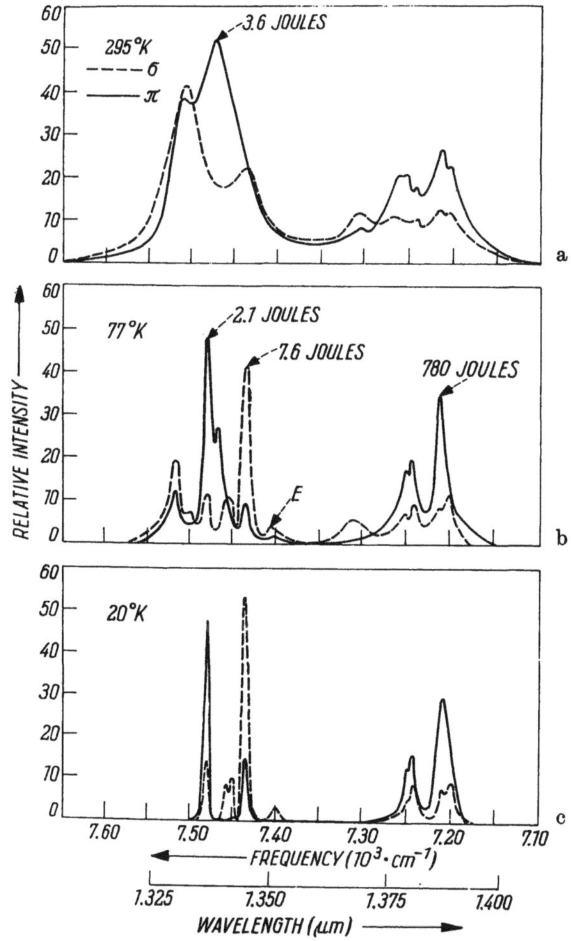

Fig. 25. Polarization of the $^4F_{3/2} \rightarrow {^4I_{13/2}}$ emission in $CaWO_4$:Nd^{3+} (Na^+ compensation) at 295, 77, and 20 °K. Laser lines and threshold energies measured in an FT 524 xenon lamp are designated (after JOHNSON and THOMAS [20])

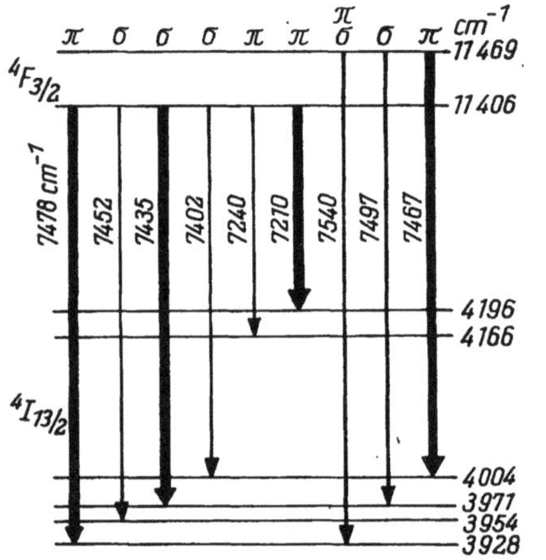

Fig. 26. Energy levels of $^4F_{3/2}$ and $^4I_{13/2}$ multiplets of Nd^{3+} in $CaWO_4$ (Na^+ compensation). Laser transitions are indicated by heavy arrows (after Johnson and Thomas [20])

295 °K. As the terminal state of the laser emission at 1.34 μm lies about 4000 cm^{-1} above the ground state (Fig. 26) and the threshold energies do not much differ in order of magnitude from those of the 1.06 μm emission, the practical realization of a continuous laser operation is within the range of possibility.

Up to now all attempts failed to achieve laser emission in the region from 5 to 7 μm by transitions between the multiplets

$$^4I_{13/2} - {}^4I_{11/2}$$

and

$$^4I_{11/2} - {}^4I_{9/2}\,.$$

CaF_2, LaF_3 as well as $CaWO_4$, for example, are absorptionfree in this region.

Both, in absorption and fluorescence, the incorporation of Nd^{3+} in CdF_2 [84] leads to the expected results characteristic of the Nd^{3+} ion.

The first references to possibilities of the incorporation of Nd^{2+} in alkaline earth fluoride crystals were recently made by Feofilov [66]. Concluding from the fact that the lowest excitation state of the divalent rare earths lies about 16 to 18% below the terms of the isoelectronic trivalent ions (verified for $Sm^{2+} - Eu^{3+}$, $Dy^{2+} - Ho^{3+}$, $Tm^{2+} - Yb^{3+}$), Feofilov supposed a luminescence of Nd^{2+} at about 7 to 8 μm [67].

4.1.2.4 Promethium, Z = 61

	Pm^{3+}	Pm^{2+}
Electron configuration	$4f^4$	—
Ground state	$^5I_4$	—

No data have so far been published on the absorption and fluorescence of Pm^{3+} or Pm^{2+} in alkaline earth fluorides or tungstates. Pm is unstable and disintegrates with a half-life of 2.6 a by emitting β-radiation.

4.1.2.5 Samarium, Z = 62

	Sm^{3+}	Sm^{2+}
Electron configuration	$4f^5$	$4f^6$
Ground state	$^6H_{5/2}$	$^7F_0$

Of all rare-earth ions used as activators in crystals, Sm^{2+} in CaF_2 may relative to its spectroscopic properties be regarded as the one most thoroughly investigated. Recently SrF_2 and BaF_2 were added as new hosts.

The fundamental investigations of Feofilov and co-authors [45] formed the basis for the research work of Sorokin and co-authors [13] as well as Kaiser and co-authors [12] who, as pointed out in our introductory remarks, reported laser

Table 3
Position of absorption bands, their half widths and oscillator strengths of Sm^{2+} in CaF_2, SrF_2, and BaF_2 at 20 °K (after Wood and Kaiser [85])

	ν (cm^{-1})	$\Delta\nu$ (cm^{-1})	f
$CaF_2:Sm^{2+}$	14118 (at 77 °K)	30	6.9×10^{-4}
	14234 (at 77 °K)	27	3.9×10^{-4}
	14497	20	3.5×10^{-5}
	15387	71	5.0×10^{-4}
	15444	71	5.2×10^{-4}
	15723		
	15823		
	16103	990	9.9×10^{-3}
	16340		
	16415		
	19677	78	1.3×10^{-4}
	20072	300	3.1×10^{-4}
	22381	200	1.1×10^{-3}
	22573		
	22676	612	2.7×10^{-3}
	22769		
	23540	715	6.3×10^{-3}
	33000	2800	2.1×10^{-2}
	35710		
	39370	1100	1.0×10^{-2}
	41841	2000	1.6×10^{-2}
$SrF_2:Sm^{2+}$	14803 (at 77 °K)		
	14925 (at 77 °K)		
	15181	18	3.6×10^{-5}
	16142	130	6.5×10^{-4}
	16460		
	16750	1070	9.6×10^{-3}
	20429	82	5.8×10^{-5}
	20790	430	4.2×10^{-4}
	23095		
	23230	210	9.2×10^{-4}
	23419		
	24154	840	7×10^{-3}
	31056		
	31646	1800	1.6×10^{-2}
	34062		
	38314		
	39682	3100	3×10^{-2}
	41494		
$BaF_2:Sm^{2+}$	17270	1110	9.6×10^{-3}
	23980		
	24691		
	24938	910	6.2×10^{-3}
	30488	1950	2.3×10^{-2}
	31949		
	33330	5500	4.6×10^{-2}
	39062		
	41670		

Table 4
Position of fluorescence lines of Sm^{2+} in CaF_2, SrF_2, and BaF_2 at 4.2 °K
(after Wood and Kaiser [85])

$CaF_2 + Sm^{2+}$		$SrF_2 + Sm^{2+}$		$BaF_2 + Sm^{2+}$	
ν (cm^{-1})	Relative intensity	ν (cm^{-1})	Relative intensity	ν (cm^{-1})	Relative intensity
14114 (sharp)	10000	14803	0.5		
13930	62	14616 (sharp)	11	14630 (sharp)	0.1
13915	64	14530	12	14465	0.36
13886	101	14470	2	14432	0.24
13875	76	14390	140	14395	0.67
13855	69	14325	357	14374 (sharp)	10
13825	71	14260	20	14331	0.13
13775	83	14353 (sharp)	10000	14186	0.16
13735	139	14125	41	14118	0.20
13725	128	14060	68	13755	0.24
13585	55	13995	2	13689	0.18
13500	55	13964 (sharp)	13	13432	0.13
13422	52	13870	16	13364	0.11
13380	49	13800	86	13166 (sharp)	0.16
12920	12	13750	131		
12890	10	13680	77		
12690	8	13625	64		
12620	5	13581 (sharp)	8		
12540	5	13490	1		
12500	3	13450	7		
		13370	50		
14497 (at 77 °K)		13290	62		
14234 (at 77 °K)		13230	7		
		13166 (sharp)*)	1		
		13140 (sharp)*)	205		
		13062 (sharp)*)	1		
		13050	2		
		12980	11		
		12950	11		
		12890	14		
		12850	14		
		12780	7		
		12580	1		
		12550	3		
		12520	2		
		12470	2		
		12450	1		
		12350 (sharp)	106		
		12276	0.5		
		12246	1		
		12225*)	0.2		
		12146	1.5		
		12096	0.2		
		12016	1		
		11589*)	0.05		
		11463*)	0.09		

*) Observed only in Ca-doped SrF_2:Sm^{2+}.

Table 5

Half width, quantum efficiency, and lifetime of the fluorescence of $CaF_2:Sm^{2+}$, $SrF_2:Sm^{2+}$, and $BaF_2:Sm^{2+}$ for several fluorescence lines at different temperatures (after Wood and Kaiser [85])

Temperature (°K)	$CaF_2:Sm^{2+}$ ν (cm^{-1})	$\Delta\nu$ (cm^{-1})	η_e	τ_{exp} (s)	$SrF_2:Sm^{2+}$ ν (cm^{-1})	$\Delta\nu$ (cm^{-1})	η_e	τ_{exp} (s)	$BaF_2:Sm^{2+}$ ν (cm^{-1})	$\Delta\nu$ (cm^{-1})
90		46				1.1				
77	14118	21	0.10	2×10^{-6}	14351.9	0.8	0.03	8×10^{-4}	14374	1.9
20	14114	1.1	0.20	2×10^{-6}	14353.0	0.8	0.4	1×10^{-2}	14375	1.5
4.2	14114	0.6	0.35		14353	0.7	0.4		14375	1.5

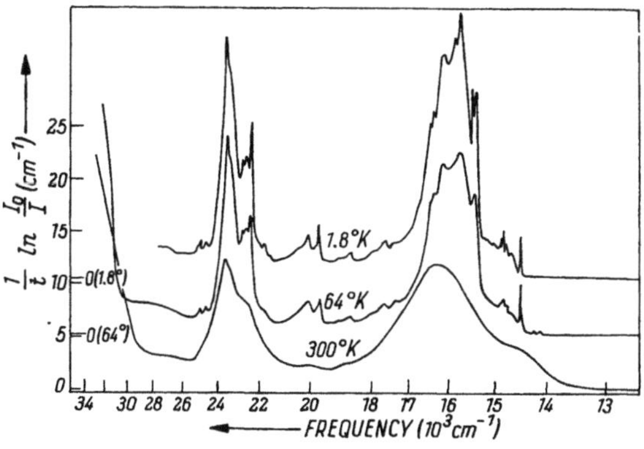

Fig. 27. Absorption spectrum for 0.05% Sm (0.01% Sm^{2+}) in CaF_2 at 1.8, 64, and 300 °K (after Kaiser and co-authors [12])

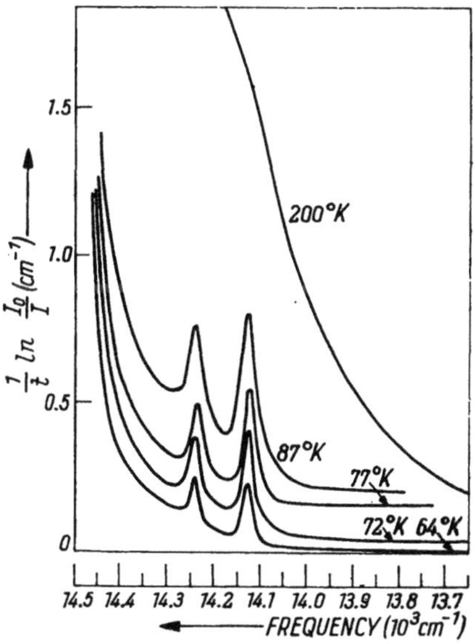

Fig. 28. Detail of absorption spectrum from Fig. 27 (after Kaiser and co-authors [12])

action of $CaF_2:Sm^{2+}$ as early as 1961. Later, detailed investigations were conducted by research teams under FEOFILOV and KAISER, who at the same time published results on the properties of Sm^{2+} in CaF_2, SrF_2, and BaF_2, which were in reasonable agreement with those of SOROKIN and KAISER.

First we shall turn to the system $CaF_2:Sm^{2+}$. By employing reducing growth conditions it is possible to incorporate Sm^{2+} predominantly in the divalent state [12, 13, 44, 45, 64, 77, 85]. An absorption spectrum typical of Sm^{2+} is shown in Fig. 27. In addition, weak but very sharp lines at 4053 cm^{-1} ($\sim$ 2.47 µm) and 2290 cm^{-1} ($\sim$ 4.37 µm) with oscillator strengths of $f = 3 \times 10^{-7}$ were observed on rather heavily doped $CaF_2:Sm^{2+}$ samples (0.75% Sm) [85].

In the visible two broad bands of several thousand cm^{-1} are dominant at room temperature, which at low temperatures split into numerous extremely narrow lines (Table 3). Fig. 28 shows a section of Fig. 27 with the absorption lines at 14128 cm^{-1}, 14241 cm^{-1}, and the considerably stronger line at 14497 cm^{-1}.

These lines among others could with minimum deviations also be detected in the fluorescence spectrum (Fig. 29) (see also Table 4). On further cooling down (4.2 °K) only the strongest of these lines at 14118 cm^{-1} is left (it showed laser emission at 20 °K). The variation of the half width of this line with the temperature may be seen from Table 5. At 4.2 °K its half width is as small as 0.6 cm^{-1}, which is most favourable for laser applications. Of disadvantage is however the continuum adjacent to the sharp line extending towards the long wavelength region with isolated weak narrow bands; at 77 °K the continuum emits about 10

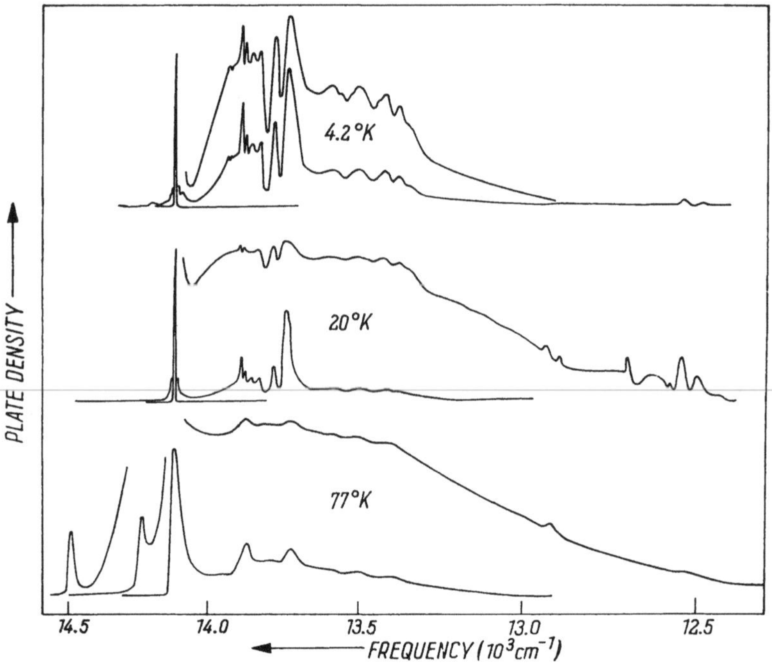

Fig. 29. Fluorescence emission for 0.05% Sm (0.01% Sm^{2+}) in CaF_2. The ordinate represents plate density (after KAISER and co-authors [12])

times the energy of the line, 5 times at 20 °K, and 3.5 times at 4.2 °K. The intensity of the main line (14118 cm^{-1}) is at 4.2 °K by a factor of 100 and at 20 °K by a factor of 3.5 greater than at 77 °K (Table 6).

Table 6

Relative intensities of the strongest fluorescence lines of $CaF_2:Sm^{2+}$, $SrF_2:Sm^{2+}$, and $BaF_2:Sm^{2+}$ (after Wood and Kaiser [85])

Temperature (°K)	$CaF_2:Sm^{2+}$ 14118 cm^{-1}	$SrF_2:Sm^{2+}$ 14353 cm^{-1}	$BaF_2:Sm^{2+}$ 14374 cm^{-1}
4.2	10000	10000	10
20	3500	10000	10
77	100	900	1

It was further pointed out by Kaiser that the fluorescence (line including continuum) can be excited in the whole wavelength region between 300 and 650 nm (Fig. 30). The excitation peaks coincide with the absorption peaks. By taking into consideration the absorption data the true quantum efficiency (quotient of the number of emitted photons and the number of absorbed photons of wavelength λ) was determined by Kaiser to be 1 for all wavelengths between 300 and 650 nm, i. e. that all bands in this region are equally effective in populating the initial level of the fluorescence.

With falling temperature the fraction of emission in the continuum decreases in favour of the emission into the line. Its fraction decreases from 90% at 77 °K via 80% at 20 °K to 65% at 4.2 °K. Accordingly the quantum efficiency of the line emission increases from 0.1 at 77 °K to 0.2 at 20 °K and to 0.35 at 4.2 °K (Table 5). The lifetime of the line emission was determined to be $\tau = 2 \times 10^{-6} \pm 0.2 \times 10^{-6}$ s for 77 °K and 20 °K [12] which is in good agreement with the value of 1.4×10^{-6} s given by Kaplyanskii and Feofilov [64]. Using the value of

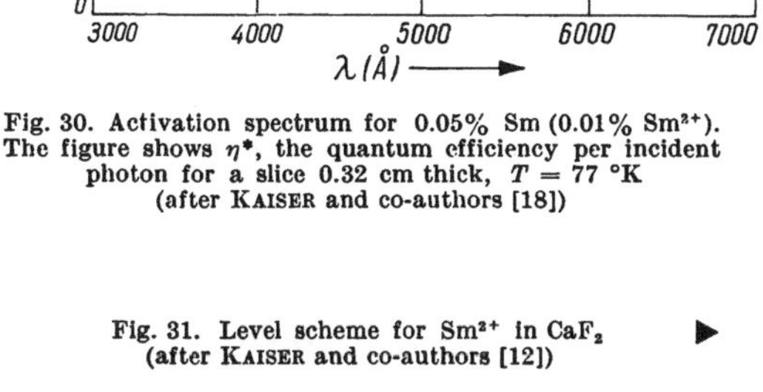

Fig. 30. Activation spectrum for 0.05% Sm (0.01% Sm^{2+}). The figure shows η^*, the quantum efficiency per incident photon for a slice 0.32 cm thick, $T = 77$ °K (after Kaiser and co-authors [18])

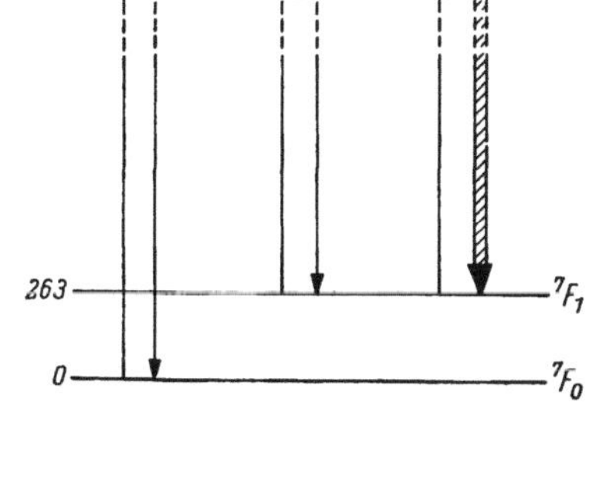

Fig. 31. Level scheme for Sm^{2+} in CaF_2 (after Kaiser and co-authors [12]) ▶

2×10^{-6} s for the radiative lifetime, KAISER and co-authors calculated an oscillator strength ($f = m\, c^3/8\, \pi^2\, e^2\, \nu^2\, \tau$) of 3.7×10^{-4} at 77 °K, which in order of magnitude is consistent with the value of $f = 6.9 \times 10^{-4}$ derived from absorption measurements. As may be seen from Table 3 the oscillator strength varies in the wavelength region 250 to 710 μm from 2.1×10^{-2} (33000 cm^{-1}) to 3.5×10^{-5} (14497 cm^{-1}).

There are at present still difficulties in interpreting the energy level scheme of Sm^{2+} in CaF_2. On the basis of the results of the absorption and fluorescence measurements and following the interpretation of the lowest energy levels suggested by BUTEMENT [69], KAISER and co-authors proposed the energy level scheme shown in Fig. 31. The terminal state $^7F_1$ of the line emission at 708.5 nm (14118 cm^{-1}) lies 263 cm^{-1} above the ground state $^7F_0$ in reasonable agreement with BUTEMENT, who had found a separation of 269 cm^{-1} for Sm^{2+} in $BaCl_2$. The $^5D_1$ and $^5D_0$ states of the 4 f shell were assumed to be excitation levels and the strongest emission at 14118 cm^{-1} was assigned to the $^5D_0 - {}^7F_1$ transition. Later investigations by FEOFILOV [64], KAPLYANSKII and PRZHEVUSKII [86] as well as observations of the Zeeman effect by ZAKHARCHENYA and RYSKIN [87] as well as KAISER [85] also led to the result that the initial level involved cannot be a level of the 4 f^6 shell, therefore excluding the $^5D_0$ level, but must be a transition in the 4 f^5 d configuration. An exact identification of the excited level has still not been presented; KAISER's suggestion is illustrated in Fig. 32.

After ruby and CaF_2:U^{3+}, CaF_2:Sm^{2+} [12, 13] was the third solid state laser substance. As the terminal state $^7F_1$ of the transition showing stimulated emission lies only 263 cm^{-1} above the ground state $^7F_0$, low temperatures ($\leqq$ 20 °K) are required for the depopulation of the $^7F_1$ level. The strong absorption bands several hundred cm^{-1} broad are excellently suited for the population of the metastable fluorescence level, where in particular an excitation in the absorption band (at about 625 nm) in the red relatively near the laser emission at 708 μm is most

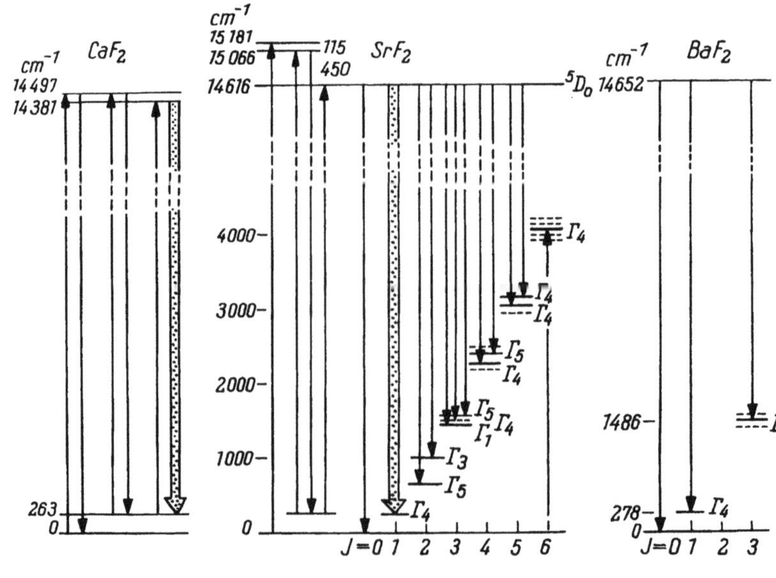

Fig. 32. Comparison of the lower energy levels (in cm^{-1}) of Sm^{2+} in CaF_2, SrF_2, and BaF_2. Transitions to dashed levels are forbidden and not observed. Transitions in absorption or fluorescence, which have been observed and identified are indicated by arrows (after WOOD and KAISER [85])

favourable. In this case the host lattice requires only about 10% of the excitation energy. Pulsed laser oscillations (Fig. 33) could be achieved with an energy flux of 20 W/cm² [12]. This value exceeds the theoretically expected value by a factor of 8. The cause of this is believed to be the optical imperfections of the $CaF_2:Sm^{2+}$ material which increase the threshold value. On further cooling down to 4.2 °K the threshold is said to be changing only slightly.

It is worthy of note that the laser oscillations of this laser system do not exhibit any spikes. The investigation of the laser output by KAISER and co-workers [12] showed that the emission consisted of 5 discrete frequencies with a width of 0.1 cm^{-1} each making a total of 0.5 cm^{-1} and that the number of the excited modes was 1000 or so.

Due to their inherent physical connection, both $SrF_2:Sm^{2+}$ and $BaF_2:Sm^{2+}$ will briefly be dealt with together. There is a strong similarity between the absorption spectra of $CaF_2:Sm^{2+}$, $SrF_2:Sm^{2+}$, and $BaF_2:Sm^{2+}$. Fig. 34 shows measurements of FEOFILOV at room temperature and Fig. 35 measurements of WOOD and KAISER at 77 °K.

The shifting of the bands (several hundred cm^{-1}) with the variation of the lattice constants in the homologous host crystals may well be seen from Fig. 36. The resolution of the narrow bands and lines at 20 °K (Table 3) and 4.2 °K [64] decreases in the order CaF_2, SrF_2, BaF_2.

With regard to the fluorescence (Fig. 37) things are different. $SrF_2:Sm^{2+}$ and $BaF_2:Sm^{2+}$ exhibit a greater similarity and are sharply contrasted with $CaF_2:Sm^{2+}$. While $CaF_2:Sm^{2+}$ has only one sharp line and several narrower bands in a continuum, several sharp lines and a reduced continuum are observed for $SrF_2:Sm^{2+}$ and $BaF_2:Sm^{2+}$ (Table 4). Again, characteristic differences in the increase of emission intensity occur for the strongest line of the $MeF_2:Sm^{2+}$ crystals (Table 6). While on cooling from 77 °K down to 4.2 °K the emission of $CaF_2:Sm^{2+}$ increases

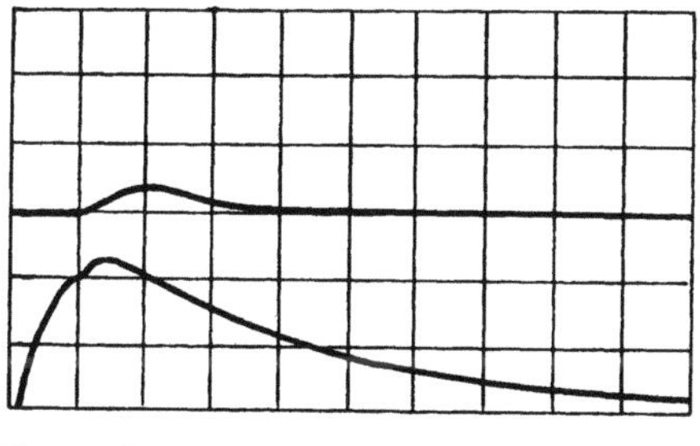

Fig. 33. Optical maser action with a short light flash. Upper beam: laser signal; lower beam: pump signal; time scale 1 μs per division (after KAISER and co-authors [12])

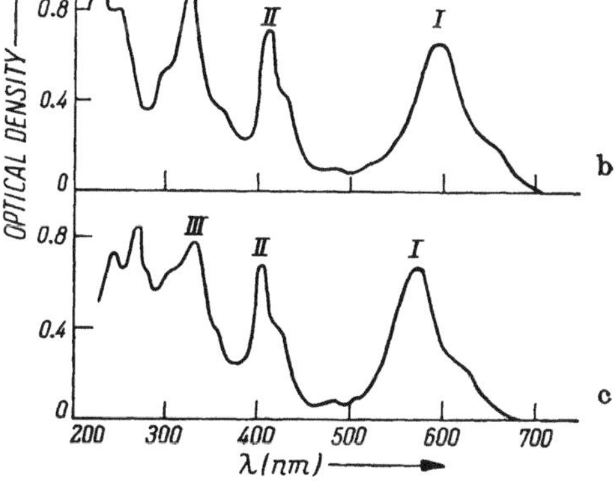

Fig. 34. Absorption spectra of $MeF_2:Sm^{2+}$ at room temperature. a) CaF_2, b) SrF_2, c) BaF_2 (after KAPLYANSKII and FEOFILOV [64])

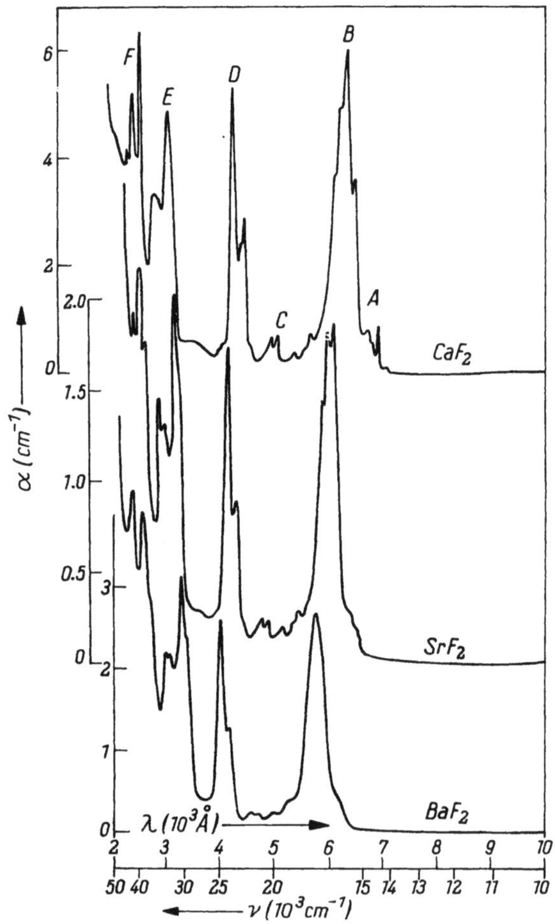

Fig. 35. Absorption curves for Sm^{2+} ion in CaF_2, SrF_2, and BaF_2 at 77 °K. The Sm^{2+} concentrations in the three samples differ by less than one order of magnitude (after Wood and Kaiser [85])

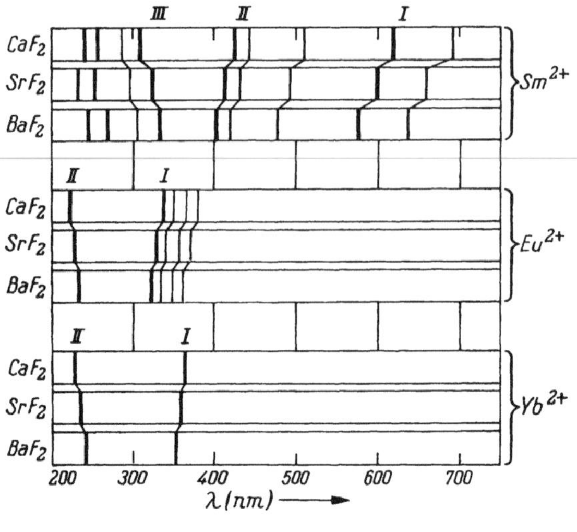

Fig. 36. Shift of the absorption band maxima in spectra of MeF_2:RE^{+2} with change in host cation (Me), (after Kaplyanskii and Feofilov [65])

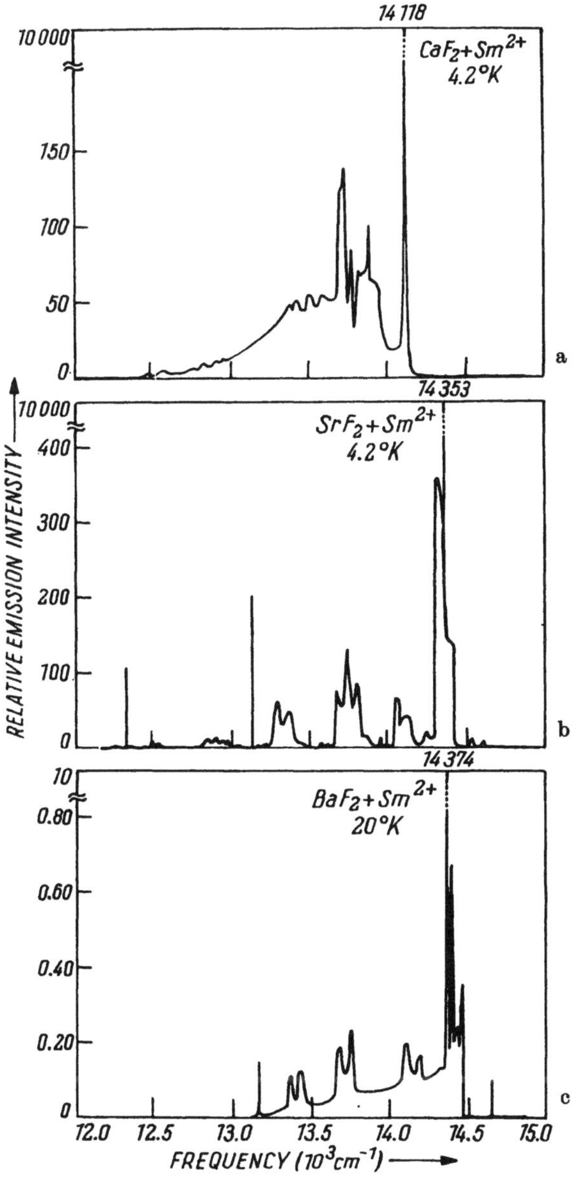

Fig. 37. Fluorescence spectra of Sm^{2+} ion in CaF_2, SrF_2, and BaF_2. a) CaF_2:Sm^{2+} at 4.2 °K; b) SrF_2:Sm^{2+} at 4.2 °K; c) BaF_2:Sm^{2+} at 20 °K (after WOOD and KAISER [85])

by a factor of 100, the increase is only 10 times for SrF_2:Sm^{2+} and BaF_2:Sm^{2+}. Furthermore, the dependence of the line width upon the temperature was found to be much stronger for the most intense fluorescence line in the case of CaF_2:Sm^{2+} than for SrF_2:Sm^{2+} and BaF_2:Sm^{2+} (Table 5). At 20 °K the lifetime of emission of SrF_2:Sm^{2+} is by a factor of 10^3 greater than for CaF_2:Sm^{2+} and decreases strongly with rising temperatures.

These basic differences in the luminescent process are also corroborated by the investigation of the Zeeman effect [85, 87] and the splitting of lines due to deformations of the crystal field [86] and lead to the assumption that in the absorption of all three hosts and in fluorescence electrical dipole transitions between the 4 f and 5 d shells are involved, in the fluorescence of $SrF_2:Sm^{2+}$ and $BaF_2:Sm^{2+}$, however, transitions of magnetic dipoles and forced electrical dipole transitions within the 4 f shell. This view was supported by FEOFILOV's and TOLSTOI's [88] investgations of the temperature dependence of the luminescence intensity and the luminescence duration for $SrF_2:Sm^{2+}$ and $BaF_2:Sm^{2+}$ (Fig. 38), which yielded and increase expected for the $^5D_0$ and $^7F_1$ transitions at falling temperatures. The energy level schemes for $CaF_2:Sm^{2+}$, $SrF_2:Sm^{2+}$, $BaF_2:Sm^{2+}$ resulting from WOOD's and KAISER's measurements are shown in Fig. 32. They do not in all details agree with those of KAPLYANSKII and FEOFILOV [64].

From the reported spectroscopic data $SrF_2:Sm^{2+}$ appears to be suited as laser substance. In 1962 SOROKIN and co-workers [24] actually detected the stimulated emission on a $SrF_2:Sm^{2+}$ (0.1%) crystal in flashlight operation at 4.2 °K. The laser wavelength is 696.9 nm and originates from the extremely sharp transition $^5D_0 - {}^7F_1$ (at 4.2 °K, line width < 0.6 cm^{-1}). About 50% of the total fluorescence intensity is concentrated in the line. In contrast to $CaF_2:Sm^{2+}$ the laser output is characterized by a multiplicity of relaxation oscillations. This is associated with the considerably greater lifetime of 15 ms ($CaF_2:Sm^{2+}$ — $\tau = 0.002$ ms) and leads to a long induction time before laser oscillation starts. Under the same operating conditions the authors obtained a lower energy output than with CaF_2: Sm^{2+}. In their opinion it would be conceivable to operate $SrF_2:Sm^{2+}$ resonators as continuous wave lasers in the visible, however at temperaturs below 77 °K.

Sm^{2+} (0.3%) was not only incorporated in CaF_2, SrF_2, and BaF_2, but also in $SrCl_2$ single crystals (having likewise fluorite structure) [89]. The absorption properties are very similar to those of SrF_2, but the superposition of electron transitions and lattice vibrations (which was also observed for $CaF_2:Sm^{2+}$, $SrF_2:Sm^{2+}$, and $BaF_2:Sm^{2+}$) emerges more and more clearly (Fig. 39). In contrast to SrF_2: Sm^{2+}, $SrCl_2:Sm^{2+}$ fluoresces already at room temperature (Fig. 40). At 4.2 °K the fluorescence spectrum consists of several very sharp lines. For the sharpest dominant line Table 7 lists the values of half width, quantum efficiency, and lifetime of $SrCl_2:Sm^{2+}$ in comparison with $SrF_2:Sm^{2+}$, which differ only slightly at 4.2 °K.

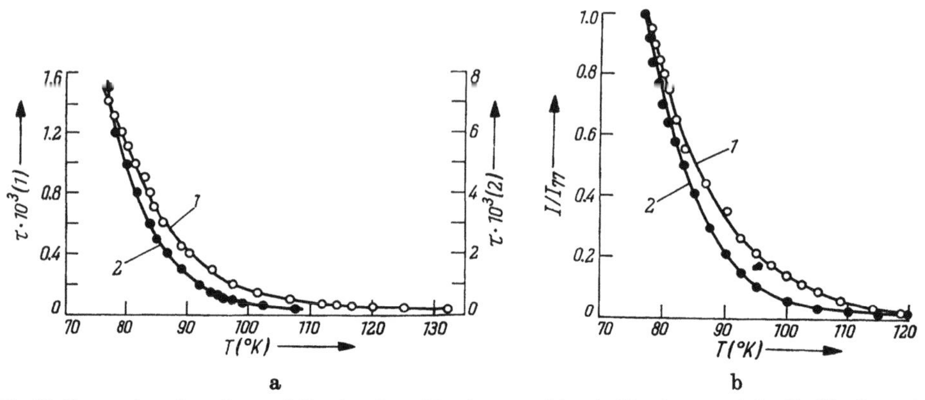

Fig. 38. Temperature dependence of the duration of luminescence (a) and of luminescence intensity (b). Curve 1: $SrF_2:Sm^{2+}$ ($\lambda = 696.74$ nm), curve 2: $BaF_2:Sm^{2+}$ ($\lambda = 695.0$ nm) (after FEOFILOV and TOLSTOI [88])

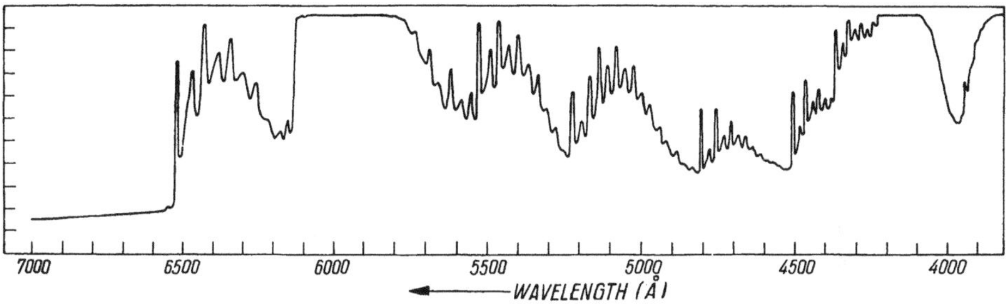

Fig. 39. Absorption spectrum at 4.2 °K of $SrCl_2:Sm^{2+}$ (after AXE and SOROKIN [89])

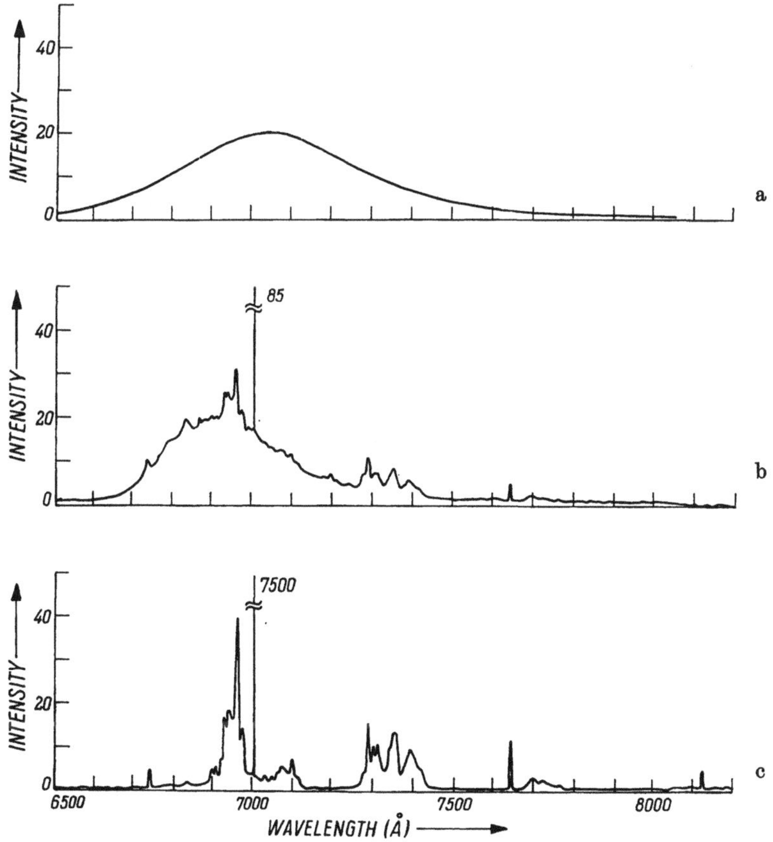

Fig. 40. Fluorescence spectrum of $SrCl_2:Sm^{2+}$. a) 300 °K, b) 77 °K, c) 4.2 °K (after AXE and SOROKIN [89])

Recently also Sm^{2+} (0.05% by weight) was incorporated in crystal lattices of the NaCl type (NaCl, NaBr, NaJ, KCl, KBr, KJ) by KARAPETYAN and co-workers [90]. The absorption spectra exhibit a high degree of similarity and are characteristic of Sm^{2+} (Fig. 41). At room temperature a broadband fluorescence occurs at 750 to 800 nm (Fig. 41, 42), which has a more detailed structure at 77 °K. The narrowest bands could be observed in $NaCl:Sm^{2+}$ crystals.

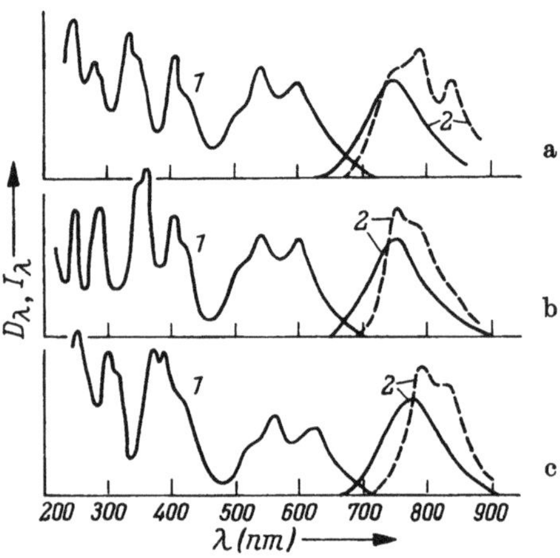

Fig. 41. Absorption (1) and luminescence (2) spectra of Sm^{2+} in potassium halide. Solid curves at 300 °K, dashed curves at 77 °K. a) KCl:Sm^{2+}, b) KBr:Sm^{2+}, c) KJ:Sm^{2+} (after Karapetyan and co-authors [90])

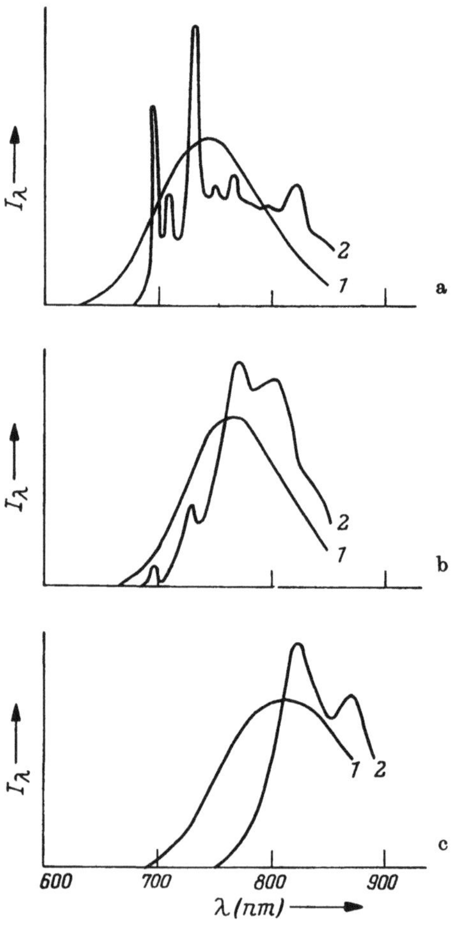

Fig. 42. Luminescence spectra of Sm^{2+} in sodium halides. 1: at 300 °K, 2: at 77 °K; a) NaCl:Sm^{2+}, b) NaBr:Sm^{2+}, c) NaJ:Sm^{2+} (after Karapetyan and co-authors [90])

Table 7

Comparison of half width, quantum efficiency, and lifetime of SrF_2:Sm^{2+} and SrCl:Sm^{2+} (after Axe and Sorokin [68])

Temperature (°K)	$SrCl_2$			SrF_2		
	$\Delta\nu$ (cm^{-1})	η_e	τ (s)	$\Delta\nu$ (cm^{-1})	η_e	τ (s)
221			5×10^{-5}			
90				1.1		
77	0.9	0.03	7.5×10^{-3}	0.8	0.03	8×10^{-4}
20				0.8	0.4	1×10^{-2}
4.2	0.5	0.64	1.4×10^{-2}	0.7	0.4	1.4×10^{-2}

Considerably less experimental data are available on the behaviour of trivalent samarium ions in crystals than in the case of the divalent ions. Owing to the very low oscillation strength the absorptive power is small. In the case of CaF_2 the absorption bands are only weakly marked at 400 nm and in the infrared at 1.060, 1.214, 1.455, and 1.540 μm [9]. The fluorescence (type I and type II) [43] has been investigated by STEPANOV and FEOFILOV [43, 55] and recently by RABBINER [91], who detected about 80 lines in the region 550 to 680 nm and established the energy level schemes. The differences between the theoretically expected and the measured frequencies were maximally 5 cm^{-1}. The author assigned the spectrum II to transitions between crystal field splitting levels of the $^4F_{5/2}$ state and those of the $^6H_{5/2}$, $^6H_{7/2}$, $^6H_{9/2}$ states and the spectrum I to transitions between the crystal field splitting levels of the $^4F_{7/2}$ and higher levels and those of the ground multiplet state $^6H_{7/2}$, $^6H_{9/2}$, $^6H_{11/2}$.

While Sm^{3+} allows readily to be incorporated in CaF_2, the incorporation of Sm^{2+} can usually only imperfectly be carried out even under reducing growth conditions [12, 24, 64]. A reduction of Sm^{3+} to Sm^{2+} of approximately 50% was achieved by O'CONNOR and BOSTICK [75] by intensive γ-irradiation. The result is shown in Fig. 43. The kinetics of this process has still not completely been clarified. The authors associate the result with a charge compensation of Sm^{3+} in CaF_2 in the form of fluorine ions in interstitial sites.

4.1.2.6 Europium, Z = 63

	Eu^{3+}	Eu^{2+}
Electron configuration	$4f^6$	$4f^7$
Ground state	$^7F_0$	$^8S_{7/2}$

Comprehensive and systematic investigations of the absorption and fluorescence behaviour of Eu^{2+} ions in host crystals of the MeF_2 type (Me = Ca, Sr, Ba) have been conducted by KAPLYANSKII and FEOFILOV [44, 65]. Concerning CaF_2:Eu^{2+} good agreement exists between the crystals of FEOFILOV [45] grown from natural CaF_2 material and the CaF_2:Eu^{2+} crystals of GÖRLICH and co-workers [77, 92] grown from highly purified synthetic substances.

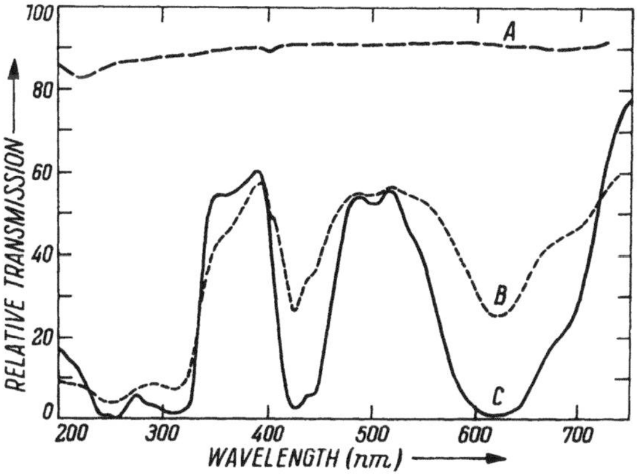

Fig. 43. The optical transmission of CaF_2:Sm^{3+} at room temperature.
A) Sample I: 1.35 × 10^{19} Sm^{3+} cm^{-3},
B) Sample I: after exposure to 1.9 × 10^6 r gamma radiation,
C) Sample II: 1.72 × 10^{19} Sm^{3+} cm^{-3}.
The thickness of sample I is 3 mm and of sample II is 0.6 mm. Curves A and B are plotted for a maximum transmission of 100%. The maximum transmission of curve C is 10% (after O'CONNOR and BOSTICK [75])

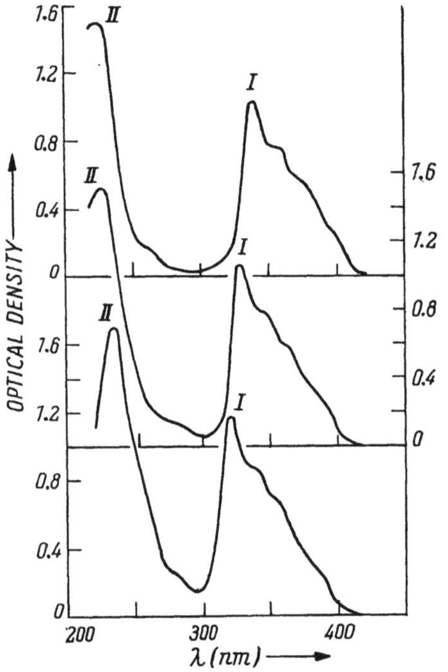

Fig. 44. Absorption spectra of $MeF_2:Eu^{2+}$ monocrystals at room temperature. a) CaF_2, b) SrF_2, c) BaF_2 (after KAPLYANSKII and FEOFILOV [65].

Great similarity exists between the absorption spectra of $CaF_2:Eu^{2+}$, $SrF_2:Eu^{2+}$ and $BaF_2:Eu^{2+}$ (Fig. 44). The shifting of the two broad strong bands I and II and of the structure observed on the long wavelength tail of band I as a result of the variation of the host lattice may be seen from Fig. 36. This structure is well resolved at 4.2 °K and allows the identification of intensive narrow analogous bands, which are shifted towards the shorter wavelengths in the order CaF_2, SrF_2, and BaF_2 (Table 8). In the same order the distinctness of splitting is reduced. Furthermore, a multiplicity of weak equidistant narrow bands become noticeable, which are generally regarded as a result of an interaction between the electron states and vibrations of different types.

Greater differences occur between the fluorescence spectra of $CaF_2:Eu^{2+}$, $SrF_2:Eu^{2+}$, and $BaF_2:Eu^{2+}$. At 4.2 °K the luminescence spectra of $CaF_2:Eu^{2+}$ and $SrF_2:Eu^{2+}$ (strong blue fluorescence at 77 °K) are splitting up into numerous extremely small lines. The most intensive emission lines lie at 413 or 401.15 nm, respectively, and are resonance lines. Towards the longer wavelengths additional rather weak narrow lines and very weak diffuse bands stand out against a continuous background. In $BaF_2:Eu^{2+}$ even at 4.2 °K no fluorescence could be stimulated.

From the spectroscopic data, especially from the considerable shifting of the bands in absorption and fluorescence as a result of the variation of the host lattice, KAPLYANSKII and FEOFILOV [65] concluded that in the case of the spectra describ-

Table 8

Position of analogous absorption bands in $MeF_2:Eu^{2+}$ (Me = Ca, Sr, Ba) (after KAPLYANSKII and FEOFILOV [65])

CaF_2			SrF_2			BaF_2	
λ (nm)	ν (cm^{-1})	Δ Ca-Sr (cm^{-1})	λ (nm)	ν (cm^{-1})	Δ Sr-Ba (cm^{-1})	λ (nm)	ν (cm^{-1})
413.0	24210	720	401.15	24930	—	—	—
407.5	24540	710	396.1	25250	—	—	—
389.0	25710	730	378.2	26440	600	369.8	27040
379.0	26390	730	368.7	27120	—	—	—
368.3	27150	750	358.4	27900	590	351.0	28490
322.5	31010	720	315.2	31730	—	—	—

Fig. 45. Microphotograms of the splitting of $\lambda = 4130$ Å line of $CaF_2:Eu^{2+}$. a) free crystal; b, d) $P \parallel C_2$; a, b) absorption; c, d) luminescence; a, b, c) $T = 4.2$ °K; d — $T > 4.2$ °K (after KAPLYANSKII and PRZHEVUSKII [86] ▶

g. 46. Absorption spectra of BaF_2:Eu, Sm monocrystals at room temperature. a) before irradiation; b) after irradiation by a spark (after FEOPILOV [94])

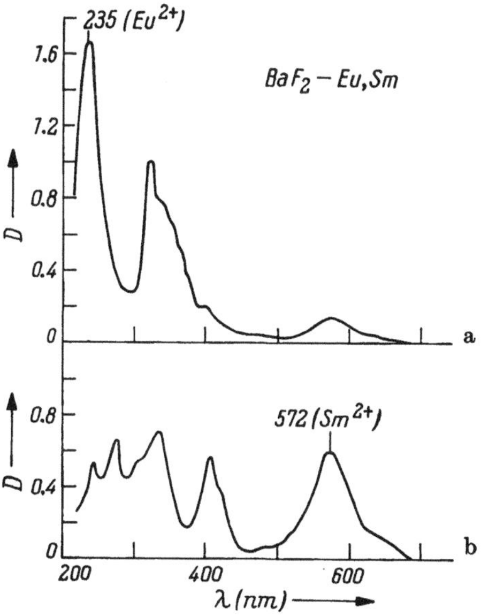

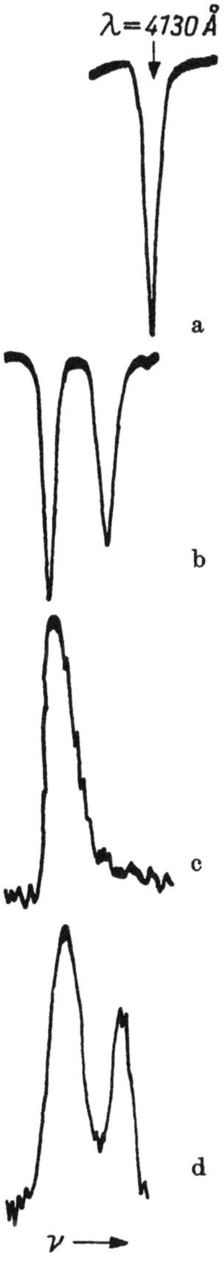

ed transitions between the ground state $^8S_{7/2}$ and the terms of a different configuration (e. g. $4f^6 5d$) must be involved. KAPLYANSKII and PRZHEVUSKII [86] corroborated this view on $CaF_2:Eu^{2+}$ for the line at 413 nm in absorption and emission with the aid of KAPLYANSKII's method of the deformation splitting of spectral lines [49]. The amount of splitting differs very distinctly for lines originating from $f - f$ and $f - d$ transitions. The splitting of 14 cm^{-1} (Fig. 45) observed in the particular case mentioned compares with a splitting of no more than about 2 cm^{-1} for an $f - f$ transition under the same operating conditions. A further confirmation of the assumption of $f - d$ transitions was recently presented by ZAKHARCHENYA and RYSKIN [93] by investigating the Zeeman effect on the line at 413 nm (absorption and luminescence) in $CaF_2:Eu^{2+}$.

FEOFILOV [94] reported the change of valency from Eu^{2+} to Eu^{3+} in double-activated crystals of CaF_2:Eu, Sm, SrF_2:Eu, Sm, and BaF_2:Eu, Sm due to the action of ionized radiation. The observation of the effect was based on the fact that even in one and the same element the spectroscopic properties of the rare earth ions in the divalent and trivalent states are fundamentally different. The crystals produced under specified growth conditions contained also Sm^{3+} in addition to Eu^{2+}. In the absorption spectrum (Fig. 46) the ultraviolet absorption of Eu^{2+} is predominant. Upon irradiation with ultraviolet light this absorption decreases, while at 572 nm the absorption being typical of Sm^{2+} is increasing. The process caused by irradiation:

$$Eu^{2+} + Sm^{3+} \underset{T}{\overset{h\nu}{\rightleftarrows}} Eu^{3+} + Sm^{2+} ,$$

can be reversed at temperatures between 400 and 600 °C and according to the author can be repeated many times with complete reproducibility. At room temperature bleaching of the Sm^{2+} band does not yet occur. The effect described is only observable in crystals exhibiting a fluorescence spectrum of type II and is associated with a particular structure near Sm^{3+} or Eu^{3+} ions which is necessary for the compensation of Sm^{3+} and Eu^{3+}.

For Eu^{3+} ions in CaF_2, Stepanov and Feofilov reported fluorescence lines in the region 525 to 670 nm (type I and type II) [43, 55]. The suggestion made by Feofilov that the deviations between the spectra of type I and II are due to the influence of different charge compensations was confirmed on $CaF_2{:}Eu^{3+}$. In the case of a compensation by an O^{2-} ion in place of an F^- ion (Fig. 5) one obtains spectra of type I. The dominant symmetry direction (3rd order axis of symmetry) was detected by Feofilov using his method of the investigation of polarized luminescence [41, 56, 57, 58]. His findings were confirmed by the study of the Zeeman effect and the method of the elastic deformation [49, 59]. The expected

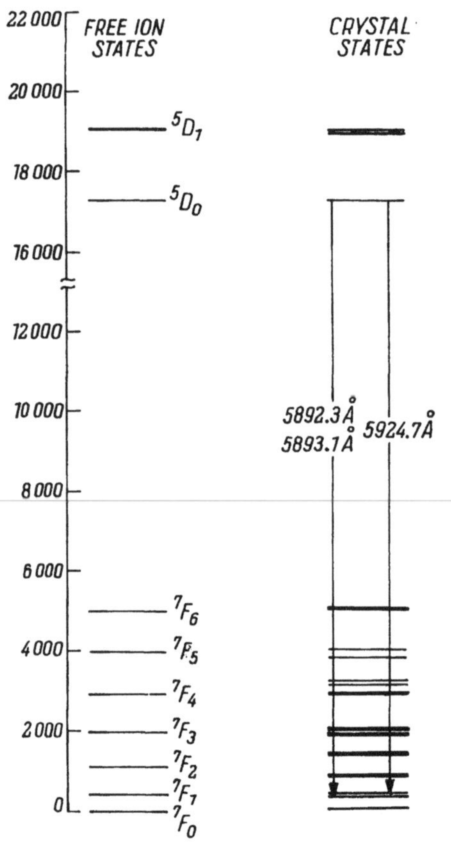

Fig. 47. The states of the Eu^{3+} as deduced from the emission spectra (after Kingsley and Prener [95])

Fig. 48. The half-width of the 5893.1 Å emission line as a function of temperature (after Kingsley and Prener [95])

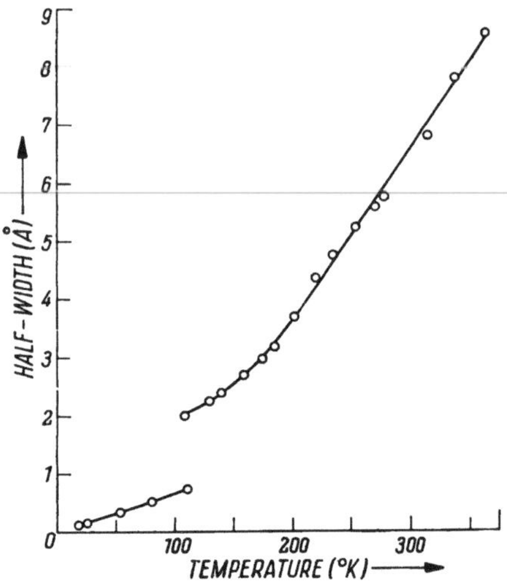

3rd order axis of symmetry was ascertained on the lines at 580.8 nm, 573.5 nm, and 528.2 nm. 20 kg/mm² compression along the G_2 axis led to a splitting of 2.5 cm^{-1} ($\triangleq$ 0.08 nm) at $\lambda = 580.8$ nm.

Eu^{3+} ions were incorporated in the CdF_2 lattice (fluorite structure) by KINGSLY and PRENER [95] and investigated relative to their suitability as laser material. From the measured values the authors calculated a threshold value of 18 W ($T = 20$ °K) for a rod having a cross section of 0.1 cm². As expected for Eu^{3+}, absorption was very small owing to the low oscillator strength. On a $CdF_2:Eu^{3+}$ crystal, which had rather heavily been doped for the absorption measurements (10 mole %), absorption lines were detected at 592.47, 589.31, 578.75, 534.23, and 525.59 nm, which except for the lines at 578.15 and 525.59 were frozen out at 85 °K.

In the wavelength region 520 to 820 nm the emission spectrum consists of about 80 lines. The main line at 589.31 nm with a lifetime of 10 ms (resonance line) concentrates 10% of the whole fluorescence intensity. The energy level scheme derived from the absorption and fluorescence measurements is shown in Fig. 47. The terminal state of the strongest emission line lies 310 cm^{-1} above the $^7F_0$ ground state. At temperatures below 100 °K this main emission proves to be consisting of two lines only 0.08 nm apart, the stronger of which is very narrow (Fig. 48), its width being only about 0.01 nm (0.32 cm^{-1}) at 15 °K. The values in Fig. 48 apply to concentrations of 10^{-2} to 10^{-1} mole %. Even at 1 mole % Eu^{3+} an essentially larger half width was measured.

For laser operation it is necessary to achieve an absorption coefficient of several cm^{-1} in the spectral emission region of the excitation light source; for this purpose sulphur was added to the crystal, thereby giving rise to a strong absorption between 250 nm and 350 nm (quantum efficiency 0.8 ± 0.2); the spectral distribution of the emission remained unaffected. To date nothing has become known on the operation of a $CdF_2:Eu^{3+}$ laser.

4.1.2.7 Gadolinium, Z = 64

	Gd^{3+}	Gd^{2+}
Electron configuration	$4f^7$	—
Ground state	$^8S_{7/2}$	—

Gd^{3+} in glass is so far known to be the only laser substance showing laser emission in the ultraviolet spectral region (312.5 nm) [35]. This value is in accord with the measurements of STEPANOV and FEOFILOV [43] on $CaF_2:Gd^{3+}$ at spontaneous emission. Absorption bands were observed on CaF_2:Gd (valency unknown) at 248 and 335 nm [77]. No further spectroscopic data have been published so far.

4.1.2.8 Terbium, Z = 65

	Tb^{3+}	Tb^{2+}
Electron configuration	$4f^8$	—
Ground state	$^7F_6$	—

Terbium belongs to those activators that have received little attention. From the investigations of FEOFILOV [43, 50, 55] it is known that the luminescence spectrum of Tb^{3+} in CaF_2 consists of two series of lines in the regions 380 to 480 nm and 490 to 660 nm. The two line groups correspond to transitions of two excited levels (Fig. 49), the lower of which is interpreted as $^5D_4$, while the upper could not be classified. For the short wavelength line group the luminescence period was determined to be 2×10^{-3} s.

4.1.2.9 Dysprosium, Z = 66

	Dy^{3+}	Dy^{2+}
Electron configuration	$4f^9$	$4f^{10}$
Ground state	$^6H_{15/2}$	$^5I_8$

In 1962 KISS and DUNCAN [25] were the first to report pulsed and continuous wave laser operation of Dy^{2+} in CaF_2. Reports on continuously operating $CaF_2:Dy^{2+}$ lasers were also given by JOHNSON [26] and YARIV [27].

Since dysprosium cannot readily be incorporated in the divalent state but rather well in the trivalent one, it is subsequently enriched by an intensive γ-irradiation of the crystals. The absorption spectrum (Fig. 50) covers the whole region from 300 nm to 1 μm. The oscillator strengths approach 10^{-2} to 10^{-3}. The absorption bands, some 1000 cm^{-1} broad ($4f - 5d$ or $4f - 6s$ transitions), compare with

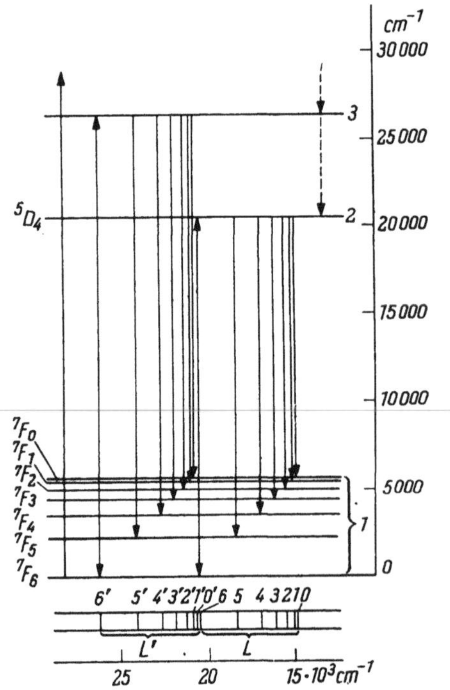

Fig. 49. Diagram of transitions which determine the luminescence of $CaF_2:Tb^{3+}$ crystals (after FEOFILOV [50])

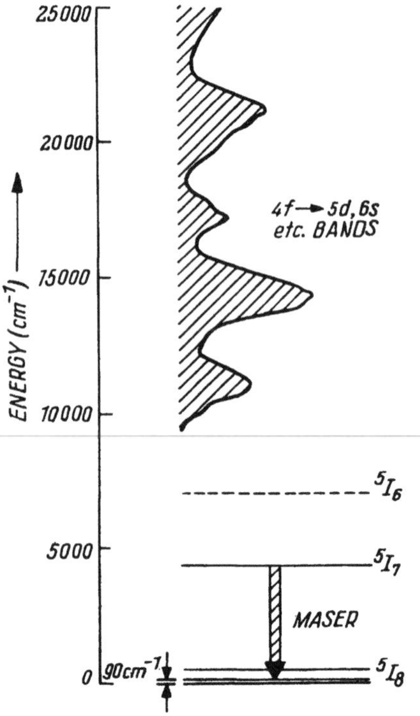

Fig. 50. Absorption spectrum and energy level diagram for $CaF_2:Dy^{2+}$ showing laser transition (after KISS and DUNCAN [25])

Fig. 51. Fluorescence spectra of CaF_2:Dy^{2+} at 77 °K. The frequency of the maser-oscillation is marked with the arrow (after KISS and DUNCAN [25])

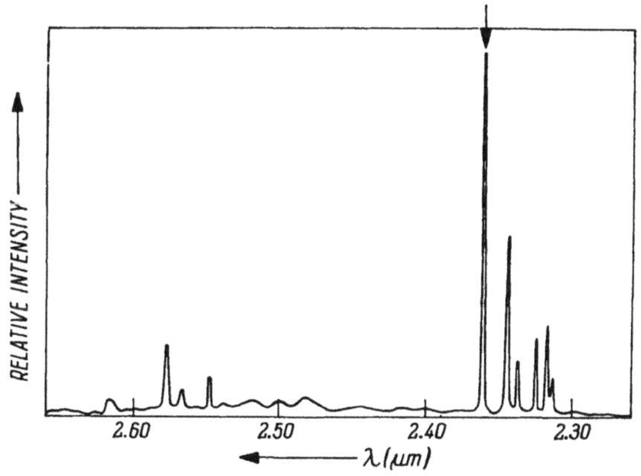

very narrow luminescence lines of about 0.5 cm^{-1} width at 77 °K [26] (Fig. 51), where transitions within the ground level multiplet ($^5I_7$ — $^5I_8$) are involved. The structure is due to the splitting of the $^5I_8$ ground state in the crystal field. At 20 °K merely the two strongest lines at 2.360 and 2.344 μm were observed [26].

The strongest line at 2.36 nm terminates merely 90 cm^{-1} above the ground state. Only this line allowed to be excited to laser oscillations at 20 °K as well as at 77 °K. The threshold value was determined to be 1 J [25]. Under the same conditions the best $CaWO_4$:Nd^{3+} and CaF_2:U^{3+} crystals had a threshold value of about 8 J. At 4.2 °K a threshold value of 0.1 J was reported for a CaF_2:Dy^{2+} crystal. The laser output does not exhibit any relaxations (Fig. 52). The very long lifetime (12 ms) of the fluorescence process (dependent upon concentration) together with the low threshold values constitute the basis for the continuous operation, whose power output of 2 mW (0.03 mole %) at 27 °K and 600 W input [25] was increased to 300 mW (0.5% Dy) at an input of 800 W [27]. A further increase is expected to be possible by using optically better CaF_2:Dy^{2+} crystals and dielectric multiple layers. YARIV also investigated the dependence of threshold values of CaF_2:Dy^{2+} in continuous operation upon the Dy^{2+} concentration (0.02 to

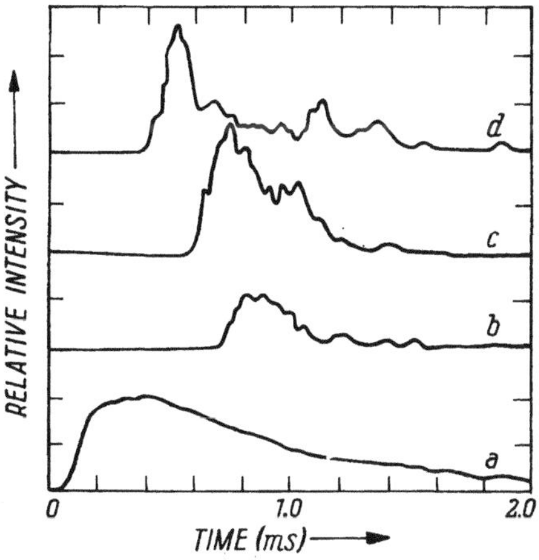

Fig. 52. Laser output of a CaF_2:Dy^{2+} crystal
a: relative lamp intensity as a function of time below threshold. Laser output: b: 2 per cent, c: 4 per cent, d: 15 per cent above the threshold voltage.
Ordinate: a: 10 mV/cm, b: 100 mV/cm, c: 100 mV/cm, d: 200 mV/cm (After KISS and DUNCAN [25])

Table 9

Threshold values of $CaF_2:Dy^{2+}$ in dependence upon the activator concentration (after YARIV [27])

	a.c. threshold	d.c. threshold
0.5 per cent Dy^{2+}	> 800 W	> 800 W
0.2 per cent Dy^{2+}	600 W	660 W
0.1 per cent Dy^{2+}	300 W	350 W
0.05 per cent Dy^{2+}	180 W	210 W
0.02 per cent Dy^{2+}	80 W	100 W

0.5%). As is apparent from Table 9, the threshold value decreases with decreasing Dy^{2+} concentration. The differences for d. c. and a. c. operation of the pumping lamp are slight. With these values (< 100 W) $CaF_2:Dy^{2+}$ of all CW solid state lasers has the lowest threshold.

According to STEPANOV and FEOFILOV [55] Dy^{3+} ions in CaF_2 give rise to fluorescence in the form of three line groups at about 450, 550, and 650 nm. A report on the more exact determination of the fluorescence spectrum of type I was recently published by RABBINER [96].

4.1.2.10 Holmium, Z = 67

	Ho^{3+}	Ho^{2+}
Electron configuration	$4f^{10}$	$4f^{11}$
Ground state	$^5I_8$	$^4I_{15/2}$

Ho^{3+} ions embedded in the $CaWO_4$ lattice [6] show an absorption spectrum, which, as is expected, consists of numerous lines (Fig. 53). The main absorption region lies in the blue spectral range. The strongest absorption lines have oscillator

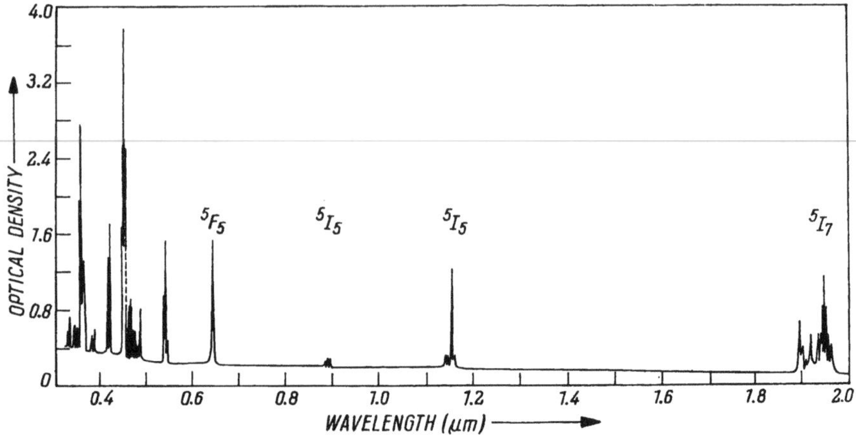

Fig. 53. Absorption spectrum of $CaWO_4:Ho^{3+}$ (0.8%) at 77 °K. Sample thickness = 6.63 mm (after JOHNSON [6])

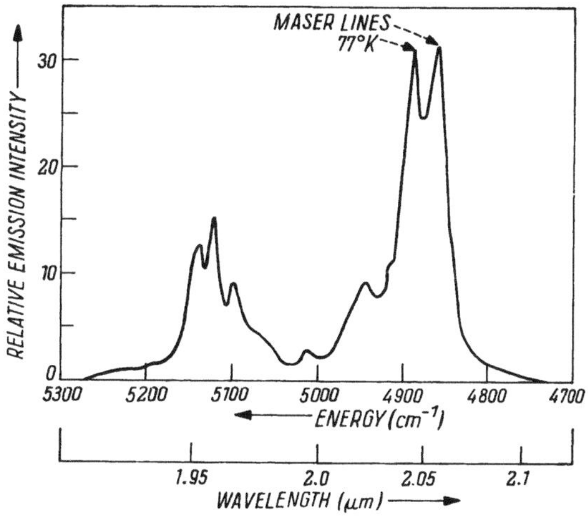

Fig. 54. Luminescence spectrum ($^5I_7 - ^5I_8$) of Ho^{3+} in $CaWO_4$ at 77 °K (after JOHNSON [6])

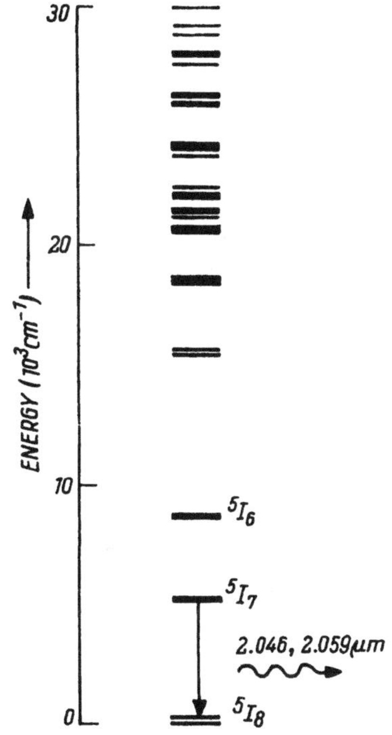

Fig. 55. Energy levels of Ho^{3+} in $CaWO_4$. Laser transitions are indicated by the arrow (after JOHNSON [6])

strengths of about 10^{-6}. Excitation in this region leads to strong infrared fluorescence at about 2 μm (Fig. 54). In addition, further fluorescence lines were observed in $CaF_2:Ho^{3+}$ at 550 nm [43], 650 nm, and 1.2 μm [55]. Aside from the splitting of the ground level, the energy level schemes of FEOFILOV [55] and JOHNSON [6] (Fig. 55) derived from the absorption and fluorescence measurements do not differ very considerably. They agree in ascribing the emission at 2 μm to a $^5I_7 - ^5I_8$ transition. The $^5I_6$ state is the initial level of the 1.2 μm emission, while the initial levels of the visible luminescence at 550 and 650 nm have yet to be identified [55]. According to JOHNSON further measurements at low temperatures are necessary for a more exact establishment of the structure within one multiplet term.

In 1962 JOHNSON, BOYD, and NASSAU [28] succeeded in achieving pulsed laser operation in $CaWO_4:Ho^{3+}$ (0.5%) at 77 °K in the lines at 2.046 and 2.059 μm. The terminal state lies 230 to 250 cm^{-1} above the ground state. The threshold value was 300 J and is assumed to decrease at 20 °K due to the lower splitting of the ground level. JOHNSON [6] pointed out that the emission lines of Ho^{3+} in CaF_2 are broader and of lower intensity. Recently BALLMAN, PORTO and YARIV [16] reported the laser emission of Ho^{3+} in calcium niobate $Ca(NbO_3)_2$. For the charge compensation of the Ho^{3+} ions (0.5% by weight) an equivalent quantity of Ti^{4+} ions was incorporated in the crystals. The threshold value for the laser emission at 2.04 μm was 90 J at 77 °K. The lifetime of the radiative transition was determined to be 2.2 ± 0.1 ms.

By resorting to reducing growth conditions and subsequent intensive γ-irradiation KARISS and FEOFILOV [67] were able to stabilize Ho^{2+} ions in CaF_2 and SrF_2.

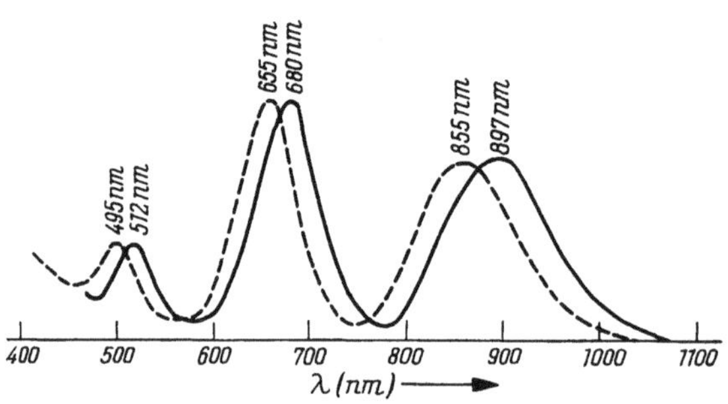

Fig. 56. Absorption spectrum of Ho^{2+} in CaF_2 (solid curve) and SrF_2 (dashed curve) at room temperature (after KARISS and FEOFILOV [67])

The broad absorption bands (Fig. 56) with half widths between 60 and 130 nm corresponding to the $4f - 5d$ transitions occur at the following wavelengths:

CaF_2	897	680	512 nm
SrF_2	855	655	495 nm

When changing from $CaF_2:Ho^{2+}$ to $SrF_2:Ho^{2+}$ all bands are shifted by about 500 to 600 cm^{-1} towards the shorter wavelengths. In the fluorescence spectrum (Fig. 57) two narrow closely spaced lines at 1.821 and 1.832 μm ($CaF_2:Ho^{2+}$) and 1.825 and 1.834 μm ($SrF_2:Ho^{2+}$), respectively, occur at 77 °K. Their half widths are expected to lie below 10 cm^{-1} (~ 5nm), while the lifetime is stated to be 30 ms. The additional bands in the fluorescence spectrum between 1.9 and 2 μm are due to Ho^{3+} (transition $^5I_7 - {^5I_8}$). The authors interpret the luminescence as transitions within the $4f$ shell ($^4I_{13/2} - {^4I_{15/2}}$) of the character of magnetic dipole radiation. JOHNSON [6] reported vain attempts to induce the afore-mentioned lines occurring in $CaF_2:Ho^{2+}$ at 1.823 and 1.834 μm to show stimulated emission.

Comprehensive measurements of the paramagnetic electron resonance on CaF_2:Ho were reported by SABISKY and LEWIS [97]. The authors believe that it should be possible to use $CaF_2:Ho^{2+}$ in powdered form as maser material.

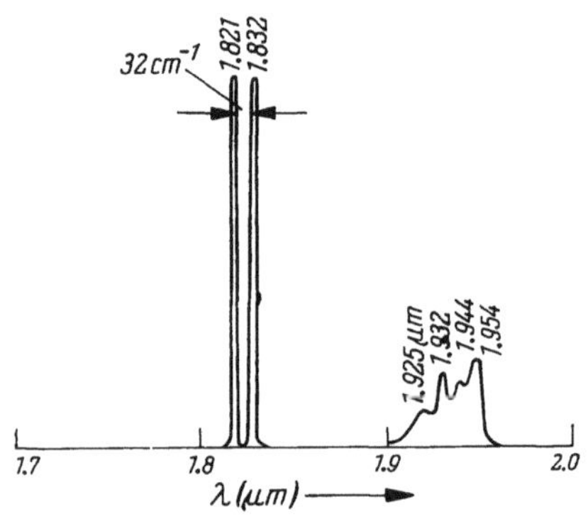

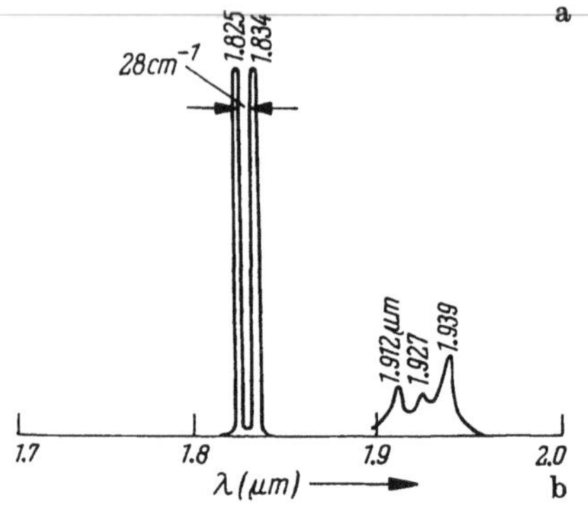

Fig. 57. Luminescence spectrum of Ho^{2+} in CaF_2 (a) and SrF_2 crystals (b) at 77 °K (after KARISS and FEOFILOV [67])

4.1.2.11 Erbium, Z = 68

	Er^{3+}	Er^{2+}
Electron configuration	$4\,f^{11}$	$4\,f^{12}$
Ground state	$^4I_{15/2}$	$^3H_6$

In 1962 KISS and DUNCAN [29] published data on the stimulated emission of Er^{3+} in $CaWO_4$. The absorption spectrum of $CaWO_4$:Er^{3+} (Fig. 58) is characteristic of the forbidden $4\,f - 4\,f$ transitions, whose strongest lines lie at about 380 and 520 nm. As the fluorescence quantum efficiency is almost unity, pumping in the lines will suffice to initiate laser oscillations.

The spontaneous fluorescence (Fig. 59) covers a range 1.45 to 1.65 μm as well as lines at about 550 and 650 nm [55] and is ascribed by the authors to transitions from the $^4I_{13/2}$ initial level to the levels of the ground state $^4I_{15/2}$ split by the crystal field (Fig. 60). The terminal state of the laser emission at 1.61 μm (77 °K) lies

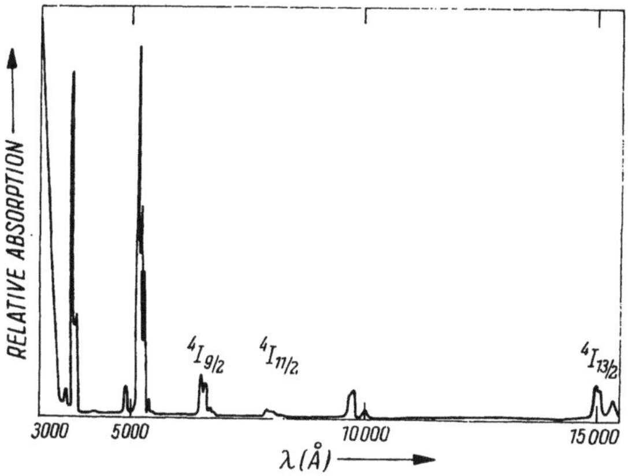

Fig. 58. Absorption spectra of $CaWO_4$:Er^{3+} at 77 °K (after KISS and DUNCAN [29])

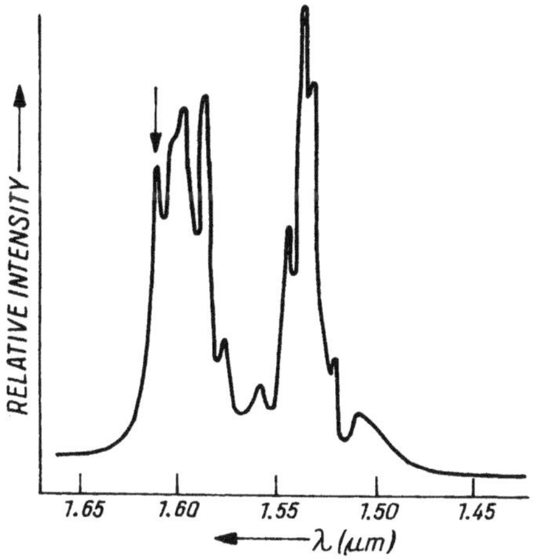

Fig. 59. Fluorescence spectra of $CaWO_4$:Er^{3+} at 77 °K. The laser line is marked with an arrow (after KISS and DUNCAN [29])

Fig. 60. Energy level diagram for $CaWO_4$:Er^{3+} showing laser transition (after KISS and DUNCAN [29])

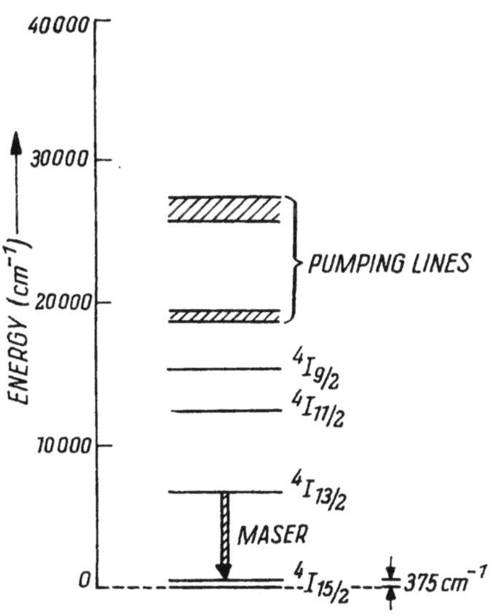

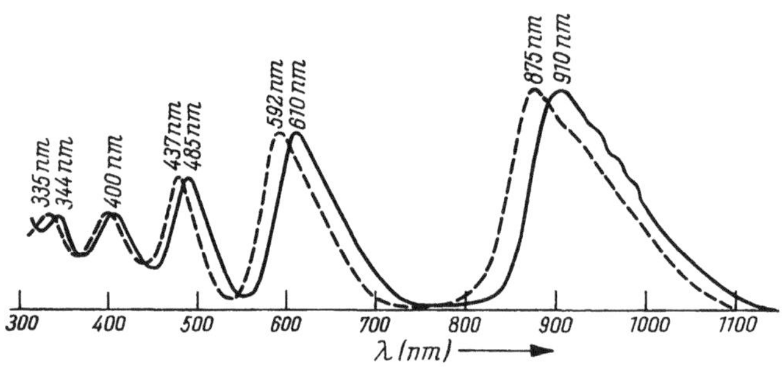

Fig. 61. Absorption spectrum of Er^{2+} in CaF_2 (solid curve) and SrF_2 (dashed curve) at room temperature (after KARISS and FEOFILOV [67])

only 375 cm^{-1} above the ground state. Further lines at about 550 and 650 nm observed by FEOFILOV [55] are due to transitions from higher pumping levels.

Charge compensation was also in this case realized by an equivalent addition of Li^+ corresponding to the Er^{3+} concentration (1 mole %). The high threshold value (800 J) is expected to be reduced by optimization of the activator concentration and the optical quality of the crystals. In calcium niobate ($Ca(NbO_3)_2$) as host lattice BALLMAN, PORTO, and YARIV [16] achieved laser emission of Er^{3+} at 1.61 μm with a threshold value of likewise 800 J.

Also KARISS and FEOFILOV [67] had to resort to reducing growth conditions and subsequent irradiation of the crystals (γ-rays) in order to incorporate Er^{2+} in CaF_2 and SrF_2. The absorption produced (300 nm to 1 μm) consists of 5 broad bands with characteristic unsymmetrical band shape (Fig. 61). In SrF_2 all bands are shifted towards the shorter wavelengths by merely several hundred cm^{-1} as compared with CaF_2. At 77 °K the authors observed a fluorescence in the region 1.9 to 2.4 μm, which could be excited in the whole absorption range (Fig. 62) and consisted of numerous lines with half widths between 5 and 10 cm^{-1}. The fluorescence spectra of Er^{2+} in CaF_2 and SrF_2 are very similar. Merely the fine structure of the fluorescence bands is somewhat more pronounced in the case of $SrF_2 : Er^{2+}$. The authors interpret the absorption as allowed $4f^{12}$–$4f^{11}5d$ transitions and the fluorescence as forbidden transitions within the $4f^{12}$ shell between the terms $^3H_5$ (or $^3H_4$) and $^3H_6$. Also the great luminescence time of 20 to 25ms pleads for magnetic dipole radiation.

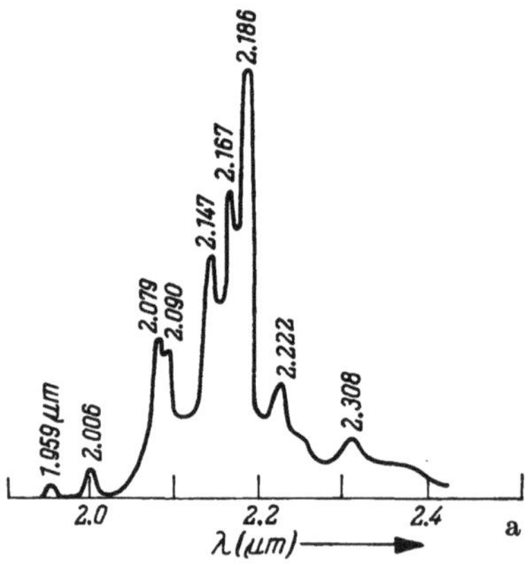

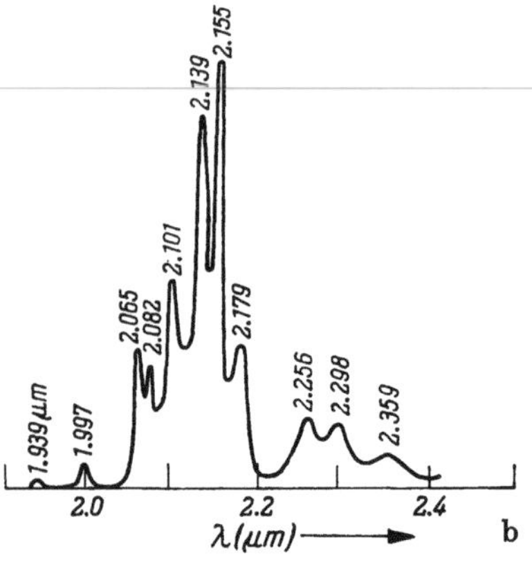

Fig. 62. Luminescence spectrum of Er^{2+} in CaF_2 (a) and SrF_2 (b) at 77 °K (after KARISS and FEOFILOV [67])

4.1.2.12 Thulium, Z = 69

	Tm^{3+}	Tm^{2+}
Electron configuration	$4f^{12}$	$4f^{13}$
Ground state	$^3H_6$	$^2F_{7/2}$

Thulium is the only element of the rare earths that produced a stimulated emission both in the trivalent state ($CaWO_4:Tm^{3+}$, $SrF_2:Tm^{3+}$) and in the divalent state ($CaF_2:Tm^{2+}$). The laser effect of Tm^{3+} was first observed by JOHNSON, BOYD, and NASSAU [32], who used $CaWO_4$ as host crystal. The absorption spectrum (Fig. 63) consists of a few sharp lines with very low oscillator strengths (10^{-7}). Irradiation in the strongest absorption lines entails a strong infrared emission (Fig. 64). Pumping in the region 460 to 480 nm (coinciding well with the

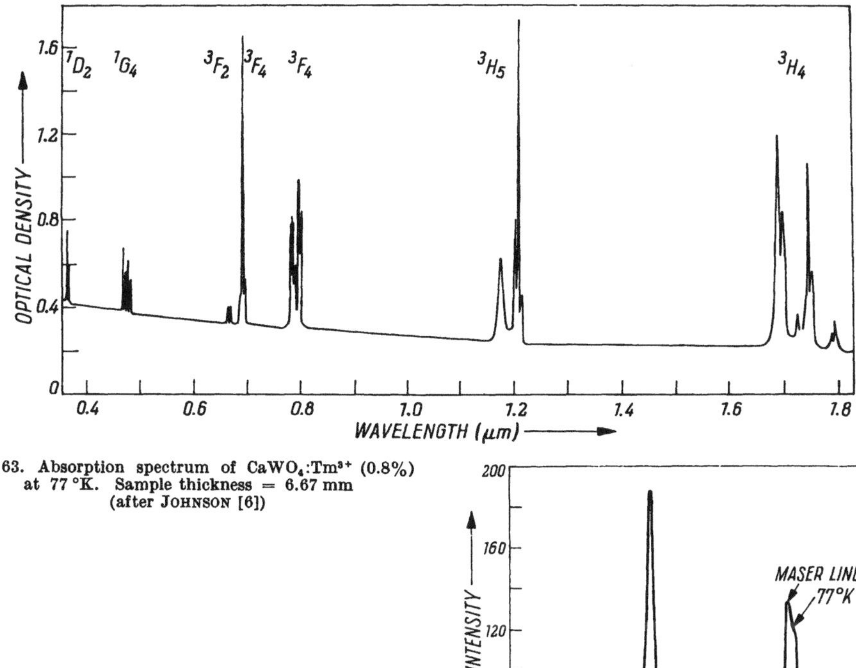

Fig. 63. Absorption spectrum of $CaWO_4:Tm^{3+}$ (0.8%) at 77 °K. Sample thickness = 6.67 mm (after JOHNSON [6])

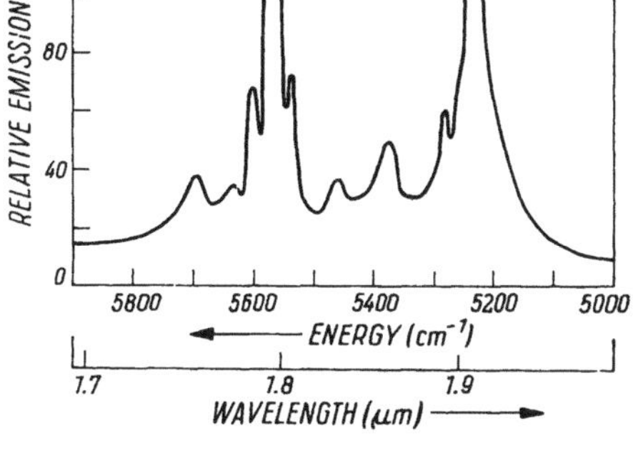

Fig. 64. Fluorescence spectrum ($^3H_4 \rightarrow {}^3H_6$) of Tm^{3+} in $CaWO_4$ at 77 °K (after JOHNSON [6])

intensive lines of a xenon lamp) led to a stimulated emission at 1.911 μm (threshold value 60 J) at 77 °K, whereas the excitation at 1.7 to 1.8 μm led to a laser emission at 1.916 (threshold value 73 J). The pertinent term scheme may be seen from Fig. 65. The laser emission corresponds to a transition from the $^3H_4$ level [6] (in [32] $^3H_5$ level) to the level lying about 325 cm^{-1} above the split $^3H_6$ ground state. Temperatures of 77 °K are necessary to achieve the inverted population. According to Johnson [6], $SrF_2:Tm^{3+}$ exhibits a strong fluorescence comparable to that of $CaWO_4:Tm^{3+}$, but has a very high threshold value (1600 J) for the laser emission at 1.972 and 1.918 [5]. In $Ca(NbO_3)_2$ the threshold value for the stimulated emission at 1.91 μm is reported by Ballman, Porto, and Yariv [16] to be 125 J.

The fluorescence spectra with several lines at 450 nm given by Stepanov and Feofilov [43, 55] for Tm^{3+} in CaF_2 are probably due to transitions from higher excitation levels.

Kiss and Duncan [30] were the first to induce Tm^{2+} to show stimulated emission in CaF_2 at 4.2 °K. At the same time they announced the publication of further data on $CaF_2:Tm^{2+}$ gained from extensive investigations. In the region between 200 and 700 nm $CaF_2:Tm^{2+}$ (0.05 mole %) has very broad absorption bands with considerable oscillator strengths of about 10^{-3} (Fig. 66). The fluorescence spectrum (Fig. 67) essentially consists of two strong lines at 1.116 μm and 1.189 μm, which in the opinion of the authors result from transitions to the crystal field splitting levels of the $^2F_{7/2}$ ground state. From the energy level diagram (Fig. 68) it might be expected that at a separation between the terminal state and the ground state of 556 cm^{-1} the system $CaF_2:Tm^{2+}$ acts as a 4-level laser and that in addition even at room temperature a considerable depopulation of the terminal

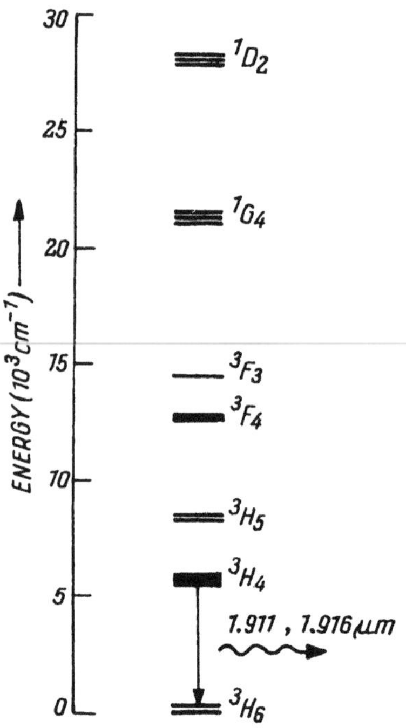

◀ Fig. 65. Energy levels of Tm^{3+} in $CaWO_4$. The arrow indicates the laser transitions (after Johnson [6])

Fig. 66. Band absorption spectra of $CaF_2:Tm^{2+}$ at 300 °K and 4.2 °K (after Kiss and Duncan [30])

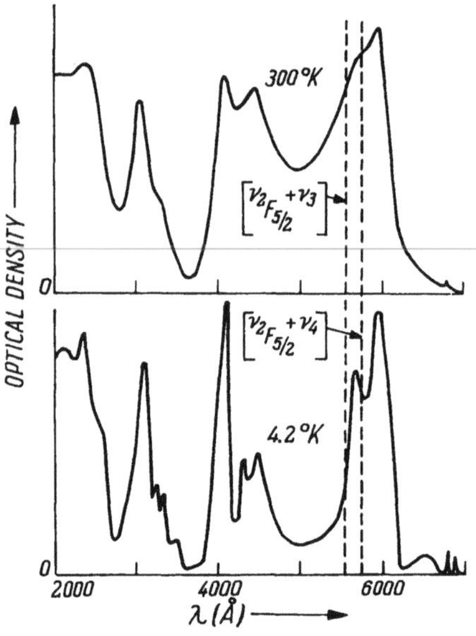

Fig. 67. Fluorescence spectrum of CaF_2: Tm^{2+} at 77 °K. The laser line (at 4.2 °K) is marked with an arrow (after KISS and DUNCAN [30])

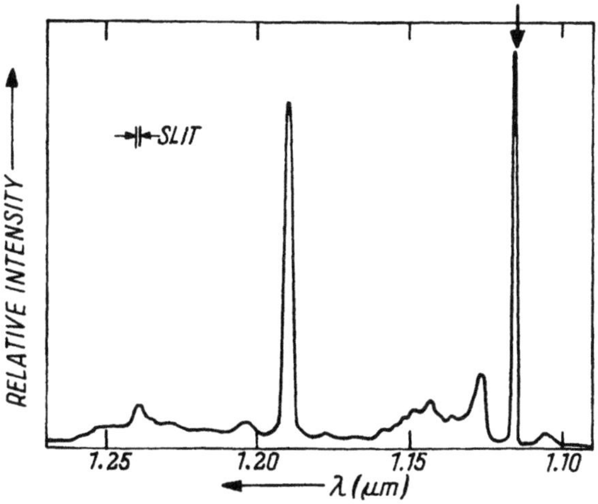

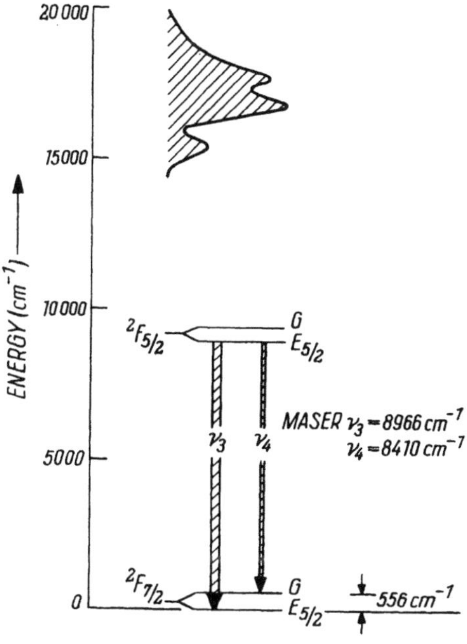

Fig. 68. Absorption spectrum and energy level diagram for CaF_2:Tm^{2+} showing the maser transition (after KISS and DUNCAN [30])

state occurs. It was, however, rather a surprise that KISS and DUNCAN did not observe the wavelength of 1.189 μm, which corresponds to the 4-level laser, but at 4.2 °K the wavelength of 1.116 μm, which corresponds to the 3-level laser. An explanation of this may be found in the absorption spectrum (Fig. 66). The 1.189-μm radiation is absorbed due to possible transitions from the excited $^2F_{5/2}$ level to the 5 d band. Only for the line at 1.116 μm and temperatures at 4.2 °K the system can oscillate owing to the lower absorption in the tail of the absorption band at 600 nm. Thus CaF_2:Tm^{2+} is the first 3-level laser that has become known to operate with an ion of the rare earths as activator.

4.1.2.13 Ytterbium, Z = 70

	Yb^{3+}	Yb^{2+}
Electron configuration	$4f^{13}$	$4f^{14}$
Ground state	$^2F_{7/2}$	$^1S_0$

The absorption and luminescence properties of Yb^{3+} in CaF_2 [47, 55] and Yb^{2+} in CaF_2, SrF_2, and BaF_2 were investigated by KAPLYANSKII and FEOFILOV [65].

Due to a $4f - 5d$ transition, Yb^{3+} in CaF_2 gives rise to a broad structureless absorption band at 255 nm and a line absorption ($H \sim 2$ to 3 Å) occurring already at room temperature in the infrared at 973.67 and 974.64 nm, which is interpreted by FEOFILOV as $4f - 4f$ transitions. Excitation in the short wavelength absorption band leads to an intensive emission in the infrared. The two strongest lines occur at 973.67 and 974.64 nm (Fig. 69) and are in resonance with the absorption lines. On cooling to 77 °K merely the luminescence doublet and the line at 1036 nm are left. The two resonance lines are interpreted as transitions between the $^2F_{7/2}$ and $^2F_{5/2}$ states which are split by the Stark effect. For the excited state a relatively large lifetime of 4 ms was found, which may be understood from the forbidden $^2F_{5/2} - {}^2F_{7/2}$ state (selection rules for n and L infringed). From the lifetime the oscillator strength of this transition was determined to be 1.7×10^{-6}, which was in an order-of-magnitude agreement with the

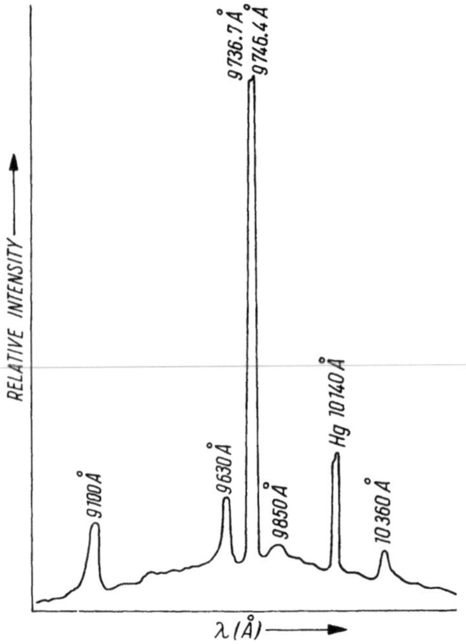

Fig. 69. Fluorescence spectrum of $CaF_2:Yb^{3+}$ in the infrared at 20 °C (after FEOFILOV [47])

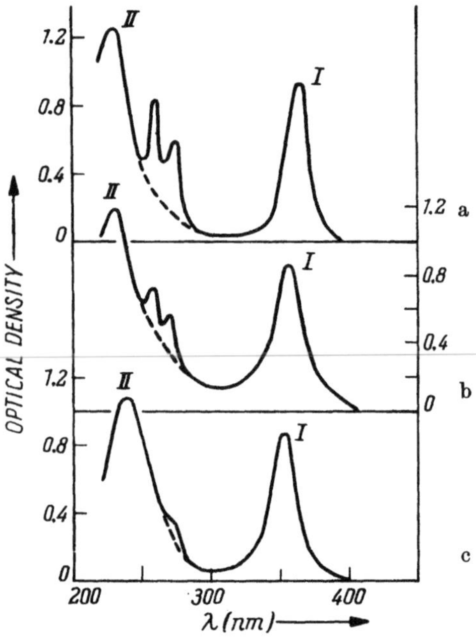

Fig. 70. Absorption spectra of Yb^{2+} in MeF_2 crystals at room temperature. a) $CaF_2:Yb^{2+}$, b) $SrF_2:Yb^{2+}$, c) $BaF_2:Yb^{2+}$ (after KAPLYANSKII and FEOFILOV [65])

values obtained from the absorption spectra. The fluorescence spectrum reported by FEOFILOV was designated by him as one of type I.

The poorly split absorption bands of Yb^{2+} in CaF_2, SrF_2, and BaF_2 (Fig. 70) lie in the ultraviolet spectral region. Even on cooling down to 77 °K the resolution of the line structure is not essentially increased. Only in the neighbourhood of the narrow bands on the long wavelength tail of band I, series of electron oscillations were observed just as in the case of Sm^{2+} and Eu^{2+}. The effect of the host lattice on the position of the absorption bands I and II may be seen from Fig. 36. On cooling to 77 °K an extremely intensive broadband luminescence peaked at about 570 nm can be observed in $CaF_2:Yb^{2+}$. Cooling down to 4.2 °K does not lead to a splitting of this band. Even at this temperature $SrF_2:Yb^{2+}$ and $BaF_2:Yb^{2+}$ could not be excited to emission. The observed relatively long lifetime of 2×10^{-4} s and the remarkably large Stokes displacement of about 10000 cm^{-1} were discussed by KAPLYANSKII and FEOFILOV on the basis of the electron configuration, the energy levels, and the known selection rules.

While Yb^{2+} and Yb^{3+} as activators in ionic crystals could not yet be excited to stimulated emission, laser emission of Yb^{3+} in silicate glass was reported as early as 1962 by ETZEL, GANDY, and GINTHER [37].

4.1.2.14 Lutecium, Z = 71

	Lu^{3+}	Lu^{2+}
Electron configuration	—	$6s\,4f^{14}$
Ground state	—	$^2S_{1/2}$

To date the authors have not found any reference in literature to the use of Lu as activator in ionic crystals.

4.1.3 Glasses

In addition to the crystalline substances described various glasses used as host substances for rare-earth ions proved to be suitable for laser applications. Pulsed laser operation was reported of Nd^{3+} in barium crown glass [33] and, later, of Yb^{3+} [37], Ho^{3+} [36], and Gd^{3+} [35] in silicate glass. Recently YOUNG [34] also reported continuous laser operation of Nd^{3+} in barium crown glass at room temperature (pumping power 900 W). While the stimulated emission of Nd^{3+} and Ho^{3+} has already become known for a crystalline host, it could so far in glass only be achieved of Yb^{3+} and Gd^{3+}. As already mentioned, Gd^{3+} in glass is the only laser system with ions of the rare earths as active medium showing stimulated emission in the ultraviolet.

Absence of crystal structure (well-defined environment for each active ion) generally manifests itself in a broadening of the emission lines and comparatively higher threshold values. The advantages of the glass result from a simpler production technology, a better optical quality, higher possible activator concentrations as well as the possibility to optimize the properties by varying both the composition and the proportions. The combination of laser technology and fibre glass technology will offer great possibilities for the future. SNITZER [98] reported laser action of a 6 m glass-Nd^{3+} fibre bundle. The system glass:Nd^{3+} is likewise suited for high energy outputs. SNITZER and WOODCOCK [99] used a fibre bundle of 18″ length and 1/4″ diameter and obtained an energy output of 113 J with 9000 J input energy.

Laser emission in quartz glass was achieved by optical pumping. As the cause of this is believed to lie in transitions of the silicon (exciton transition), the result is discussed in Section 4.4.

References

[76] G. MANDEL, R. P. BAUMAN, and E. BANKS, J. chem. Phys. **33**, 192 (1960).
[77] P. GÖRLICH, H. KARRAS, and R. REIMANN, Sitzungsber. Dtsch. Akad. Wiss., Kl. Math. Phys. Techn., Nr. 3, Berlin 1960.
[78] P. GÖRLICH, H. KARRAS, A. KÖTHE, and K. KÜHNE, phys. stat. sol. **1**, 366 (1961).
[79] A. A. KAPLYANSKII, V. N. MEDVEDEV, and P. P. FEOFILOV, Optika i Spektroskopiya **14**, 664 (1963).
[80] G. H. DIEKE and R. SARUP, J. chem. Phys. **29**, 741 (1958).
[81] E. SNITZER, Phys. Rev. Letters **7**, 444 (1961).
[82] E. H. CARLSON and H. G. DIEKE, J. chem. Phys. **24**, 224 (1958).
[83] E. H. CARLSON, Johns Hopkins Univ. Spectroscopic Rept. **16**, March (1960).
[84] P. GÖRLICH, H. KARRAS, G. KÖTITZ, and W. LÜDKE, phys. stat. sol. in press.
[85] D. L. WOOD and W. KAISER, Phys. Rev. **126**, 2079 (1962).
[86] A. A. KAPLYANSKII and A. K. PRZHEVUSKII, Optika i Spektroskopiya **13**, 882 (1962).
[87] B. P. ZAKHARCHENYA and A. YA. RYSKIN, Optika i Spektroskopiya **13**, 875 (1962).
[88] P. P. FEOFILOV and M. N. TOLSTOI, Optika i Spektroskopiya **13**, 294 (1962).
[89] J. D. AXE and P. P. SOROKIN, Phys. Rev. **130**, 945 (1963).
[90] V. E. KARAPETYAN, B. I. MAKSAKOV, and P. P. FEOFILOV, Optika i Spektroskopiya **14**, 441 (1963).
[91] N. RABBINER, Phys. Rev. **130**, 502 (1963).
[92] P. GÖRLICH, H. KARRAS, and K. KÜHNE, Sitzungsber. Dtsch. Akad. Wiss., Kl. Math. Phys. Techn., Nr. 2, Berlin 1959;
P. GÖRLICH, Monatsber. Dtsch. Akad. Wiss. Berlin **1**, 228 (1959).
[93] B. P. ZAKHARCHENYA and A. YA. RYSKIN, Optika i Spektroskopiya **14**, 309 (1963).
[94] P. P. FEOFILOV, Optika i Spektroskopiya **12**, 531 (1962).
[95] J. D. KINGSLEY and J. S. PRENER, Phys. Rev. **126**, 458 (1962).
[96] N. RABBINER, Phys. Rev. **132**, 224 (1963).
[97] E. S. SABISKY and H. R. LEWIS, Proc. IEEE **51**, 53 (1963).
[98] E. SNITZER, Paper read at the Symposium on Optical Processing of Information October 23, 1962, at Washington D.C.
[99] E. SNITZER and R. F. WOODCOCK, Paper read at the Sixth International Congress on Glass, July 1962, Washington D.C.

(Received May 23, 1964)

Original Papers

phys. stat. sol. 6, 319 (1964)

Институт физики Академии Наук Украинской ССР, Киев

Влияние ангармонизма колебаний атомов на фотопереходы в примесных центрах твердого тела

А. Ф. Лубченко

Проведено квантовомеханическое изучение влияния ангармонизма кристаллических колебаний на оптические свойства примесных центров в твердых телах. Исследовано влияние ангармонизма на форму спектров поглощения и излучения света примесными центрами. Показано, что зависимость квазиупругих постоянных твердого тела от электронного состояния примесного центра и ангармонизм кристаллических колебаний приводит к нарушению зеркальной симметрии спектров излучения и поглощения света.

A quantum mechanical calculation is given of the effect of lattice anharmonicity on the optical properties of impurity centres in solids. The optical properties considered are the shape and temperature dependence of the absorption and emission bands of an impurity. It is shown that, due to the inter-dependence of the electronic state of an impurity and the quasi-elastic constants of the solid, the anharmonicity of the lattice vibrations destroys the "mirror symmetry" of the absorption and emission bands.

1. Введение

Исследованию фонон-фононного взаимодействия в монокристаллах и его роли в различных физических процессах последнее время приделяется довольно значительное внимание. В работах [1, 2, 3] была развита квантовая теория ширины линии, которая явилась обобщением классической теории Борна-Блэкмана [4] и Уоллеса-Джепсена [5]. В работах [6, 7, 8] исследовано вклад ангармонизма колебаний атомов идеального кристалла в его термодинамические функции; роль ангармонизма при рассеянии рентгеновских лучей и нейтронов монокристаллами исследовалась Ханом [9] и другими авторами [10, 11, 12, 13].

При переходе от идеальных кристаллов к кристаллам, содержащим примеси (в дальнейшем мы будем называть их твердыми растворами) возникает новый механизм рассеяния фононов, а также появляется ряд особенностей в фонон-фононном рассеянии, связанным с отсутствием трансляционной симметрии системы. В идеальном кристалле при рассеянии фононов на фононах закон сохранения импульса выполняется; в твердых растворах, благодаря отсутствию трансляционной симметрии, строго говоря, теряет смысл понятие фонона, как плоской монохроматической волны с квазиимпульсом $\boldsymbol{k}$. Это обстоятельство значительно усложняет исследование роли ангармонизма в неидеальном кристалле. Учет ангармонизма при изучении фотопереходов в примесных центрах осложняется также тем, что в адиабатическом приближении [14] положения равновесия атомов твердого тела при различных электронных состояниях примеси оказываются различными. В связи с этим в настоящей работе мы ограничимся исследованием фотопереходов в примесных центрах при достаточно низких температурах, когда можно считать, что роль ангармонизма в состоянии, из которого совершается фотопереход, пренебрежимо мала, и будем учитывать ангармонизм лишь того состояния, в которое совершается фотопереход. Как будет показано ниже, уже при таком упрощающем предположении „закон зеркальной симметрии" [15] для спектров излучения и поглощения нарушается, в то время, как учет одной лишь зависимости квазиупругих постоянных от электронного состояния примеси, вопреки распространенному мнению [16, 17, 18], в гармоническом приближении, не приводит [19] к нарушению закона зеркальной симметрии.

2. Вероятность фотопереходов в примесных центрах при учете ангармонизма

Рассмотрим систему, состоящую из твердого тела — растворителя, примесного центра и поля излучения. Будем предполагать, что область поглощения света примесными центрами лежит далеко от области поглощения растворителя и концентрация их настолько мала, что взаимодействием между ними можно пренебречь. Тогда для вычисления волновых функций стационарных состояний такой системы можно воспользоваться адиабатическим методом [14]. Считая оптические электроны примесного центра и ядра всех атомов системы „медленной" подсистемой, а все прочие электроны-„быстрой" подсистемой, в адиабатическом приближении найдем [17] волновую функцию φ_n — описывающую состояния этих электронов. Собственные значения, соответствующие функции φ_n являются, как известно, потенциалом, в котором движется „медленная" подсистема. В последней мы опять выделим электронную подсистему, включающую оптические электроны примеси, как „быструю" подсистему по отношению к подсистеме, состоящей из всех ядер твердого раствора. Пользуясь адиабатическим методом, вычислим волновую функцию оптических электронов примеси $\psi_{nl}(\boldsymbol{r}\,\boldsymbol{R})$. Она будет зависить от координат $\boldsymbol{r}$ оптических электронов примеси и координат $\boldsymbol{R}$ всех ядер раствора; индекс l означает полный набор квантовых чисел, описывающих состояния этих электронов, причем, поскольку в дальнейшем рассматриваются лишь фотопереходы с возбужденнием нижайших состояний примесного центра, без возбуждения остальных электронов системы, то индекс n у φ_n и ψ_{nl} можно опустить. При этом колебания ядер системы будет описываться гамильтонианом

$$H_l = T_{\boldsymbol{R}} + V_l(\boldsymbol{R}),$$

где $T_{\boldsymbol{R}}$ оператор кинетической энергии ядер, $V_l(\boldsymbol{R})$ оператор потенциальной энергии ядер в адиабатическом смысле. Разлагая далее $V_l(\boldsymbol{R})$ в ряд по степеням смещения $\boldsymbol{u}_j = \boldsymbol{R}_j - \boldsymbol{R}_j^\circ$ ядер от положений равновесия $\boldsymbol{R}_j^\circ$, получим

$$H_l = V_l(\boldsymbol{R}_0) + T_{\boldsymbol{R}} + \sum_{j\alpha} \frac{\partial V_l}{\partial R_{j\alpha}} u_{j\alpha} + \frac{1}{2} \sum_{\substack{j, j' \\ \alpha, \alpha'}} \frac{\partial^2 V_l}{\partial R_{j\alpha}\, \partial R_{j'\alpha'}} u_{j\alpha}\, u_{j'\alpha'} +$$
$$+ \frac{1}{3!} \sum_{\substack{j\, j'\, j'' \\ \alpha\, \alpha'\, \alpha''}} V_{3l}(j\,\alpha, j'\,\alpha', j''\,\alpha'')\, u_{j\alpha}\, u_{j'\alpha'}\, u_{j''\alpha''} +$$
$$+ \frac{1}{4!} \sum_{\substack{j\, j'\, j''\, j''' \\ \alpha\, \alpha'\, \alpha''\, \alpha'''}} V_{4l}(j\,\alpha, j'\,\alpha', j''\,\alpha'', j'''\,\alpha''')\, u_{j\alpha}\, u_{j'\alpha'}\, u_{j''\alpha''}\, u_{j'''\alpha'''} + \cdots,$$

где V_{3l}, V_{4l} соответственно третяя и четвертая производные от V_l по $R_{j\alpha}$ взятые в точке $\boldsymbol{R}_j^\circ$, причем $V_{3l} \sim a_0\, V_{4l}$, где a_0 среднее значение межатомных расстояний [20]. Если в последнем выражении положить

$$u_{j\alpha} = \sum_s \left(\frac{\hbar}{2\, M_j \cdot \omega_s^{l_0}}\right)^{1/2} q_{j\alpha s}\, (b_s^+ + b_s), \qquad p_{j\alpha} = -\, i \sum_s \left(\frac{M_j\, \hbar\, \omega_s^{l_0}}{2}\right)^{1/2} q_{j\alpha s}\, (b_s - b_s^+),$$

где M_j масса ядра, $\omega_s^{l_0}$ частоты нормальных колебаний раствора, когда примесь находится в основном состоянии ($l = l_0$), b_s^+, b_s операторы рождения и уничтожения элементарных возбуждений для s го нормального колебания, то оператор Гамильтона H_l, с точностью до N^{-1}, где N число атомов системы, запишется в виде

$$H = V_l(\boldsymbol{R}_0) + \sum_s \hbar\, \omega_s^l \left(b_s^+\, b_s + \frac{1}{2}\right) + \sum_s V_s^l\, (b_s^+ + b_s) + \sum_s \Delta\omega_s\, (b_s^+\, b_s^+ + b_s\, b_s) +$$
$$+ \hbar \sum_{s_3 s_1 s_2} Y^l_{s_3 s_1 s_2} \prod_{i=1}^{3} (b_{s_i}^+ + b_{s_i}) + \hbar \sum_{s_4 s_1 s_2 s_3} Y^l_{s_4 s_1 s_2 s_3} \prod_{i=1}^{4} (b_{s_i}^+ + b_{s_i}) + \cdots. \tag{1}$$

Здесь

$$\left.\begin{aligned}
\Delta\omega_s &= \omega_s^l - \omega_s^{l_0}, \ V_s^l = \sum_{j\alpha} \frac{\partial V_l}{\partial R_{j\alpha}} q_{j\alpha s} \left(\frac{\hbar}{2\, M_j \omega_s^{l_0}}\right)^{1/2}, \\
Y^l_{s s_1 s_2} &= \frac{\hbar^{1/2}}{3!} \sum_{\substack{j\, j'\, j'' \\ \alpha\, \alpha'\, \alpha''}} \frac{V_{3l}\, q_{j\alpha s}\, q_{j'\alpha' s_1}\, q_{j''\alpha'' s_2}}{\sqrt{M_j\, M_{j'}\, M_{j''} \omega_s^{l_0} \omega_{s_1}^{l_0} \omega_{s_2}^{l_0}}}, \\
Y^l_{s s_1 s_2 s_3} &= \frac{\hbar}{4!} \sum_{\substack{j\, j'\, j''\, j''' \\ \alpha\, \alpha'\, \alpha''\, \alpha'''}} \frac{V_{4l}\, q_{j\alpha s}\, q_{j'\alpha' s_1}\, q_{j''\alpha'' s_2}\, q_{j'''\alpha''' s_3}}{\sqrt{M_j\, M_{j'}\, M_{j''}\, M_{j'''} \omega_s^{l_0} \omega_{s_1}^{l_0} \omega_{s_2}^{l_0} \omega_{s_3}^{l_0}}}.
\end{aligned}\right\} \tag{2}$$

Легко убедиться, пользуясь (2), что величины $Y^l_{s s_1 s_2}$, $Y^l_{s s_1 s_2 s_3}$ симметричны относительно перестановки индексов s, причем, поскольку $q_{j\alpha s}$ нормированы так, что

$$\sum_s q_{j'\alpha' s}\, q_{j\alpha s} = \delta_{j j'}\, \delta_{\alpha\alpha'}, \qquad \sum_{j\alpha} q_{j\alpha s}\, q_{j\alpha s'} = \delta_{s s'},$$

то они имеют соответственно порядок не ниже $N^{-3/2}$ и N^{-2}. Третье слагаемое в (1) обусловлено зависимостью квазиупругих постоянных от электронного состояния примеси; как было показано в [19] для кристаллических колебаний оно не вносит вклада в функцию, определяющую форму спектра, поскольку $\Delta\omega_s \sim N^{-1}$; для локальных и внутримолекулярных колебаний (если примесным центром является молекула) это слагаемое необходимо учитывать. Взаимодействие внутримолекулярных (локальных) колебаний с кристаллическими, которое учитывается в (1), обуславливает ширину уровней; мы будем учитывать ее феноменологически (обозначая полную ширину уровня Γ), а оператор H_l перепишем в виде

$$H_l = H_{lm} + H_{lk},$$

где H_{lm} описывает локальные и внутримолекулярные колебания в примесном центре, а

$$H_{lk} = V_l(\boldsymbol{R}_0) + \sum_s \hbar\,\omega_s^l\left(b_s^+ b_s + \frac{1}{2}\right) + \sum_s V_s^l\,(b_s^+ + b_s) + \\ + \hbar \sum_{s_3 s_1 s_2} Y^l_{s_3 s_1 s_2} \prod_{i=1}^{3} (b_{s_i}^+ + b_{s_i}) + \sum_{s_4 s_1 s_2 s_3} Y^l_{s_4 s_1 s_2 s_3} \prod_{i=1}^{4} (b_{s_i}^+ + b_{s_i}) + \cdots \tag{3}$$

описывает колебания решетки растворителя вместе с примесями, как целым (без локальных колебаний).

Обозначим далее через $|v_\tau\rangle$, ε_{v_τ} волновые функции и собственные значения оператора H_{lm}, а через $|i\rangle$, E_i и $|f\rangle$, E_f волновые функции и собственные значения оператора H_{lk} для начального и конечного состояний и рассмотрим переход $l_1\,|v\rangle\,|i\rangle \to l_0\,|v_\tau\rangle\,|f\rangle$ при котором примесный центр переходит из возбужденного состояния $l_1\,|0\rangle$ в основное $l_0\,|v_\tau\rangle$ с возбуждением внутримолекулярного колебания номера τ и излучает квант света энергии $\hbar\,\nu$. Вероятность такого перехода можно записать в виде [17, 19]

$$w_r = \frac{1}{\hbar^2} \sum_{\{i\}\{f\}} w(i)\,|L_{l_1 l_0}|^2 \frac{|\langle i|\,f\rangle|^2}{\left(\nu - \Omega_1 + \frac{1}{\hbar} E_f - \frac{1}{\hbar} E_i\right)^2 + \frac{\Gamma^2}{4}}, \tag{4}$$

где $\hbar\,\Omega_1 = V_{l_1}(\boldsymbol{R}_0) - V_{l_0}(\boldsymbol{R}_0) + \varepsilon_0 - \varepsilon_{v_\tau}$, $L_{l_1 l_0}$ матричный элемент оператора энергии взаимодействия электромагнитного поля с примесью, вычисленный на функциях $\psi_{l_1}\,|0\rangle$, $\psi_{l_0}\,|v_\tau\rangle$, $w(i)$ функция распределения Гиббса для элементарных возбуждений, описывающих колебания решетки раствора с примесными центрами при $l = l_1$. Выберем в качестве равновесных положений ядер системы положения, соответствующие $l = l_0$; тогда $V_s^{l_0} = 0$; при этом последнее выражение можно привести к виду

$$w_r = \frac{|L_{l_1 l_0}|^2}{\hbar^2} F_r(\nu)\,.$$

Здесь

$$F_r(\nu) = \frac{2}{Z_0\,\Gamma} \operatorname{Re} \int_0^\infty d\mu \exp\left\{i\,\mu\,(\Omega_1 - \nu) - \frac{\mu\,\Gamma}{2}\right\} \operatorname{Sp}\left\{e^{-\beta\hbar H_i + i\mu H_i}\, e^{-i\mu H_f}\right\}, \tag{5}$$

$$H_i = \sum_s \omega_s^{l_1}\left(b_s^+ b_s + \frac{1}{2}\right) + \frac{1}{\hbar} \sum_s V_s^{l_1}\,(b_s^+ + b_s)\,, \quad H_f = \frac{H_{k\,l_0}}{\hbar},$$

причем в соответствии с вышесказанным в выражении для H_i мы пренебрегли ангармоническими членами, Z_0 статистическая сумма.

Умножим далее выражение, стоящее под знаком шпура на $S_{l_1}^+ S_{l_1}$, где $S_{l_1} = \exp\left\{\sum_s \xi_{s l_1}(b_s^+ - b_s)\right\}$, $\xi_{s l_1} = -\frac{V_s^{l_1}}{\hbar \omega_s^{l_1}}$. При таком унитарном преобразовании в операторах H_i исчезнет линейный по $(b_s^+ + b_s)$ член и мы получим

$$Z_0^{-1}\,\mathrm{Sp}\,\{S_{l_1}^+ S_{l_1} e^{-\beta\hbar H_i + i\mu H_i} e^{-i\mu H_f}\} = Z^{-1}\,\mathrm{Sp}\,\{e^{-\beta\hbar H_1 + i\mu H_1} S_{l_1}^+ e^{-i\mu H_f} S_{l_1}\} \times$$
$$\times\, e^{i\mu \sum_s \frac{\omega_s^{l_1}}{2} - i\mu \sum_s \omega_s^{l_1} \xi_{s l_1}^2}, \tag{6}$$

где

$$H_1 = \sum_s \omega_s^{l_1} b_s^+ b_s\,, \qquad Z = \mathrm{Sp}\,\{e^{-\beta\hbar H_1}\}\,.$$

Выполняя далее в (6) унитарное преобразование $S_{l_1}^+$, S_{l_1} и вводя обозначения

$$\sum_{s s_1 s_2} \xi_{s l_1} \xi_{s_1 l_1} \xi_{s_2 l_1} \{8\, Y_{s s_1 s_2}^{l_0} + 16 \sum_{s_3} Y_{s s_1 s_2 s_3}^{l_0} \xi_{s_3 l_1} + \cdots\} = \Phi^{l_1 l_0}\,,$$
$$\sum_{s_1 s_2} \xi_{s_1 l_1} \xi_{s_2 l_2} \{12\, Y_{s s_1 s_2}^{l_0} + 32 \sum_{s_3} Y_{s s_1 s_2 s_3}^{l_0} \xi_{s_3 l_1} + \cdots\} = \Phi_s^{l_1 l_0}\,,$$
$$\sum_{s_2} \xi_{s_2 l_1} \{6\, Y_{s s_1 s_2}^{l_0} + 24 \sum_{s_3} Y_{s s_1 s_2 s_3}^{l_0} \xi_{s_3 l_1} + \cdots\} = \Phi_{s s_1}^{l_1 l_0}\,,$$
$$Y_{s s_1 s_2}^{l_0} + 8 \sum_{s_3} Y_{s s_1 s_2 s_3}^{l_0} \xi_{s_3 l_1} + \cdots = \Phi_{s s_1 s_2}^{l_1 l_0}\,,$$

получим

$$F_r(\nu) = \frac{2}{\Gamma}\,\mathrm{Re} \int_0^\infty d\mu \exp\left\{i\mu(\Omega_2 - \nu) - \frac{\mu\Gamma}{2}\right\} A(\mu)\,.$$

Здесь

$$A(\mu) = Z^{-1}\,\mathrm{Sp}\,\{e^{-\beta\hbar H_1 + i\mu H_1}\; e^{-i\mu H_0 - i\mu U}\}\,, \tag{7}$$

$$\left.\begin{aligned}
&\beta = (kT)^{-1}\,,\\
&\Omega_2 = \Omega_1 + \frac{1}{2}\sum_s \Delta\tilde{\omega}_s - \Phi^{l_1 l_0} - \sum_s (\omega_s^{l_1} + \omega_s^{l_0})\,\xi_{s l_1}\,, \quad \Delta\tilde{\omega}_s = \omega^{l_1} - \omega_s\,,\\
&\omega_s = \omega_s^{l_0} + 2\,\Phi_{ss}^{l_1 l_0}\,, \quad H_0 = \sum_s \omega_s b_s^+ b_s\,,\\
&U = \sum_s (\omega_s^{l_0} \xi_{s l_1} + \Phi_s^{l_1 l_0})(b_s^+ + b_s) + \sum_s \Phi_{ss}^{l_1 l_0}(b_s^+ b_s^+ + b_s b_s) +\\
&\quad + \sum_{\substack{s s_1\\(s \neq s_1)}} \Phi_{s s_1}^{l_1 l_0}(b_s^+ + b_s)(b_s^+ + b_s) + \sum_{s_1 s_2 s_3} \Phi_{s s_1 s_2}^{l_1 l_0} \prod_{i=1}^{3} (b_{s_i}^+ + b_{s_i}) +\\
&\quad + \sum_{s_1 s_2 s_3 s_4} \Phi_{s_1 s_2 s_3 s_4}^{l_1 l_0} \prod_{i=1}^{4} (b_{s_i}^+ + b_{s_i}) + \cdots\,.
\end{aligned}\right\} \tag{8}$$

Последнюю экспоненту (7) можно представить в виде

$$e^{i\mu H_0} S_r(\mu)\,,$$

где $S_r(\mu) = T \exp\left\{-\int_0^\mu U(t)\,dt\right\}$. $U(t)$ оператор определяемый (8) в представлении взаимодействия, построенном с помощью H_0, T оператор хронологического упорядочения. Если положить теперь $\varkappa_s = \beta \hbar \omega_s^{l_1} - i\mu \Delta\widetilde{\omega}_s$, то выражение для $A(\mu)$ можно представить в виде

$$A(\mu) = \langle S_r(\mu) \rangle \frac{\mathrm{Sp}\left\{e^{-\sum\limits_s \varkappa_s b_s^+ b_s}\right\}}{Z}, \tag{9}$$

где

$$\langle \cdots \rangle = \frac{\mathrm{Sp}\left\{e^{-\sum\limits_s \varkappa_s b_s^+ b_s} \cdots\cdots\right\}}{\mathrm{Sp}\left\{e^{-\sum\limits_s \varkappa_s b_s^+ b_s}\right\}}.$$

Последний множитель (9) легко вычислить, если воспользоваться теоремой Вика, обобщенной на случай отличных от нуля температур [21]; при этом учитывая, что $\Delta\widetilde{\omega}_s \sim N^{-1}$, получим

$$Z^{-1}\,\mathrm{Sp}\left\{e^{-\sum\limits_s \varkappa_s b_s^+ b_s}\right\} = \exp\left\{i\mu \sum_s \bar{n}_s \Delta\widetilde{\omega}_s\right\},$$

где

$$\bar{n}_s = \{e^{\beta\hbar\omega_s^{l_1}} - 1\}^{-1}.$$

Таким образом, для функции, определяющей спектральное распределение люминесценции, получаем выражение

$$F_r(\nu) = \frac{2}{\Gamma}\,\mathrm{Re}\int_0^\infty d\mu \exp\left\{i\mu(\Omega_3 - \nu) - \frac{\mu\Gamma}{2}\right\} \langle S_r(\mu)\rangle, \tag{10}$$

где

$$\Omega_3 = \Omega_2 + \sum_s \bar{n}_s \Delta\omega_s.$$

Совершенно аналогично можно найти выражение для функции, определяющей спектральное распределение поглощаемого света. Обозначим в этом случае через E_i, $|i\rangle$ и E_f, $|f\rangle$ собственные значения и волновые функции, описывающие колебания ядер кристалла вместе с примесными центрами (как целым), когда последние находятся соответственно в состояниях $\psi_{l_0}|0\rangle$ и $\psi_{l_1}|v_\tau\rangle$. Тогда вероятность w_a перехода $l_0|0\rangle|i\rangle \to l_1|v_\tau\rangle|f\rangle$ с поглощением кванта света частоты ν, при сделанных выше предположениях, можно записать в виде

$$w_a = \frac{\bar{N}_\nu |L_{l_1 l_0}|^2}{\hbar^2} F_a(\nu),$$

где

$$F_a(\nu) = \sum_{\{i\}\{f\}} w(i) \frac{|\langle i|f\rangle|^2}{\left(\Omega' - \nu + \frac{1}{\hbar}E_f - \frac{1}{\hbar}E_i\right)^2 + \frac{\Gamma^2}{4}},$$

$\bar{N}_\nu$ среднее число фотонов на один радиационный осциллятор; все остальные обозначения те же, что и в (4). Также как и выше выражение для $F_a(\nu)$ можно привести к виду

$$F_a(\nu) = \frac{2}{Z\Gamma}\,\mathrm{Re}\int_0^\infty d\mu\, e^{i\mu(\nu - \Omega') - \frac{\mu\Gamma}{2}}\,\mathrm{Sp}\left\{e^{-\beta\hbar H_2 + i\mu H_2}\, e^{-i\mu(H_1 + U_1)}\right\}.$$

Здесь

$$\left.\begin{aligned} &H_2 = \sum_s \omega_s^{l_0} b_s^+ b_s\,, \quad \hbar\,\Omega' = V_{l_1}(\boldsymbol{R}_0) - V_{l_0}(\boldsymbol{R}_0) + \varepsilon_{v_\tau} - \varepsilon_0 + \frac{\hbar}{2}\sum_s \Delta\omega_s\,,\\ &Z = \mathrm{Sp}\,\{e^{-\beta\hbar H_2}\}\,,\\ &U_1 = -\sum_s \omega_s^{l_1}\,\xi_{s\,l_1}\,(b_s^+ + b_s) + \sum_{s_1 s_2 s_3} Y^{l_1}_{s_1 s_2 s_3}\prod_{i=1}^{3}(b_{s_i}^+ + b_{s_i}) +\\ &\qquad + \sum_{s_1 s_2 s_3 s_4} Y^{l_1}_{s_1 s_2 s_3 s_4}\prod_{i=1}^{4}(b_{s_i}^+ + b_{s_i})\,. \end{aligned}\right\} \tag{11}$$

Полагая далее

$$e^{-i\mu(H_1 + U_1)} = e^{-i\mu H_1}\, S_a(\mu)\,,$$

где $S_a(\mu) = T\exp\left\{-i\,\mu\int\limits_0^\mu U_1(t)\;dt\right\}$, $U_1(t)$ оператор U_1, определяемый (11), в представлении взаимодействия, построенном с помощью H_1, так же как и выше, получим

$$F_a(\nu) = \frac{2}{\Gamma}\,\mathrm{Re}\int\limits_0^\infty d\mu\,\exp\left\{i\,\mu\,(\nu - \Omega' - \sum_s \bar{n}_s\,\Delta\omega_s) - \frac{\mu\,\Gamma}{2}\right\}\langle S_a(\mu)\rangle\,. \tag{12}$$

Таким образом, вычисление формы спектра излучения и поглощения света примесными центрами твердого тела сводится, в первую очередь, к вычислению выражений $\langle S_r(\nu)\rangle$ и $\langle S_a(\nu)\rangle$.

3. Форма спектров излучения и поглощения света примесными центрами твердого тела при учете ангармонизма

Для вычисления средних $\langle S_i(\mu)\rangle$, $i = r, a$ рассмотрим S матрицу

$$S_\alpha(\mu) = T\,e^{-i\,\alpha\int\limits_0^\mu U(t)\,dt}\,.$$

Очевидно, что при $\alpha = 0$

$$\langle S_0(\mu)\rangle = 1$$

поэтому продифференцировав $\langle S_\alpha(\mu)\rangle$ по α и разделив правую и левую части полученного выражения на $\langle S_\alpha(\mu)\rangle$, после интегрирования его по α', получим (при $\alpha = 1$)

$$\langle S_1(\mu)\rangle = \exp\left\{-\int\limits_0^1 d\alpha'\int\limits_0^\mu \frac{\langle T\;U(t)\;S_{\alpha'}(\mu)\rangle}{\langle S_{\alpha'}(\mu)\rangle}\,dt\right\}. \tag{13}$$

В этом выражении будем рассматривать $S_\alpha(\mu)$ как функцию параметра η, определяемую соотношением

$$S_{\alpha'\eta} = T\;\exp\left\{-\,i\,\alpha'\,\eta\int\limits_0^\mu U(t)\,dt\right\}.$$

Подставляя его в (13), найдем явное выражение для $\langle S_{1\eta}(\mu)\rangle$. Очевидно, что

$$\langle S_{1\,0}(\mu)\rangle = \exp\left\{-\,i\int\limits_0^\mu \langle U(t)\rangle\,dt\right\}.$$

Пользуясь этим выражение, как начальным условием, продифференцируем $\langle S_{1\eta}(\mu)\rangle$ по η разделим полученное выражение на $\langle S_{1\eta}(\mu)\rangle$ и проинтегрируем его по η. При этом легко определить $\langle S_{11}(\mu)\rangle$, которое выражается через $U(t)$ и $S_{\alpha'\eta'}(\mu)$. В полученном выражении будем рассматривать $S_{\alpha'\eta'}(\mu)$ как функцию параметра λ, т. е. положим

$$S_{\alpha'\eta'\lambda} = T \exp\left\{-i\,\alpha'\,\eta'\,\lambda \int_0^\mu U(t)\,dt\right\}.$$

Тогда легко показать, что

$$\langle S_{110}(\mu)\rangle = \exp\left\{-i\int_0^\mu \langle U(t)\rangle\,dt - \frac{1}{2}\int\int_0^\mu dt\,dt'\,(\langle T\,U(t)\,U(t')\rangle - \langle U(t)\rangle\,\langle U(t')\rangle)\right\}. \tag{14}$$

Выражение (14) служит начальным условием для определения $\langle S_{111}(\mu)\rangle$, которое вычисляется указанным выше способом. При этом для $\langle S_{111}(\mu)\rangle$ находим выражение

$$\langle S_{111}(\mu)\rangle = \langle S_{110}(\mu)\rangle \exp\left\{\int_0^1 A_{\lambda'}(\mu)\,d\lambda'\right\},$$

где $A_{\lambda'}(\mu)$ составлено из операторов $U(t)$ и $S_{\alpha'\eta'\lambda'}(\mu)$ по тем же правилам, что и (13). Вводя в матрицу $S_{\alpha'\eta'\lambda'}(\mu)$ еще один параметр, легко найти выражение для $\langle S_{111\varrho}(\mu)\rangle$, причем

$$\langle S_{1110}(\mu)\rangle = \langle S_{110}(\mu)\rangle \exp\left\{i\frac{1}{6}\int\int\int_0^\mu [\langle T\,U(t)\,U(t')\,U(t'')\rangle - \right.$$
$$\left. - 3\,\langle U(t)\,U(t')\rangle\,\langle U(t'')\rangle - 2\,\langle U(t)\rangle\,\langle U(t')\rangle\,\langle U(t'')\rangle]\,dt\,dt'\,dt''\right\}, \tag{15}$$

а

$$\langle S_{1111}(\mu)\rangle = \langle S_{1110}(\mu)\rangle \exp\left\{-\int_0^1 B_{\varrho'}(\mu)\,d\mu\right\},$$

где $B_{\varrho'}(\mu)$ также составлено из операторов $U(t)$ и $S_{\alpha'\eta'\lambda'\varrho'}(\mu)$. Очевидно, что эту процедуру можно продолжить до бесконечности, однако уже из (15) видно, что все множители при $\langle S_{110}(\mu)\rangle$ содержат все высшие корреляции элементарных возбуждений, начиная с тройных. Ввиду малости ангармонических членов в разложении потенциальной энергии колебания ядер растворителя мы можем ограничиться лишь парными корреляциями; тогда для $\langle S(\mu)\rangle$ получим выражение

$$\langle S(\mu)\rangle = \exp\left\{-i\int_0^\mu \langle U(t)\rangle\,dt - \frac{1}{2}\int\int_0^\mu dt\,dt'\,(\langle T\,U(t)\,U(t')\rangle - \langle U(t)\rangle\,\langle U(t')\rangle)\right\}. \tag{16}$$

При вычислении его мы будем учитывать третий и четвертый ангармонические члены в разложении $V_l(\boldsymbol{R})$, пренебрегая квадратом четвертого члена. Если теперь положить

$$K_1(\mu) = 2\,i\,\mu \sum_{\substack{s,s_1\\(s\neq s_1)}} (\Phi_{ss_1}^{l_1 l_0})^2 \left\{\frac{\bar n_s \bar n_{s_1} - (\bar n_s + 1)(\bar n_{s_1} + 1)}{\omega_s + \omega_{s_1}} - \frac{2(\bar n_s + 1)\,\bar n_{s_1}}{\omega_s - \omega_{s_1}}\right\} -$$
$$- 2 \sum_{\substack{s,s_1\\(s\neq s_1)}} (\Phi_{ss_1}^{l_1 l_0})^2 \left\{\frac{\bar n_s \bar n_{s_1}[e^{i\mu(\omega_s+\omega_{s_1})} - 1] + (\bar n_s + 1)(\bar n_{s_1} + 1)[e^{-i\mu(\omega_s+\omega_{s_1})} - 1]}{(\omega_s + \omega_{s_1})^2} + \right.$$
$$\left. + \frac{2(\bar n_s + 1)\,\bar n_{s_1}}{(\omega - \omega_{s_1})^2}[e^{-i\mu(\omega_s-\omega_{s_1})} - 1]\right\}, \tag{17a}$$

$$K_2(\mu) = 6\,i\,\mu \sum_{s s_1 s_2} (\Phi^{l_1 l_0}_{s s_1 s_2})^2 \left\{ \frac{\bar n_s\, \bar n_{s_1}\, \bar n_{s_2} - (\bar n_s + 1)(\bar n_{s_1} + 1)(\bar n_{s_2} + 1)}{\omega_s + \omega_{s_1} + \omega_{s_2}} - \right.$$

$$\left. - \frac{3(\bar n_s + 1)(\bar n_{s_1} + 1)\,\bar n_{s_2}}{\omega_s + \omega_{s_1} - \omega_{s_2}} - \frac{3(\bar n_s + 1)\,\bar n_{s_1}\,\bar n_{s_2}}{\omega_s - \omega_{s_1} - \omega_{s_2}} \right\} -$$

$$- 6 \sum_{s s_1 s_2} (\Phi^{l_1 l_0}_{s s_1 s_2})^2 \left\{ \frac{3(\bar n_s + 1)\,\bar n_{s_1}\,\bar n_{s_2}}{(\omega_s - \omega_{s_1} - \omega_{s_2})^2} [e^{-i\mu(\omega_s - \omega_{s_1} - \omega_{s_2})} - 1] + \right.$$

$$+ \frac{\bar n_s\,\bar n_{s_1}\,\bar n_{s_2} [e^{i\mu(\omega_s + \omega_{s_1} + \omega_{s_2})} - 1] + (\bar n_s + 1)(\bar n_{s_1} + 1)(\bar n_{s_2} + 1)[e^{-i\mu(\omega_s + \omega_{s_1} + \omega_{s_2})} - 1]}{(\omega_s + \omega_{s_1} + \omega_{s_2})^2} +$$

$$\left. + \frac{3(\bar n_s + 1)(\bar n_{s_1} + 1)\,\bar n_{s_2}}{(\omega_s + \omega_{s_1} - \omega_{s_2})^2} [e^{-i\mu(\omega_s + \omega_{s_1} - \omega_{s_2})} - 1] \right\}, \qquad (17б)$$

$$\zeta^2_{s l_1 l_0}\,\omega_s = \omega^{l_0}_s \left(\xi_{s l_1} + \frac{\Phi^{l_1 l_0}_s}{\omega^{l_0}_s} \right)^2 + 6 \sum_{s_2} \left(\xi_{s l_1} + \frac{\Phi^{l_1 l_0}_s}{\omega^{l_0}_s} \right) \Phi^{l_1 l_0}_{s s_2 s_2} (2\,\bar n_{s_2} + 1) +$$

$$+ \frac{9}{\omega_s} \left(\sum_{s_2} \Phi^{l_1 l_0}_{s s_2 s_2} (2\,\bar n_{s_2} + 1) \right)^2, \qquad (18)$$

$$g_r(\mu) = \sum_s \zeta^2_{s l_1 l_0} [\bar n_s\, e^{i\mu\omega_s} + (\bar n_s + 1)\, e^{-i\mu\omega_s} - (2\,\bar n_s + 1)], \qquad (19)$$

то для $\langle S_r(\mu)\rangle$ получим выражение

$$\langle S_r(\mu)\rangle = \exp \left\{ i\,\mu \sum_s \zeta^2_{s l_1 l_0}\,\omega_s + g_r(\mu) - K_1(\mu) - K_2(\mu) - \right.$$

$$\left. - 3\,i\,\mu \sum_{s s_2} \Phi^{l_1 l_0}_{s s s_2 s_2} (2\,\bar n_s + 1)(2\,\bar n_{s_2} + 1) \right\}. \qquad (20)$$

Для вычисления $\langle S_a(\mu)\rangle$ необходимо в (16) подставить $U_1(t)$; выполнив соответствующее интегрирование, получим

$$\langle S_a(\mu)\rangle = \exp \left\{ i\,\mu \sum_s \zeta^2_{s l_1}\,\omega^{l_1}_s + g_a(\mu) - K_{2a}(\mu) - \right.$$

$$\left. - 3\,i\,\mu \sum_{s s_2} Y^{l_1}_{s s s_2 s_2} (2\,\bar n_s + 1)(2\,n_{s_2} + 1) \right\}, \qquad (21)$$

где $\omega^{l_1}_s \zeta^2_{s l_1}$ получается из (18), если в нем положить $\Phi^{l_1 l_0}_s = 0$ и выполнить замену $\omega^{l_0}_s \to \omega^{l_1}_{s_1}$, $\Phi^{l_1 l_0}_{s s_2 s_2} \to - Y^{l_1}_{s s_2 s_2}$; $g_a(\mu)$ получается из (19) заменой $\zeta^2_{s l_1 l_0}$ на $\xi^2_{s l_1}$ и ω_s на $\omega^{l_1}_s$; $K_{2a}(\mu)$ получается из (17б) заменой $\Phi^{l_1 l_0}_{s s_1 s_2}$ на $Y^{l_0}_{s s_1 s_2}$ и $\omega_s, \omega_{s_1}, \omega_{s_2}$ соответственно на $\omega^{l_1}_s, \omega^{l_1}_{s_1}, \omega^{l_1}_{s_2}$; кроме того при выводе (20), (21) мы полагаем

$$\bar n_s = \left\{ e^{\beta \hbar \omega^{l_1}_s} - 1 \right\}^{-1} = \left\{ e^{\beta \hbar \omega^{l_0}_s} - 1 \right\}^{-1}$$

поскольку $\Delta\omega_s \sim \frac{1}{N}$.

Зная $\langle S_r(\mu)\rangle$ и $\langle S_a(\mu)\rangle$ можно перейти теперь к вычислению формы спектра излучения и поглощения света примесными центрами. Предварительно необходимо, однако, отметить одно важное обстоятельство: выражения (10) и (12) вместе с (20) и (21) определяют соответственно весь спектр излучения и поглощения света примесным центром твердого тела, т. е. как его фононную часть, так и бесфононную линию, которая является аналогом линии Мессбауэра [22] при γ переходах в ядре. Существование этой линии в спектрах изучения и поглощения впервые отмечалось в [23], а затем и другими авторами; ей, очевидно, отвечают те фотопереходы, при которых колебательное состояние раствора для начального и конечного состояния примесного центра будет одним и тем же. В рассматриваемом приближении считается, что фотопереходы совершаются из состояний, в которых ангармонизм несущественный, поэтому бесфононной линии будут соответствовать только те из них, которые совершаются в состояния, для которых ангармонизм также несущественный. В связи с этим при вычислении бесфононной части спектра излучения и поглощения необходимо в выражениях (20) и (21) положить все $\Phi^{l_1 l_0}$, Y^{l_1} равными нулю.

Для вычисления формы спектра излучения света примесными центрами разобъем область интегрирования $[0, \infty]$ в (10) на две области $[0, \mu_0]$ и $[\mu_0, \infty]$, где μ_0 определяется условием $\mu_0 \omega_0 < 1$, ω_0 максимальное значение частоты ω_s. Так как в интеграле с границами $[0, \mu_0]$ $\mu \omega_s < 1$ для всех ω_s, то мы можем разложить экспоненты, которые входят в выражения для $g_r(\mu)$, $K_1(\mu)$, $K_2(\mu)$ в ряд, ограничиваясь первыми членами разложения. Тогда вводя обозначения

$$\left.\begin{aligned}
\hbar\,\Omega &= V_{l_1}(\boldsymbol{R}_0) - V_{l_0}(\boldsymbol{R}_0) + \varepsilon_0 - \varepsilon_{v_\tau} + \frac{\hbar}{2}\sum_s \Delta\omega_s - \sum_s \hbar\,\omega_s^{l_1}\,\xi_{s l_1}\,,\\
A_r &= \sum_s \omega_s^{l_0}\,\xi_{s l_1}^2 + \Phi^{l_1 l_0} + \sum_s \Phi_s^{l_1 l_0}(2\bar{n}_s + 1) + 3\sum_{s,s_2}\Phi^{l_1 l_0}_{s s s_2 s_2}(2\bar{n}_s + 1)(2\bar{n}_{s_2} + 1)\,,\\
B_r^2 &= \frac{1}{2}\sum_s \zeta_{s l_1 l_0}^2\,\omega_s^2\,(2\bar{n}_s + 1) + \sum_{\substack{s, s_1\\(s\neq s_1)}}\left(\Phi^{l_1 l_0}_{s s_1}\right)^2(2\bar{n}_s + 1)(2\bar{n}_{s_1} + 1) + \\
&\quad + 3\sum_{s s_1 s_2}\left(\Phi^{l_1 l_0}_{s s_1 s_2}\right)^2(2\bar{n}_s + 1)(2\bar{n}_{s_1} + 1)(2\bar{n}_{s_2} + 1)\,,\\
C_r &= \frac{1}{3!}\sum_s \zeta_{s l_1 l_0}^2\,\omega_s^3\,, \qquad G = \sum_s \xi_{s l_1}^2\,(2\bar{n}_s + 1)\,,
\end{aligned}\right\} \tag{22}$$

запишем интеграл по области $[0, \mu_0]$ в виде

$$i_1 = \operatorname{Re}\int_0^{\mu_0} d\mu\,\exp\left\{i\,\mu\left(\Omega + \sum_s \bar{n}_s\,\Delta\omega_s - \nu - A_r\right) - \mu^2\,B_r^2 + i\,\mu^3\,C_r + \cdots\right\}. \tag{23}$$

Если предположить теперь, что выполняется условие

$$\omega_0^{-2}\,B_r^2 > 1 \tag{24}$$

которое мы будем называть в дальнейшем условием большего тепловыделения, то интегрирование выражения (23) легко выполнить, пользуясь методом перевала [17]. Интеграл по области $[\mu_0, \infty]$ при условии (24), большой дисперсии частот нормальных колебаний и плавной функции распределения их, дает заметный вклад лишь при $\mu = 0$; он определяет бесфононную линию в спектре излучения при вычислении которой, в соответствии с вышесказанным, необходимо пренебречь всеми ангармоническими членами. При этом находим

$$F_r(\nu) = \frac{\sqrt{\pi}}{\Gamma B_r} e^{-\frac{\alpha_r^2}{4 B_r^2}} \left[1 + \frac{C_r \alpha_r^3}{8 B_r^6} + \cdots\right] + \frac{\exp\{-G\}}{\left(\Omega + \sum_s \bar{n}_s \Delta\omega_s - \nu\right)^2 + \frac{\Gamma^2}{4}}, \quad (25)$$

где $\alpha_r = \Omega + \sum_s \bar{n}_s \Delta\omega_s - A_r - \nu$.

Функция $F_r(\nu)$ при $C_r \ll B_r^6$ т. е. при достаточно высоких температурах определяет гауссову кривую с полушириной

$$\Delta_r = 2 B_r \sqrt{\ln 2} \quad (26)$$

и максимумом в точке $\nu_r = \Omega + \sum_s \bar{n}_s \Delta\omega_s - A_r$; величину A_r обычно называют стоксовым смещением, а Ω частотой фотоперехода при $T = 0$ (в частности, при $v_r = 0$ частотой чисто-электронного перехода). При учете ангармонических членов величина стоксового смещения, как следует из (22), существенно зависит от температуры, притом если принятое в работе приближение выполняется до $T > \frac{\hbar\omega_s}{k}$, где k постоянная Больцмана, то

$$A_r = A_1 + A_2 T + A_3 T^2 , \quad (27)$$

где A_1, A_2, A_3 независящие от температуры величины. Направление смещения всего спектра в шкале частот с изменением температуры определяется соотношением $\sum_s \bar{n}_s \Delta\omega_s$ и A_r. Если роль ангармонизма пренебрежимо мала, то $A_2 = A_3 = 0$ и величина стоксового смещения не зависит от температуры. В этом случае смещение спектра с изменением температуры определяется только величиной $\sum_s \bar{n}_s \Delta\omega_s$, появление которой в (25) обусловлено зависимостью квазиупругих постоянных раствора от электронного состояния примесного центра.

Пользуясь (22) легко также найти, что при $\hbar\omega_s < k T$

$$\Delta_r = \sqrt{a_1 T + a_2 T^3 + a_3 T^3} , \quad (28)$$

где a_1, a_2, a_3 независящие от температуры величины, причем a_2 и a_3 связаны только с ангармоническими членами, так что при отсутствии ангармонизма $a_2 = a_3 = 0$.

Установить в общем случае связь между константами A_1, A_2, A_3 а также a_1, a_2, a_3 невозможно; можно лишь утверждать, что в случае большого тепловыделения $A_1 > A_2$, A_3, $a_1 > a_2$, a_3.

Из (25) следует далее, что с понижением температуры фононная часть спектра излучения становится асимметричной: в сторону больших частот интенсивность ее убывает более резко, чем в сторону меньших частот: максимум полосы при этом смещается в шкале частот, полуширина ее уменьшается, интенсивность растет. Если условие (24) выполняется вплоть до $T = 0$, то понижение температуры до нуля вызывает лишь рост асимметрии, интенсивности и смещения максимума кривой; интенсивность бесфононной линии с понижением температуры быстро возрастает. Если условие (24) выполняется лишь до некоторой температуры, то начиная с нее на коротковолновой части спектра появляются отдельные пики, тогда как длинноволновая часть спектра остается все еще плавной и может быть с некоторым приближением апроксимирована (25). Дальнейшее снижение температуры ведет к тому, что пики на коротковолновой части спектра становятся более резкими, смещаются в шкале частот, кроме того они начинают появляться и на плавной части кривой. Таким образом, понижение температуры приводит к тому, что пики появляются на все большей части спектра, пока не охватывают всю полосу. Общий вид аналитического выражения для формы полосы в этом случае будет таким же, как и при пренебрежении ангармонизмом [17], только с несколько измененными параметрами за счет ангармонизма.

Не останавливаясь больше на особенностях спектров излучения света примесными центрами твердого тела (так как они подробно исследованы в [17]), займемся вычислением формы спектров поглощения. Пользуясь (12) и (21) и вводя обозначения

$$\left.\begin{aligned} \hbar\,\Omega_a &= V_{l_1}(\boldsymbol{R}_0) - V_{l_0}(\boldsymbol{R}_0) + \varepsilon_{v_\tau} - \varepsilon_0 + \frac{\hbar}{2}\sum_s \Delta\omega_s - \hbar\sum_s \omega_s^{l_1}\xi_{sl_1}^2\,, \\ A_a &= \sum_s \omega_s^{l_1}\xi_{sl_1}^2 + 3\sum_{ss_2} Y^{l_1}_{sss_2s_2}\,(2\,\bar{n}_s+1)\,(2\,\bar{n}_{s_2}+1)\,, \\ B_a &= \frac{1}{2}\sum_s \zeta_{sl_1}^2\left(\omega_s^{l_1}\right)^2(2\,\bar{n}_s+1) + 3\sum_{ss_1s_2}\left(Y^{l_1}_{ss_1s_2}\right)^2(2\,\bar{n}_s+1)\,(2\,\bar{n}_{s_1}+1)\,(2\,\bar{n}_{s_2}+1)\,, \\ C_a &= \frac{1}{3!}\sum_s \zeta_{sl_1}^2\left(\omega_s^{l_1}\right)^3\,, \end{aligned}\right\} \tag{29}$$

также как и в случае излучения находим

$$i_1 = \operatorname{Re}\int_0^\mu d\mu\,\exp\left\{i\,\mu\left(\nu - \Omega_a - \sum_s \bar{n}_s\,\Delta\omega_s - A_a\right) - \mu^2\,B_a^2 + i\,\mu^3\,C_a + \cdots\right\}.$$

Если как и выше предположить, что выполняется условие

$$\left(\omega_0^{l_1}\right)^{-2} B_0^2 > 1\,, \tag{30}$$

то при большой дисперсии частот нормальных колебаний и плавной функции распределения их для функции $F_a(\nu)$ получим выражение

$$F_a(\nu) = \frac{\sqrt{\pi}}{\Gamma\,B_a}\,e^{-\frac{\alpha_a^2}{4\,B_a^2}}\left[1 + \frac{C_a\,\alpha_a^3}{8\,B_a^6} + \cdots\right] + \frac{\exp\{-G\}}{\left(\Omega_a + \sum_s \bar{n}_s\,\Delta\omega_s - \nu\right)^2 + \frac{\Gamma^2}{4}}\,, \tag{31}$$

где

$$\alpha_a = \nu - \Omega_a - \sum_a \bar{n}_s\,\Delta\omega_s - A_a\,.$$

Отсюда следует, что в случае большого тепловыделения, когда выполняется условие (30), и при $C_a \ll B_a^6$, полоса поглощения будет иметь форму гауссовой кривой с полушириной

$$\Delta_a = 2\, B_a \sqrt{\ln 2}$$

и максимумом в точке $\nu = \Omega_a + \sum_s \bar{n}_s \Delta\omega_s + A_a$. Величина стоксового смещения A_a зависит от температуры, причем при высоких температурах, когда $\hbar\, \omega_s^{l_1} < k\, T$

$$A_a = A_{1a} + A_{3a}\, T^2\,, \tag{32}$$

при этом температурная зависимость полуширины полосы будет определяться выражением

$$\Delta_a = \sqrt{b_1\, T + b_2\, T^2 + b_3\, T^3}\,, \tag{33}$$

где, как и выше, A_{1a}, A_{3a}, b_1, b_2, b_3, не зависят от температуры, причем $A_{1a} > A_{3a}$ и $b_1 > b_2, b_3$. Из выражений (22), (27), (28) и (29), (32), (33) следует, что в общем случае соответственные полосы в спектрах излучения и поглощения не будут зеркально-симметричными не только в смысле положения их максимумов в шкале частот ($\Omega_a \neq \Omega$, $A_r \neq A_a$), но и по полуширине. Если фотопереход совершается без возбуждения внутренних колебательных степеней свободы примеси (когда примесью является молекула) или же в случае примесных центров без внутренних колебательных степеней свободы — $\Omega = \Omega_a$, однако в силу различия A_r и A_a полоса поглощения и полоса излучения не будут зеркально-симметричными относительно частоты $\Omega + \sum_s \bar{n}_s \Delta\omega_s$; в общем случае они будут по-разному смещаться в шкале частот с изменением температуры, полуширины полос будут разными. Это действительно наблюдается для F центров в кристаллах KCl [24]. Только при пренебрежительно малом вкладе ангармонизма, когда $a_2, a_3 \to 0$, $b_2, b_3 \to 0$ эти полосы будут зеркально-симметричными относительно частоты $\Omega + \sum_s \bar{n}_s \Delta\omega_s$ как в смысле положения максимумов, так и полуширины, причем с изменением температуры их максимумы будут смещаться в шкале частот одинаковым образом (температурное смещение определяется величиной $\sum_s \bar{n}_s \Delta\omega_s$).

Генезис полосы поглощения с температурой будет происходить, в общем счучае, таким же образом, как и полосы излучения; все отличие состоит в том, что асимметричная кривая поглощения более резко спадает в сторону меньших частот и с понижением температуры пики начинают появляться на длинноволновом крыле кривой. Однако в силу различия B_a и B_r условия (24) и (30) перестают выполняться, вообще говоря, при разных температурах, поэтому отдельные пики на крыльях кривой поглощения и излучения начинают появляться при разных температурах; при одной и той же температуре на одной из полос эти пики могут появиться, а на другой нет. При пренебрежительно малом вкладе ангармонизма условия (24) и (30) совпадают; в этом случае можно ожидать, что генезис спектров поглощения и излучения с температурой будет происходить одинаково.

Литература

[1] A. A. Maradudin and R. F. Wallis, Phys. Rev. **123**, 777 (1961).
[2] R. F. Wallis and A. A. Maradudin, Phys. Rev. **125**, 1277 (1962).
[3] Л. Г. Гуревич и И. П. Ипатова, ЖЭТФ **45**, 231 (1963).
[4] M. Born and M. Blackman, Z. Phys. **82**, 551 (1933).
[5] D. F. Jepsen and R. F. Wallis, Phys. Rev. **125**, 1496 (1962).
[6] Г. Лейбфрид и В. Людвиг, Теория ангармонических эффектов в кристаллах, Москва 1963.
[7] A. A. Maradudin, P. A. Flinn, and R. A. Coldwell-Horsfall, Ann. Phys. **15**, 337, 360 (1961).
[8] А. Е. Маринчук и В. А. Москаленко, ФТТ **5**, 575 (1963).
[9] H. Hahn, Z. Phys. **165**, 569 (1961); Proceedings of the symposium on inelastic scattering of neutrons in solids and liquids (Intern. Atomic Energy Agency, Vienna 1963).
[10] В. Н. Кащеев и М. А. Кривоглаз, ФТТ **3**, 1528 (1961).
[11] S. S. S. Kokkedee, Physica **28**, 374 (1962).
[12] A. A. Maradudin and A. E. Fein, Phys. Rev. **128**, 2589 (1962).
[13] A. A. Maradudin, A. E. Fein, and G. H. Vineyard, phys. stat. sol. **2**, 1479 (1962).
[14] M. Born and J. Oppenheimer, Ann. Phys. **84**, 453 (1927).
[15] W. L. Lewschin, Z. Phys. **72**, 368 (1931).
[16] А. С. Давыдов, Наукови записки КДУ **14**, 5 (1955).
[17] А. Ф. Лубченко, Изв. АН СССР (сер. физ.) **18**, 718 (1954); Укр. физ. журн. **1**, 120, 265, 285 (1956).
[18] R. Kubo and I. Toyazawa, Progr. theor. Phys. **13**, 160 (1955).
[19] А. Ф. Лубченко и Б. И. Павлик, Укр. физ. Журн. (в печати).
[20] Р. Пайерлс, Квантовая теория твердых тел, Москва 1956.
[21] D. J. Thouless, Phys. Rev. **107**, 1162 (1957).
[22] R. L. Mössbauer, Z. Phys. **151**, 124 (1958); Naturwissenschaften **45**, 538 (1958).
[23] И. П. Дзюб и А. Ф. Лубченко, ФТТ **3**, 3602 (1961).
[24] F. Lüty and W. Gebhardt, Z. Phys. **169**, 475 (1962).

(Received April 7, 1964)

phys. stat. sol. 6, 333 (1964)

Arbeitsstelle für Tieftemperaturphysik der Deutschen Akademie der Wissenschaften zu Berlin, Forschungsgemeinschaft, Dresden

Die Feldstärkeabhängigkeit der Dielektrizitätskonstanten von $SrTiO_3$-Einkristallen im Temperaturbereich von 15 bis 80 °K

Von

E. HEGENBARTH

An einem $SrTiO_3$-Einkristall wurde im Temperaturbereich von 15 bis 80 °K die Dielektrizitätskonstante in Abhängigkeit von einer überlagerten Gleichspannung untersucht. Dabei wurden Maxima der Dielektrizitätskonstanten festgestellt, wie sie auch bei normalen Ferroelektrika, die einen Phasenübergang zweiter Ordnung besitzen, auftreten. Die Ergebnisse werden mit Hilfe der thermodynamischen Theorie der Ferroelektrika behandelt.

The dielectric constant of a $SrTiO_3$ monocrystal is measured between 15 and 80 °K as a function of an applied d. c. field. Maxima are observed in the dielectric constant. This behaviour is analogous to that of a normal ferroelectric, with a second order ferroelectric transition. A theoretical explanation of the results is given on the basis of the thermodynamic theory of ferroelectrics.

1. Einleitung

Wenn man die DK eines ferroelektrischen Kristalls bei konstantem Gleichspannungsfeld E_a in Abhängigkeit von der Temperatur mißt, so beobachtet man in der Nähe des Phasenübergangs eine Verschiebung des DK-Maximums nach höheren Temperaturen. Wie aus der phänomenologischen Theorie von DEVONSHIRE [1] hervorgeht, hängt dieser Effekt außerdem noch davon ab, ob der Phasenübergang von 1. oder 2. Ordnung ist.

Bei einem Übergang 1. Ordnung ergibt sich ein linearer Zusammenhang zwischen der Temperatur des DK-Maximums und der überlagerten Feldstärke E_a, und außerdem bleibt bei wachsendem E_a das Maximum scharf, wie KÄNZIG und MAIKOFF [2] experimentell an $BaTiO_3$ gezeigt haben. Im Gegensatz dazu ergibt sich für einen Übergang 2. Ordnung ein nichtlinearer Zusammenhang zwischen der Temperatur des DK-Maximums und der überlagerten Feldstärke E_a, wobei das DK-Maximum mit wachsender Feldstärke immer stärker verwaschen wird. Obwohl man bei $SrTiO_3$ bis zu Temperaturen von $\sim 1{,}5$ °K kein Maximum der DK findet [3, 4, 5], haben wir an $SrTiO_3$ unter Anlegen eines äußeren Feldes derartige Maxima festgestellt, die sich so verhalten, wie es einem Übergang 2. Ordnung entspricht.

2. Theoretische Betrachtungen

Für einen „normalen" ferroelektrischen Kristall mit einem Phasenübergang 2. Ordnung kommt man über den Ansatz von DEVONSHIRE für die freie Energie

$$F - F_0 = A\,(T - T_0)\,P^2 + B\,P^4 + \cdots - E_a\,P \tag{1}$$

(A und B sind Konstanten, P ist die Polarisation) mit dem Zusatzglied $- E_a\,P$, welches die Einwirkung des äußeren Feldes E_a berücksichtigt, zu folgender Beziehung zwischen der Temperatur des DK-Maximums und der angelegten Feldstärke E_a:

$$T - T_0 = \left(\frac{3}{4}\right)^{2/3} \cdot \left(\frac{3}{2}\frac{B}{A^3}\right)^{1/3} \cdot E_a^{2/3} . \tag{2}$$

Dabei ist T_0 die Curietemperatur der betreffenden Substanz. Diese Beziehung ist jedoch nicht auf $SrTiO_3$ anwendbar, da man hier keinen eigentlichen Curiepunkt findet. Das zeigt sich auch darin, daß keine Übereinstimmung mit den experimentellen Werten der Fig. 1 festgestellt werden kann.

Eine recht interessante Beziehung für $SrTiO_3$ ergibt sich, wenn man $A\ (T - T_0)$ in Gleichung (1) durch

$$A\left(\frac{T_1}{2} \coth\frac{T_1}{2\,T} - T_0\right) = \beta/2 \tag{3}$$

ersetzt. Dabei ist β die dielektrische Impermeabilität. Diese Beziehung (3) wurde von BARRETT [6] abgeleitet, als er die Slatersche Theorie auf quantenmechanischer Grundlage behandelte.

Geht man also von dem Ansatz

$$F - F_0 = A\left(\frac{T_1}{2} \coth\frac{T_1}{2\,T} - T_0\right) P^2 + B\,P^4 + \cdots - E_a\,P \tag{4}$$

aus und sucht nach dem Minimum der freien Energie, so ergibt sich

$$\frac{\partial(F - F_0)}{\partial P} = E = 2\,A\left(\frac{T_1}{2} \coth\frac{T_1}{2\,T} - T_0\right) P + 4\,B\,P^3 + \cdots - E_a = 0\,. \tag{5}$$

Daraus erhält man eine Beziehung $P = f(E_a, T)$.

Weiter ergibt sich aus (5) für die dielektrische Impermeabilität β — sie entspricht in unserem Fall $\varepsilon \gg 1$ der reziproken Dielektrizitätskonstanten — die Beziehung

$$\frac{\partial^2(F - F_0)}{\partial P^2} = \frac{\partial E}{\partial P} = \beta = 2\,A\left(\frac{T_1}{2} \coth\frac{T_1}{2\,T} - T_0\right) + 12\,B\,P^2 + \cdots . \tag{6}$$

β ist also nur noch eine Funktion von P und T: $\beta = f(T, P)$. Fragt man nun nach dem Extremwert von β, so ergibt sich

$$\frac{d\beta}{dT} = \frac{\partial\beta}{\partial T} + \frac{\partial\beta}{\partial P} \cdot \frac{\partial P}{\partial T} . \tag{7}$$

Sieht man zunächst von einer Temperaturabhängigkeit der Konstanten B ab, so erhält man aus (6)

$$\frac{\partial\beta}{\partial T} = A\,\frac{T_1^2}{T^2} \cdot \frac{1}{\sinh^2 \frac{T_1}{2\,T}} \qquad \text{und} \qquad \frac{\partial\beta}{\partial P} = 24\,B\,P\,. \tag{8}$$

Mit der Forderung, daß E_a während der Messung konstant bleiben soll, ergibt sich aus (5)

$$dE_a = 0 = A\,\frac{T_1^2}{T^2} \cdot \frac{P}{\sinh^2 \frac{T_1}{2\,T}}\,dT + 2\,A\left(\frac{T_1}{2} \coth\frac{T_1}{2\,T} - T_0\right) dP + 12\,B\,P^2\,dP \tag{9}$$

und damit

$$\frac{dP}{dT} = -A\frac{T_1^2}{T^2}\cdot\frac{P}{\left|2A\left(\frac{T_1}{2}\coth\frac{T_1}{2T}-T_0\right)+12BP^2\right|}. \tag{10}$$

Setzt man die Ausdrücke (8) und (10) in (7) ein, so ergibt sich

$$P = \left[\frac{A}{6B}\left(\frac{T_1}{2}\coth\frac{T_1}{2T}-T_0\right)\right]^{1/2}. \tag{11}$$

Wenn man diesen Ausdruck (11) in (5) einsetzt, ergibt sich

$$\coth\frac{T_1}{2T} = \frac{3}{T_1}\left(\frac{B}{4A^3}\right)^{1/3}\cdot E_a^{2/3} + \frac{2T_0}{T_1}. \tag{12}$$

Mit dieser Beziehung kann man für eine bestimmte Feldstärke E_a die Temperatur des DK-Maximums errechnen.

Aus Gleichung (12) läßt sich noch eine interessante Folgerung ziehen. Da $\coth\frac{T_1}{2T}$ mit sinkender Temperatur T gegen den Wert 1 strebt, kann man die Feldstärke ausrechnen, bei der das DK-Maximum bei $T = 0$ °K liegen würde:

$$E_{ak} = \left(\frac{T_1-2T_0}{3}\right)^{3/2}\cdot\left(\frac{4A^3}{B}\right)^{1/2}. \tag{13}$$

Bei einer Feldstärke, die kleiner ist als diese kritische Feldstärke E_{ak}, kann man demnach keinen Extremwert für die DK mehr erwarten. Um die Temperaturen des DK-Maximums bzw. die kritische Feldstärke E_{ak} berechnen zu können, müssen A, B, T_1 und T_0 bekannt sein. Die Konstante A ist gleich $\frac{4\pi}{2C}$ bzw. gleich $\frac{1}{2\varepsilon_0 C}$ je nachdem ob man im el. stat. CGS-System oder im Giorgischen System rechnet; C ist die Curiekonstante, die man experimentell bestimmen kann; die Konstante B kann aus den im Maximum gemessenen DK-Werten mit Hilfe der Beziehung

$$\beta_{\min}^{3/2} = (54\,B)^{1/2}\cdot E_a^{2/3} \tag{14}$$

berechnet werden. Diese Beziehung erhält man aus Gleichung (6), wenn man dort die Gleichungen (11) und (12) einsetzt.

3. Experimentelle Ergebnisse und Diskussion

Die Messungen wurden an einem Strontiumtitanateinkristall (Dicke $d =$ $= 0{,}408$ mm, Fläche $F = 0{,}468$ cm²) in der Weise durchgeführt, daß die Kapazität in Abhängigkeit von der Temperatur gemessen wurde, wobei dem Meßspannungswechselfeld ein Gleichspannungsfeld überlagert wurde.

Die Ergebnisse der Messungen sind in Fig. 1 zu sehen, wo $\varepsilon = f(T)$ mit der angegebenen Feldstärke als Parameter zu sehen ist. Wie man sieht, tritt ein Maximum der DK auf, welches sich mit steigender Feldstärke nach höheren Temperaturen verschiebt und dabei immer flacher wird.

Aus den DK-Werten im Maximum kann man die dielektrische Impermeabilität $\beta_{\min}$ berechnen und $\beta_{\min}^{3/2}$ gegen E_a auftragen. Das Ergebnis in Fig. 2 kann als Bestätigung der Gleichung (14) angesehen werden.

Aus der Neigung der Geraden entnimmt man die Werte für B:

$$B = 2{,}00\cdot 10^{-12}\ \text{el. stat. Einh.} \quad \text{bzw.} \quad B = 0{,}162\cdot 10^{10}\,\frac{\text{Vm}^5}{\text{C}^3}.$$

Dieser Wert ist etwas größer als der früher gemessene Wert $B = 0{,}137 \times 10^{10}\,\frac{\mathrm{Vm^5}}{\mathrm{C^3}}$ [7]. Als Ursache für diese Differenz kann man mehrere Möglichkeiten in Betracht ziehen, z. B. kann sowohl der elektrokalorische Effekt als auch eine Randschicht zwischen dem Inneren des Dielektrikums und den Elektroden zu einem niedrigeren DK-Wert im Maximum und damit zu einem größeren B-Wert führen.

Fig. 1. Die Feldabhängigkeit der DK von $SrTiO_3$-Einkristallen im Temperaturbereich von 15 bis 80 °K

Fig. 2. $\beta_{\min}^{3/2}$ als Funktion von E_a

Fig. 3. Theoretisch berechnete und experimentell gefundene Temperaturen der DK-Maxima in Abhängigkeit vom überlagerten Feld

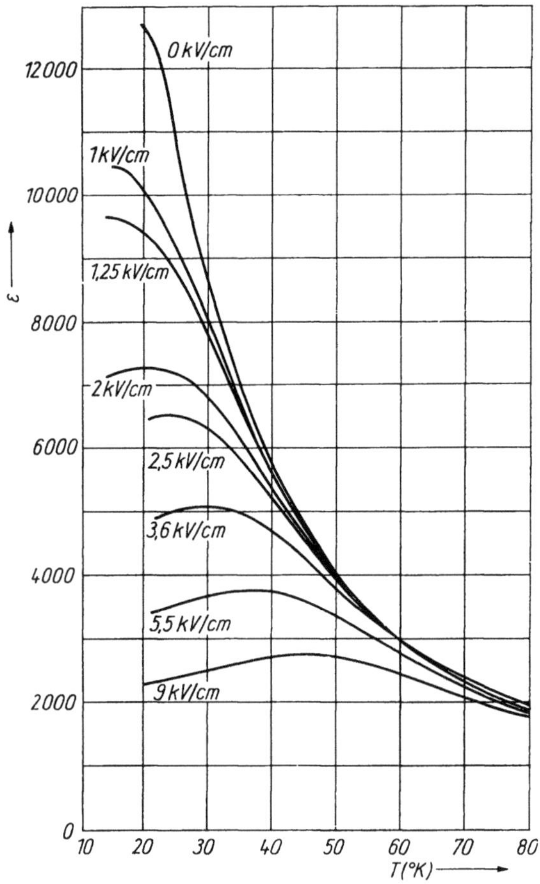

Fig. 1

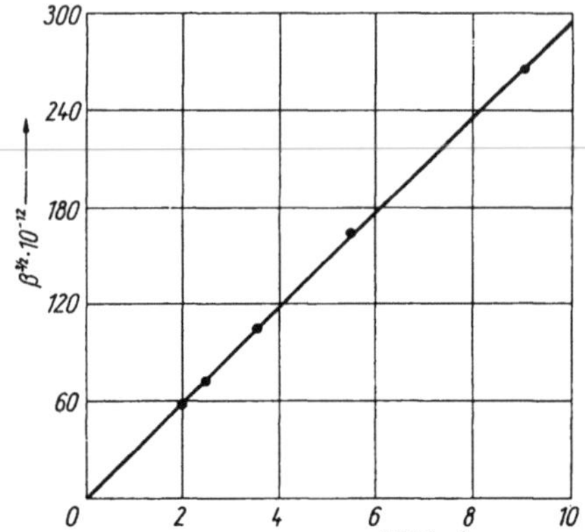

Fig. 2

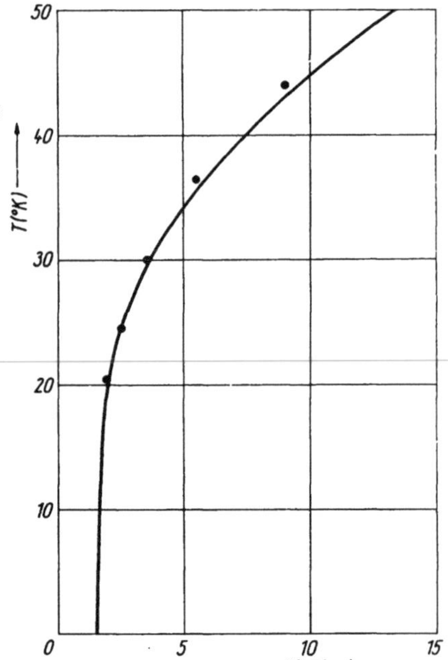

Fig. 3

In Fig. 3 sind die Temperaturen des DK-Maximums gegen die überlagerte Feldstärke E_a aufgetragen.

Die ausgezogene Kurve in Fig. 3 ist nach Werten gezeichnet, die aus der Gleichung (12) mit $A = \frac{4\pi}{2C}$ bzw. $A = \frac{1}{2\varepsilon_0 C}$ und $C = 8{,}7 \cdot 10^4$ °K, den oben angegebenen Werten von B und mit $T_1 = 100$ °K und $T_0 = 45$ °K nach Weaver [4] berechnet wurden. Wie man sieht, ist die Übereinstimmung zwischen berechneten Werten und experimentell gefundenen Werten recht gut. Das Ergebnis kann als Bestätigung der von Barrett gefundenen Gleichung (3) angesehen werden.

Setzt man die oben angegebenen Werte von A, B, T_1 und T_0 in Gleichung (13) ein, so bekommt man für die kritische Feldstärke

$$E_{ak} = 1{,}58 \text{ kV/cm} .$$

Nach Fig. 1 scheint auch dieses Ergebnis in dem dort angegebenen Temperaturbereich experimentell bestätigt zu sein. Eine endgültige Klärung sollen aber die noch im Gang befindlichen Messungen bei Temperaturen des flüssigen Heliums bringen.

Diese Ergebnisse können auch als eine indirekte Bestätigung der Gültigkeit der Slaterschen Theorie angesehen werden, denn der Ansatz des modifizierten Curie-Weissschen Gesetzes wurde ja von Barrett [6] mit Hilfe der Slaterschen Theorie abgeleitet. Nach unserer Meinung deuten die unter dem Einfluß eines äußeren Feldes zu beobachtenden Maxima der DK darauf hin, daß $SrTiO_3$ bei tiefen Temperaturen ferroelektrisch wird.

Herrn Dr. G. Schmidt bin ich für die Überlassung der Kristalle und viele wertvolle Ratschläge zu Dank verpflichtet. Herrn Prof. Bewilogua möchte ich ebenfalls für viele wertvolle Anregungen danken.

Literatur

[1] A. F. Devonshire, Adv. Phys. **3**, 85 (1954).
[2] W. Känzig und N. Maikoff, Helv. phys. Acta **24**, 343 (1951).
[3] N. Gränicher, Helv. phys. Acta **29**, 211 (1956).
[4] N. E. Weaver, J. Phys. Chem. Solids **11**, 274 (1959).
[5] E. Hegenbarth, Monatsber. Dtsch. Akad. Wiss., **6**, 309 (1964).
[6] I. H. Barrett, Phys. Rev. **86**, 118 (1952).
[7] G. Schmidt und E. Hegenbarth, phys. stat. sol. **3**, 329 (1963).

(Received April 10, 1964)

phys. stat. sol. 6, 339 (1964)

Physikalisches Institut der Universität Münster

Über die Ermittlung der Verfestigungskurve durch den Torsionsversuch an zylindrischen Vollstäben und das Verhalten von vielkristallinem Kupfer bei sehr hoher plastischer Schubverformung

Von

H. G. Grewe und E. Kappler

Durch Torsionsversuche an zylindrischen Vollstäben wurde das Verfestigungsverhalten von vielkristallinem Kupfer bei Schubverformung bis zu Scherungen an der Probenoberfläche von $\gamma = 11$ (bei Raumtemperatur) bzw. $\gamma = 8$ (bei 77 °K) untersucht. Es kann abgeschätzt werden, daß diese Verformung mit einem Zugversuch vergleichbar ist, bei dem die Probe auf mehr als das zehnfache ihrer Ausgangslänge gedehnt wird. Aus den gemessenen Drehmoment-Drehwinkel-Kurven läßt sich die Schubspannung an der Probenoberfläche ermitteln, vorausgesetzt, daß die Scherung in einem Volumelement im Innern der Probe proportional zu seiner Entfernung von der Probenachse ist. Es konnte gezeigt werden, daß dies der Fall ist. Die Schubspannungs-Scherungskurven erreichten bei Raumtemperatur einen Grenzwert $\tau_m = 26{,}6$ kp/mm² bei $\gamma = 7{,}5$. Die Versuche lassen vermuten, daß dieser Grenzwert durch eine konstante Fließspannung bedingt ist. Bei 77 °K bzw. 90 °K gelang es nicht, einen konstanten Wert der Schubspannung nachzuweisen, da die Proben vorher brachen. Aus Temperatur- bzw. Geschwindigkeitswechselversuchen ergab sich, daß τ_m nur von der Temperatur und der Verformungsgeschwindigkeit abhängt, nicht aber von der Vorgeschichte der Probe.

By torsion tests on cylindrical specimens, the work-hardening behaviour of polycrystalline copper is investigated for shear strains up to $\gamma = 11$ (at 293 °K) and $\gamma = 8$ (at 77 °K). It is estimated that for such a deformation the test is equivalent to a tensile test in which the specimen is strained to between ten and twenty times its initial length. From the measured torque-twist curves, the shear stress at the surface of the specimens can be calculated by assuming that the shear of a volume element inside the specimen is proportional to its distance from the axis. It is shown that this assumpion is justified. The shear stress-strain curves at room temperature reach a limiting value of $\tau_m = 26.6$ kp/mm² at $\gamma = 7.5$. This limiting value is thought to be due to a constant yield stress. At 77 and 90 °K a constant shearing stress could not be obtained, due to premature fracture of the specimens. From temperature and speed change tests it is found that the limiting value is a function of temperature and speed only, and does not depend on the previous treatment of the specimen.

1. Einleitung

Unter der Verfestigungskurve eines vielkristallinen Materials versteht man die Abhängigkeit der Fließspannung von der plastischen Verformung. Dabei ist die Fließspannung — sie sei im Folgenden mit τ_0 bezeichnet — proportional der Invarianten der Fließbedingung. Im Falle eines plastisch isotropen Materials ist sie auch proportional einer Invarianten des Spannungszustandes. Da es verschiedene mögliche Fließbedingungen gibt, gibt es auch verschiedene mögliche Definitionen der Verfestigungskurve. Wir benutzen im Folgenden, sofern es sich um isotropes plastisches Fließen handelt, die Fließbedingung von Huber und von Mises.

Danach tritt an einer Stelle des Körpers plastisches Fließen ein, wenn dort die zweite Invariante des Spannungsdeviators einen bestimmten Wert erreicht hat, den wir mit $3/2\,\tau_0^2$ bezeichnen, wobei gilt

$$\tau_0^2 = \frac{1}{9}\left[(\sigma_x-\sigma_y)^2+(\sigma_y-\sigma_z)^2+(\sigma_z-\sigma_x)^2+6\,(\tau_{xy}^2+\tau_{yz}^2+\tau_{zx}^2)\right], \quad (1\,\mathrm{a})$$

$$\tau_0^2 = \frac{1}{9}\left[(\sigma_1-\sigma_2)^2+(\sigma_2-\sigma_3)^2+(\sigma_3-\sigma_1)^2\right], \quad (1\,\mathrm{b})$$

$$\tau_0^2 = \frac{4}{9}\,(\tau_1^2+\tau_2^2+\tau_3^2)\,. \quad (1\,\mathrm{c})$$

Gleichung (1 a) gilt bei Benutzung eines zur Probe beliebig orientierten Koordinatensystem ($x\,y\,z$), die Gleichungen (1 b) und (1 c) für ein Koordinatensystem, dessen Richtungen körperfest sind und mit den Richtungen der Hauptspannungen $\sigma_1, \sigma_2, \sigma_3$ zusammenfallen. $\tau_1 = \frac{\sigma_2-\sigma_3}{2}$, $\tau_2 = \frac{\sigma_3-\sigma_1}{2}$, $\tau_3 = \frac{\sigma_1-\sigma_2}{2}$ sind die zugehörigen Hauptschubspannungen.

Der Ausdruck auf der rechten Seite der Gleichung (1) ist das Quadrat der Oktaederschubspannung. Beim Zugversuch ist die Zugspannung σ, bei der das plastische Fließen einsetzt, gleich $3\sqrt{2}\,\tau_0$, beim Torsionsversuch ist die Schubspannung beim Fließbeginn $\tau = \tau_{z\Theta} = \sqrt{3/2}\,\tau_0$ (siehe Fig. 1).

Die Fließbedingung (1) gilt für nur ein plastisch isotropes Material. Wenn das Material anisotrop ist, ist die rechte Seite der Gleichung (1 a) durch einen allgemeineren Ausdruck zu ersetzen, in dem außer den Spannungskomponenten noch gewisse Anisotropiefaktoren auftreten, welche die Textur der Probe berücksichtigen [1]. Der Ausdruck hat dann nicht mehr die Bedeutung der Oktaederschubspannung.

Unter Zugrundelegung der Fließbedingung (1) versteht man unter der Verfestigungskurve den Zusammenhang zwischen der Oktaederschubspannung τ_0 und der plastischen Deformation, wenn die letztere durch die sogenannte Oktaederscherung γ_0 beschrieben wird, d. h. also die Funktion $\tau_0(\gamma_0)$. Unter gewissen

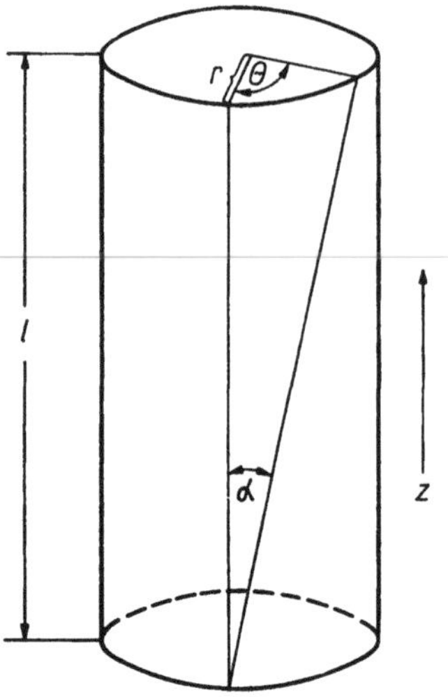

Fig. 1

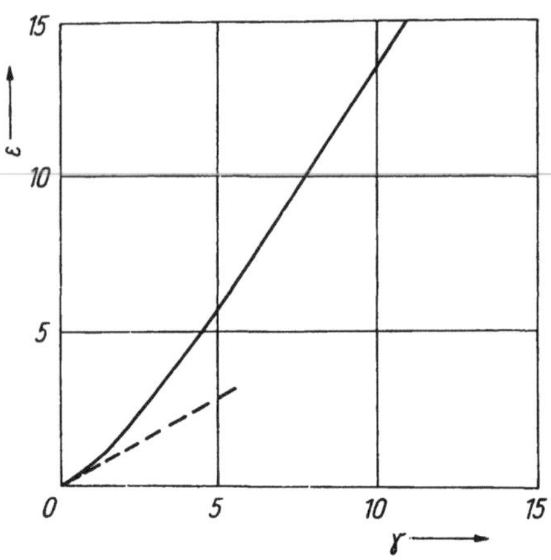

Fig. 2

Voraussetzungen läßt sich γ_0 durch die „natürlichen Hauptdehnungen" $\delta_i = \ln(1 + \varepsilon_i)$ in folgender Weise ausdrücken [2]:

$$\gamma_0 = \frac{2}{3}\sqrt{\delta_1^2 + \delta_2^2 + \delta_3^2}\,. \tag{2}$$

Diese Voraussetzungen sind:

1. Es ist angenommen, daß die Verformung ohne Änderung des Volumens erfolgt, was bei dem plastischen Fließen gut erfüllt ist, d. h. es gilt $\delta_1+\delta_2+\delta_3=0$.
2. Die infinitesimalen Änderungen der drei Hauptdehnungen müssen einander proportional sein. Dies ist beim Zug- und beim Torsionsversuch erfüllt.

Beim Zugversuch ist

$$\delta_1 = \ln(1+\varepsilon)\,, \qquad \delta_2 = \delta_3 = -\frac{1}{2}\ln(1+\varepsilon)\,, \tag{3}$$

wobei $\varepsilon = (l - l_0)/l_0$ die „konventionelle Dehnung" ist. Für $\varepsilon \ll 1$ wird dann $\gamma_0 = \sqrt{3/2}\,\varepsilon$.
Beim Torsionsversuch gilt

$$\delta_1 = -\delta_2 = \ln\left(\sqrt{1+\gamma^2/4} + \gamma/2\right)\,, \qquad \delta_3 = 0\,, \tag{4}$$

wobei γ die Scherung ist. Für $\gamma \ll 1$ folgt $\gamma_0 = \gamma/\sqrt{2}$.

Die Untersuchung der Verfestigung bei sehr hohen Verformungsgraden ist mit dem Zugversuch, bei dem wegen der Einschnürung der Proben in günstigen Fällen homogene Dehnungen von maximal $\varepsilon \approx 0{,}6$ erreichbar sind, nicht möglich. Dagegen sind im Torsionsversuch je nach Temperatur und Vorbehandlung Scherungen von γ bis etwa 7 erreichbar, was im Zugversuch Dehnungen von $\varepsilon \approx 10$ entspricht.[1])

Der Zusammenhang zwischen der Scherung γ beim Torsionsversuch und der Dehnung ε, die beim Zugversuch zur gleichen Oktaederscherung führen würde, ist entsprechend den Gleichungen (3) und (4) in Fig. 2 dargestellt. Bei kleinen Verformungen ist nach den obigen Beziehungen $\varepsilon = \gamma/\sqrt{3}$, d. h. die Neigung der Kurve $\varepsilon(\gamma)$ ist im Nullpunkt $1/\sqrt{3}$. (Gestrichelte Tangente im Nullpunkt der Fig. 2).

Eine Schwierigkeit liegt bei der Torsion von Vollstäben insofern vor, als im Gegensatz zum Zugversuch die Verformung inhomogen ist. Bei vorgegebenem Torsionswinkel Θ nimmt der Verformungsgrad von innen nach außen zu. Auf diese Schwierigkeit wird in Abschnitt 2.1 näher eingegangen.

Eine nahezu homogene Verformung liegt bei der Torsion dünnwandiger Rohre vor. Da man mit Rohren aber nur kleine Verformungsgrade erreichen kann (wir erreichten maximal $\gamma = 0{,}3$), war es notwendig, mit Vollstäben zu arbeiten.

Eine weitere Komplikation, die bei allen Verformungsversuchen mit hohen Verformungsgraden auftritt, ergibt sich dadurch, daß die Probe eine mit wachsender Verformung mehr und mehr sich ausprägende Verformungstextur bekommt. Die Probe wird plastisch anisotrop, und die Fließbedingung (1) ist nicht mehr anwendbar, und muß, wie oben erwähnt, durch eine allgemeinere ersetzt werden.

---

[1]) Noch höhere Verformungsgrade als die hier erreichten sind zuerst von TAYLOR und QUINNEY [5] durch Stauchversuche erzielt worden. Bei diesen Versuchen wurden die Proben mehrfach entlastet und neu geschmiert. Einige Male wurde außerdem der Querschnitt der Proben verringert. Dieses Verfahren hat gegenüber dem Torsionsversuch den Nachteil, daß hierbei durch die Entlastung der Zustand der Probe geändert werden kann gegenüber demjenigen bei kontinuierlicher Belastung und konstanter Verformungsgeschwindigkeit. Dann ist die so gewonnene Verfestigungskurve abhängig von der Zahl der Entlastungen.

Im Falle des Torsionsversuches äußert sich die Anisotropie, die für die einzelnen Spannungskomponenten jeweils durch einen bestimmten mittleren Orientierungsfaktor $\overline{m}$ der Kristallite beschrieben werden kann, darin, daß die Schubspannung $\tau = \tau_{z\Theta}$ sich mit ihr ändert. τ ändert sich dann also mit wachsender Scherung aus zwei Gründen, einmal infolge der Verfestigung, zum anderen infolge der Textur, d. h. die Schubspannung wäre durch eine Funktion $\tau = \tau[\tau_0(\gamma), \overline{m}(\gamma)]$ darzustellen, wobei die Größe $\tau_0(\gamma)$ nach wie vor die Invariante der Fließbedingung bedeutet, die den Verfestigungszustand kennzeichnet. Zur Aufstellung der Verfestigungskurve müßte danach der Einfluß der Textur eliminiert werden. Das kann im Prinzip durch Messungen der Verformungstextur geschehen. Bei den hier mitgeteilten Messungen ist dies allerdings nicht geschehen, da nur vereinzelte Texturmessungen gemacht worden sind. Aus diesem Grunde muß darauf verzichtet werden, die reine Verfestigungskurve $\tau_0(\gamma_0)$ zu bestimmen. In den meisten Abbildungen des Abschnitts 5 ist deswegen die gemessene Schubspannung $\tau = \tau_{z\Theta}$ gegen die gemessene Scherung γ aufgetragen.

Es soll in dieser Untersuchung vielmehr ein Teilproblem der Verfestigungskurve behandelt werden, nämlich die Frage, ob die Fließspannung τ_0 bei sehr hohen Verformungsgraden einen konstanten Sättigungswert erreicht. Wie in Abschnitt 5 gezeigt wird, beobachtet man im Bereich höherer Temperaturen — bei Cu ab etwa 196 °K — bei hohen Verformungsgraden einen konstanten Wert der Schubspannung τ. Daraus kann jedoch noch nicht eindeutig entschieden werden, ob dies auch eine konstante Fließspannung τ_0 bedeutet. Es wäre noch möglich, daß hier zwei entgegengesetzt wirkende Vorgänge sich in ihrem Einfluß so aufheben, daß eine Zunahme der Schubspannung τ infolge Verfestigung durch eine entsprechende Abnahme infolge Texturänderung kompensiert wird.[2]) Systematische Texturuntersuchungen zur Entscheidung dieser Frage sind im Gange. Nach den bisher vorliegenden Untersuchungen ändert sich bei den hohen Verformungsgraden die Textur im Rahmen der Meßgenauigkeit nicht mehr, woraus dann in der Tat auf einen konstanten Grenzwert der Fließspannung geschlossen werden kann.

In einer zweiten Arbeit, die demnächst in dieser Zeitschrift veröffentlicht werden soll, werden noch andere Beobachtungen mitgeteilt — u. a. elektronenmikroskopische Gefügeuntersuchungen — welche ebenfalls für die Ansicht sprechen, daß die Fließspannung τ_0 einen Grenzwert erreicht und $\frac{d\tau_0}{d\gamma} = 0$ ist.

2. Der Torsionsversuch

2.1 Schubspannung und Scherung

Der Torsionsversuch an zylindrischen Vollstäben wurde zuerst von Ludwik und Scheu [3] zur Ermittlung der Schubspannungs-Scherungskurve $\tau = \tau(\gamma)$ von vielkristallinen Metallen herangezogen.

Gemessen wird beim Torsionsversuch die Abhängigkeit des zum Aufrechterhalten des Fließens erforderlichen Drehmomentes M vom Torsionswinkel Θ.

---

[2]) Wenn bei hohen Verformungsgraden die Schubspannung $\tau[\tau_0(\gamma), \overline{m}(\gamma)]$ konstant gefunden wird, würde also gelten:

$$d\tau = 0 = \left(\frac{\partial\tau}{\partial\tau_0}\right)_{\overline{m}} d\tau_0 + \left(\frac{\partial\tau}{\partial\overline{m}}\right)_{\tau_0} d\overline{m}$$

oder mit

$$d\tau_0 = \frac{d\tau_0}{d\gamma}\, d\gamma \qquad \text{und} \qquad d\overline{m} = \frac{d\overline{m}}{d\gamma}\, d\gamma\,, \qquad \frac{d\tau_0}{d\gamma} = -\frac{(\partial\tau/\partial\overline{m})_{\tau_0}}{(\partial\tau/\partial\tau_0)_{\overline{m}}} \cdot \frac{d\overline{m}}{d\gamma}\,. \tag{5}$$

LUDWIK und SCHEU haben gezeigt, daß man unter gewissen Voraussetzungen aus dem gemessenen Drehmoment die Schubspannung τ_a an der Oberfläche der zylindrischen Probe bestimmen kann. Da die in der Probenoberfläche vorliegende Scherung γ_a sich aus dem Torsionswinkel Θ direkt ergibt — sie beträgt

$$\gamma_a = \frac{R\,\Theta}{l}, \tag{6}$$

wobei R der Radius und l die Länge der Probe ist — kann man so für die Probenoberfläche die Schubspannungs-Scherungskurve $\tau = \tau(\gamma)$ ermitteln.

Diese Voraussetzungen sind, daß die Scherung γ_r eines im Inneren der Probe im Abstand r von der Probenachse gelegenen Volumelementes nach einer Torsion um den Winkel Θ durch $\gamma_r = \frac{r\Theta}{l}$ gegeben ist, d. h. daß γ_r wie bei der elastischen Torsion proportional zu r ist, und daß ferner zu Beginn der Torsion alle Volumenelemente der Probe im gleichen physikalischen Zustand sind. Dann kann man nach LUDWIK und SCHEU aus der Momentenkurve die Schubspannung τ_r an der Oberfläche der Probe ermitteln. Sie beträgt

$$\tau_a = \frac{1}{2\,\pi\,R^3}\left(3\,M + \Theta\,\frac{dM}{d\Theta}\right). \tag{7}$$

(Die Herleitung der Formel (7) findet sich auch in dem Lehrbuch von NADAI [4], S. 347ff.)

Bei dieser Ableitung ist angenommen, daß während der Torsion die Probenlänge und der Probenradius konstant bleiben. Dies ist nicht streng erfüllt. Die Berücksichtigung der Längen- und Radienänderung erfolgt in Abschnitt 2.2.

Zur experimentellen Prüfung der Frage, ob die der Formel (7) zugrundeliegende Voraussetzung der Proportionalität zwischen γ_r und r erfüllt ist, braucht man ein Kriterium, an dem man nach Beendigung des Versuches erkennen kann, welche plastische Scherung die in verschiedenen Entfernungen r von der Probenachse liegenden Volumenelemente erfahren haben. Ein solches Kriterium ergibt sich aus der Beobachtung von Ätzriefen, die an tordiertem Kupfer nach Anätzen mit Chromschwefelsäure entstehen. Diese haben eine Vorzugsrichtung. Sie sind umso

Fig. 3. Ätzriefen in verschiedenen Zonen einer tordierten Probe

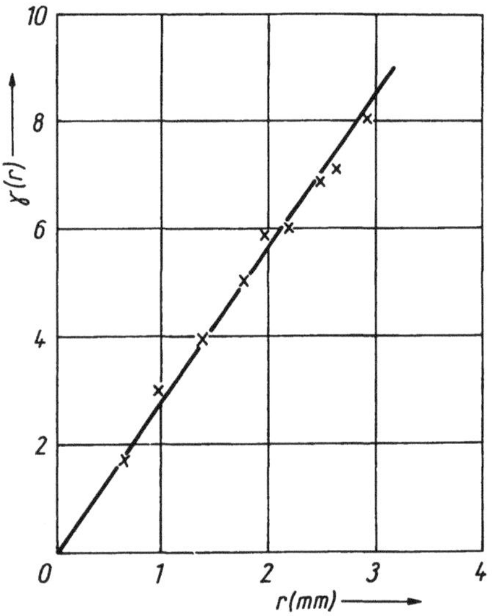

Fig. 4. Verteilung der Scherung im Probeninnern

stärker gegen die Probenachse geneigt, je höher der Torsionsgrad ist. Es erweist sich, daß der Tangens dieses Neigungswinkels im Mittel gleich der Scherung γ ist. Dies wurde an der Probenoberfläche dadurch geprüft, daß auf die Probe vor Beginn der Torsion eine Mantellinie eingezeichnet wurde, die nach der Torsion in eine gleichmäßig geneigte Schraubenlinie der Steigung $1/\gamma$ verwandelt wird (Fig. 7). Die Messung der Neigung der Ätzriefen für verschiedene Entfernungen r von der Probenachse erfolgte nach Anätzen einer an verschiedenen Stellen verschieden stark abgedrehten Probe. Vor dem Anätzen wurden die Proben elektrolytisch poliert. Das Ergebnis zeigt Fig. 3. — Die Abhängigkeit der Neigung vom Radius der Zonen ist in Fig. 4 aufgezeichnet, womit innerhalb der Meßgenauigkeit (ca. $\pm 3\%$) die Berechtigung zur Anwendung von (7) erwiesen ist.[3])

2.2 Längen- und Radiusänderungen und ihr Einfluß auf das Schubspannungs-Scherungsdiagramm

Beim Tordieren eines zylindrischen Vollstabes beobachtet man außer der Verdrillung des Stabes auch eine geringe Änderung der Probenlänge sowie des Probenradius. Dieses Verhalten wurde zuerst von SWIFT [8] genauer untersucht. Die Versuche von SWIFT ergaben, daß die Längenänderung nicht nur vom Torsionsgrad abhängt, sondern daß sie auch bei verschiedenen Metallen unterschiedlich ist.

Die Längenänderung einiger im Rahmen dieser Untersuchung tordierten Proben zeigt Fig. 5. Die Dehnung verlief bei Tieftemperaturversuchen wesentlich

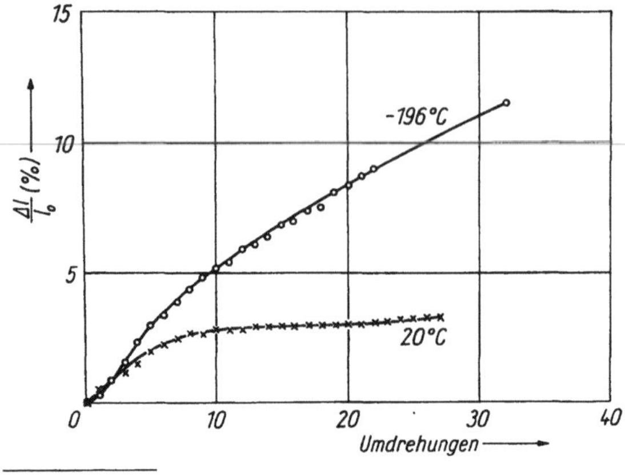

Fig. 5. Längenänderung von in Längsrichtung entlasteten Torsionsproben

[3]) Die Proportionalität zwischen γ und r wurde noch durch andere Verfahren geprüft, auf die hier nicht eingegangen werden soll. Vergleiche hierzu [6] und [7].

anders als bei Raumtemperaturversuchen. Die Radien änderten sich bei Raumtemperatur so, daß das Volumen der Proben konstant blieb. Bei der Temperatur des flüssigen Stickstoffs hingegen wurde aus den Längen- und Radiusänderungen eine Volumzunahme bis zu 1,7 % (bei $\gamma = 8$) errechnet.

Die Ursachen für diese Erscheinung sind noch weitgehend ungeklärt. Es sei aber darauf hingewiesen, daß bereits rein elastisch tordiertes Material geringe Längenänderungen zeigt. Dies wurde zuerst von POINTING [9] beobachtet.

Aus der Tatsache, daß mit der Torsion eine Längenänderung verknüpft ist, folgt, daß bei starrer Einspannung eine Axialspannung σ_z auftreten muß. Aus diesem Grunde war die Torsionsapparatur so konstruiert, daß die Probe sich frei ausdehnen konnte.[4])

Die Berücksichtigung der Längen- und Radiusänderungen bei der Herleitung von (7) führt nach einer Überlegung von TURCK [6] zu der Gleichung

$$\tau_a = \frac{1}{2\,\pi\, R^3 \left(1 + \frac{\varphi}{R} \cdot \frac{dR}{d\varphi}\right)} \left(3\, M + \varphi\, \frac{dM}{d\varphi}\right) \tag{7a}$$

mit $\varphi = \frac{\Theta}{l}$. Das Korrekturglied ist bei Raumtemperatur und höheren Temperaturen praktisch gleich 1. Nur bei den Tieftemperaturversuchen (77 °K) wird bei $\gamma_a \approx 8$ der Klammerausdruck im Nenner gleich 0,97. Diese Korrektur ist bei der Ermittlung der Kurven der Fig. 12 berücksichtigt worden.

3. Die Torsionsapparatur und die Meßverfahren

3.1 Messung des Drehmomentes M

Die Messung der Momentenkurve erfolgte mit einer Torsionsmaschine, die nach dem Prinzip der von SAUTTER, DEHLINGER und KOCHENDÖRFER [12] beschriebenen Apparatur gebaut worden war. Die Maschine besteht im wesentlichen aus zwei gegenüberliegenden Spannköpfen K, von denen der untere, der in vertikaler Richtung in dem Rahmen R sich frei bewegen kann, mit konstanter Geschwindigkeit gedreht wird, während der obere über einen Meßstab St mit dem gemeinsamen Rahmen R verbunden ist (Fig. 6).

Das Drehmoment wird durch die elastische Verdrillung des geeichten Meßstabes St bestimmt, die optisch mit Hilfe des durch den Meßstab bewegten Spiegels S auf einem mit konstanter Geschwindigkeit bewegten Film registriert wird.

3.2 Messung des Torsionswinkels Θ und die Bestimmung von γ_a

Der Torsionswinkel Θ der Proben wurde durch Verdrehung der Spannköpfe gemessen. Die Anzeige von Θ erfolgte dadurch, daß nach einer vollen Umdrehung des Spannkopfes die Registrierlampe kurzzeitig ausgeschaltet und so die photographisch aufgenommene Registrierkurve unterbrochen wurde.

Für die Berechnung von γ_a nach (6) muß wegen der Längen- und Querschnittsänderung der Proben (siehe Abschnitt 2.2) der jeweilige Wert von R und l ein-

[4]) Eine Abschätzung für die bei starrer Einsparung auftretenden Axialspannung σ_z mit Hilfe der Pragerschen Fließtheorie [10, 11, 12] und auf Grund der beobachteten Längenänderungen (Fig. 5) ergab im übrigen bei den Raumtemperaturversuchen bei $\gamma \approx 1$ einen Wert von 1,17 kp/mm², dessen Vernachlässigung bei der Berechnung von τ_0 aus (1) nur einen Fehler von etwa 1‰ zur Folge hat.

Fig. 6. Die Torsionsmaschine

gesetzt werden. Die Änderung von l wurde aus der vertikalen Verschiebung des unteren Spannkopfes mit einer Meßuhr bestimmt. Die Änderung des Radius wurde mit einer Mikrometerschraube gemessen.

Dieses Verfahren zur Bestimmung von γ_a ist aus zwei Gründen nicht unbedenklich. Die Messung von Θ könnte dadurch verfälscht werden, daß sich mit zunehmender Verfestigung der Probe die Verformung in einem nicht mehr vernachlässigbaren Grade in die Einspannfüße hineinzieht. Dies würde einer Vergrößerung der effektiven Meßlänge entsprechen und es würden sich für γ_a aus Formel (6) zu große Werte ergeben. Daß dies tatsächlich nicht der Fall ist, wurde auf zweierlei Weise geprüft. Einmal zeigte sich dies bei einer Untersuchung des Gefüges im Übergangsgebiet zwischen Meßstrecke und Einspannfuß (Fig. 8).

Außerdem wurde bei einigen Proben γ_a mit Hilfe zweier auf der Meßstrecke im Abstand l' angebrachter Marken aus $\gamma_a = \frac{R\,\Theta'}{l'}$ bestimmt, wo Θ' der Torsionswinkel zwischen den beiden Marken ist. Die Ergebnisse waren nach beiden Methoden die gleichen. — In den Ergebnissen des Abschnitts 5 sind die Größen τ_a und γ_a einfach mit τ und γ bezeichnet.

3.3 Messungen bei verschiedenen Temperaturen

Außer bei Raumtemperatur wurden Messungen bei höheren Temperaturen (bis zu 600 °C) durchgeführt, die mit Hilfe eines zwischen den Spannköpfen angebrachten elektrischen Ofens hergestellt wurden. Ferner erfolgten Messungen unterhalb Raumtemperatur, und zwar bei den Temperaturen des flüssigen Stickstoffs und Sauerstoffs, sowie in einem Gemisch von festem CO_2 und Alkohol (196 °K).

4. Die Proben und die Materialien

Die meisten der verwendeten Proben hatten eine zylindrische Meßstrecke l von 60 mm Länge und $2\,R = 6$ mm Durchmesser, an die sich auf beiden Seiten die zylindrischen Einspannfüße von 10 mm Durchmesser und 20 mm Länge anschlossen. Fig. 7 zeigt eine Probe nach einer Torsion bis $\gamma_a = 7$, auf deren Oberfläche vor Beginn der Torsion mit Ölfarbe eine Mantellinie gezeichnet worden war, die sich in eine gleichmäßig geneigte Schraubenlinie verwandelt hat. Eine Vermessung der Neigung der Schraubenlinie an 15 verschiedenen Stellen der Oberfläche ergab für γ_a eine mittlere Schwankung von $\pm 1\%$, womit die Homogenität der Torsion längs der Probe nachgewiesen ist. Die Versuche oberhalb Raumtemperatur wurden mit kleineren Proben (Meßstrecke $l = 10$ mm) durchgeführt.

Als Material wurde vorwiegend Elektrolytkupfer von 99,95% Reinheit verwendet, das als gezogene Stange angeliefert war. Vor der Torsion wurden die Proben 2 Stunden bei 800 °C im Vakuum

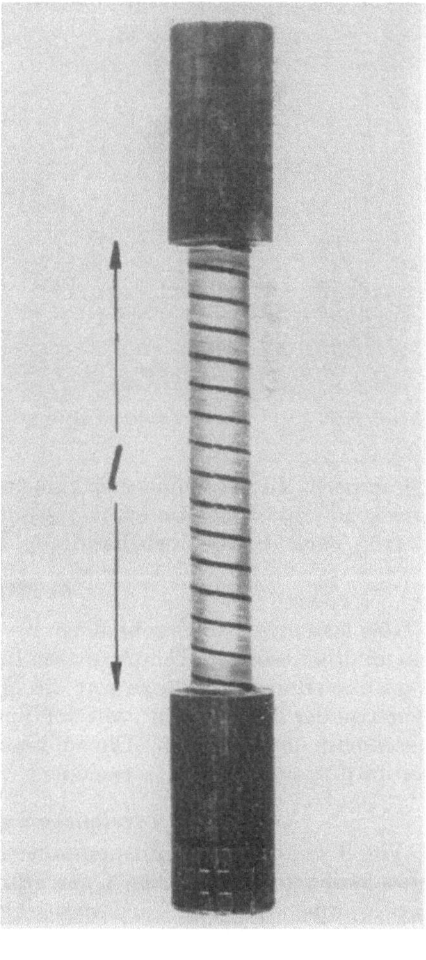

Fig. 7. Tordierte Probe. Eine vor dem Versuch eingezeichnete Mantellinie hat sich in eine gleichmäßig geneigte Schraubenlinie verwandelt

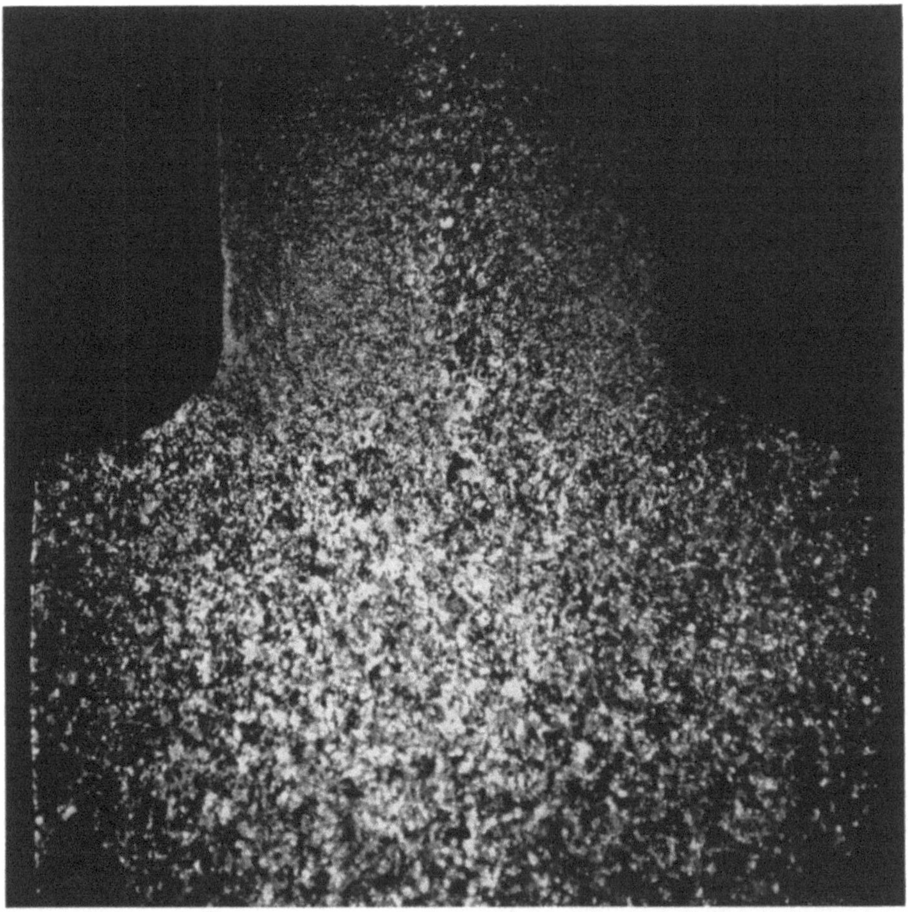

Fig. 8. Schliffbild von der Übergangszone zwischen tordiertem Material und Spannfuß. Die Probe ist längs der Achse durchgesägt

getempert. Zu der Glühdauer kamen noch die Anheizzeit von etwa 2 Stunden sowie die Abkülzeit von etwa 8 Stunden. Die mittlere Korngröße des Materials betrug nach dieser Vorbehandlung etwa 50 μm.

5. Meßergebnisse

Das wesentlichste Ergebnis der Versuche ist, daß bei Kupfer bei Raumtemperatur und bei höheren Temperaturen die Schubspannung einen Grenzwert erreicht. Von besonderem Interesse war die Messung dieses Grenzwertes, seine Abhängigkeit von der Temperatur, von der Verformungsgeschwindigkeit und von der Vorgeschichte des Materials. Die zu dieser Grenzfestigkeit gehörende Schubspannung sei im Folgenden mit τ_m bezeichnet.

5.1 Torsionsversuche bei Raumtemperatur

Fig. 9 zeigt eine Schubspannungs-Scherungskurve von Elektrolytkupfer, welches zwei Stunden bei 800 °C im Vakuum in der oben angegebenen Weise getempert worden war. Die Torsionsgeschwindigkeit betrug $\Theta = 0{,}15\ \mathrm{s}^{-1}$.

Fig. 9. Verfestigung von ausgeglühtem Kupfer. Die Breite der Kurve veranschaulicht die Grenzen der Reproduzierbarkeit

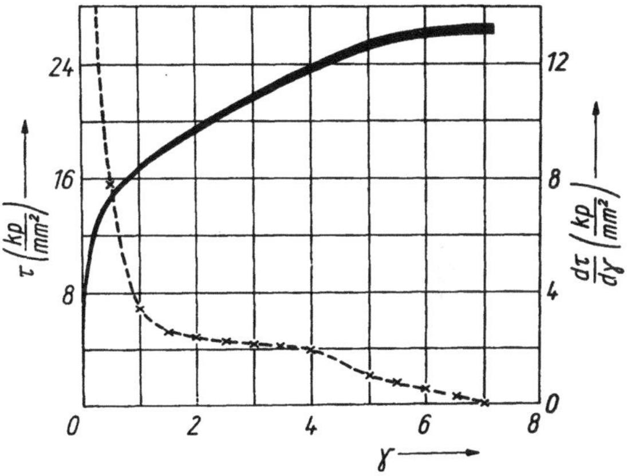

Die zunächst immer flacher werdende Schubspannungs-Scherungskurve weist nach einer Scherung von $\gamma \approx 1{,}5$ einen annähernd linearen Bereich auf, der sich etwa bis $\gamma = 4{,}5$ erstreckt. Von da ab nimmt die Neigung der Kurve weiterhin ab bis zum Bruch, wo sie nahezu Null geworden ist. Die gestrichelte Kurve in Fig. 9 (rechte Skala!) stellt den Verlauf des Anstiegs $d\tau/d\gamma$ dar. Der Bruch der Probe tritt in der Regel nach einer Scherung von $\gamma \approx 7{,}5$ ein. Die Schubspannung beim Bruch beträgt etwa $26 \frac{\text{kp}}{\text{mm}^2}$.

Eine aus kalt gezogenem Material (Anlieferungszustand) hergestellte Probe brach bei $\tau = 26{,}6 \frac{\text{kp}}{\text{mm}^2}$ (siehe Fig. 10, Kurve „0").

Die Schubspannungs-Scherungskurven von getempertem und ungetempertem Material lassen sich für Spannungen über $\tau = 20 \frac{\text{kp}}{\text{mm}^2}$ annähernd durch eine Parallelverschiebung in γ-Richtung ineinander überführen (Fig. 10).

Im Folgenden soll die Frage nach der Ursache des Bruches diskutiert werden. Die Oberfläche einer ursprünglich glatten Probe zeigt nach der Torsion eine mehr oder weniger große Rauhigkeit. Diese ist größer bei grobkörnigen Proben als bei feinkörnigen und hat die Größenordnung des ursprünglichen Korndurch-

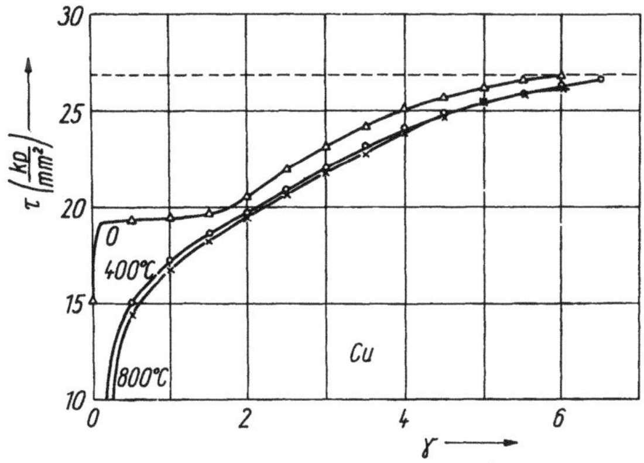

Fig. 10. Verfestigung von ungetempertem (Kurve „0") und 2 Stunden bei 400 °C bzw. 800 °C getempertem Kupfer. Der Grenzwert der Verfestigung, der aus anderen Versuchen ermittelt wurde, ist durch die gestrichelte Gerade angedeutet

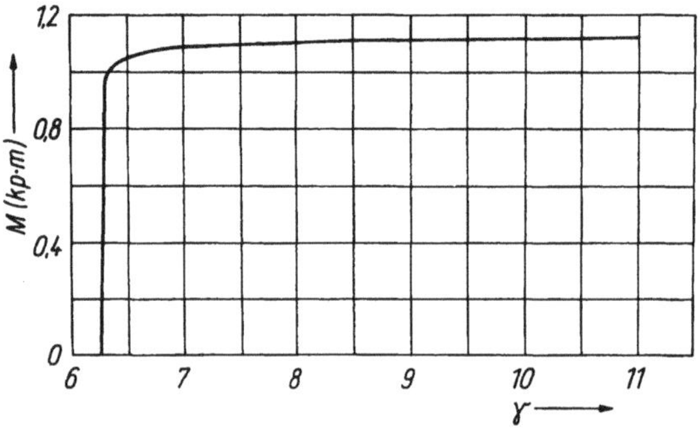

Fig. 11. Verfestigung einer kurz vor dem bei $\gamma \approx 7$ zu erwartenden Bruch glattgedrehten und dann weitertordierten Cu-Probe

messers vor der Verformung. Die Ursache für das Eintreten des Bruches dürfte in der Ausbildung von Rissen zu suchen sein, die sich an solchen in der Oberfläche verlaufenden Korngrenzen bilden, wo infolge der Oberflächenrauhigkeit eine besonders hohe Kerbwirkung auftritt. Danach sollte der Bruch verhindert werden können, wenn man eine bis kurz vor den normalen Bruch (etwa bei $\gamma = 7$) tordierte Probe glattdreht. Tatsächlich lassen sich auf diese Weise wesentlich höhere Torsionsgrade erreichen. Eine Momentenkurve, die mit einer im Anschluß an eine Vortorsion von $\gamma = 6{,}3$ glattgedrehten Probe erhalten wurde, ist in Fig. 11 wiedergegeben. Das Drehmoment erreichte einen Grenzwert bei einer Gesamttorsion von $\gamma = 8$, und es war möglich, die Probe nach Erreichen dieses Grenzwertes noch bis $\gamma = 11$ weiter zu tordieren. Aus dem konstanten Enddrehmoment ergibt sich nach (7a) ein Grenzwert der Schubspannung von $\tau_m = 26{,}2 \frac{\text{kp}}{\text{mm}^2}$, also innerhalb der Fehlergrenzen von $\pm 1{,}5\%$ der gleiche Wert wie er als Bruchspannung an Proben gefunden wird, die nicht nach einer Vortorsion glattgedreht worden waren.

Bemerkenswert ist, daß die glattgedrehten Proben bei einer Weiterverformung im wesentlichen glatt bleiben. Dies ist ein Hinweis darauf, daß bei sehr hohen plastischen Verformungen die Kristallite kleiner werden, denn die Oberflächenrauhigkeit nach der Verformung ist von der Größenordnung der bei Beginn der Verformung vorliegenden Kristallitgröße. Es tritt also eine Kornzerkleinerung[5]) auf. Diese konnte auch durch licht- und elektronenmikroskopische Untersuchungen nachgewiesen werden (vergleiche Teil II dieser Arbeit).[6])

5.2 Versuche bei höheren Temperaturen

Bei Torsionsversuchen oberhalb Raumtemperatur ergaben sich ab etwa 600 °K Schubspannungs-Scherungskurven, deren Verlauf sich qualitativ von dem Verlauf der bei Raumtemperatur gemessenen Kurven dadurch unterschied, daß die Schubspannung nach einer gewissen Verformung wieder absinkt und erst dann einem Grenzwert zustrebt (Fig. 12). Das Maximum liegt bei umso kleineren Verformungen, je höher die Temperatur ist. Ein solches Verhalten wurde auch von Kartwick, Tegart, Rossard und Blain [13] sowie von Turck [6] beobachtet.

---

[5]) Damit ist gemeint, daß der Kristallit eine Aufteilung in viele kleine Bereiche mit größeren Orientierungsunterschieden erfährt, wobei aber möglicherweise die Orientierungsänderung von Bereich zu Bereich nicht unstetig wie bei Kristallitgrenzen erfolgt, sondern innerhalb einer schmalen Zone stetig (Zellbildung).

[6]) Erscheint demnächst in dieser Zeitschrift.

Fig. 12. Verfestigung von ausgeglühtem Kupfer bei verschiedenen Temperaturen

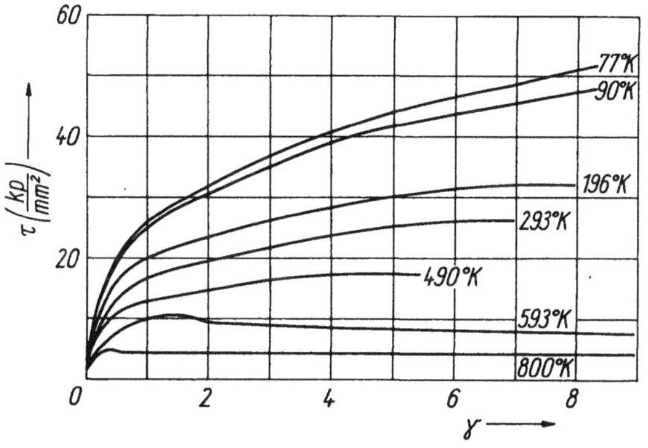

Möglicherweise hängt diese Erscheinung damit zusammen, daß entsprechend den Überlegungen in Abschnitt 1 in diesem Verformungsbereich die Abnahme der Schubspannung infolge der sich ausbildenden Verformungstextur die Zunahme der Schubspannung infolge Verfestigung überwiegt. Bei den Versuchen oberhalb 600 °K wurden wesentlich höhere Verformungen erreicht als bei Raumtemperatur. Bei 800 °K betrug die Scherung beim Bruch der Probe $\gamma = 17{,}5$.

5.3 Versuche bei tiefen Temperaturen

Die Versuche bei 196 °K, 90 °K und 77 °K (siehe Fig. 12) ergaben gut reproduzierbare Bruchspannungen, die höher sind als bei Raumtemperatur — bei 77 °K fast der doppelte Wert, bei 196 °K etwa 32 $\frac{\text{kp}}{\text{mm}^2}$. Es tritt zwischen $\gamma \approx 1{,}5$ und $\gamma \approx 3{,}0$ ebenfalls ein linearer Bereich auf.

Der Anstieg der Kurven ist bei 77 °K und 90 °K vor dem Eintreten des Bruches nicht gleich Null. Es gelang bei diesen Temperaturen auch nicht in der oben beschriebenen Weise durch Glattdrehen und Weitertordieren der Proben einen Grenzwert der Festigkeit festzustellen. Sie brachen bei der gleichen Schubspannung wie die nicht abgedrehten Proben.

5.4 Temperaturwechselversuche

Unterbricht man einen bei der Temperatur des flüssigen Stickstoffs durchgeführten Torsionsversuch nach einer Scherung von $\gamma \approx 3{,}5$ und setzt die Torsion

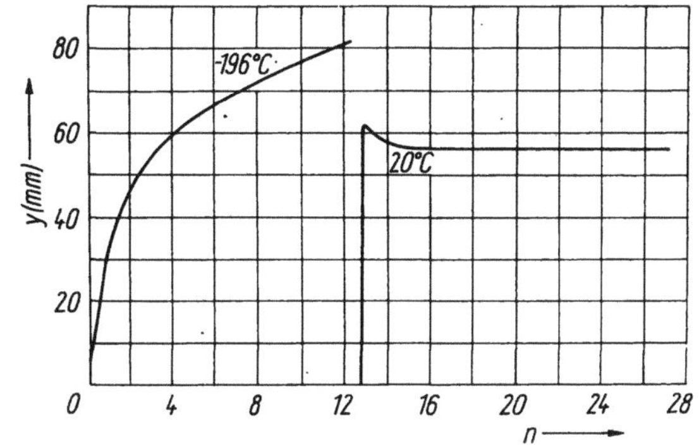

Fig. 13. Momentenkurve bei Temperaturwechsel von 77 °K auf Raumtemperatur. n = Zahl der Umdrehungen. Als Ordinate ist die Schreibhöhe der Registrierkurve aufgetragen

nach Erwärmung der Probe auf Raumtemperatur fort, so ergibt sich eine Momentenkurve, wie sie Fig. 13 zeigt. Das Drehmoment steigt bei Fortsetzung des Versuches bei Raumtemperatur zunächst auf einen Maximalwert an und fällt dann um etwa 10% dieses Maximalwertes ab, um dann konstant zu bleiben. Aus diesem konstanten Drehmoment errechnet man nach (7a) eine konstante Schubspannung, die gleich dem Grenzwert der Festigkeit bei reinen Raumtemperaturversuchen ist.

Durch solche Temperaturwechselversuche zwischen 77 °K und 196 °K konnte auch noch bei 196 °K ein Grenzwert der Festigkeit nachgewiesen werden. Die erforderliche Vortorsion bei 77 °K betrug ca. $\gamma = 4{,}5$. Die Proben wurden in diesem Falle vor der Weitertorsion durch elektrolytisches Polieren geglättet. Es ergaben sich Grenzfestigkeiten zwischen $31{,}2\ \frac{\text{kp}}{\text{mm}^2}$ und $32{,}3\ \frac{\text{kp}}{\text{mm}^2}$, welche im Rahmen der Meßgenauigkeit mit der Bruchspannung in Fig. 12 übereinstimmen.

Die vorstehend geschilderten Beobachtungen legen die Vermutung nahe, daß sich zumindest bei Temperaturen oberhalb 196 °K und bei konstanter Verformungsgeschwindigkeit ein Zustand einstellt, der nicht von der Vorgeschichte abhängt und der durch eine konstante Fließspannung gekennzeichnet ist.

Diese Vorstellung wird nicht nur durch die oben erwähnten vorläufigen Ergebnisse der Texturbeobachtungen nahegelegt, sondern es spricht dafür besonders auch der folgende Versuch: Eine Probe, die bis $\gamma_a \approx 3{,}5$ in flüssigem Stickstoff vortordiert worden war, wurde abwechselnd bei Raumtemperatur bzw. in flüssigem Stickstoff in folgender Weise weitertordiert: Zunächst erfolgten drei Umdrehungen bei Raumtemperatur. Danach war das Drehmoment konstant geworden, und die Schubspannung hatte den für Raumtemperatur charakteristischen Wert τ_m erreicht. Sodann erfolgten zwei Umdrehungen bei 77 °K, dann wieder eine Umdrehung bei Raumtemperatur usw. bis zum Bruch der Probe. Es zeigte sich, daß die Momentenkurven, die bei tiefer Temperatur gemessen wurden, nach jeder Zwischentorsion bei Raumtemperatur praktisch den gleichen Verlauf hatten (siehe Fig. 14). Dieses Ergebnis deutet darauf hin, daß durch die Zwischentorsion bei Raumtemperatur stets der gleiche Ausgangszustand mit gleicher Festigkeit τ_0 und gleicher Textur für die nachfolgende Torsion in flüssigem Stickstoff geschaffen wird, und daß damit der letzte Bereich der Verfestigungskurve, der durch eine konstante Grenzschubspannung gekennzeichnet ist, einen von der Vorgeschichte unabhängigen Zustand mit konstanter Festigkeit und konstanter Textur darstellt. Würde nämlich bei Raumtemperatur in dem Bereich konstanter Schubspannung entsprechend den Überlegungen von Abschnitt 1 die Textur sich ändern und damit auch die Fließspannung, so sollte sich dies in den anschließenden (τ, γ)-Kurven bei der tiefen Temperatur bemerkbar machen, die mit zunehmender Textur sich systematisch im Sinne kleinerer Schubspannungen ändern müßten.

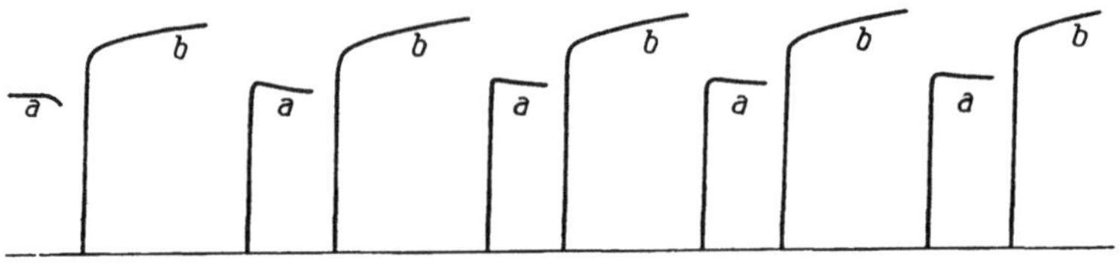

Fig. 14. Verlauf der Momentenkurve bei abwechselnder Torsion bei Raumtemperatur (a) bzw. in flüssigem Stickstoff (b)

5.5 Der Einfluß der Torsionsgeschwindigkeit

Der Grenzwert τ_m der Schubfestigkeit hängt außer von der Temperatur noch von der Verformungsgeschwindigkeit ab. Die Geschwindigkeitsabhängigkeit des Drehmomentes beim Torsionsversuch ist in verschiedenen Arbeiten [14 bis 17] untersucht worden. Sie läßt sich gut durch ein Potenzgesetz der Form $M = \text{const} \cdot \dot{\Theta}^n$ beschreiben. Die Geschwindigkeitsabhängigkeit ist umso größer, je höher die Temperatur ist.

Unsere Versuche ergaben ein ähnliches Verhalten für den Grenzwert τ_m. In dem Geschwindigkeitsbereich zwischen $\dot{\Theta} = 9{,}0\ \text{min}^{-1}$ und $\dot{\Theta} = 4 \cdot 10^{-2}\ \text{min}^{-1}$ findet man für n den Wert 0,0168. Die Geschwindigkeitsabhängigkeit von τ_m wurde vorwiegend durch Geschwindigkeitswechselversuche ermittelt.

5.6 Versuche mit Wechsel der Torsionsrichtung

Eine dem Betrage nach viel größere Verformung als bei den bisher geschilderten Versuchen läßt sich erreichen, wenn man die Probe abwechselnd rechts- bzw. linkssinnig tordiert. Es wurde versucht, auf diese Weise die Grenzfestigkeit bei tiefer Temperatur zu messen.

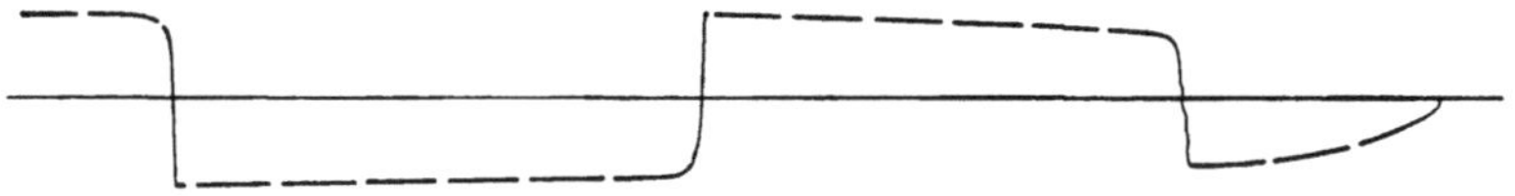

Fig. 15. Registrierkurve eines Torsionsversuches mit Wechsel der Torsionsrichtung (Raumtemperatur)

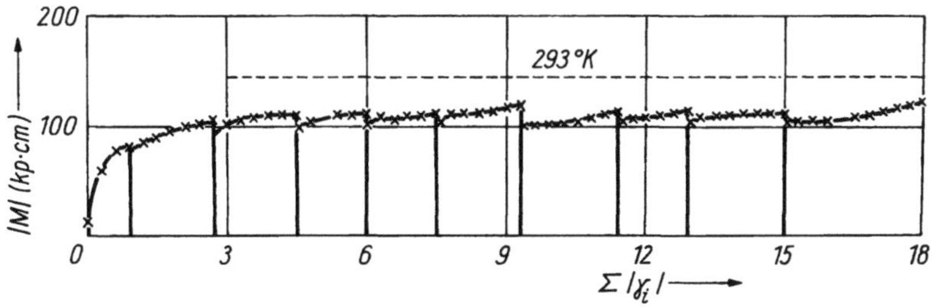

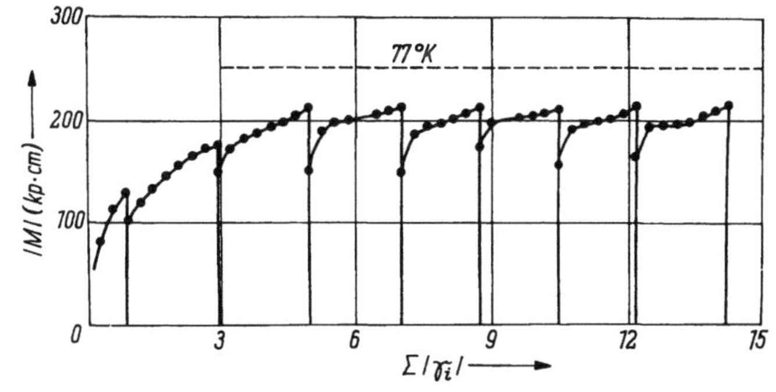

Fig. 16. Abhängigkeit des Betrages des Drehmomentes von der Summe der Beträge der Einzelscherungen bei Wechsel der Torsionsrichtung. Gestrichelt ist das Grenz- bzw. Bruchdrehmoment bei einsinniger Torsion (auf gleiche Endradien der Proben umgerechnet) eingetragen

Fig. 15 gibt eine Registrierkurve wieder, die bei einem Richtungswechselversuch an ausgeglühtem Kupfer bei einer Versuchstemperatur von 77 °K gewonnen wurde. In dem Diagramm Fig. 16 ist der Betrag des Drehmomentes gegen den Betrag der Scherung $\sum_i |\gamma_i| = \sum_i \frac{R}{l} |\Theta_i|$ für je einen Versuch bei Raumtemperatur und bei 77 °K aufgetragen. Aus den Messungen ergibt sich, daß das Drehmoment bei einem Richtungswechselversuch kleiner bleibt, als bei einsinniger Verformung.

Die Deutsche Forschungsgemeinschaft hat die Untersuchung durch eine Beihilfe wirksam gefördert, wofür die Verfasser bestens danken.

Literatur

[1] R. Hill, The Mathematical Theory of Plasticity, Clarendon Press Oxford 1960 (S. 317ff.).
[2] A. Nadai, J. appl. Phys. 8, 205 (1937).
[3] P. Ludwik und R. Scheu, Stahl und Eisen **45**, 373 (1925).
[4] A. Nadai, Theory of Flow and Fracture of Solids, McGraw-Hill Book Comp., New York 1950 (S. 347ff.).
[5] G. I. Taylor und H. Quinney, Proc. Roy. Soc. **A 143**, 307 (1933/34).
[6] H. Turck, Diplomarbeit, T.H. Aachen 1963.
[7] H. G. Grewe, Dissertation, Münster 1962.
[8] H. W. Swift, Engineering, Apr., 253 (1947).
[9] J. H. Pointing, Proc. Roy. Soc. **A 82**, 546 (1909); **A 86**, 534 (1912).
[10] W. Prager, J. appl. Phys. **20** 235 (1949).
[11] D. Drucker, J. appl. Mech. **16**, 349 (1949).
[12] W. Sautter, U. Dehlinger und A. Kochendörfer, Z. Metallk. **44**, 442 und 553 (1953).
[13] D. Hartwick, W. J. Mc.G. Tegart, C. Rossard und P. Blain, C. R. Acad. Sci. (France) **252**, 2881 (1961).
[14] P. Blain und C. Rossard, Mem. Sci. Rev. Met. **56**, 285 (1959).
[15] P. Blain und C. Rossard, Rev. Met. **55**, 573 (1958).
[16] H. Ormerod und W. J. Mc.G. Tegart, J. Inst. Met. **89**, 94 (1960/61).
[17] D. S. Fields und W. A. Backofen, Trans. Amer. Soc. Metals **51**, 946 (1959).

(Received April 16, 1964)

phys. stat. sol. 6, 355 (1964)

Department of Physics, Battersea College of Technology, London

The Determination of the Orientation of Rhombohedral and Cubic Crystals from Traces on a Single Surface

By

M. BEVIS, F. HECKSCHER, and A. G. CROCKER[1])

The determination of the orientation of rhombohedral and cubic crystals from the angles between traces produced on a single surface of a specimen by three crystallographically equivalent planes is considered. It is shown that the direction normal to the crystal surface may be determined directly by taking readings from simple physical models which may be readily constructed. A new algebraic formulation of the problem which leads directly to the possible orientations, is also presented. Degenerate cases which arise in this analysis are discussed and a table of results is given for the case of traces produced by $\{11\bar{1}\}$ planes in mercury crystals. A convenient method of plotting these results is described, and it is shown that the results obtained using the models are accurate to within $1^1/_2°$ when compared with those obtained numerically.

Es werden Methoden diskutiert, mit deren Hilfe die Orientierung von rhomboedrischen und kubischen Kristallen aus den Winkeln zwischen Spuren bestimmt werden kann, die auf einer einzigen Oberfläche durch drei kristallografisch äquivalente Ebenen hervorgerufen werden. Es wird gezeigt, daß die Richtung senkrecht zur Oberfläche direkt auf der Grundlage einfacher, leicht konstruierbarer Modelle bestimmt werden kann. Es wird eine neue algebraische Formulierung des Problems angegeben, die direkt zu den möglichen Orientierungen führt. Ferner werden Entartungsfälle diskutiert und Resultate für Spuren angegeben, die auf $\{11\bar{1}\}$-Flächen in Quecksilberkristallen auftreten. Eine bequeme Methode zur Auftragung dieser Resultate wird beschrieben. Es zeigt sich, daß die mit Hilfe von Modellen erhaltenen Ergebnisse auf $\pm 1^1/_2°$ mit den numerisch gewonnenen übereinstimmen.

1. Introduction

The problem of determining the orientation of a crystal from a knowledge of the angles between traces produced on a flat surface of a specimen by given crystallographic planes has been discussed by several authors. The earliest treatment is that due to BARRETT [1] who briefly mentions a trial and error method of solution involving the manipulation of tracing paper on a stereographic net. This method is in principle valid for any crystal lattice, but is often lengthy and can give rise to most inaccurate results. It has been described in more detail by MYKURA [2] who has applied it to determine the orientation of face centered cubic crystals from traces produced by three or four $\{111\}$ planes. Another approach is to use the charts which have been constructed for certain special cases (for references see [3]), but again these are often difficult to use and for some orientations prove to be rather unreliable. Finally, the orientation may be determined analytically, in which case the accuracy is clearly limited only by the measurements of

[1]) On leave of absence at RIAS, Baltimore, Maryland USA.

the angles themselves. An analysis of this kind has been given recently by DRAZIN and OTTE [3] for traces produced by three {111} planes of cubic crystals. The algebra involved in this analysis is extremely lengthy and tedious and although the authors describe three procedures for obtaining numerical solutions, one mainly graphical, a second suitable for desk computation and a third for solution using an electronic digital computer, it is clear that this formulation of the problem could not be used extensively in practice. Tables have, however, been prepared by DRAZIN and OTTE [4] from their analysis and these will enable rapid accurate determinations of the orientations of cubic crystals to be made.

The complexity of the algebra of the DRAZIN and OTTE method [3] is essentially due to the fact that a quartic equation has to be solved in order to deduce the vector normal to the plane surface containing the three traces. For different sets of initial data this equation may have four, two or no real solutions, corresponding to four, two or no possible normals. It should, however, be noted that two orientations related by reflection in the surface have the same surface normal and thus there are in general eight, four or no full orientations for a given set of data.

In section 2 of the present paper we shall describe simple physical models which have been used to determine the orientations of mercury crystals from surface traces produced by slip on the three $\{11\bar{1}\}$ planes of the face centered rhombohedral cell. Although these models were made specifically for orienting mercury crystals, the principles involved are quite general and in particular may be readily applied to the determination of the orientation of face centered cubic crystals from {111} traces. The accuracy which can be obtained using the models is discussed, results being compared with numerical solutions obtained using a new analysis of the problem which is presented in section 3. This analysis is valid for any three crystallographically equivalent symmetrically disposed planes and thus includes as special cases traces produced by {111} and {100} planes of cubic crystals and by $\{11\bar{1}\}$ and {100} planes of mercury. Also, unlike the presentation of DRAZIN and OTTE [3], in this treatment the quartic equation, giving the possible surface normals, is derived directly. Degenerate solutions of this equation are examined and, as an illustrative example, a specimen table is presented which enables the orientation of mercury crystals to be determined. Finally, a convenient method of plotting these results graphically is described and the ranges of the sets of data which give rise to different numbers of solutions are discussed.

2. Orientation Determination Using Models

This section contains a description of the use of simple physical models in determining the orientation of crystals from a knowledge of the angles between three surface traces. Let the three crystallographically equivalent planes which produce the traces meet at a point O and intersect along the lines AOA', BOB', and COC' as shown in Fig. 1. We restrict ourselves to cases in which the three angles AOB, BOC, and COA are all equal and represent the magnitude of this angle by β. Let the surface of the crystal intersect OA, OB, and OC at points L, M, and N distant a, b, and c respectively from O. The triangle LMN thus has the shape of the observed set of similar triangles produced on the crystal surface by the three sets of planes, and the orientation of the crystal is fixed by the magnitudes of the parameters a, b, and c. The problem of determining the orientation thus reduces to finding a, b, and c for given values of the angles L, M, and N. This may be carried out conveniently by constructing two models in the form of tetrahedra for the

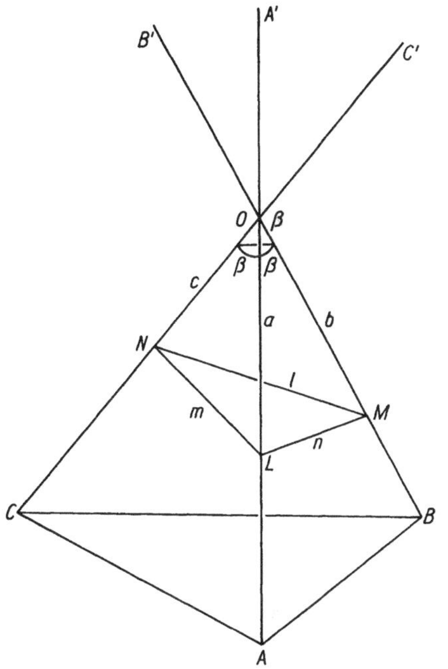

Fig. 1. The tetrahedron $OABC$ formed by three crystallographically equivalent planes meeting symmetrically at the point O, and the triangles of traces LMN produced at the surface of the crystal

two essentially different situations which may arise. These models consist of the tetrahedra $OABC$ and $OAB'C'$ with the faces ABC and $AB'C'$ open. A triangular template with angles L, M, and N is then prepared and placed in each tetrahedron in turn so that the edges LM, MN, and NL lie in the faces of the tetrahedron, as illustrated for the tetrahedron $OABC$ in Fig. 1. The magnitudes of a, b, and c can then be read off directly from scales along the edges OA, OB, OC and OA, OB', OC' for the two models respectively, where intercepts on OB' and OC' are, of course, negative.

The size of the template is arbitrary and it is thus only the relative magnitudes of a, b, and c with which we are concerned. Also, as the three planes meeting at O are equivalent, the labelling of the corners L, M, and N of the triangle and of the edges AOA', BOB', and COC' of the tetrahedra is clearly arbitrary. The two models described above may thus be taken as representative of the total of eight tetrahedra which can be defined by these three planes, tetrahedron $OA'B'C'$ being equivalent to $OABC$ and tetrahedra $OBCA'$, $OA'B'C$, etc. being equivalent to $OAB'C'$. There are several ways of attempting to place the triangular template in each of the two models, as each corner may be associated with each edge of the model in two different ways by inverting the template, and each of these associations may give more than one solution. However, because of the symmetry of the models, the maximum number of solutions is always found to be four, although for certain ranges of template shape the number of solutions obtained is two or even zero. The case of no solutions should, however, never occur, as it indicates that the original assumption that all three traces are produced by known crystallographically equivalent planes was false. For isosceles and equilateral triangle shapes degeneracies arise which result in some of the solutions being crystallographically equivalent, so that in special cases one or three independent solutions may also be obtained. However, these degeneracies and the way in which the number of solutions is divided between the two models depends critically on the angle β and the shape of the template and is more conveniently studied analytically.

The parameters a, b, c fix the orientation of the surface of the crystal, the Miller indices of the surface relative to the tetrahedron axes OA, OB, OC being $\{a^{-1}\ b^{-1}\ c^{-1}\}$. It will, however, be necessary in some applications to convert these indices to those referred to a basis consisting of the $\langle 100 \rangle$ edges of the standard cell.

This may be readily carried out using the relation

$$(h_i)_s = (h_i)_t \, {}_tC_s\,,$$

where $(h_i)_s$ and $(h_i)_t$ $(i = 1, 2, 3)$ represent the Miller indices relative to standard and tetrahedron axes, and the rows of the 3×3 symmetric matrix ${}_tC_s$ are the Miller indices of the edges of the standard cell relative to the tetrahedron axes. For example, for traces produced by three $\{111\}$ planes in cubic structures or three $\{11\bar{1}\}$ planes in mercury we have

$${}_tC_s = \begin{pmatrix} \bar{1} & 1 & 1 \\ 1 & \bar{1} & 1 \\ 1 & 1 & \bar{1} \end{pmatrix}.$$

The Miller indices h_i give only the orientation of the normal to the surface of the crystal and not the full orientation of the crystal. The latter may, however, be readily deduced from h_i and the positions of the surface traces. In particular the Miller indices of a given trace, which will clearly lie in the surface h_i and the plane m_i producing the trace, are given by

$$[m_2 h_3 - m_3 h_2,\, m_3 h_1 - m_1 h_3,\, m_1 h_2 - m_2 h_1]\,.$$

The two model tetrahedra used in determining the orientation of mercury crystals from $\{11\bar{1}\}$ traces were constructed of transparent synthetic plastic, the edges of the triangular faces being joined with adhesive. The angle β is in this case $70°45'$ and the edges were graduated every 0.1 cm.[2])

In using models for determining crystallographic orientations errors can arise in essentially three different ways. The construction of the models may be faulty, the shape of the template may be incorrect and readings of the intercepts a, b, and c may be in error. The effective error in an orientation determined from the models is thus difficult to assess and therefore, in order to determine the accuracy of the model method, a set of test templates covering the whole range of triangle shapes with angles at $10°$ intervals was prepared. The orientations deduced from these templates were compared with those determined numerically using the analysis to be described in section 3. In no case were the model results found to be in error by more than $1^1/_2°$. These errors are in fact often less than those which have been found to arise from uncertainties in the true magnitudes of the angles between the surface traces themselves. It is thus considered that the use of these models is a valuable technique in determining the orientation of crystals from surface traces of known crystallographic planes.

3. Analytic Determination of Orientations

In this section we present an analysis of the problem of orienting any rhombohedral crystal, including as a special case cubic crystals, from surface traces produced by three crystallographically equivalent planes which intersect each other in crystallographically equivalent directions. We again refer to Fig. 1 and the problem reduces to determining the intercepts a, b, c when the angles L, M, N are known. As it is only the relative magnitudes of a, b, and c with which we are con-

---

[2]) If desired scales graduated in terms of the recipocal of the distance from the vertex can be placed along the edges, enabling the Miller indices, relative to the tetrahedron axes, to be read off directly.

cerned we may put $a = 1$ and let the corresponding magnitudes of MN, NL and LM be l, m, and n respectively. Then, using the sine and cosine rules we obtain

$$b^2 + 1 - 2\,b\cos\beta = r\,l^2\,, \tag{1}$$

$$c^2 + 1 - 2\,c\cos\beta = s\,l^2\,, \tag{2}$$

$$b^2 + c^2 - 2\,b\,c\cos\beta = l^2\,, \tag{3}$$

where $r = (\sin N/\sin L)^2$ and $s = (\sin M/\sin L)^2$. Eliminating b and l^2 from equations (1) to (3) we arrive at the following quartic equation for c

$$C_0 + C_1\,c + C_2\,c^2 + C_3\,c^3 + C_4\,c^4 = 0\,, \tag{4}$$

where

$$C_4 = \{(-r + s + 1)^2 - 4\,s\cos^2\beta\}\,,$$

$$C_1 = 4\cos\beta\,\{(-r + s + 1)(r - 1) + s\,(r - s + 1)\cos\beta + 2\,s\cos^2\beta\}\,,$$

$$C_2 = 2\,\{-(-r + s + 1)(r + s - 1) + 2\,[(r - 1)^2 - (r + 1)\,s + 2\,s^2] \times \\ \times \cos^2\beta - 4\,(r + 1)\,s\cos^3\beta\}\,,$$

$$C_3 = 4\cos\beta\,\{-(r + s - 1)(r - 1) + s\,(r - s + 1)\cos\beta + 2\,r\,s\cos^2\beta\}\,,$$

$$C_4 = \{(r + s - 1)^2 - 4\,r\,s\cos^2\beta\}\,.$$

This equation gives four, two or no real solutions for c depending on the values of the parameters β, r and s. When c is known, b may be determined from the relationship

$$b = \frac{(r - 1)(c^2 + 1 - 2\,c\cos\beta) + s\,(c^2 - 1)}{2\,s\,(c - 1)\cos\beta}\,, \tag{5}$$

and thus, as a is fixed at unity, the vector normal to the surface is known relative to the tetrahedron axes.

Certain degenerate solutions of equations (4) and (5) should be noted. In particular when one of the angles of the triangle LMN is β or $\pi - \beta$ the equations produce a solution with either b or c zero. This case is not observable physically as the surface of the crystal is then parallel to one of the planes producing the traces. Also equation (5) for b becomes indeterminate when $c = 1$, because, as we have already let $a = 1$, the triangle LMN is then isosceles so that $r = 1$. In this case equation (4) reduces to

$$(c^2 + C_1'\,c + 1)(c - 1)^2 = 0\,, \tag{6}$$

where $C_1' = 2\,[s\,(1 - 2\cos^2\beta) + 4\cos^3\beta]/(s - 4\cos^2\beta)$.

The two solutions $c = 1$ of equation (6) are associated with values of b given by the quadratic equation

$$b^2 - 2\,b\cos\beta + [1 - 2\,(1 - \cos\beta)/s] = 0\,. \tag{7}$$

The other two solutions of equation (6) are clearly reciprocal to each other and on substituting these values into equation (5) for b, which when $r = 1$ reduces to $b = (c + 1)/2\cos\beta$, one finds that they result in only one independent solution for the orientation. We may thus obtain, depending on the values of s and β, one, two or three possible solutions for the orientation when the triangle LMN is isosceles. In the case of the equilateral triangle, however, only two independent solutions arise for the triplet of parameters a, b, c. By putting $s = 1$ in equations

Table 1

The orientation of mercury crystals as calculated from three $\{11\bar{1}\}$ traces

M	N	I $10^4 k$	I $10^4 l$	II $10^4 k$	II $10^4 l$
40	40	−6110	+1052	+1052	−6110
	35	−6788	+3445	+2903	−6980
45	45	−5488	−1386	−1386	−5488
	40	−5495	−0085	−0589	−5945
	35	−5771	+1509	+0494	−6481
	30	−6626	+3909	+2248	−7212
50	50	−5323	−3080	−3080	−5323
	45	−5071	−2025	−2578	−5672
	40	−4977	−0924	−1982	−6040
	35	−5070	+0322	−1235	−6440
	30	−5452	+1894	−0216	−6902
	25	−6485	+4319	+1420	−7503
55	55	−5455	−4525	−4525	−5455
	55	+ 10^4	+3740	−9417	−9417
	50	−4981	−3520	−4171	−5747
	50	+9732	−5754	+9380	−0986
	45	−4674	−2566	−3777	−6041
	40	−4508	−1600	−3324	−6343
	35	−4488	−0556	−2783	−6661
	30	−4657	+0658	−2102	−7006
	25	−5142	+2227	−1172	−7398
	20	−6361	+4690	+0307	−7881
60	60	+ 10^4	+ 10^4	−5924	−5924
	55	−5166	−4859	−5670	−6168
	55	+8581	+6543	−7299	−7130
	50	−4639	−3918	−5399	−6406
	50	+3601	+7950	−9292	−9199
	45	−4284	−3044	−5103	−6643
	45	+9222	−5654	+8171	−0654
	40	−4060	−2178	−4772	−6884
	35	−3961	−1269	−4389	−7134
	30	−4003	−0261	−3930	−7399
	25	−4244	+0939	−3350	−7688
	20	−4833	+2521	−2560	−8012
	15	−6250	+5033	−1324	−8395
	10	+7880	−9658	+1718	−8944

M	N	I $10^4 k$	I $10^4 l$	II $10^4 k$	II $10^4 l$
65	50	−6952	−7465	−6952	−7465
	50	+6080	+6080	−4287	−4287
	45	−4813	−5170	−6763	−7629
	45	+4013	+6444	−9479	−9330
	40	−3618	−2692	−6557	−7796
	40	+8751	−5604	+6871	−0046
	35	−3457	−1884	−6325	−7968
	30	−3411	−1014	−6059	−8149
	25	−3507	−0027	−5740	−8344
	20	−3817	+1173	−5341	−8561
	15	−4517	+2782	−4806	−8808
	10	−6149	+5357	−4004	−9101
	5	−9864	+8406	−2252	−9464
70	40	−3166	−3166	−9305	−9526
	40	+4048	+4048	−9305	−9526
	35	−2953	−2435	−9263	−9565
	35	+5263	+0928	+8319	−5593
	30	−2842	−1670	−9218	−9606
	25	−2840	−0826	−9168	−9650
	20	−2985	+0151	−9111	−9699
	15	−3363	+1363	−9045	−9755
	10	−4183	+3016	−8964	−9820
	5	−6057	+5668	−8863	−9899
75	30	−2270	−2270	+7923	−5610
	30	+2690	+2690	+7923	−5610
	25	−2193	−1534	+7557	−5645
	20	−2226	−0707	+7214	−5695
	15	−2414	+0270	+6890	−5753
	10	−2861	+1505	+6578	−5820
	5	−3817	+3224	+6273	−5893
80	20	−1482	−1482	+1650	+1650
	15	−1541	−0668	+3403	−1214
	10	−1766	+0315	+3582	−2240
	5	−2283	+1588	+3534	−2899
85	10	−0740	−0740	+0780	+0780
	5	−0991	+0251	+1392	−0642

Columns M and N give the measured angles in degrees between the traces, using the convention $L \geqq M \geqq N$. Columns I and II give the components k and l of the possible normals $\{h, k, l\}$ to the crystal surface relative to the standard face centered rhombohedral cell. The largest component h is put equal to unity in all cases.

(6) and (7) these may be shown to be 1, 1, 1 and 1, 1, 2 cos β − 1. Further degeneracies also arise when the angle β takes on certain values. For {111} traces in cubic crystals for example we have $\beta = \pi/3$ and the two orientations corresponding to $c = 1$ for the isosceles triangle are crystallographically equivalent. Thus, in this case we may only have one or two independent solutions for the orientation. Again when $\beta = \pi/2$, corresponding to {100} traces in cubic crystals, all four solutions of the quartic equation are equivalent for all possible sets of data.

Having determined a triplet of parameters a, b, c corresponding to a given set of data L, M, N, the Miller indices of the surface of the crystal, referred to standard axes, can now be obtained using the procedure described in section 2.

The whole analysis described above has been programmed for an electronic digital computer, the method suggested by LANCZOS [5] and BUCKINGHAM [6] being used to solve the quartic equation. Results have been obtained for surface traces arising from {111} and {100} planes of cubic crystals and from {$11\bar{1}$} and {100} planes of mercury crystals. As an example of the way in which the results may be concisely presented a selection of surface normals of mercury crystals as calculated from {$11\bar{1}$} slip traces is given in the table.

As described in the introduction to this paper, these results may also be presented on charts [7]. This method is particularly advantageous in indicating clearly the ranges of shapes for the triangle $L\ M\ N$ of Fig. 1 which give rise to different numbers of solutions for the surface normal. As the labelling of the corners L, M, and N of the triangle of surface traces is arbitary, it is convenient in this connection to make use of a plot which has three-fold symmetry. Thus we have used the principle utilized in the plotting of ternary phase diagrams i.e. the sum of the three perpendiculars from any interior point of an equilateral triangle to the three edges is invariant. Thus, if the three perpendiculars represent the three angles of a triangle of surface traces, their sum will always be 180°. This is illustrated in Fig. 2 where the ranges for four, two or no solutions for triangles of traces produced on {111} cubic planes are also shown.

The symmetry of Fig. 2 makes it clear that it is not necessary to make use of the whole ternary diagram; the plot consists of six crystallographically equivalent right angled triangles. Thus, in the present work the restriction $L \geqq M \geqq N$ was imposed limiting results to the shaded triangle JKL of Fig. 2. In Fig. 3 we make use of this triangle to plot the ranges of four, two and no solutions for values

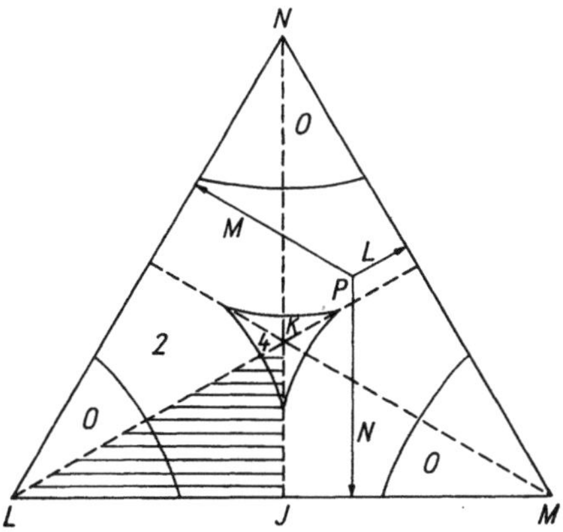

Fig. 2. Representation of triangles of surface traces on a ternary diagram. The triangle reprensented by the point P has angles L, M, and N. The plot consists of six equivalent right angled triangles one of which is shown shaded. The boundaries of the regions in which four, two and no solutions for traces produced by cubic {111} planes are also indicated

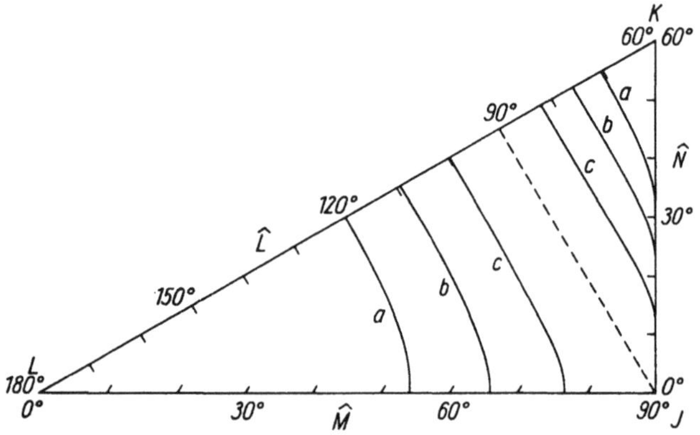

Fig. 3. The number of solutions for the surface normal for differently shaped triangles of slip traces produced by cubic {111} planes and mercury {11$\bar{1}$} and {100} planes, plotted on the region of the ternary diagram defined by $L \geqq M \geqq N$. The boundaries of the regions for these three sets of planes are represented by curves *a*, *b*, and *c* respectively, the number of solutions in each case decreasing from four to two to zero as one moves from right to left. To the right of the broken line cubic {100} traces give one solution, no solutions arising for this system over the remainder of the diagram

of β equalling 60°, 70.75° and 98.37°, corresponding to {111} traces in cubic crystals and {11$\bar{1}$} and {100} traces in mercury crystals respectively. Also shown is the region in which the unique solution arises for the degenerate case of {100} cubic traces. Use has been made of the ranges for the {11$\bar{1}$} and {100} mercury planes to determine the operative slip plane of this metal [8]. Triangles of slip traces were observed experimentally in the range lying outside the {100} no solution boundary but inside the {11$\bar{1}$} no solution boundary indicating that the latter set of planes is operative. A detailed account of this work has been given elsewhere [8].

Although a unique crystal orientation cannot normally be obtained directly from three surface traces, additional experimental information is often available which enables some of the possible solutions to be eliminated. For example, in the case of {111} cubic traces, a fourth {111} trace may be present enabling a unique choice of orientation to be made [4]. Again in orienting mercury crystals from surface traces [9] the correct orientation has to be consistent with the positions of traces produced by deformation twins and their associated surface tilts and displacements. Thus, once the limited number of possible solutions which are consistent with the angles between three surface traces are known, a unique choice of orientation can often be made.

4. Summary

This paper considers the problem of determining the orientation of rhombohedral crystals, and therefore as a special case cubic crystals, from traces produced on a flat surface of a specimen by three crystallographically equivalent planes symmetrically disposed about a triad axis. A method is described whereby the possible orientations for a given set of angles may be determined using two simple physical models in the form of tetrahedra. Results obtained using the models to determine the orientation of mercury crystals from {11$\bar{1}$} traces have been found to be accurate to within $1^1/_2$° on comparing them with numerical results calculated using a new algebraic analysis of the problem. The analysis is more general than an earlier treatment, which dealt only with {111} traces of cubic crystals, and leads directly to the quartic equation which defines the possible normals to the crystal surface. The equation has four, two or no real solutions depending on the crystal structure and the angles between the traces. Degenerate solutions of this equation have been examined and numerical results obtained for {111} and {100} traces in cubic crystals and $\{11\bar{1}\}$ and {100} traces in mercury. A specimen table of results for traces produced by $\{11\bar{1}\}$ slip in mercury is presented. A convenient method for plotting the results on triangular charts is described and in particular this method is used to illustrate the ranges of data giving rise to different numbers of solutions for the orientation.

Acknowledgements

We are indebted to Mr. F. Bristow, who constructed the models, and to Drs. H. M. Otte and M. P. Drazin for their comments on the manuscript. Financial assistance from Battersea College is acknowledged by F. H., from The States of Jersey by M. B., and from the U.S. Atomic Energy Commission by A. G. C.

References

[1] C. S. Barrett, Structure of Metals, McGraw-Hill, New York 1942.
[2] H. Mykura, Bull. Inst. Metals **4**, 102 (1958).
[3] M. P. Drazin and H. M. Otte, phys. stat. sol. **3**, 824 (1963).
[4] M. P. Drazin and H. M. Otte, Tables for Determining Cubic Crystal Orientations Harrod, Towson, Maryland 1964.
[5] C. Lanczos, Applied Analysis, Prentice Hall, Englewood Cliffs N. J. 1956.
[6] R. A. Buckingham, Numerical Methods, Pitman, London 1962.
[7] M. Bevis, unpublished work.
[8] A. G. Crocker, F. Heckscher, and M. Bevis, Phil. Mag. **8**, 1863 (1963).
[9] F. Heckscher and A. G. Crocker, unpublished work.

(Received April 17, 1964)

phys. stat. sol. 6, 365 (1964)

*Forschungsinstitut für metallische Spezialwerkstoffe Dresden
der Deutschen Akademie der Wissenschaften zu Berlin*

Spontane Remanenzmagnetisierung von Cu-Ni-Fe-Legierungen

Von

O. HENKEL

Nach einer feldfreien Abkühlung unterhalb 160 °K tritt an Cu–Ni–Fe-Legierungen im homogenisierten und abgeschreckten Zustand eine spontane Remanenzmagnetisierung auf, deren Größe und Richtung von dem vor der Erwärmung angelegten Feld abhängt. Diese auch als Gedächtniseffekt bezeichnete Erscheinung kann durch Aushärtung und plastische Verformung beseitigt werden. Die spontane Remanenz wird nur in dem Vorstadium der wahren Ausscheidung, wo die Gitterkohärenz noch gewahrt ist, beobachtet.

A spontaneous remanent magnetization ("memory effect") occurs in homogenized and quenched Cu–Ni–Fe alloys when cooled to temperatures below 160 °K in the absence of a magnetic field. The magnitude and direction of this magnetization depends upon the magnetic field which is applied before heating. It is shown that this magnetization may be destroyed by hardening treatments and plastic deformation, and is only observed when a coherent phase exists, i. e., before the beginning of segregation processes.

1. Einleitung

Obwohl die aushärtungsfähigen Cu–Ni–Fe-Dauermagnetlegierungen mit 50 bis 75% Cu, ca. 20% Ni und dem Rest Fe nur eine geringe technische Bedeutung haben, sind sie seit 1935 oft Gegenstand umfangreicher Untersuchungen gewesen [1]. Zur Deutung des Mechanismus des heterogenen Zerfalls und der Ursache der magnetischen Härte haben besonders die Untersuchungen von BIEDERMANN und KNELLER [2] und MÜLLER und SCHÜLER [3] beigetragen. Hiernach bilden diese Legierungen bei genügend hoher Temperatur homogene Mischkristalle, die durch eine übliche Wasserabschreckung in eine Ni-Fe-reiche ferromagnetische und eine Cu-reiche nichtferromagnetische Phase zerfallen. Mit zunehmender Aushärtung ändert sich die Sättigungsmagnetisierung nicht mehr, d. h., die bei der Abschreckung ausgeschiedenen kugelförmigen Teilchen — teilweise unterhalb der kritischen Größe mit superparamagnetischem Verhalten — ändern ihre chemische Zusammensetzung nicht, sondern es findet nur eine Art Umkristallisation zu Paketen ziemlich regelmäßig parallel geschichteter, tetragonal verzerrter Plättchen statt [4]. Die magnetische Härte ist vorwiegend durch die Formanisotropie bedingt.

Wenn aber der Querschnitt der Probenkörper verkleinert und damit die Abkühlgeschwindigkeit aus dem homogenen Mischkristallgebiet vergrößert wird, ist die Sättigungsmagnetisierung nicht mehr unabhängig vom Aushärtungszustand. Sie nähert sich beträchtlich dem theoretischen Wert für homogene Legierungen ($4\pi I_{s20} = 2950$ G gegenüber 4400 G für heterogene Legierungen mit 66% Cu und 21% Ni). Darüber hinaus ließen Untersuchungen über das Remanenzproblem und die anhysteretische Suszeptibilität vermuten (siehe z. B. [5] bzw. noch unveröffentlichte Arbeit), daß der Übergang von dem Zustand nach der Abschreckbehandlung zu dem der optimalen magnetischen Härte diskontinuierlich ist.

In dem Vorstadium der wahren Ausscheidung kann thermoremanente Magnetisierung beobachtet werden. Besondere Aufmerksamkeit verdient aber eine Erscheinung, die z. B. an ferrimagnetischen Stoffen Gedächtniseffekt genannt wird [6]. Die nach einer feldfreien Abkühlung von einer Temperatur, bei der keine remanente Magnetisierung vorhanden ist, auftretende Remanenz ist in Richtung und Größe von dem vor der Erwärmung angelegten Feld abhängig. Auch nach einer Lagerung von mehreren Monaten gewinnt das Material durch Abkühlung unterhalb einer bestimmten Temperatur spontan einen Teil seines vergangenen remanenten Zustandes wieder. Im folgenden soll über diesen Effekt an aushärtbaren Cu-Ni-Fe-Legierungen berichtet werden.

2. Spontane Remanenzmagnetisierung

Die Cu-Ni-Fe-Meßproben in Drahtform mit einem Durchmesser von etwa 0,3 mm (in Ausnahmefällen wurden Meßproben mit Durchmessern bis 3,5 mm benutzt) wurden etwa 10 Stunden dicht unterhalb des Schmelzpunktes homogenisiert und in Wasser abgeschreckt. Die Abhängigkeit der Magnetisierung I von der magnetischen Feldstärke H ist bei Raumtemperatur S-förmig und zeigt fast keine Hysterese. Somit ist die remanente Magnetisierung annähernd Null und die Koerzitivfeldstärke $_IH_c < 5$ Oe. An Meßproben mit relativ großem Querschnitt stellt man bei Abkühlung auf tiefere Temperaturen — z. B. der des flüssigen Stickstoffs — fest, daß sowohl die Koerzitivfeldstärke als auch die Remanenzmagnetisierung größer werden. Man spricht dann von einem nicht vollständigen Superparamagnetismus. Das Härte- bzw. Größenspektrum der Teilchen reicht bis ins Gebiet der Einbereichsteilchen. Überhaupt liegt bei diesen Ausscheidungslegierungen auch nach einer optimalen Anlaßbehandlung immer ein sehr breites Härtespektrum vor, d. h., die Teilchen überstreichen einen breiten Bereich vom ausgeprägten Vielbereichs- bis zum Einbereichsverhalten oder gar superparamagnetischem Verhalten.

Wenn die Abkühlungsgeschwindigkeit aus dem Gebiet der homogenen Mischkristalle groß genug ist, dann tritt beim feldfreien Abkühlen ziemlich übergangslos eine thermoremanente Magnetisierung auf. (Punkt A in Fig. 1; die Richtung ist willkürlich gewählt.) Durch Anlegen eines Feldes wird die Remanenzmagnetisierung I_r erhöht und nach der Sättigung der Punkt C erreicht. Ein Gegenfeld verändert die Magnetisierung entlang der Kurven von $+I$ und $+I_r$. Der Vollständigkeit halber sind auch die Kurven für ein gleichgerichtetes Feld mit $-I$ bzw. $-I_r$ (entsprechend der Wahl der Ausgangsrichtung) eingetragen. Nach Erwär-

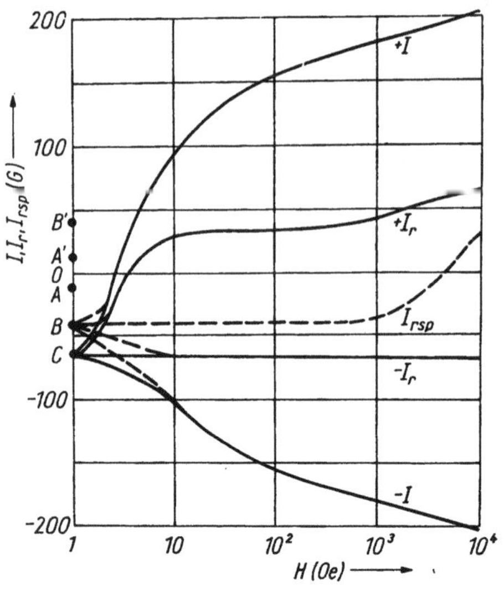

Fig. 1. Magnetisierung I, normale Remanenz I_r und spontane Remanenz I_{rsp} einer Legierung mit 70% Cu, 21% Ni, Rest Fe als Funktion des magnetischen Feldes H bei 77 °K (nähere Erläuterungen siehe Text)

mung auf Raumtemperatur stellt sich reversibel der remanenzfreie Zustand ein. Durch eine erneute Abkühlung tritt eine thermoremanente Magnetisierung auf (Punkt B), die wesentlich größer als die jungfräuliche ist. Die Richtung ist stets identisch mit der des vorher angelegten Feldes. Dagegen kann C und somit B zu A reziprok sein (A′). Wird vom Punkt B ausgehend ein gleich- oder gegengerichtetes Feld angelegt, so münden die I- und I_r-Kurven (gestrichelt eingezeichnet) bald in die I- und I_r-Kurven nach vorheriger Sättigung (ohne Zwischenerwärmung) ein. Der Punkt B wird nach Erwärmung reversibel erreicht, sofern vorher ein Feld gleicher Größe und Richtung angelegt war. Wird aber ein sehr großes Feld in der entgegengesetzten Richtung angelegt, so wird auch diese remanente Magnetisierung (sie soll im folgenden spontane Remanenz genannt werden) mit entgegengesetzten Vorzeichen erhalten (B′). Bei einem bestimmten Gegenfeld wird nach der Abkühlung keine spontane Remanenz gemessen. Es ist dies der Punkt, wo I_{rsp} von negativen zu positiven Werten übergeht. Zu bemerken ist noch, daß die I_{rsp}-Kurve von Gegenfeldern bis etwa 1000 Oe praktisch nicht beeinflußt wird, während die I_r-Kurve bereits bei Feldern um 10 Oe den Nulldurchgang hat. Trotzdem ist eine Kopplung beider Kurven vorhanden, wie an dem Anstieg der $+ I_r$-Kurve bei höheren Feldern zu sehen ist. Dieser Anstieg kann zu einer deutlichen Stufe werden, wie später noch gezeigt wird. Er ist charakteristisch für das Vorhandensein spontaner Remanenzmagnetisierung.

3. Versuchsergebnisse

3.1 Einfluß der Temperatur

In Fig. 2 ist die Magnetisierung als Funktion der Aussteuerungsfeldstärke dargestellt. Parameter ist die Meßtemperatur. Um speziell bei den tieferen Temperaturen reproduzierbare Werte zu erhalten, wurde stets von der spontanen Remanenz (Punkt B in Fig. 1) ausgegangen. Für die Temperaturen oberhalb Raumtemperatur ist das belanglos. Von $T = 475$ °K etwa bis 300 °K herunter sind die Magnetisierungskurven in bezug auf die Feldrichtung symmetrisch. Aber schon bei $T = 200$ °K ist, obwohl noch keine spontane Remanenz beobachtet wird, eine deutliche Verzerrung wahrnehmbar. In der Feldrichtung, wo vor der Messung bei $T = 77$ °K ein Feld von $-$ 4000 Oe angelegt wurde, ist bei kleinen Feldstärken ein stärkerer Anstieg der Magnetisierung als in der entgegen-

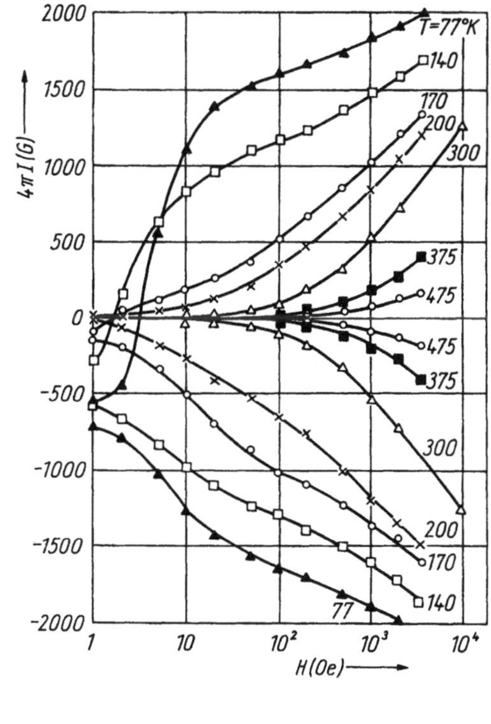

Fig. 2. Magnetisierung als Funktion der Feldstärke bei verschiedenen Meßtemperaturen (70% Cu, 21% Ni, Rest Fe). Ausgangspunkt ist die spontane Remanenz nach Anlegung eines Feldes von -4000 Oe bei 77 °K

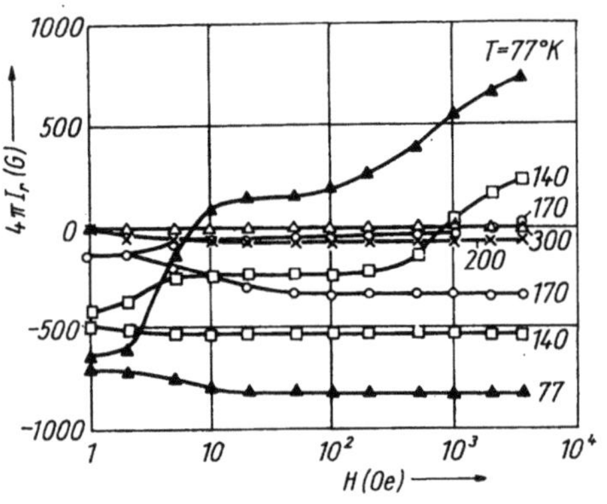

Fig. 3. Remanenz als Funktion der Feldstärke (im übrigen wie Fig. 2)

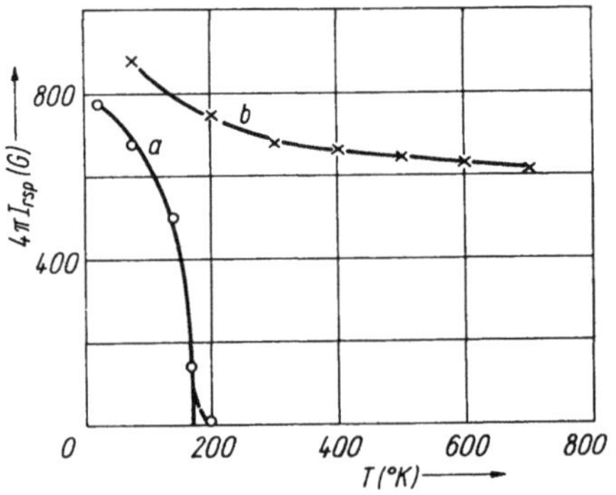

Fig. 4. Spontane Remanenz einer Legierung mit 70% Cu, 21% Ni, Rest Fe als Funktion
a: der Meßtemperatur, nachdem vorher bei 77 °K ein Feld von − 4000 Oe angelegt und dann die Probe auf Raumtemperatur erwärmt war;
b: der Erwärmungstemperatur zwischen der Feldanlegung von − 4000 Oe bei 77 °K und der Messung bei ebenfalls 77 °K

gesetzten Richtung (hier als positiv bezeichnet) vorhanden. Von $T = 170$ °K an tritt dann eine spontane Remanenz ein. Besonders bei $T = 140$ und 77 °K hat es den Anschein, als ob bei kleinen Feldstärken ein ferromagnetischer Anstieg vorhanden ist, der nach Erreichung eines bestimmten Wertes der Magnetisierung in beiden Richtungen durch einen superparamagnetischen oder antiferromagnetischen Bereich abgelöst wird. Leider war es nicht möglich, die Abhängigkeit von H und I zu noch höheren Feldstärken zu verfolgen.

Analog zu Fig. 2 ist in Fig. 3 die normale Remanenz bei verschiedenen Meßtemperaturen aufgetragen. Bei 77 °K ist der Unterschied zwischen dem positiven und negativen Kurvenast beträchtlich. Auch der auf der oberen Kurve bei einem Feld von etwa 100 Oe beginnende Anstieg ist deutlich zu sehen. Mit zunehmender Meßtemperatur werden die spontane Remanenz, die Differenz beider Kurven und der zweite Anstieg auf der oberen Kurve kleiner. Bei einer Temperatur von 300 °K, wo die Magnetisierungskurven in Fig. 2 symmetrisch sind, sind alle drei genannten Anomalitäten verschwunden.

Die Kurve a in Fig. 4 zeigt die spontane Remanenz als Funktion der Meßtemperatur. Unterhalb 200 °K steigt die spontane Remanenz steil an. Bis zur Temperatur des flüssigen Neons[1]) ist keine Sättigung festzustellen. Die Proben wurden vor der Messung stets bei 77 °K in einem Feld von − 4000 Oe aufmagnetisiert, auf Raumtemperatur erwärmt und sodann bei der angegebenen Temperatur gemessen. Kurve b zeigt die Abhängigkeit der spontanen Remanenz von der Erwärmungstemperatur. Hier wurde stets bei 77 °K die spontane Remanenz gemessen. Verfolgt wurde die Abhängigkeit nur in dem Temperaturbereich (kurzzeitige Erwärmung), wo noch keine irreversiblen Strukturänderungen auftreten. Oberhalb Raumtemperatur ist eine lineare Abhängigkeit vorhanden. Die Veränderung der spontanen Remanenz durch die Erwärmungstemperatur ist gering. Unterhalb Raumtemperatur ist eine Abweichung vom linearen Verlauf

[1]) Das flüssige Neon wurde uns freundlicherweise von der Arbeitsstelle für Tieftemperaturphysik der DAW — Leiter Prof. Dr. L. Bewilogua — zur Verfügung gestellt.

vorhanden. Es muß aber beachtet werden, daß sich die Erwärmungstemperatur immer mehr der Meßtemperatur nähert und bei 77 °K, wo praktisch keine Erwärmung mehr stattgefunden hat, letztlich die normale Remanenz ermittelt wurde, die um etwa 30% größer als die spontane Remanenz ist. Der Anstieg unterhalb Raumtemperatur kann somit auf den allmählichen Übergang von spontaner zu normaler Remanenz zurückgeführt werden.

Bezüglich Fig. 4 wurde als oberste Grenze des Bereiches für reversible Strukturänderungen $T = 700$ °K angegeben. Das ist aber nur zutreffend, wenn es sich um die kurze Zeit von wenigen Minuten handelt, die zum Erwärmen der Probe auf diese Temperatur notwendig ist. Allgemein sind die Strukturänderungen auf Diffusionsvorgänge zurückzuführen. Es können sich auch bereits bis $T = 700$ °K irreversible Änderungen bemerkbar machen, wenn die Haltezeit groß genug ist. Im folgenden sollen die Änderungen betrachtet werden, die durch eine Anlaßbehandlung von 2 Stunden auftreten. Berücksichtigt wurde der Bereich bis zu einer Anlaßtemperatur $T_A = 825$ °K. Nach den bisherigen Untersuchungsergebnissen treten in diesem Bereich noch keine merklichen Änderungen der magnetischen Eigenschaften auf. Sobald aber die Abschreckwirkung so intensiv war, daß spontane Remanenz vorliegt, ist diese Anlaßtemperatur von $T_A = 825$ °K bereits die obere Grenze des Existenzbereiches, wie die folgenden Ergebnisse zeigen werden. In Fig. 5 ist zunächst die Magnetisierung als Funktion der Feldstärke bei 300 °K dargestellt. Mit zunehmender Anlaßtemperatur wird der Anstieg der I-Kurve immer steiler, bis dann bei $T_A = 700$ °K bereits in der Nähe der Raumtemperatur spontane Remanenz auftritt (hier ausnahmsweise als positiv eingezeichnet). Der Existenzbereich der spontanen Remanenz (Kurve a in Fig. 4) wird durch zunehmende Aushärtung zu höheren Temperaturen verschoben. Durch eine Erhöhung der Anlaßtemperatur um nur 50 Grad verschwindet dann die spontane Remanenz. Zugleich wird die Erreichung einer Sättigungsmagnetisierung angedeutet, und die Koerzitivfeldstärke wird wesentlich größer.

Fig. 6 zeigt die Magnetisierung I als Funktion der Feldstärke H bei 77 °K. Es wurde von der normalen Remanenz ausgegangen und ein Gegenfeld angelegt. Bis zu einer Anlaßtemperatur von 775 °K wird die Remanenz größer. Eine wesentlich andere Charakteristik zeigt dann die I-Kurve nach der Anlaßbehandlung bei einer Temperatur von 825 °K. Trotz aller Verschiedenheit ist ein gewisser Übergang — was bei einem Diffusionsvorgang auch zu erwarten ist — bezüglich des Nulldurchganges und der Neigung der Kurven bei höheren Feldern zu erkennen. Normalerweise ist zu erwarten, daß die Remanenzkurven nach der Anlaß-

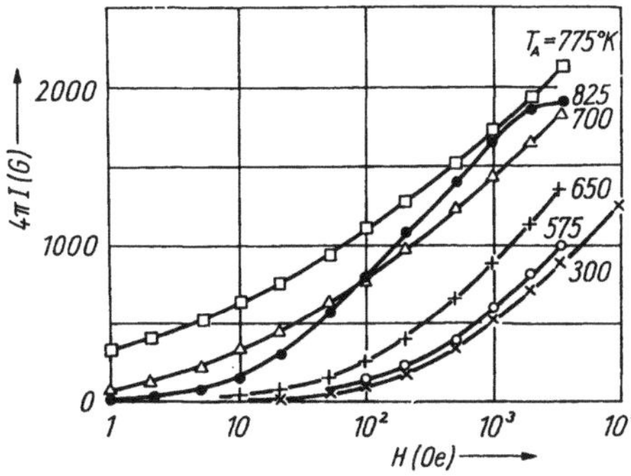

Fig. 5. Magnetisierung in Abhängigkeit vom angelegten Feld bei 300 °K. Parameter ist die Anlaßtemperatur T_A nach zweistündiger Haltezeit (70% Cu, 21% Ni, Rest Fe)

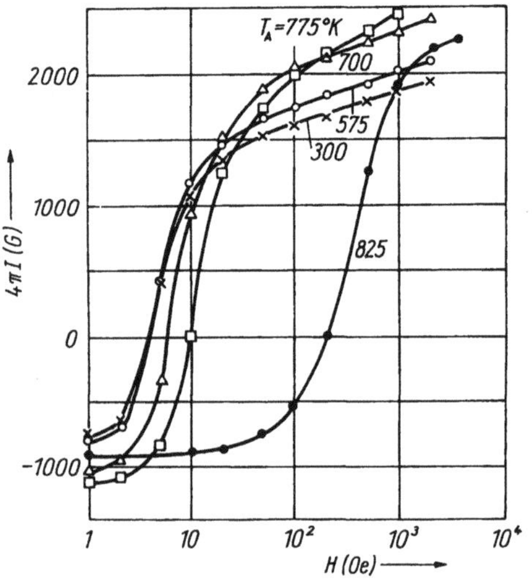

Fig. 6

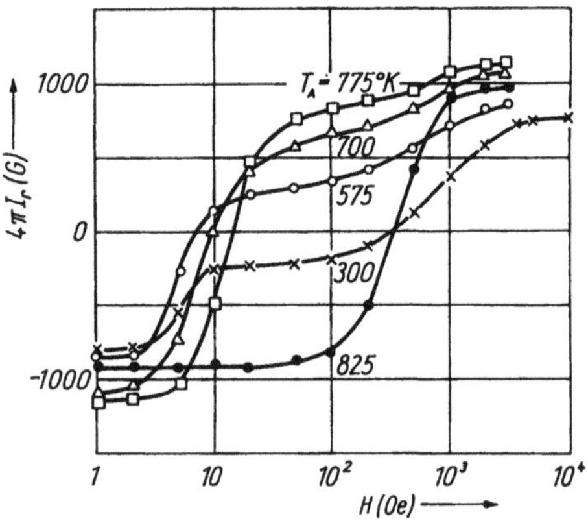

Fig. 7

Fig. 6. Magnetisierung in Abhängigkeit vom angelegten Feld bei 77 °K. Ausgangspunkt ist die normale Remanenz nach Anlegung eines Feldes von − 7000 Oe (im übrigen wie Fig. 5)

Fig. 7. Normale Remanenz in Abhängigkeit vom angelegten Feld bei 77 °K (im übrigen wie Fig. 6)

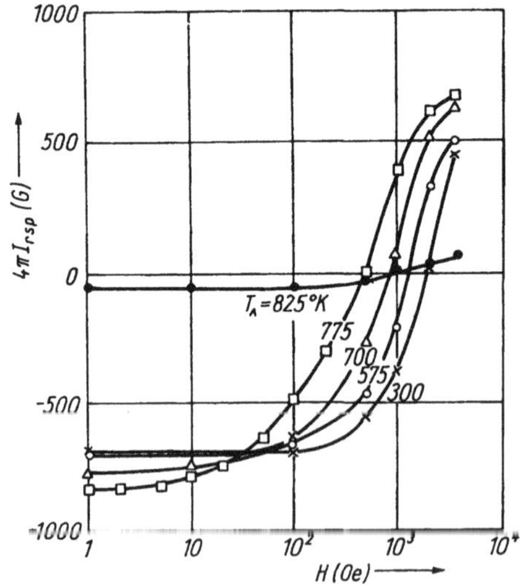

Fig. 8

Fig. 8. Spontane Remanenz in Abhängigkeit vom angelegten Feld bei 77 °K nach einer Zwischenerwärmung auf Raumtemperatur (im übrigen wie Fig. 6)

behandlung qualitativ dem Verlauf der Magnetisierungskurven entsprechen. Fig. 7 zeigt aber, daß zusätzlich beachtenswerte Abweichungen noch vor dem wieder deutlichsichtbaren Übergang von T_A = 775 °K bis T_A = 825 °K auftreten. Nach allen Anlaßtemperaturen zeigen die I_r-Kurven im Gegensatz zu den I-Kurven bei höheren Feldstärken keinen linearen Verlauf. Bei T_A = 300 °K ist die Trennung in die zwei Bereiche besonders ausgeprägt. Mit zunehmender Anlaßtemperatur wird diese Stufe in der Remanenzkurve kleiner.

Der Nulldurchgang der spontanen Remanenz wird mit zunehmender Anlaßtemperatur zu kleineren Feldstärken verschoben (Fig. 8). Bis T_A = 775 °K wird der Effekt vergrößert, so daß auch das Auftreten bei Raumtemperatur in Fig. 5 verständlich wird. Zugleich wird aber der lineare Anfangsbereich immer kleiner.

3.2 Einfluß der Zusammensetzung

Es wurde bei einem konstanten Ni-Gehalt von 21 Gew.-% der Cu-Gehalt von 60 bis 72% variiert. Bekanntlich haben die Legierungen mit etwa 60% Cu, 21% Ni und dem Rest Fe die größte Energiedichte $(B\,H)_{max}$ und diejenigen mit dem höchsten Cu-Gehalt die größte Koerzitivfeldstärke. Fig. 9 zeigt für vier Legierungen innerhalb dieses Bereiches die spontane Magnetisierung als Funktion des Gegenfeldes bei einer Meßtemperatur von 77 °K. Die Unterschiede sind nicht allzu groß. Mit zunehmendem Cu-Gehalt wird entsprechend der Abnahme des Ni–Fe-Anteiles die spontane Remanenz kleiner. Zu kleineren Cu-Gehalten hin überlagert sich dann eine andere Erscheinung, die sich durch die Verschiebung des Nulldurchganges bemerkbar macht. Es hat den Anschein, als ob mit abnehmendem Cu-Gehalt die kritische Abkühlgeschwindigkeit von der Homogenisierungstemperatur immer größer wird und bei den zu diesen Untersuchungen verwendeten Drahtdicken nicht mehr völlig erreicht wurde. Außerdem dürfte eine Zwischenerwärmung auf höhere Temperaturen erforderlich sein.

Im Zusammenhang mit der chemischen Zusammensetzung sollen noch weitere Untersuchungen durchgeführt werden. Es soll überprüft werden, ob durch die Homogenisierungsglühung dicht unterhalb des Schmelzpunktes und während der Wasserabschreckung nicht eventuell eine Oxydschicht an der Oberfläche entsteht, die eine Austauschkopplung zwischen antiferromagnetischem Mantel und ferromagnetischem Kern bewirkt. Dann soll festgestellt werden, ob bei Proben mit verschiedenem Querschnitt durch andere Abkühlwirkungen strukturelle und damit magnetische Eigenschaftsänderungen auftreten.

Zunächst wurde als Untersuchungsobjekt ein Probekörper mit einer Querschnittsfläche von 8 mm² benutzt. Er zeigte nach der Wasserabschreckung bei einer Meßtemperatur von 77 °K eine spontane Remanenz, die in der gleichen Größe wie bei den Drähten mit einer Querschnittsfläche $F < 1$ mm² war. Das Verhältnis von normaler zu spontaner Remanenz ist gleich 0,6. Der Probekörper wurde dann stufenweise abgedreht. Sowohl die Magnetisierung bei 7000 Oe Feldstärke als auch die spontane Remanenz werden hierdurch nicht verändert. Wenn eine Oxydschicht an der Oberfläche vorhanden ist, dann ist sie so dünn, daß sie

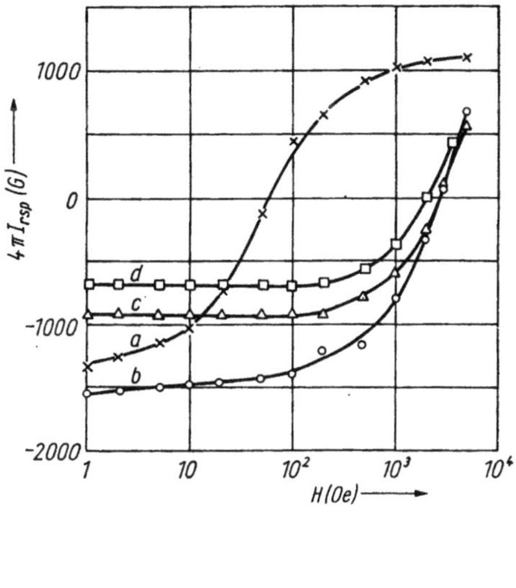

Fig. 9. Spontane Remanenz als Funktion des Gegenfeldes bei 77 °K. Parameter ist die chemische Zusammensetzung in Gew.-% a) 60% Cu, b) 66,5% Cu, c) 70% Cu, d) 72% Cu und jeweils 21% Ni, Rest Fe

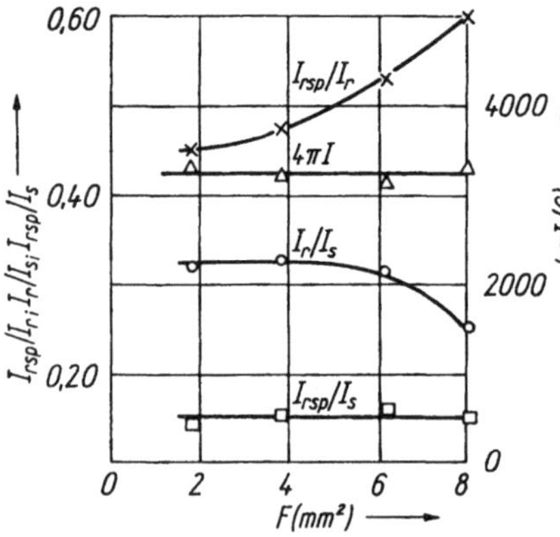

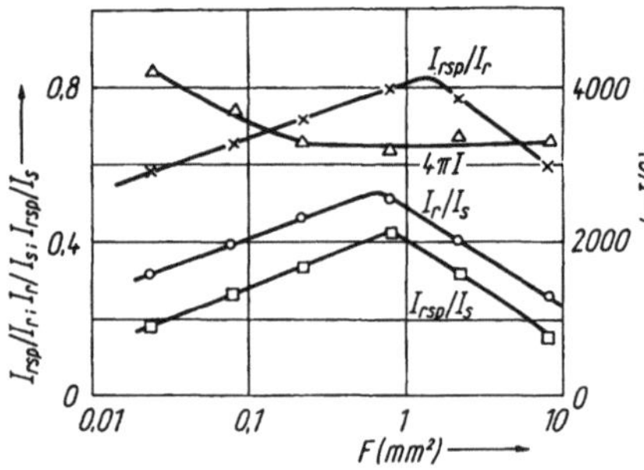

Fig. 11. Magnetisierung und Remanenzverhältnisse ($T = 77\,°K$, $H = -7000$ Oe) als Funktion des Probenquerschnittes bei der Abschreckbehandlung (71% Cu, 20% Ni, Rest Fe)

Fig. 10. Magnetisierung und Remanenzverhältnisse ($T = 77\,°K$, $H = -7000$ Oe) in Abhängigkeit vom Probenquerschnitt. Die Meßprobe wurde bei $F = 8\,mm^2$ homogenisiert und abgeschreckt und dann auf die angegebenen Querschnitte abgedreht (71% Cu, 20% Ni, Rest Fe)

die magnetischen Eigenschaften praktisch nicht beeinflußt. In Fig. 10 sind zusätzlich die Werte für I_r/I_s und I_{rsp}/I_r eingetragen. Über die Veränderung von I_r/I_s, die sicher auf die Spannungsänderung durch die Querschnittsverminderung zurückzuführen ist, wird als Sekundäreffekt auch der Wert von I_{rsp}/I_r erniedrigt.

Den Einfluß des während der Abschreckung von der Homogenisierungstemperatur vorliegenden Probenquerschnittes auf die magnetischen Eigenschaften zeigt Fig. 11. Hier konnte ein großer Bereich von $F = 0{,}01$ bis $10\,mm^2$ berücksichtigt werden. Alle charakteristischen Größen sind von dem Probenquerschnitt bei der Abschreckbehandlung abhängig. Bis zu einem Querschnitt von $0{,}2\,mm^2$ herunter ist die bei 77 °K gemessene Magnetisierung bei 7000 Oe noch querschnittsunabhängig. Aber dann steigt die Magnetisierung auf Werte, die eine Entmischung in eine Cu-reiche und eine Fe-Ni-reiche Phase vermuten lassen. Die normale und spontane Remanenz durchlaufen in diesem Bereich ein Maximum. Es kann z. B. abgelesen werden, wie groß bei einem bestimmten Querschnitt das Verhältnis von spontaner zu normaler Remanenz ist. Sowohl bei einer Drahtdicke von 0,3 mm als auch bei einer solchen von 3 mm wird ein Verhältniswert von etwa 0,65 erhalten. Dazwischen kann der Wert auf 0,8 ansteigen. Es ist aber sehr schwer, über die zu einer bestimmten Einstellung des Zustandes führenden Abkühlbedingungen auszusagen. Es kann festgestellt werden, daß eine Querschnittsabhängigkeit vorhanden ist, aber im gesamten Bereich spontane Remanenz auftritt.

3.3 *Einfluß der Verformung*

Es soll hier der Einfluß der Verformung nach der Homogenisierung und Abschreckbehandlung ermittelt werden. Fig. 12 zeigt die Magnetisierung als Funktion des Gegenfeldes bei 77 °K. Ausgangspunkt ist die remanente Magnetisierung nach vorheriger Aufmagnetisierung bis 7000 Oe. Durch eine plastische Ziehverformung, charakterisiert durch das Verhältnis von End- zu Ausgangslänge, wird die remanente Magnetisierung praktisch zu Null. Die Abhängigkeit von $I(H)$ bei 77 °K entspricht dann etwa dem unverformten Zustand bei 300 °K.

Fig. 12. Magnetisierung als Funktion des Gegenfeldes (T = 77 °K, H = −7000 Oe). Ausgangspunkt ist die normale Remanenz. Parameter ist das Längenverhältnis L/L_0 der Ziehverformung (70% Cu, 21% Ni, Rest Fe)

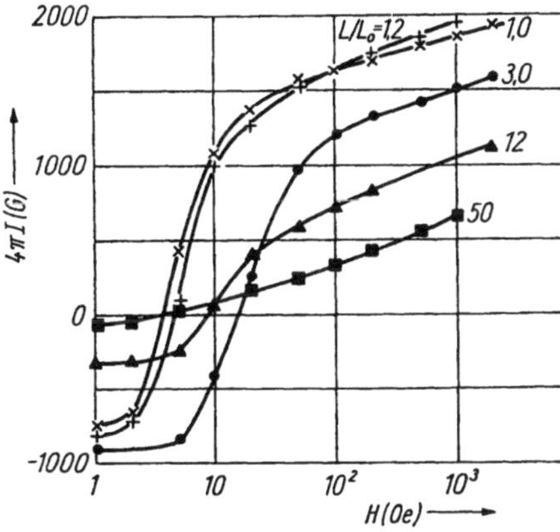

Fig. 13 zeigt den Einfluß der plastischen Verformung auf die remanente Magnetisierung. Bei kleinen Verformungsgraden wird zunächst die Stufenausbildung ähnlich wie nach einer Anlaßbehandlung verändert. Erst im Anschluß hieran wird dann die remanente Magnetisierung erniedrigt. Den Einfluß der plastischen Verformung auf die spontane Remanenz zeigt Fig. 14. Nur der Wert der spontanen Remanenz nach vorhergehender Aufmagnetisierung auf 7000 Oe wird wieder bis praktisch auf Null erniedrigt, aber die Charakteristik der Kurven und besonders der Nulldurchgang bleiben konstant. Durch eine plastische Verformung wird der Zustand, der zu einer spontanen Remanenz führt, beseitigt. Zugleich wird aber auch die remanente Magnetisierung in gleichem Maße erniedrigt, so daß die beiden Größen nur im Zusammenhang betrachtet werden können.

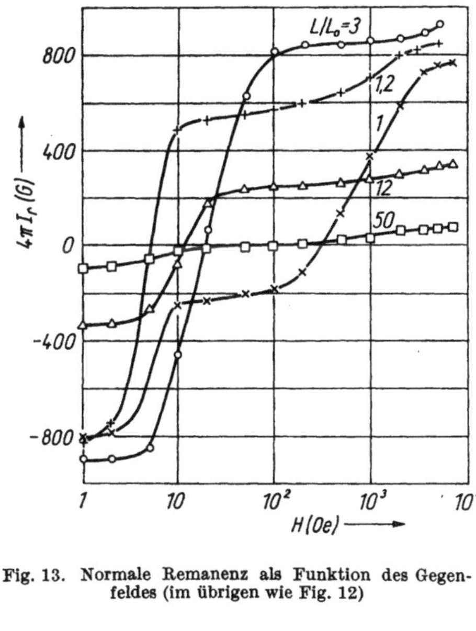

Fig. 13. Normale Remanenz als Funktion des Gegenfeldes (im übrigen wie Fig. 12)

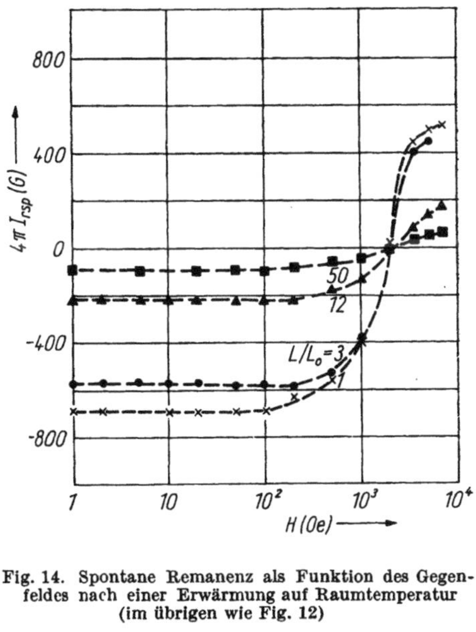

Fig. 14. Spontane Remanenz als Funktion des Gegenfeldes nach einer Erwärmung auf Raumtemperatur (im übrigen wie Fig. 12)

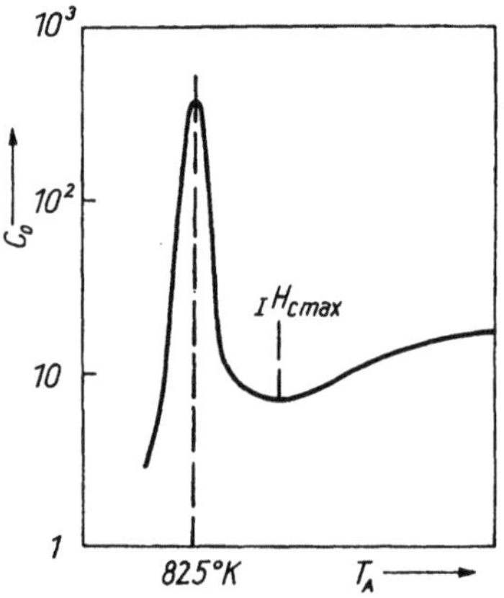

Fig. 15. Schematische Darstellung der wahren anhysteretischen Suszeptibilität von aushärtbaren, kupferreichen Cu-Ni-Fe-Legierungen als Funktion der Anlaßtemperatur

4. Diskussion der Ergebnisse

Die Ergebnisse sollen vorwiegend unter folgenden Aspekten diskutiert werden: Sind die Vorstellungen über die Ausscheidungskinetik dieser Legierungen durch ein besonderes Vorstadium der wahren Ausscheidung zu erweitern, und was kann als Ursache für die spontane Remanenz bzw. den Gedächtniseffekt angesehen werden?

Zunächst deuten die Ergebnisse im Zusammenhang mit der spontanen Remanenz selbst darauf hin (Fig. 5 bis 8), daß bei Anlaßtemperaturen um 825 °K eine Diskontinuität in den magnetischen Eigenschaftsänderungen vorhanden ist. Diese Ergebnisse sollen durch zwei weitere spezifisch magnetische Eigenschaften ergänzt werden (bisher unveröffentlicht). Die wahre anhysteretische Suszeptibilität c_0, die an diesen Legierungen experimentell gut bestimmbar ist, durchläuft bei etwa 825 °K ein ausgeprägtes Maximum. Erst oberhalb dieser Temperatur ist eine Zunahme der Koerzitivfeldstärke zu beobachten bzw. ist eine solche erreicht, die im üblichen Sinne nicht mehr als weichmagnetisch bezeichnet werden kann. Fig. 15 zeigt schematisch die anhysteretische Suszeptibilität als Funktion der Temperatur. Die c_0-Kurve durchläuft bei der maximalen Koerzitivfeldstärke ein Minimum und steigt dann nur noch wenig an. Das ausgeprägte Maximum ist sicher nicht mit einem kontinuierlichen Kornwachstum vereinbar und tritt bei der Anlaßtemperatur auf, bei der der Gedächtniseffekt verschwindet.

Als weiteres Charakteristikum bezüglich der kritischen Anlaßtemperatur sollen die Remanenzkurven betrachtet werden. Trotz vieler ungeklärter Probleme auf diesem Gebiet [5] können mit großer Sicherheit die Beziehungen von Wohlfahrth [7] auf Quasi-Einbereichsteilchen angewandt werden. In Fig. 16 sind für die Anlaßtemperaturen 775 und 825 °K die Wechselfeldremanenzkurven I'_D und die aus der Gegenfeldremanenz errechneten (Kurve b) aufgetragen. (Gleichfeldremanenzkurven sind bei Vorhandensein von spontaner Remanenz nicht sinnvoll

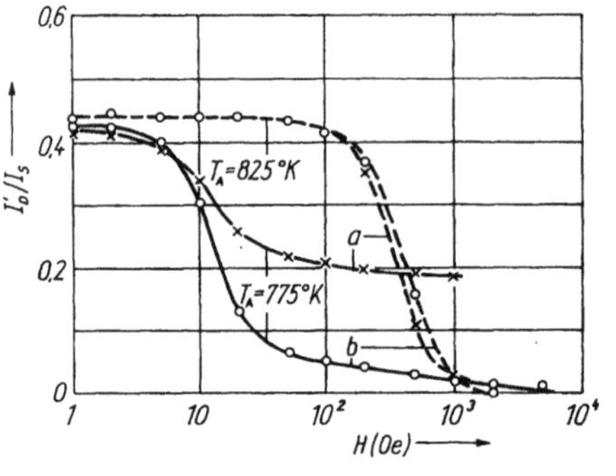

Fig. 16. Wechselfeld-Remanenzverhältnisse nach einer zweistündigen Anlaßbehandlung bei 775 und 825 °K (T = 77 °K, Meßfrequenz = 50 Hz, Zusammensetzung 70% Cu, 21% Ni, Rest Fe). Die Kurven a sind direkt gemessen, während die Kurven b aus der Gegenfeldremanenz errechnet wurden [7]

zu ermitteln.) Der Unterschied ist deutlich sichtbar. Bei einer Anlaßtemperatur von 825 °K sehen die Remanenzkurven wie die von Einbereichsteilchen aus. Die Wechselfeldremanenzkurve liegt bereits unterhalb der Gegenfeldremanenzkurve. Nach den bisherigen Ergebnissen ist anzunehmen, daß die bestimmende Komponente in diesem Zustand bereits überkritische Teilchen sind. Ganz anders ist das Aussehen beider Remanenzkurven nach einer Anlaßbehandlung bei 775 °K, wo noch der Gedächtniseffekt vorliegt. Es kann angenommen werden, daß die ausnehmend schlechte Wechselfeld-Entmagnetisierbarkeit direkt im Zusammenhang mit dem Gedächtniseffekt steht. Hier soll aber nur gezeigt werden, daß innerhalb dieser kleinen Anlaßtemperaturdifferenz von 50 Grad eine wesentliche Veränderung im Remanenzverhalten auftritt, die nicht mit den bisherigen Vorstellungen über die Ausscheidungskinetik an Cu–Ni–Fe-Dauermagnetlegierungen befriedigend erklärt werden kann. Es muß ein Vorstadium der wahren Ausscheidung vorhanden sein. Der Übergang von einem Zustand zum anderen ist magnetisch gut verfolgbar. Er beschränkt sich auf ein relativ enges Anlaßtemperaturgebiet.

Das elektrische Widerstandsverhalten als Funktion der Meßtemperatur ist in Fig. 17 dargestellt. Nach der Homogenisierungsglühung und dem Abschrecken in Wasser nimmt der elektrische Widerstand ohne Kaltverformung ($L/L_0 = 1$) bis etwa 650 °K linear zu. In dem unteren Temperaturgebiet führen kurzzeitige Änderungen zu reversiblen Widerstandsänderungen. Mit zunehmender Temperatur zeigt der elektrische Widerstand irreversibles Verhalten. Bis etwa 850 °K tritt ein beträchtlicher Widerstandsabfall ein. Erst oberhalb dieser Temperatur, wo auch die Koerzitivfeldstärke ansteigt und ferromagnetische, formanisotrope Teilchen vorhanden sind, liegt wieder ein normaler Widerstandsverlauf vor. Nach einer Luftabkühlung ist bei Raumtemperatur der elektrische Widerstand nur noch halb so groß. Durch eine Ziehverformung von z. B. $L/L_0 = 50$, durch die die spontane Remanenz bereits stark erniedrigt ist (siehe Fig. 14), ist der Anfangswiderstand fast doppelt so groß wie ohne Kaltverformung. Der Widerstandsabfall wird mit zunehmender Kaltverformung größer und der Beginn zu tieferen Temperaturen verschoben. Allgemein ist nach dem Widerstandsverhalten der Gedächtniseffekt nicht mit Gunier-Preston-Zonen — wie sie an anderen aushärtbaren Legierungen vor der wahren Ausscheidung gefunden werden [8] — zu erklären. Das zeigt auch noch eindringlich der Vergleich der mechanischen

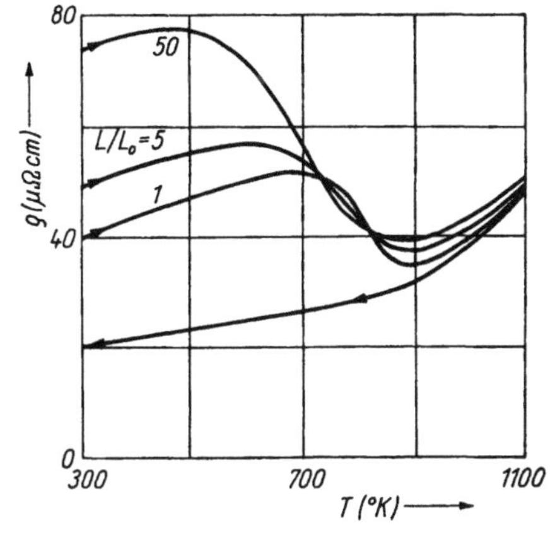

Fig. 17. Spezifischer elektrischer Widerstand einer Legierung mit 70% Cu, 21% Ni, Rest Fe als Funktion der Meßtemperatur. Parameter ist das Längenverhältnis der Ziehverformung

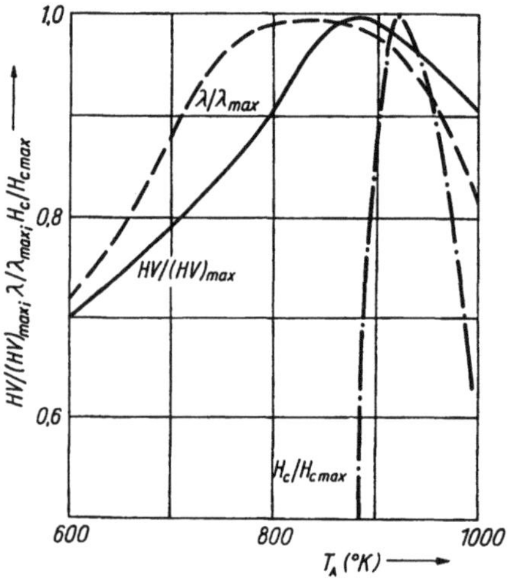

Fig. 18. Mechanische Härte HV elektrische Leitfähigkeit λ, und Koerzitivfeldstärke, jeweils im Verhältnis zu den Maximalwerten, als Funktion der Anlaßtemperatur (70% Cu, 21% Ni, Rest Fe)

Härte, der Koerzitivfeldstärke und der elektrischen Leitfähigkeit nach verschiedenen Anlaßtemperaturen in Fig. 18. Das Leitfähigkeitsmaximum befindet sich etwa bei der Anlaßtemperatur, bei der der Gedächtniseffekt verschwindet. Das Maximum der mechanischen Härte fällt mit dem Beginn des Koerzitivfeldstärke-Anstieges zusammen, während das Koerzitivfeldstärke-Maximum noch um etwa 50 Grad zu höheren Anlaßtemperaturen verschoben ist. Wenn also der Gedächtniseffekt durch eine Zonenbildung bedingt ist, so zeigt er aber keine Besonderheiten hinsichtlich der mechanischen Härte oder elektrischen Leitfähigkeit. Beide Eigenschaften, die als charakteristisch für Ausscheidungsvorgänge betrachtet werden können, nehmen erst bei einer Anlaßtemperatur oberhalb des Existenzbereiches des Gedächtniseffektes Extremwerte an.

Die magnetischen Untersuchungen lassen eine antiferromagnetische Austauschkopplung vermuten. Eine Hystereseschleifenverschiebung wird aber bei einer Temperatur von 77 °K nicht beobachtet, so daß der Antiferromagnetismus nur leicht an das Gitter gebunden sein könnte [9]. Die Drehmomentenkurve hat die dementsprechende sin 2 Θ-Form, aber Rotationshystereseverluste sind nicht feststellbar. Wenn das auch kein absolutes Kriterium gegen eine Austauschanisotropie in einem antiferromagnetischen-ferromagnetischen System ist, weil die Rotationshysterese bei einem großen Wert von Austausch-/Kristall-Anisotropiekonstanten sehr klein wird, so ist bei der Annahme ferromagnetischer Körner eine antiferromagnetische Cu-reiche Phase hypothetisch.

TOURNIER und WEIL [10] haben die Magnetisierung von Kupfer-Kobalt-Legierungen mit Kobaltgehalten von 0,6 bis 3% im abgeschreckten Zustand bis zu Temperaturen unterhalb 1 °K verfolgt. Sie fanden, daß sich solche Legierungen mit zunehmendem Co-Gehalt und steigender Temperatur immer mehr wie Antiferromagnete verhalten. Bei niedrigen Temperaturen wurde Thermoremanenz gefunden. Die Ergebnisse können nicht mit Superparamagnetismus erklärt werden, wohl aber mit dem Néelschen Modell feiner antiferromagnetischer Körner [11]. Gemäß der Konzeption eines Antiferromagneten mit zwei Untergittern genügt eine ungerade Anzahl netzartiger Ebenen, um ein resultierendes Moment zu erhalten. Wenn die Ordnung im Korn verschwindet (Néel-Temperatur), so verschwindet auch diese „ferrimagnetische" Magnetisierung. Durch Erniedrigung der Temperatur tritt eine thermoremanente Magnetisierung auf.

Nach der Methode von TOURNIER und WEIL [10] kann eine Temperatur T_N (positiv) bestimmt werden, die mit zunehmendem Anteil an Co-Atomen größer wird. Eine solche Temperatur kann auch durch Extrapolation des geradlinigen Teiles der Magnetisierungskurven von Cu-Ni-Fe-Legierungen erhalten werden. Diese Temperatur T_N, die wesentlich größer ist als die aus Kurve a in Fig. 4

ablesbare, scheint aber keine große Bedeutung zu haben. Darüber hinaus kann der Gedächtniseffekt in gleicher Größe auch an binären Cu-Ni- und sowohl an geordneten als auch an ungeordneten Ni_3Mn-Legierungen beobachtet werden. Es ist kaum anzunehmen, daß die fest fundierten Anschauungen, wonach in dem Cu-Ni-Legierungssystem bei Ni-Gehalten über 50% Mischkristalle vorliegen, lückenhaft sind. Vielmehr muß angenommen werden, daß der Gedächtniseffekt nicht an die Existenz von Teilchen im üblichen Sinne gebunden ist. Sicher sind sowohl in den binären als auch in den ternären Cu-Legierungen nach dem Abschrecken bestimmte Konzentrationsschwankungen vorhanden, die durch Anlaßbehandlungen vergrößert werden. Der Gedächtniseffekt tritt in dem Bereich von homogenen Mischkristallen bis zu solchen heterogenen auf, wo die Kohärenz nicht mehr gewahrt ist. Erst ab hier kann aber für hartmagnetische Betrachtungen von einer wahren Ausscheidung und von superpara-, antiferro- und ferromagnetischen Teilchen gesprochen werden. Das Ende des Vorstadiums der wahren Ausscheidung ist durch die Ermittlung der spontanen Remanenz, anhysteretischen Suszeptibilität und Wechselfeldremanenz als Funktion der Anlaßtemperatur gut verfolgbar.

Die Problemstellung ist, aus dem Vorhandensein bestimmter „weichmagnetischer" Mischkristalle die Existenz des Gedächtniseffektes abzuleiten. Es muß weiteren Untersuchungen vorbehalten bleiben, ob der Effekt auf die Änderung der Spinvektoren antiferromagnetischer Clusters zurückzuführen ist. Es ist aber auch nicht ausgeschlossen, daß die Ursache des Effektes in Richtung der von Wohlfahrth und Rhodes [12] entwickelten Vorstellungen über den Kollektivelektronen-Metamagnetismus oder Stacey [13] aufgestellten Theorie der Pseudo-Einbereiche zu suchen ist.

Literatur

[1] E. P. Wohlfarth, Adv. Phys. (Phil. Mag. Suppl.) 8, 196 (1959).
[2] E. Biedermann und E. Kneller, Z. Metallk. **47**, 289, 760 (1956).
[3] H.-G. Müller und K. Schüler, Wiss. Z. HfV **7**, 27 (1959/60).
[4] E. Kneller, Ferromagnetismus, Springer-Verlag 1962 (S. 447).
[5] O. Henkel, in: Metallische Spezialwerkstoffe, Akademie-Verlag, Berlin 1963.
[6] I. S. Jacobs und P. E. Lawrence, J. appl. Phys. **31**, 1388 (1960).
[7] E. P. Wohlfarth, J. appl. Phys. **29**, 595 (1958).
[8] U. Dehlinger, in: Ausscheidungsvorgänge in Legierungen, Akademie-Verlag, Berlin 1964.
[9] W. H. Meiklejohn, J. appl. Phys. **33**, Suppl., 1328 (1962).
[10] R. Tournier und L. Weil, J. Phys. Radium **23**, 522 (1962).
[11] L. Néel, C. R. Acad. Sci. (France) **253**, 203 (1961).
[12] E. P. Wohlfarth und P. Rhodes, Phil. Mag. **7**, 1817 (1962).
[13] F. D. Stacey, Phil. Mag. **7**, 1887 (1962).

(Received April 27, 1964)

phys. stat. sol. 6, 379 (1964)

Institute of Physical Chemistry, Polish Academy of Sciences, Warsaw

Exchange Correlations and Compressibilities of Mg, Al, Si and Zn

By

S. OLSZEWSKI and A. WIERZBICKI

The compressibilities of Mg, Al, Si, and Zn are calculated on the basis of the energy data of the quantum defect method. The calculations are made with and without the inclusion of the exchange volume correction in the equation of state of the Wigner-Seitz cell. The relative importance of this correction for compressibilities at zero pressure decreases with an increase in the number of electrons in the cell.

Die Kompressibilitäten von Mg, Al, Si und Zn werden auf der Grundlage der mit Hilfe der Quantendefekt-Methode erhaltenen Energiewerte berechnet, unter Berücksichtigung und Vernachlässigung der Volumenkorrektur, die von der Austausch-Wechselwirkung herrührt. Der relative Beitrag dieser Korrektur zur Kompressibilität beim Druck Null nimmt mit wachsender Elektronenzahl pro Gitterzelle ab.

The purpose of the present paper is twofold. First we like to investigate how the energy curves provided by the Quantum Defect Method (QDM) for valence electrons of the metal cells [1] can reproduce the compressibilities of several nonalkali metals (solids). Secondly we like to see how the data for metal compressibilities obtained from the QDM and the BARDEEN theory [2] are influenced by introducing the volume effect of the exchange into the equation of state of the Wigner-Seitz cell.

The meaning of the volume effect lies in considering the total volume V occupied by the electron gas in the metal cell as being equal to the sum of the volume occupied actually by the electron charge, and of the empty space, v_a, created by the exchange holes of the gas [3]. Then the corresponding correction, analogous to the volume correction entering in the Van der Waals equation of state of the classical gas, should be introduced into the equation of state of the electron gas of the metal cell. This correction equals to a replacement of the total volume V of the cell, entering in the product $p\,V$ for the electron gas, by the effective volume of the electron charge in the cell, $V - v_a$. Thus, if $\varepsilon(r_e)$ represents the energy of the valence electrons in the cell at the equilibrium radius r_e (the pressure of electron gas at the boundary of the cell is then equal to zero), we have for the bulk modulus

$$\frac{1}{\beta} = \frac{1}{12\,\pi\,r_e}\,\frac{\partial^2\varepsilon}{\partial r_e^2}\,\frac{V}{V - v_a}, \tag{1}$$

$\left(\beta \text{ compressibility, } V = \dfrac{4\,\pi\,r_e^3}{3}\right)$ instead of the formula of BARDEEN

$$\frac{1}{\beta^{\text{Bar}}} = \frac{1}{12\,\pi\,r_e}\,\frac{\partial^2\varepsilon}{\partial r_e^2}\,. \tag{2}$$

The volume v_a can be calculated from the Poisson equation as the volume of the exchange charge created by the potential

$$V_a^e(\boldsymbol{r}) = -e \int \frac{\varrho_a(\boldsymbol{r}, \boldsymbol{r}')}{|\boldsymbol{r} - \boldsymbol{r}'|} dv' , \tag{3}$$

where $\varrho_a(\boldsymbol{r}, \boldsymbol{r}')$ is the exchange charge density

$$\varrho_a(\boldsymbol{r}, \boldsymbol{r}') = -\frac{\varrho(\boldsymbol{r}')}{2} \times$$
$$\times \left[3 \frac{\sin\{[3\pi^2 \varrho(\boldsymbol{r}')]^{1/3} |\boldsymbol{r} - \boldsymbol{r}'|\} - [3\pi^2 \varrho(\boldsymbol{r}')]^{1/3} |\boldsymbol{r} - \boldsymbol{r}'| \cos\{[3\pi^2 \varrho(\boldsymbol{r}')]^{1/3} |\boldsymbol{r} - \boldsymbol{r}'|\}}{3\pi^2 \varrho(\boldsymbol{r}') |\boldsymbol{r} - \boldsymbol{r}'|^3}\right]^2 \tag{4}$$

(WIGNER and SEITZ [4]), and $\varrho(\boldsymbol{r}') = \varrho = z/V$. (The number of electrons in the cell is denoted by z.) In general we have

$$V - v_a = c\, V , \tag{5}$$

where

$$c = \frac{1}{1 + d} \tag{6}$$

and

$$d = \frac{v_a}{V - v_a} = \frac{Q_a}{z\, e} , \tag{7}$$

where Q_a is the exchange charge produced by z electrons in the cell.

c in (5) is the proportionality coefficient which is independent of V and depends only on the number of electrons enclosed in the cell. The values of c, d for different z are given in Table 1. We have neglected in (5) the effect of screening of the exchange potential due to the Coulomb correlations between the electrons, since we estimate the value of the corresponding screening correction to v_a to be still smaller for high electron densities (large z) than it was in case of the screening in Li and Na (cf. [3]).

We see from Table 1 that the value of d decreases with the increase of z, thus we have the result that the relative diminution of the electron density due to the exchange decreases with increase of the number of electrons. This result can be compared with the analogous behaviour of the screening correction in dependence on the changes of z, as it was discussed above.

Table 1

The ratio d between the exchange charge and the electron charge in the spherical cell calculated for different numbers of electrons enclosed in the cell. c is determined by equations (5) and (6)

z	d	c
1	0.53	0.65
2	0.48	0.67_5
3	0.44	0.69_5
4	0.41	0.71
5	0.39	0.72
6	0.37_5	0.73

Table 2

Cell radii, r_e, and the cohesive energies (in Ry) calculated from equations (8) and (9). $r_0 = 1.8$ for Zn has been estimated from the QDM data for r_e and for the cohesive energy of [1]. I are the experimental ionization potentials (in Ry) for the valence electrons of the free atoms [7] (1 Ry = 13.62 eV)

	z	r_e (calcul.)	r_e (observ.)	cohes. energy (calcul.)	cohes. energy (observ.)	I
Mg	2	3.46	3.30	0.124	0.116	1.658
Al	3	3.10	2.96	0.205	0.237	3.894
Si	4	2.80	3.15	0.127	0.280	7.540
Zn	2	2.86	2.91	0.091	0.100	2.000

The energy ε of electrons in the polyvalent metal (crystal) cell is given by Brooks [1]:

$$\varepsilon = z^2\left(\frac{r_0^2}{r_e^3} - \frac{3}{r_e}\right) + \frac{1.2\, z^2}{r_e} + \frac{2.21\, z^{5/3}}{r_e^2} - \frac{0.916\, z^{4/3}}{r_e} + z\, g\left(\frac{r_e}{z^{1/3}}\right) \tag{8}$$

(given in Rydberg units) basing on the simplified theory of Raimes [5]. All valence electrons are considered here as free electrons. The value r_e fulfills the relation $\frac{\partial \varepsilon}{\partial r_e} = 0$; r_0 is the integration constant determined by QDM (see [1]). The expression g is the correlation energy per electron in the free electron approximation as given for a metal with one valence electron (Nozières and Pines [6] and equation (35) of [3])[1]):

$$g\left(\frac{r_e}{z^{1/3}}\right) = -\,0.115 + 0.031 \ln\left(\frac{r_e}{z^{1/3}}\right). \tag{9}$$

Equation (9) is different than the older expression for correlation energy used by Brooks. The values obtained presently for the cohesive energy and for r_e are given in Table 2. Comparing these data with corresponding data of [1], we see that especially the cohesive energies are quite sensible on the form of the expression chosen for $g\left(\frac{r_e}{z^{1/3}}\right)$.

The data for compressibilities are given in Table 3. The major part of discrepancy remaining still between calculated and observed $1/\beta$ values for Mg and Al or Zn, is believed to be mainly due to the inaccuracy of the energy equation (8), since the effect of volume of Coulomb correlation holes is expected to be small. (The last point will be a subject of separate investigation.) The approximations involved in (8) were discussed in [1]. In view of the very existence of these appro-

[1]) There were errors in the reproduction of (9) of the present paper in equation (35) of [3]. However, the corresponding numerical data for correlation energy in [3] were calculated with the correct formula (9), with $z = 1$.

Table 3

Bulk moduli (expressed in 10^{11} dyn · cm^{-2}) calculated with (1) or (2) and with (8), (9). r_0 for Zn is 1.8. The experimental data for Mg, Al, and Zn (without asterisk) are estimated from the shock waves measurements of [8] and the data for Si are taken from the graph of [9]. The quantities with asterisk are the older experimental data obtained at hydrostatic pressures [10]

	z	$\frac{1}{\beta^{\text{Bar}}}$ (eq. (2))	$\frac{1}{\beta}$ (eq. (1))	$\frac{1}{\beta}$ (observed)	
Mg	2	1.15	1.7	4.1	3.4*
Al	3	3.7	5.3	9.0	7.5*
Si	4	9.2	12.9	10.5	
Zn	2	2.3	3.4	8.2	5.9*

ximations, the coincidence of experimental and theoretical data of Table 3 for the $1/\beta$ values of Si should be considered as accidental. In particular, the free electron assumptions underlying equation (8) are much less justified for Si than for Mg and Al (cf. here also the greatest discrepancy between the observed energy of cohesion for Si and the value for this energy calculated by means of equation (8); Table 2).

References

[1] H. BROOKS, Nuovo Cimento Suppl. **7**, No. 2, 165 (1958).
[2] J. BARDEEN, J. chem. Phys. **6**, 367 (1938).
[3] S. OLSZEWSKI, phys. stat. sol. **3**, 1358 (1963).
[4] E. WIGNER and F. SEITZ, Phys. Rev. **43**, 804 (1933); **46**, 509 (1934).
[5] S. RAIMES, Phil. Mag. **43**, ser. 7, 327 (1952).
[6] P. NOZIÈRES and D. PINES, Phys. Rev. **111**, 455 (1958).
[7] LANDOLT/BÖRNSTEIN, Zahlenwerte und Funktionen, **1**, Berlin 1950.
[8] M. H. RICE, R. G. MCQUEEN, and J. M. WALSH, Solid State Physics, **6**, New York 1958.
[9] G. BORELIUS, Solid State Physics, **6**, New York 1958.
[10] G. BORELIUS, Handbuch der Metallphysik, **1**, Leipzig 1935 (p. 220).

(Received May 4, 1964)

phys. stat. sol. 6, 383 (1964)

Department of Metallurgy, Cambridge

The Effect of Dislocation Configuration on the Superconducting Properties of Niobium and Vanadium

By

A. V. Narlikar and D. Dew-Hughes

Magnetization measurements have been made on various annealed and cold worked niobium and vanadium specimens in the superconducting state. The microstructure of these specimens was studied by transmission electron microscopy. The observed variation in magnetization has been correlated with the dislocation configuration present in the material. The results are explained on the assumption that flux lines are trapped by an interaction with dislocation tangles, the interaction arising through a local reduction in the range of coherence.

Magnetisierungsmessungen wurden an mehreren erwärmten und kaltverformten Niobium- und Vanadium-Proben im supraleitenden Zustand durchgeführt. Die Mikrostruktur dieser Proben wurde mit Hilfe eines Transmissions-Elektronenmikroskops untersucht. Die beobachteten Veränderungen der Magnetisierung werden mit der Versetzungskonfiguration des Materials in Zusammenhang gebracht. Die Ergebnisse werden unter der Annahme erklärt, daß die Flußlinien infolge von Wechselwirkung mit Versetzungsverflechtungen anhaften. Die Wechselwirkung entsteht durch eine lokale Reduktion des Kohärenzgebietes.

1. Introduction

Hard superconductors are currently of interest because of their ability to support high d. c. current densities without losses. Hard superconductors are Abrikosov or type II materials [1] in which the motion of flux lines in the mixed state is interrupted by some pinning force. This pinning force arises from the interaction between the flux lines and microstructural defects in the material. The superconducting properties of cold worked materials [2] suggest that dislocations may be one type of defect which can interact with flux lines. This paper describes some preliminary results of an investigation in to the role played by dislocations in pinning flux in type II superconductors.

2. Experimental Procedure

The materials chosen for this investigation were polycrystalline niobium and vanadium, as these are elements which appear to be intrinsically type II, and it was desired that the results be not confused with possible alloying effects.

The materials were subjected to various cold-working and annealing procedures, their final form being sheets 100 to 200 μm in thickness. Thin foils were prepared from a part of each sample for examination by transmission electron microscopy. In the case of niobium, specimens were reduced chemically in a mixture of seven parts of concentrated nitric acid to three parts of 40% hydrofluoric acid to a thickness of 20 μm. The 20 μm foils were then electropolished in a mixture of nine parts of concentrated sulphuric acid to one part of hydrofluoric acid, at a current density of 1 Acm^{-2} and a potential drop of 12 V. Vanadium foils were electropolished directly in a four to one mixture of methyl alcohol and concentrated sulphuric acid at a current density of 0.5 Acm^{-2} and a potential drop of 20 V. Stainless steel cathodes were used and all polishing solutions were maintained at 0 °C.

Specimens for the magnetization measurements were in the form of strips, 2 cm long and 2 mm wide cut from the 100 to 200 μm thick sheets. These measurements were made at 4.2 °K by recording on a ballistic galvanometer the signal generated as the specimen was moved rapidly from the centre of one search coil to the centre of a second coil. The external field was provided by a niobium-zirkonium wound superconducting solenoid capable of generating a field up to 45000 Oe. The results of each specimen were normalized by adjusting each *M-H* curve, below $H c_1$, to unit slope. From the specimen shape corrections for demagnetizing effects would be small and have been ignored.

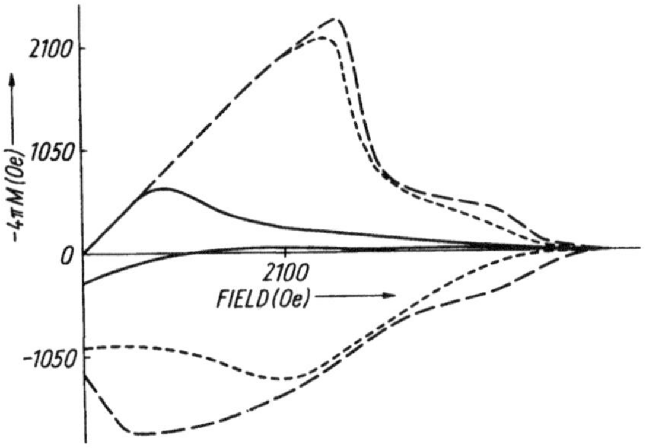

Fig. 1. Magnetization curves for annealed niobium
— — — as received specimen
—·–·– annealed in 10^{-5} Torr vacuum
- - - - - annealed in argon

Fig. 2

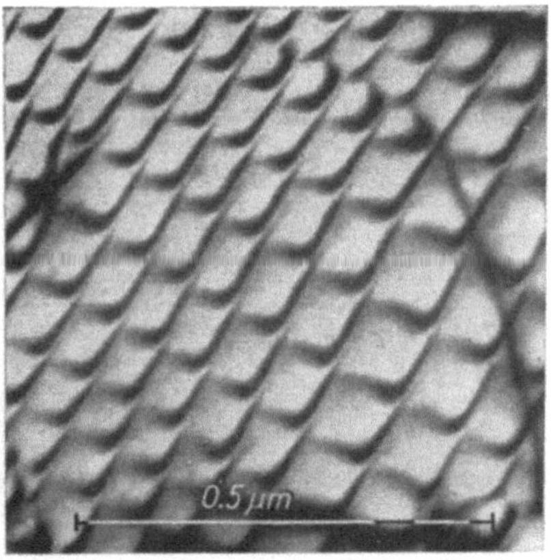

Fig. 3

Fig. 2. Electron micrograph of as received niobium specimen, cold worked to 95% reduction

Fig. 3. Cold worked niobium specimen annealed at 1200 °C in a vacuum 10^{-5} Torr; showing a regular dislocation network

3. Results

Fig. 1 shows the effect of annealing upon the magnetization behaviour of niobium. The as received material, cold worked ($\sim$ 95% reduction) with a resistivity ratio of 16, showed considerable hysteresis and flux trapping. The structure of this material (Fig. 2) is the cellular structure typical of coldworked body centred cubic metals [3]. Using the method suggested by HIRSCH and BAILEY [4] the dislocation density was estimated as 10^{12} lines per cm². Annealing this material at 1200 °C in a vacuum of 10^{-5} Torr produced a regular network of dislocations (Fig. 3). The dislocation density was estimated at 10^{8} dislocations per cm². Magnetization curves (Fig. 1) of this annealed specimen show a smaller hysteresis and flux trapping. Specimens annealed at 1200 °C in an argon atmosphere which is known to contain some oxygen showed an interesting effect. As for the vacuum anneal, the cold worked structure was initially replaced by a hexagonal network of dislocations. After about 15 minutes annealing, the network began to break-up and was replaced by the structure shown in Fig. 4. The fringes indicate the presence of some kind of extended faults which were found to lie in {112} planes. After two hours annealing these faults were fully developed and no trace of the original dislocation network was observed. The exact nature of these faults is currently being investigated, and a full report of this effect will be published later. A similar contrast effect has also been recently observed by AMELINCKX and VAN LANDUYT [5] who ascribe it to the presence of a kind of domain boundary, but in this case it is felt more likely to be due to a double, one intrinsic and the other extrinsic, stacking fault. Stacking faults have been observed in other body centred cubic materials, such as vanadium [6] and tungsten [7]; it is reasonable that niobium should also show the presence of stacking faults. The magnetization curve for the specimen containing these extended faults shows (Fig. 1) nearly as much hysteresis and flux trapping as the cold worked material. $H c_2$, the field of complete flux penetration, for all these specimens was constant at 5250 Oe.

The effect of cold work was investigated by the following experiments. Niobium foil was annealed for two hours at 1750 °C in vacuum better than 10^{-7} Torr; referred to as

Fig. 4. Cold worked niobium specimen annealed at 1200 °C in argon; showing some kind of extended defects giving fringe contrast

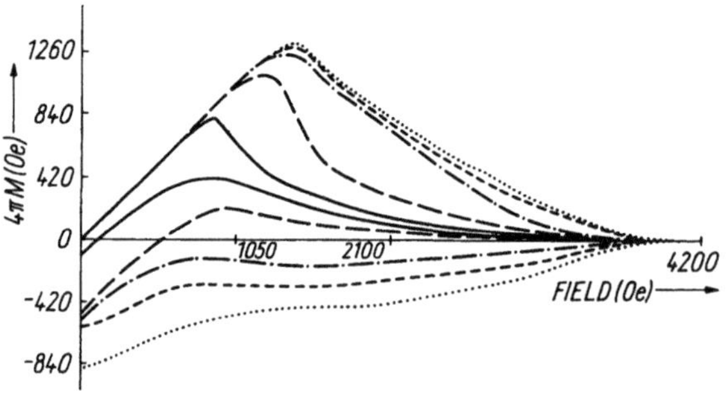

Fig. 5. Magnetization curves for cold rolled niobium specimens

——— degassed specimen, — — — 14% deformed, — · — · 35% deformed, - - - - - 62.5% deformed, ... 62.5% deformed and repeatedly bent

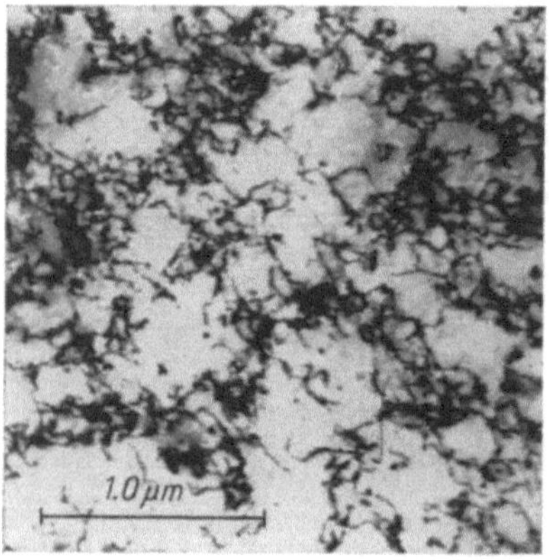

Fig. 6

Fig. 6. Electron micrograph of well annealed niobium deformed to 14% reduction. Note the formation of cellular structure

Fig. 7. Electron micrograph of well annealed niobium deformed to 35% reduction. Note the increase in dislocation density in the cell walls.

Fig. 8. Electron micrograph of well annealed niobium deformed to 62.5% reduction

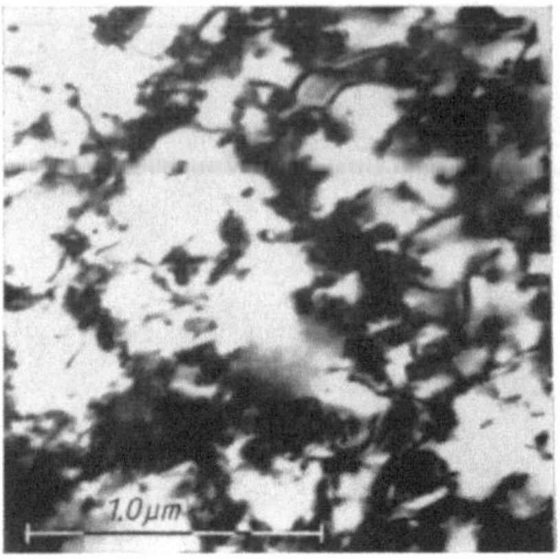

Fig. 7

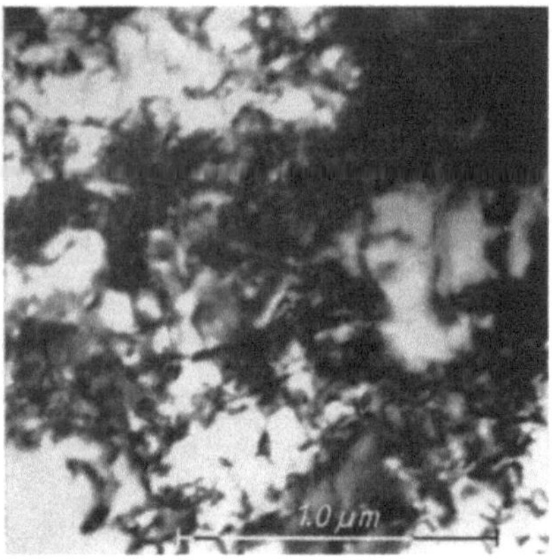

Fig. 8

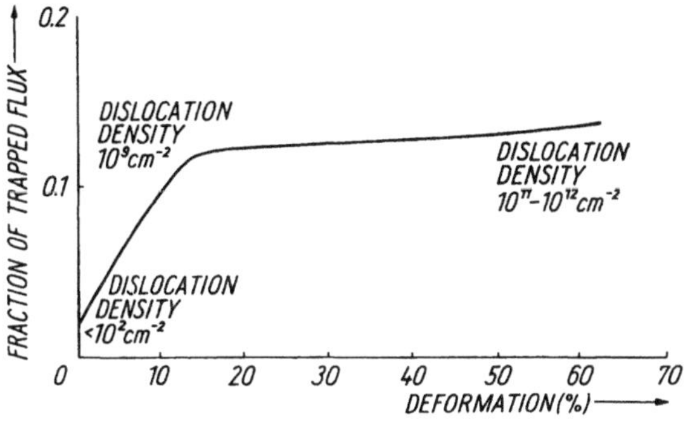

Fig. 9. A curve showing the variation of trapped flux at zero field, expressed as a fraction of $H c_2$, with percentage deformation

the degassing treatment. Electron microscopy of this material revealed that essentially all the dislocations had been removed by this treatment, the dislocation density being less than 100 lines per cm². The magnetization curve for this specimen (Fig. 5) is very close to being reversible. This annealed material was subsequently cold rolled and reduced 14%, 35% and 62.5%. Fig. 5 shows the increasing hysteresis in the magnetization curves for these deformed specimens; their respective microstructures are shown in Fig. 6, 7, and 8. $H c_2$ remained constant at 4200 Oe for all the specimens. Fig. 9 is a polt of the trapped flux at zero field, expressed as a fraction of $H c_2$, versus percentage deformation. The knee of the curve coincides with the formation of the cellular dislocation structure. Deformation beyond this point increases the dislocation density in the cell walls without changing the cell size, as can be seen from Fig. 6, 7 and 8. Trapped flux rises but slightly with increasing deformation once the cellular structure has been formed. The dislocation density at the knee of the curve is 10^8 to 10^9 lines per cm². The trapped flux corresponds to an induction of about 500 G; $2.5 \cdot 10^8$ flux quanta per cm² are required to give this induction. The close agreement

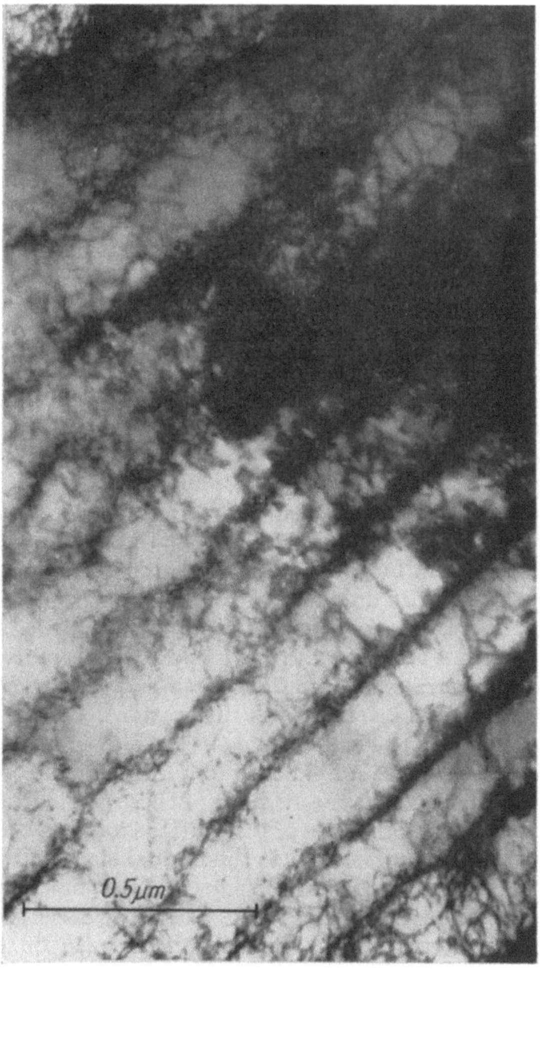

Fig. 10. Polygonized diclocation walls in a repeatedly bent cold worked niobium foil

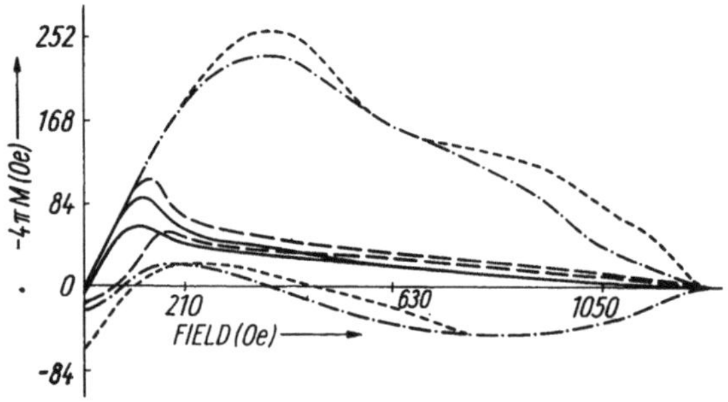

Fig. 11. Magnetization curves for cold worked vanadium specimens

—— · —— degassed specimen
— — — 14% deformed
— · — · as received specimen
- - - - - repeatedly bent

between the number of flux quanta and the dislocation density is interesting but probably coincidental.

If a cold worked niobium foil is repeatedly bent back and forth, dislocations become aligned into walls parallel to {100} planes (Fig. 10). This effect has been found in other b. c. c. metals [8]. The magnetization curve for the repeatedly bent material is also shown in Fig. 5. The curve for the field increasing is very close to that for the 62.5% cold worked material; the flux trapped on decreasing the field is, however, much greater.

Vanadium has not been examined in as great detail as niobium; the results so far obtained are qualitatively similar. Fig. 11 shows magnetization curves for vanadium specimens after various treatments. Electron micrographs revealed no significant difference from niobium specimens after similar treatments.

4. Discussion

The shape of the magnetization curves in Fig. 5 is in agreement with the strong pinning model developed by CAMPBELL, EVETTS, and DEW-HUGHES [9]. The dislocations introduced by the deformation process must interact strongly with the flux lines. The lowering of $H\,c_2$ for niobium after the degassing treatment indicates that a purification accompanied the reduction in dislocation density. The shape of the magnetization curves in Fig. 1, showing a plateau after the sharp decrease in the magnetization, suggests the presence of two types of pinning centre [9]. As the plateau is not present in the degassed specimens (Fig. 5) it may be associated with an impurity effect acting separately to the dislocation pinning. The magnetization curves for the vanadium specimens in Fig. 11 seem to be intermediate between those for impure and degassed niobium.

The results on niobium suggest that groups of dislocations, or "tangles", such as are observed in the cell walls of cold worked material or after repeated bending, are more effective in flux trapping than individual dislocations or regular dislocation arrays. There have been several suggestions as to the way in which flux lines might interact with dislocations. FLEISCHER [10] has calculated an interaction due to local changes in the elastic moduls, and WEBB [11] has shown that the elastic strain field around a dislocation would account for the observed superconducting behaviour. Both of these calculations were for individual dislocations; the effect of bundles of dislocations has not been calculated. BARBEE and LITTLE [12] have suggested that the normal electron mean free path, and hence the range of coherence ξ [13] of the superconducting electrons, must be much shorter within the dislocation walls than within the essentially dislocation free cells of a cold rolled structure. As the self energy of a flux line within a superconductor is a function of the local value of ξ, the resulting spatial variations in ξ will inhibit the motion of flux. The extended faults poduced upon annealing in argon presumably also give rise to local changes in the range of coherence.

Any pinning mechanism in a type II superconductor must involve a spatial variation in the self energy ϵ of a flux line. This self energy, ignoring any interaction with the flux lines, is given by SILCOX and ROLLINS [14] as

$$\epsilon = \frac{\varphi_0 H c_1}{4 \pi}, \tag{1}$$

where φ_0 is one flux quantum.

For a type II superconductor [15],

$$H c_1 = H c_0 \left(\frac{\xi}{\lambda}\right)^{1/2}, \tag{2}$$

where $H c_0$ is the thermodynamic critical field and λ the penetration depth.

PIPPARD [13] gives $\frac{1}{\xi} = \frac{1}{\xi_0} + \frac{1}{\alpha l}$ and $\lambda = \lambda_L \left(1 + \frac{\xi_0}{\alpha l}\right)^{1/2}$ where ξ_0 is the range of coherence, and λ_L is the London penetration depth for the pure material, and l is the normal electron mean free path. Substituting in equation (1), gives

$$\epsilon = \frac{\varphi_0 H c_0}{4 \pi} \left(\frac{\xi_0}{\lambda_L}\right)^{1/2} \left\{\frac{\alpha l}{\xi_0 + \alpha l}\right\}^{3/4}. \tag{3}$$

For a given superconductor the only parameters which may be varied locally by the presence of dislocations are $H c_0$ and l. Any effect on $H c_0$, through the elastic strain field, will be small. l also will not change much in the vicinity of a single dislocation, but as mentioned above can be reduced considerably in a region of high dislocation density. Thus the large local variations in dislocation density in cold worked or repeatedly bent body centred cubic metals are expected to produce very effective trapping centres for flux. This conclusion is borne out by the experiments of HEATON and ROSE-INNES [16] on a Nb-Ta alloy. Bending or twisting of the annealed material, which would produce a uniform dislocation distribution, resulted in little flux trapping; drawing operations, which are expected to give rise to dislocation tangles, however produced specimens with very high flux trapping.

5. Conclusions

It is concluded that the increase in flux trapping and current carrying capacity on cold working a type II superconductor is due to a nonuniform distribution of dislocations. The resulting local variations in the range of coherence interfere with the motion of flux lines through a variation in their self energy.

Acknowledgements

The authors wish to thank Professor A. H. COTTRELL and the Department of Metallurgy for the provision of facilities to pursue this work, and their colleagues for stimulating discussion. One of them (A. V. N.) is indebted to the Master and Fellows of Peterhouse for the award of a research studentship. Finally, the authors are grateful to Central Electricity Generating Board for financially supporting the project.

References

[1] J. D. LIVINGSTON, Report No. 62-RL-3155 M, General Electric Research Laboratory, New York, Oct. 1962.
[2] R. W. SHAW and D. E. MAPOTHER, Phys. Rev. **118**, 1474 (1960).
J. J. HAUSER and E. BUEHLER, Phys. Rev. **127**, 142 (1962).
J. J. HAUSER and E. HELFLAND, Phys. Rev. **127**, 386 (1962).
[3] W. CARRINGTON, K. F. HALE, and D. McLEAN, Proc. Roy. Soc. A **259**, 203 (1960).
[4] P. B. HIRSCH and J. E. BAILEY, Phil. Mag. **5**, 485 (1960).
C. S. SMITH and L. GUTTMAN, J. Metals **5**, 81 (1953).
[5] S. AMELINCKX and J. VAN LANDUYT, Appl. Phys. Letters **4**, 15 (1964).
[6] T. C. LINDLEY and R. E. SMALLMAN, Acta metall. **11**, 361 (1963).
[7] Y. NAKAYAMA, S. WEISSMAN, and T. IMURA, Direct Observation of Imperfections in Crystals. Interscience Publishers, John Wiley & Sons, New York 1962 (p. 573).
[8] A. V. NARLIKAR and D. DEW-HUGHES, to be presented at the "Third European Regional Conference on Electron Microscopy, Prague, August 1964".
[9] A. M. CAMPBELL, J. E. EVETTS, and D. DEW-HUGHES, to be published.
[10] R. L. FLEISCHER, Phys. Rev. Letters **3**, 111 (1962).
[11] W. W. WEBB, Phys. Rev. Letters **11**, 191 (1963).
[12] T. W. BARBEE and W. A. LITTLE, private communication.
[13] A. B. PIPPARD, Proc. Roy. Soc. A **203**, 210 (1950).
A. B. PIPPARD, Physica **19**, 765 (1953).
A. B. PIPPARD, Phil. Trans. Roy. Soc. A **248**, 97 (1955).
[14] J. SILCOX and R. W. ROLLINS, Appl. Phys. Letters **2**, 231 (1963).
[15] C. P. BEAN and R. W. SCHMITT, Science **140**, 26 (1963).
[16] J. W. HEATON and A. C. ROSE-INNES, private communication.

(Received May 6, 1964)

phys. stat. sol. 6, 391 (1964)

FIAT Research Laboratories, Torino

Vacancy as a Phonon Field Perturbation

I. Lattice Distorsion around a Vacancy

By

V. GALLINA and M. OMINI

The second quantization of the phonon field allows a theory to be developed which gives the reticular centre displacements, produced by vacancy formation. The theory is based on Bravais lattices, and a detailed calculation is carried out for the Debye approximation. Numerical results for argon at 80 °K are compared with those of previous theories.

Par un calcul de seconde quantization on étudie la distorsion d'un réseau cristallin autour d'une vacance. La théorie est appliquée aux cristaux de Bravais en employant l'approximation de Debye. Les résultats numériques pour un cristal d'argon à 80 °K sont comparés à ceux des théories précédentes.

1. Introduction

The knowledge of reticular distorsion around a vacancy is a fundamental problem in modern solid state physics. In fact, all experimental methods to measure the vacancy concentration, are based on the determination of those macroscopic physical parameters (such as the solid dilatation, the variation of transport properties, etc.) that are, *a priori*, connected with the effective number of vacancies. On theoretical grounds, this connection may be easily made only if lattice centre displacements around the defect are sufficiently well known. Their knowledge allow, for example, to calculate the scattering cross-section in electron-vacancy interactions (and then the contribution to electrical resistivity), as well as to evaluate the so-called "relaxation energy", which gives a not negligeable contribution to the vacancy creation energy.

The problem was first tackled by several authors [1, 2, 3, 4, 5] in the light of the elasticity theory, which leads to the following solution:

$$\eta(r) = \alpha \frac{v}{r^2}, \tag{1}$$

where r is the distance of the atom from the defect point, v the atomic volume, and α a dimensionless constant, related to the material properties. However, equation (1) is an unsatisfactory solution, because elasticity theory fails in the immediate neighbourhood of the vacancy, i.e. just in the region which gives the most important contribution to scatterings and relaxation energy. The first attempt to account for the lattice *discrete* nature, in a static model, was made by KANZAKI [6], for a f.c.c. lattice. His method, based on expansion of the atom displacements in normal lattice coordinates, can be used both for vacancies and interstitials. An iterative procedure, leading to successive solutions of linear algebraic equation systems, has been formulated by TEWORDT [7]. It requires the use of a large electronic computer, and, moreover, may be regarded as an attempt to give *numerical* rather than analytical solutions.

Later on, NARDELLI and CHIAROTTI [8], taking into account lattice vibrations, have used KANZAKIS method and calculated the distortion of an argon crystal at various temperatures.

In the present and the next papers, we wish to give a solution of the problem, starting from a different theoretical point of view. Our calculation is made using the second quantization theory. A vacancy will be considered as a perturbation of the phonon state related to the perfect crystal. In this scheme displacements are shown to be independent of phonon distribution, and will prove our theory to be applicable both for static and vibrating lattice, provided that thermal dilatation be duly accounted for. A general formula can be obtained, which, in principle, allows calculation of displacements in any Bravais lattice, providing dispersion curves to be known. A detailed calculation is carried out for a f.c.c. crystal, in the Debye approximation, and the solution is expressed in terms of universal functions, whose tabulations are given.

2. General Considerations

Let $V(\boldsymbol{l} + \boldsymbol{\eta}_l - \boldsymbol{l}' - \boldsymbol{\eta}_{l'})$ be the potential energy between lattice centres denoted by $\boldsymbol{l}$ and $\boldsymbol{l}'$, and let $\boldsymbol{\eta}_l$ and $\boldsymbol{\eta}_{l'}$ be respective displacements from their equilibrium positions. By power expansion we have

$$V(\boldsymbol{l} + \boldsymbol{\eta}_l - \boldsymbol{l}' - \boldsymbol{\eta}_{l'}) = V(\boldsymbol{l} - \boldsymbol{l}') + \frac{\partial V(\boldsymbol{h})}{\partial \boldsymbol{h}} \cdot \{\boldsymbol{\eta}_l - \boldsymbol{\eta}_{l'}\} + \\ + \frac{1}{2}\left\{\boldsymbol{\eta}_l \frac{\partial^2 V(\boldsymbol{h})}{\partial \boldsymbol{h}^2} \boldsymbol{\eta}_l + \boldsymbol{\eta}_{l'} \cdot \frac{\partial^2 V(\boldsymbol{h})}{\partial \boldsymbol{h}^2} \boldsymbol{\eta}_{l'} - \boldsymbol{\eta}_l \frac{\partial^2 V(\boldsymbol{h})}{\partial \boldsymbol{h}^2} \boldsymbol{\eta}_{l'} - \boldsymbol{\eta}_{l'} \frac{\partial^2 V(\boldsymbol{h})}{\partial \boldsymbol{h}^2} \boldsymbol{\eta}_l\right\} + \cdots,$$

where $\boldsymbol{h} = \boldsymbol{l} - \boldsymbol{l}'$. Introducing the tensors

$$\boldsymbol{f}_h = \frac{\partial V(\boldsymbol{h})}{\partial \boldsymbol{h}}, \qquad g_h = -\frac{\partial^2 V(\boldsymbol{h})}{\partial \boldsymbol{h}^2}, \tag{2}$$

we can also write

$$V(\boldsymbol{h} + \boldsymbol{\eta}_l - \boldsymbol{\eta}_{l'}) = V(\boldsymbol{h}) + \boldsymbol{f}_h \cdot \{\boldsymbol{\eta}_l - \boldsymbol{\eta}_{l'}\} - \frac{1}{2}(\boldsymbol{\eta}_l - \boldsymbol{\eta}_{l'})\, g_h\, (\boldsymbol{\eta}_l - \boldsymbol{\eta}_{l'}) + \cdots. \tag{3}$$

Definitions (2) require $\boldsymbol{h} \neq 0$; besides, being $V(\boldsymbol{h}) = V(-\boldsymbol{h})$, we have $\boldsymbol{f}(-\boldsymbol{h}) = -\boldsymbol{f}(\boldsymbol{h})$, and $g(-\boldsymbol{h}) = g(\boldsymbol{h})$. Then the total potential energy of centre $\boldsymbol{l}$ will be

$$W_l = \sum_{\boldsymbol{h} \neq 0} \left\{V(\boldsymbol{h}) + \boldsymbol{f}_h \cdot [\boldsymbol{\eta}_l - \boldsymbol{\eta}_{l'}] - \frac{1}{2}[\boldsymbol{\eta}_l - \boldsymbol{\eta}_{l'}]\, g_h\, [\boldsymbol{\eta}_l - \boldsymbol{\eta}_{l'}]\right\} \tag{4}$$

and that of the whole crystal

$$W = \frac{1}{2} \sum_{\boldsymbol{l}} \sum_{\boldsymbol{h} \neq 0} \left\{V(\boldsymbol{h}) + \boldsymbol{f}_h \cdot [\boldsymbol{\eta}_l - \boldsymbol{\eta}_{l'}] - \frac{1}{2}[\boldsymbol{\eta}_l - \boldsymbol{\eta}_{l'}]\, g_h\, [\boldsymbol{\eta}_l - \boldsymbol{\eta}_{l'}]\right\}. \tag{5}$$

Hence, in a perfect crystal, it results that the linear term in the potential energy is identically vanishing as $\boldsymbol{f}_h$ is an antisymmetric vector. Therefore, the Hamiltonian of a perfect crystal is

$$H = \sum_{\boldsymbol{l}} \frac{\boldsymbol{p}_l^2}{2m} - \frac{1}{4} \sum_{\boldsymbol{l}} \sum_{\boldsymbol{h} \neq 0} [\boldsymbol{\eta}_l - \boldsymbol{\eta}_{l'}]\, g_h\, [\boldsymbol{\eta}_l - \boldsymbol{\eta}_{l'}], \tag{6}$$

where $\boldsymbol{p}_l$ is $\boldsymbol{l}$-centre momentum canonically conjugated with $\boldsymbol{\eta}_l$, the energy origin having been taken so that $\sum_{l} \sum_{l \neq 0} V(\boldsymbol{h}) = 0$. Hamiltonian (6) can be easily identified with that of phonon field

$$H = \sum_{l} \frac{\boldsymbol{p}_l^2}{2\,m} + \frac{1}{2} \sum_{l,l'} \boldsymbol{\eta}_l\, g\,(\boldsymbol{l} - \boldsymbol{l}')\, \boldsymbol{\eta}_{l'}\,, \tag{6'}$$

providing we define

$$g(\vec{0}) = - \sum_{\boldsymbol{h} \neq 0} g(\boldsymbol{h})\,. \tag{7}$$

It is well known that (6′), and hence (6), results in the phonon field quantization if applied to a reticular lattice with periodicity properties. Calling $\boldsymbol{q}$ the wave vector, p the branch, $\boldsymbol{\epsilon}_{\boldsymbol{q}p}$ the polarization vector, $\omega_{\boldsymbol{q}p}$ the angular frequency of a generical vibration mode, we have the following eigenvalue equation[1])

$$\frac{1}{m}\, E_{\boldsymbol{q}}\, \boldsymbol{\epsilon}_{\boldsymbol{q}p} = \omega_{\boldsymbol{q}p}^2\, \boldsymbol{\epsilon}_{\boldsymbol{q}p}\,, \tag{8}$$

where $E_{\boldsymbol{q}}$ is defined by

$$E_{\boldsymbol{q}} = \sum_{\boldsymbol{h}} e^{-i\boldsymbol{q}\cdot\boldsymbol{h}}\, g(\boldsymbol{h})\,. \tag{8'}$$

The operational expansions of $\boldsymbol{\eta}_l$ and $\boldsymbol{p}_l$ in the Schrödinger scheme are given by

$$\left.\begin{aligned} \boldsymbol{\eta}_l &= i \sqrt{\frac{\hbar}{2\,m\,N}} \sum_{\boldsymbol{q}p} \frac{1}{\sqrt{\omega_{\boldsymbol{q}p}}}\, \boldsymbol{\epsilon}_{\boldsymbol{q}p}\, [e^{i\boldsymbol{q}\cdot\boldsymbol{l}}\, a_{\boldsymbol{q}p} - e^{-i\boldsymbol{q}\cdot\boldsymbol{l}}\, a_{\boldsymbol{q}p}^{+}]\,, \\ \boldsymbol{p}_l &= \sqrt{\frac{\hbar\, m}{2\,N}} \sum_{\boldsymbol{q}p} \sqrt{\omega_{\boldsymbol{q}p}}\, \boldsymbol{\epsilon}_{\boldsymbol{q}p}\, [e^{i\boldsymbol{q}\cdot\boldsymbol{l}}\, a_{\boldsymbol{q}p} + e^{-i\boldsymbol{q}\cdot\boldsymbol{l}}\, a_{\boldsymbol{q}p}^{+}]\,, \end{aligned}\right\} \tag{9}$$

being N the total number of lattice cells and $a_{\boldsymbol{q}p}$, $a_{\boldsymbol{q}p}^{+}$ the absorption and creation operators of the $(\boldsymbol{q}\,p)$ phonon.

3. Vacancy as a Perturbation

Consider a vacancy at $\boldsymbol{l} = 0$. In this case, the kinetic and potential contribution of the missing centre must be subtracted from the crystal energy (6). This contribution, with sign changed, can be considered as the perturbation Hamiltonian to the perfect crystal. Hence one has

$$H'_{\text{vac}} = - \frac{\boldsymbol{p}_0^2}{2\,m} - W_0 = \sum_{\boldsymbol{h} \neq 0} \boldsymbol{f}_{\boldsymbol{h}} \cdot \boldsymbol{\eta}_{\boldsymbol{h}} - \frac{\boldsymbol{p}_0^2}{2\,m} + \frac{1}{2} \sum_{\boldsymbol{h} \neq 0} (\boldsymbol{\eta}_{\boldsymbol{h}} - \boldsymbol{\eta}_0)\, g_{\boldsymbol{h}}\, (\boldsymbol{\eta}_{\boldsymbol{h}} - \boldsymbol{\eta}_0)\,. \tag{10}$$

Expression (10) can immediately be considered as a quantum operator, if we use the expansions (9). Thus we are assuming, as in the field theory, that perturbation does not appreciably modify the field operator structure. In our case this seems a reasonable assumption because the perturbation is small, and it will be confirmed — a posteriori — proving that reticular displacements result much smaller than lattice constants.

Finally, it must be pointed out that (10) is actually the perturbation Hamiltonian associated with a lattice centre *removed* from the crystal, while a vacancy results in a reticular centre transferred to the surface. However, as long as we

---

[1]) ZIMAN, Electrons and Phonons, Oxford Clarendon Press, 1962 (p. 28f.).

are interested in local properties, concerning the immediate neighbourhood of the vacancy (scatterings, displacement distribution, etc.), expression (10) is obviously true. In the case one studies properties concerning the crystal as a whole (e.g. vibrational energy variations), expression (10) should be duly modified.

In this paper, we shall be exclusively interested in the perturbation term

$$H'^{(1)}_{\text{vac}} = \sum_{\boldsymbol{h}} \boldsymbol{f_h} \cdot \boldsymbol{\eta_h}$$

which describes, as later on shown, the atom rearrangement in new equilibrium positions, i.e. after the vacancy formation. To this purpose, let us indicate by the symbol $|n_q\rangle$ the perfect crystal state having $n_{q_1}, n_{q_2}, \ldots$ phonons in the modes $q_1, q_2, \ldots$ respectively, and by $|n_{q_r} \pm 1\rangle$ the previous state increased or reduced of one phonon in the mode q_r.[2]) Then, the state $|n_q\rangle_{\text{vac}}$ perturbed by $H'^{(1)}_{\text{vac}}$ results to be

$$|n_q\rangle_{\text{vac}} = |n_q\rangle + \sum_r \frac{\langle n_{q_r}+1|H'^{(1)}_{\text{vac}}|n_q\rangle}{-\hbar\,\omega_{q_r}}\,|n_{q_r}+1\rangle + $$

$$+ \sum_r \frac{\langle n_{q_r}-1|H'^{(1)}_{\text{vac}}|n_q\rangle}{\hbar\,\omega_{q_r}}\,|n_{q_r}-1\rangle + \cdots \tag{11}$$

because $H'^{(1)}_{\text{vac}}$ has not-vanishing matrix elements between states differing by one phonon. Let us now observe that, in the perfect crystal, $\langle n_q|\boldsymbol{\eta_l}|n_q\rangle = 0$, $\langle n_q|\boldsymbol{p_l}|n_q\rangle = 0$, as it can be easily verified. This is related to the fact that $\boldsymbol{l}$ points are actually equilibrium position for the lattice centres. By analogy, we can think that vectors $\boldsymbol{l} + \langle\boldsymbol{\eta_l}\rangle_{\text{vac}} = \boldsymbol{l} + \langle n_q|\boldsymbol{\eta_l}|n_q\rangle_{\text{vac}}$ indicate the new equilibrium position in the perturbed crystal, providing $\langle n_q|\boldsymbol{p_l}|n_q\rangle_{\text{vac}} = 0$. Therefore let us calculate the expectation value of $\boldsymbol{\eta_l}$ operator in the state (11). We need the following matrix elements:

$$\left.\begin{aligned}
\langle n_{q_r}+1|H'^{(1)}_{\text{vac}}|n_q\rangle &= \sqrt{\frac{\hbar}{2mN}}\sqrt{\frac{n_{q_r}+1}{\omega_{q_r}}}\,(\boldsymbol{\epsilon}_{q_r}\cdot\boldsymbol{R}_{q_r}),\\
\langle n_{q_r}-1|H'^{(1)}_{\text{vac}}|n_q\rangle &= \sqrt{\frac{\hbar}{2mN}}\sqrt{\frac{n_{q_r}}{\omega_{q_r}}}\,(\boldsymbol{\epsilon}_{q_r}\cdot\boldsymbol{R}_{q_r}),\\
\langle n_{q_r}+1|\boldsymbol{\eta_l}|n_q\rangle &= -\,i\sqrt{\frac{\hbar}{2mN}}\sqrt{\frac{n_{q_r}+1}{\omega_{q_r}}}\,e^{-i\,\boldsymbol{q}_r\cdot\boldsymbol{l}}\,\boldsymbol{\epsilon}_{q_r},\\
\langle n_{q_r}-1|\boldsymbol{\eta_l}|n_q\rangle &= i\sqrt{\frac{\hbar}{2mN}}\sqrt{\frac{n_{q_r}}{\omega_{q_r}}}\,e^{i\,\boldsymbol{q}_r\cdot\boldsymbol{l}}\,\boldsymbol{\epsilon}_{q_r},
\end{aligned}\right\} \tag{12}$$

having introduced the vector

$$\boldsymbol{R}_{q_r} = \sum_{\boldsymbol{h}} \boldsymbol{f_h} \sin \boldsymbol{q}_r\cdot\boldsymbol{h}\,. \tag{13}$$

---

[2]) Index q_r is here used instead of the pair $(\boldsymbol{q}_r, p_r)$.

One easily gets

$$\langle \boldsymbol{\eta}_{\boldsymbol{l}} \rangle_{\text{vac}} = \sum_r \frac{\langle n_{q_r}+1 |\, H'^{(1)}_{\text{vac}} \,| n_q \rangle \langle n_q |\, \boldsymbol{\eta}_{\boldsymbol{l}} \,| n_{q_r}+1 \rangle}{-\hbar\,\omega_{q_r}} +$$

$$+ \sum_r \frac{\langle n_{q_r} - 1 |\, H'^{(1)}_{\text{vac}} \,| n_q \rangle \langle n_q |\, \boldsymbol{\eta}_{\boldsymbol{l}} \,| n_{q_r} - 1 \rangle}{\hbar\,\omega_{q_r}} + \text{c.c.} =$$

$$= \frac{1}{m\,N} \sum_r \frac{(\boldsymbol{R}_{q_r} \cdot \boldsymbol{\epsilon}_{q_r})}{\omega^2_{q_r}} \boldsymbol{\epsilon}_{q_n} \sin \boldsymbol{q}_r \cdot \boldsymbol{l}\,. \tag{14}$$

The equation (14) proves $\langle \boldsymbol{\eta}_{\boldsymbol{l}} \rangle_{\text{vac}}$ to be independent of the lattice vibrational state. Moreover, by the matrix elements

$$\left.\begin{aligned} \langle n_{q_r} + 1 \,|\boldsymbol{p}_{\boldsymbol{l}}|\, n_q \rangle &= \sqrt{\frac{\hbar\,m}{2\,N}} \sqrt{\omega_{q_r}\,(n_{q_r} + 1)}\; e^{-i\,\boldsymbol{q}_r \cdot \boldsymbol{l}} \boldsymbol{\epsilon}_{q_r}\,, \\ \langle n_{q_r} - 1 \,|\boldsymbol{p}_{\boldsymbol{l}}|\, n_q \rangle &= \sqrt{\frac{\hbar\,m}{2\,N}} \sqrt{\omega_{q_r}\, n_{q_r}}\, e^{-i\,\boldsymbol{q}_r \cdot \boldsymbol{l}} \boldsymbol{\epsilon}_{q_r}\,, \end{aligned}\right\} \tag{15}$$

we readily obtain

$$\langle \boldsymbol{p}_{\boldsymbol{l}} \rangle_{\text{vac}} = -\frac{1}{2\,N} \sum_r \frac{(\boldsymbol{R}_{q_r} \cdot \boldsymbol{\epsilon}_{q_r})}{\omega_{q_r}} \boldsymbol{\epsilon}_{q_r}\, e^{i\,\boldsymbol{q}_r \cdot \boldsymbol{l}}\, (2\, n_{q_r} + 1) + \text{c.c.}\,.$$

Being $\boldsymbol{R}_{q_r}$ an odd function of $\boldsymbol{q}_r$, the sum $\sum_r$ is a purely imaginary expression, hence

$$\langle \boldsymbol{p}_{\boldsymbol{l}} \rangle_{\text{vac}} = 0\,. \tag{16}$$

These results enable us to consider $\langle \boldsymbol{\eta}_{\boldsymbol{l}} \rangle_{\text{vac}}$ as the displacement of $\boldsymbol{l}$-centre in the perturbed lattice. Expression (14) allows, at least in principle, to evaluate the reticular rearrangement due to the vacancy in any Bravais lattice, provided the dispersion curves of the crystal be known. However, in order to obtain a simpler result, we wish to specify (14) for a Debye model, which represents a suitable approximation for all f.c.c. lattices.

4. Displacements in a Debye Model

In this approximation one merely has $\omega_{qp} = s\,q$, where s is the sound velocity assumed independent of $(\boldsymbol{q}, p)$. We get

$$\langle \boldsymbol{\eta}_{\boldsymbol{l}} \rangle_{\text{vac}} = \frac{1}{m\,N} \sum_{qp} \frac{(\boldsymbol{R}_{\boldsymbol{q}} \cdot \boldsymbol{\epsilon}_{qp})}{s^2\,q^2} \boldsymbol{\epsilon}_{qp} \sin \boldsymbol{q} \cdot \boldsymbol{l} = \frac{1}{m\,N\,s^2} \sum_{\boldsymbol{q}} \frac{\boldsymbol{R}_{\boldsymbol{q}}}{q^2} \sin \boldsymbol{q} \cdot \boldsymbol{l} \tag{17}$$

since $\boldsymbol{\epsilon}_{qp}$ vectors make an orthonormal reference frame in a Bravais lattice. Because our model is spherically symmetrical, vector $\langle \boldsymbol{\eta}_{\boldsymbol{l}} \rangle_{\text{vac}}$ is parallel to $\boldsymbol{l}$; therefore, we shall account only for the scalar quantity $\xi_l = \langle \boldsymbol{\eta}_{\boldsymbol{l}} \rangle_{\text{vac}} \cdot \boldsymbol{u}_{\boldsymbol{l}}$, being $\boldsymbol{u}_{\boldsymbol{l}}$ the versor of $\boldsymbol{l}$. Substituting the sum on $\boldsymbol{q}$ by an integral, we readily obtain

$$\xi_l = \frac{v}{8\,\pi^3\,m\,s^2} \sum_{\boldsymbol{h}} (\boldsymbol{f}_{\boldsymbol{h}} \cdot \boldsymbol{u}_{\boldsymbol{l}}) \int_0^Q dq \int d\Omega \sin \boldsymbol{q} \cdot \boldsymbol{l} \sin \boldsymbol{q} \cdot \boldsymbol{h}\,,$$

where $v = V/N$ is the lattice cell volume, Q the radius of the Debye sphere and $d\Omega$ the differential solid angle in the reciprocal space. The above expression writes

$$\xi_l = \frac{v}{4\pi^2 m s^2} \sum_{\boldsymbol{h}} (\boldsymbol{f_h} \cdot \boldsymbol{u}_l) \left\{ \frac{\mathrm{Si}\, Q\, |\boldsymbol{l} - \boldsymbol{h}|}{|\boldsymbol{l} - \boldsymbol{h}|} - \frac{\mathrm{Si}\, Q\, |\boldsymbol{l} + \boldsymbol{h}|}{|\boldsymbol{l} + \boldsymbol{h}|} \right\} =$$

$$= \frac{v\, Q}{2\pi^2 m s^2} \sum_{\boldsymbol{h}} (\boldsymbol{f_h} \cdot \boldsymbol{u}_l) \frac{\mathrm{Si}\, Q\, |\boldsymbol{l} - \boldsymbol{h}|}{Q\, |\boldsymbol{l} - \boldsymbol{h}|},$$

where $\mathrm{Si}\, x = \int\limits_0^x \frac{\sin z}{z}\, dz$ is the integral sine function. Assuming the atoms around the vacancy to be uniformely distributed on successive spherical shells of radii $\varrho_1, \varrho_2, \ldots$ and occupation numbers $\nu_1, \nu_2, \ldots$ respectively, the last sum can be easily evaluated.

As a consequence of the spherical symmetry, vectors $\boldsymbol{f}_r$ must be radially directed, i.e. $\boldsymbol{f}_r = f_r\, \boldsymbol{u}_r$, being $\boldsymbol{u}_r = \frac{\boldsymbol{\rho}_r}{\varrho_r}$, so that

$$\xi_l = \frac{v\, Q}{2\pi^2 m s^2} \sum_r f_r \frac{\nu_r}{4\pi} \int d\Omega (\boldsymbol{u}_r \cdot \boldsymbol{u}_l) \frac{\mathrm{Si}\, Q\, |\boldsymbol{l} - \boldsymbol{\rho}_r|}{Q\, |\boldsymbol{l} - \boldsymbol{\rho}_r|}.$$

Finally it easily follows

$$\xi_l = \frac{v\, Q}{8\pi^2 m s^2} \sum_r f_r\, \nu_r\, F_{\lambda_r}\!\left(\frac{l}{\varrho_r}\right), \tag{18}$$

where the following definitions have been introduced:

$$\lambda_r = Q\, \varrho_r, \tag{19}$$

$$F_\lambda(x) = 2 \int\limits_{-1}^{1} du \cdot u\, \frac{\mathrm{Si}\, \lambda \sqrt{1 + x^2 - 2xu}}{\lambda \sqrt{1 + x^2 - 2xu}} = \frac{1}{\lambda x^2} \int\limits_{x-1}^{x+1} (1 + x^2 - z^2)\, \mathrm{Si}\, \lambda z\, dz. \tag{20}$$

The λ_r's given by (19) are pure numbers depending only on the geometry of the crystal lattice. To show this, let us observe that, if a^3 is the elementary cell volume, one has[3] $Q\, a = (6\pi^2)^{1/3}$; moreover, in a f.c.c. lattice, $a^3 = \frac{1}{4}(\varrho_1 \sqrt{2})^3$ so that $\lambda_1 = Q\, \varrho_1 = \sqrt{2}\, (3\pi^2)^{1/3} = 4.3749$. In a similar way, we can proceed for the other λ_r's which are listed in Table 1, together with the radii and the occupation numbers, for the first sixteen shells.

For large values of l the asymptotic behaviour of (18) is easily obtained substituting in (20) $\mathrm{Si}\, \lambda z$ by $\pi/2$, i.e.

$$F_\lambda(x) \underset{(x \to \infty)}{\approx} \frac{1}{\lambda x^2} \frac{\pi}{2} \int\limits_{x-1}^{x+1} (1 + x^2 - z^2)\, dz = \frac{2\pi}{3\lambda x^2}.$$

---

[3]) See, e.g., ZIMAN, l.c., page 53.

Table 1

Radii, occupation numbers and λ_r parameters up to the 16th shell in a f.c.c. lattice (d = cube edge)

Shell	ϱ_r/d	ν_r	λ_r
1	$1/\sqrt{2}$	12	4.3749
2	1	6	6.1871
3	$\sqrt{3}/\sqrt{2}$	24	7.5777
4	$\sqrt{2}$	12	8.7498
5	$\sqrt{5}/\sqrt{2}$	24	9.7827
6	$\sqrt{3}$	8	10.7164
7	$\sqrt{7}/\sqrt{2}$	48	11.5752
8	2	6	12.3742
9	$3/\sqrt{2}$	36	13.1247
10	$\sqrt{5}$	24	13.8347
11	$\sqrt{11}/\sqrt{2}$	24	14.5100
12	$\sqrt{6}$	24	15.1553
13	$\sqrt{13}/\sqrt{2}$	72	15.7741
14	$\sqrt{15}/\sqrt{2}$	48	16.9440
15	$2\sqrt{2}$	12	17.4999
16	$\sqrt{17}/\sqrt{2}$	48	18.0385

Then

$$\xi_l \gtrsim \frac{v}{12\,\pi\, m\, s^2}\left(\sum_r f_r\,\nu_r\,\varrho_r\right)\Big/ l^2 \tag{21}$$

which has just the asymptotic form of the elastic solution (1) with coefficient α expressed in terms of the microscopic properties of the lattice:

$$\alpha = \frac{1}{12\,\pi\, m\, s^2}\left(\sum_r f_r\,\nu_r\,\varrho_r\right).$$

ESHELBY [9] has shown that a reticular distorsion like that given by (1) corresponds, in a freely bounded crystal, to the volume change

$$\delta V = 4\,\pi\,\alpha\, v\cdot 3\,\frac{1-\sigma}{1+\sigma},$$

where σ is the Poisson ratio. Then the relative volume variation for a crystal with a vacancy concentration $c = n/N$ is given by

$$\frac{\delta V}{V} = c\,\frac{1-\sigma}{1+\sigma}\,\frac{1}{m\, s^2}\sum_r f_r\,\nu_r\,\varrho_r\,. \tag{22}$$

Finally, we want to point out that the presence of the reticular mass m in (18) is indeed apparent (as required by the solution of a static problem) because of

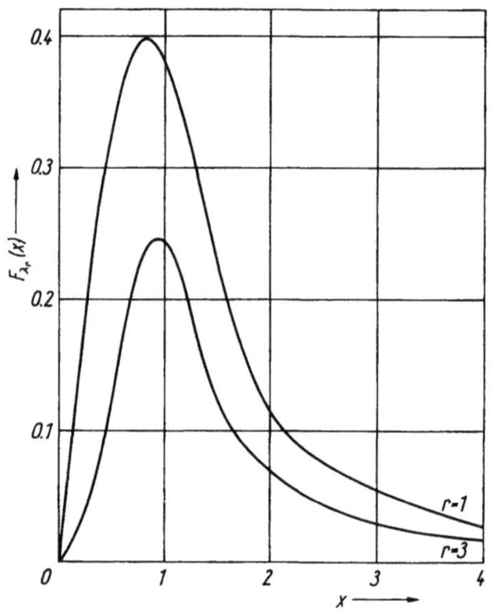

Fig. 1. $F_{\lambda_r}(x)$ functions plotted against x for $r = 1; 3$ in a f.c.c. lattice

the relation

$$-\frac{1}{2}\sum_{\boldsymbol{h}} (\boldsymbol{u}_q \cdot \boldsymbol{h})^2\, g_{\boldsymbol{h}} = m\, \kappa_q^2$$

deriving from equations (8) and (8′) when we take into account the scalar nature of the $g_{\boldsymbol{h}}$'s in a spherically symmetric crystal.

5. Numerical Results

In Tables 3 and 4 $F_\lambda(x)$ functions for a f.c.c. lattice up to the 16th shell are given; in Fig. 1 we have plotted, as an example, $F_{\lambda_1}(x)$ and $F_{\lambda_3}(x)$.

For an argon crystal, assuming true the (12,6) interatomic potential function [8]

$$V(r) = A\, r^{-12} - B\, r^{-6}$$

with $A = 1.63 \cdot 10^{-7}$ erg Å^{12}, $B = 1.05 \cdot 10^{-10}$ erg Å^6, we have calculated the displacement distribution around

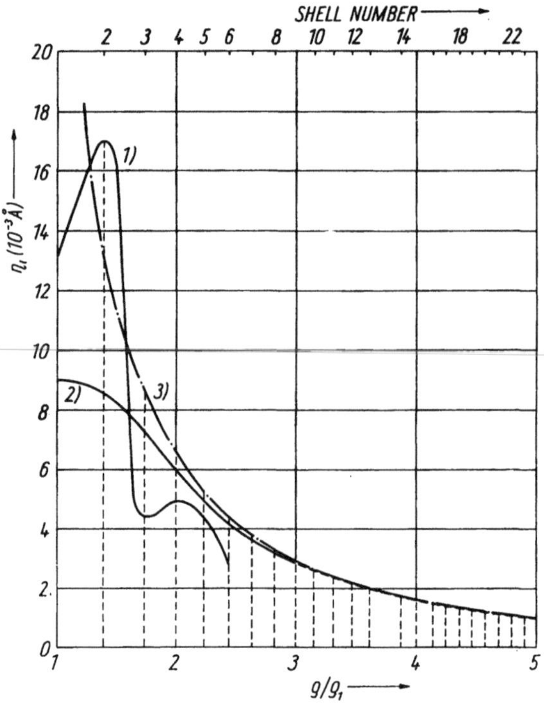

Fig. 2. Lattice displacements in solid argon (80 °K)

Curve *1* : NARDELLIS and CHIAROTTIT results
Curve *2* : Present calculation
Curve *3* : Elastic extrapolation of the present work

a vacancy at 80 °K, taking $\varrho_1 = 3.8555$ [10]. The results are quoted in Table 2 together with the corresponding values obtained by NARDELLI and CHIAROTTI [8]. In Fig. 2 the two solutions and the elastic one, given by (21), are plotted.

Our solution (curve 1) appears to be more regular than NARDELLIS and CHIAROTTIS result (curve 2). This is partially due to the isotropic model we have used: a little effect of lattice geometry could be taken into account be evaluating the rigorous solution (14). However, considering that a Debyes model is a good approximation for f.c.c. lattices, we must not expect any way a result so irregular as curve 1.

Table 2

Displacements in 10^{-3} Å for an argon crystal lattice at 80 °K

Shells	1	2	3	4	5	6
Present calculation	8.94	8.62	7.33	5.99	4.90	4.12
NARDELLI and CHIAROTTI	13.20	17.00	4.42	4.93	4.45	2.54

Table 3

$F_\lambda(x)$ functions up to the 8th shell of neighbours, in a f.c.c. lattice, for values of x in the range (0—15)

x \ r	1	2	3	4	5	6	7	8
0.000	0.0000	0.0000	0.0000	0.0000	0.0000	0.0000	0.0000	0.0000
0.125	0.0988	0.0414	0.0142	0.0203	0.0328	0.0370	0.0321	0.0228
0.250	0.1908	0.0860	0.0390	0.0447	0.0588	0.0620	0.0552	0.0453
0.375	0.2701	0.1349	0.0790	0.0764	0.0791	0.0755	0.0696	0.0652
0.500	0.3318	0.1855	0.1310	0.1152	0.1018	0.0908	0.0851	0.0826
0.625	0.3732	0.2322	0.1844	0.1564	0.1327	0.1181	0.1098	0.1032
0.750	0.3936	0.2681	0.2260	0.1918	0.1672	0.1523	0.1405	0.1295
0.875	0.3943	0.2869	0.2463	0.2121	0.1915	0.1768	0.1632	0.1520
1.000	0.3780	0.2856	0.2423	0.2117	0.1939	0.1783	0.1653	0.1560
1.125	0.3488	0.2657	0.2188	0.1920	0.1736	0.1577	0.1465	0.1377
1.250	0.3115	0.2323	0.1849	0.1610	0.1418	0.1280	0.1184	0.1098
1.375	0.2704	0.1932	0.1505	0.1290	0.1123	0.1023	0.0944	0.0876
1.500	0.2297	0.1556	0.1221	0.1038	0.0923	0.0854	0.0790	0.0746
1.625	0.1924	0.1251	0.1019	0.0875	0.0804	0.0742	0.0690	0.0652
1.750	0.1605	0.1037	0.0884	0.0775	0.0714	0.0646	0.0600	0.0556
1.875	0.1348	0.0905	0.0787	0.0696	0.0620	0.0557	0.0513	0.0474
2.000	0.1152	0.0826	0.0703	0.0614	0.0531	0.0483	0.0446	0.0420
2.250	0.0904	0.0705	0.0547	0.0462	0.0420	0.0388	0.0362	0.0335
2.500	0.0764	0.0552	0.0435	0.0383	0.0348	0.0311	0.0285	0.0268
3.000	0.0546	0.0368	0.0309	0.0260	0.0241	0.0217	0.0199	0.0186
4.000	0.0302	0.0206	0.0171	0.0150	0.0131	0.0121	0.0113	0.0104
5.000	0.0187	0.0132	0.0107	0.0094	0.0086	0.0077	0.0072	0.0067
6.000	0.0133	0.0089	0.0076	0.0065	0.0058	0.0054	0.0050	0.0046
8.000	0.0074	0.0050	0.0042	0.0036	0.0033	0.0030	0.0028	0.0026
10.000	0.0046	0.0033	0.0027	0.0024	0.0021	0.0020	0.0018	0.0017
15.000	0.0023	0.0015	0.0012	0.0011	0.0009	0.0009	0.0008	0.0007

Table 4

$F_\lambda(x)$ functions from the 9th up to the 16th shell of neighbours, in a f.c.c. lattice, for values of x in the range (0—15)

x \ r	9	10	11	12	13	14	15	16
0.000	0.0000	0.0000	0.0000	0.0000	0.0000	0.0000	0.0000	0.0000
0.125	0.0145	0.0103	0.0104	0.0133	0.0170	0.0212	0.0205	0.0182
0.250	0.0374	0.0333	0.0320	0.0321	0.0323	0.0313	0.0302	0.0293
0.375	0.0624	0.0600	0.0570	0.0534	0.0496	0.0442	0.0428	0.0420
0.500	0.0805	0.0774	0.0737	0.0703	0.0676	0.0635	0.0614	0.0592
0.625	0.0968	0.0913	0.0872	0.0843	0.0820	0.0771	0.0746	0.0725
0.750	0.1204	0.1136	0.1083	0.1035	0.0990	0.0912	0.0883	0.0858
0.875	0.1437	0.1368	0.1304	0.1243	0.1191	0.1110	0.1074	0.1040
1.000	0.1483	0.1411	0.1346	0.1292	0.1248	0.1167	0.1130	0.1098
1.125	0.1295	0.1224	0.1168	0.1121	0.1075	0.0996	0.0964	0.0938
1.250	0.1026	0.0971	0.0925	0.0882	0.0843	0.0784	0.0759	0.0734
1.375	0.0828	0.0790	0.0753	0.0721	0.0696	0.0652	0.0631	0.0614
1.500	0.0711	0.0676	0.0645	0.0621	0.0598	0.0552	0.0535	0.0519
1.625	0.0612	0.0577	0.0551	0.0525	0.0500	0.0464	0.0450	0.0436
1.750	0.0518	0.0491	0.0468	0.0445	0.0429	0.0403	0.0391	0.0380
1.875	0.0449	0.0429	0.0409	0.0394	0.0380	0.0352	0.0341	0.0331
2.000	0.0402	0.0380	0.0363	0.0347	0.0331	0.0307	0.0297	0.0288
2.250	0.0312	0.0297	0.0283	0.0272	0.0263	0.0245	0.0236	0.0229
2.500	0.0257	0.0242	0.0231	0.0220	0.0210	0.0198	0.0191	0.0186
3.000	0.0178	0.0168	0.0160	0.0152	0.0147	0.0137	0.0133	0.0128
4.000	0.0100	0.0094	0.0090	0.0087	0.0083	0.0077	0.0075	0.0072
5.000	0.0064	0.0060	0.0058	0.0055	0.0053	0.0049	0.0048	0.0046
6.000	0.0044	0.0042	0.0040	0.0038	0.0037	0.0034	0.0033	0.0032
8.000	0.0025	0.0023	0.0022	0.0021	0.0021	0.0019	0.0019	0.0018
10.000	0.0016	0.0015	0.0014	0.0014	0.0013	0.0012	0.0012	0.0012
15.000	0.0007	0.0007	0.0006	0.0006	0.0006	0.0005	0.0005	0.0005

Acknowledgements

Thanks are due to FIAT Laboratories Management for permission to publish this work and to Dr. C. P. GALOTTO for his useful suggestions and discussions.

References

[1] H. B. HUNTINGTON and F. SEITZ, Phys. Rev. **61**, 315 (1942).
[2] F. SEITZ, Rev. mod. Phys. **18**, 384 (1946).
[3] C. W. TUCKER and J. B. SAMPSON, Acta metall. **2**, 433 (1954).
[4] H. B. HUNTINGTON, Acta metall. **2**, 554 (1954).
[5] G. J. DIENES, Phys. Rev. **86**, 228 (1956).
[6] H. KANZAKI, J. Phys. Chem. Solids **2**, 24 (1957).
[7] L. TEWORDT, Phys. Rev. **109**, 61 (1958).
[8] G. F. NARDELLI and A. CHIAROTTI, Nuovo Cimento **18**, 1053 (1960).
[9] J. D. ESHELBY, J. appl. Phys. **25**, 255 (1954).
[10] I. J. ZUCKER, Phil. Mag. **3**, 987 (1958).

(Received May 11, 1964)

phys. stat. sol. 6, 401 (1964)

Институт физики металлов АНСССР, Свердловск

К квантовой теории термогальваномагнитных явлений в металлах и полупроводниках (I)

П. С. Зырянов

Найдены формулы для недиссипативных плотностей потоков тепла и тока проводимости в квантующем магнитном поле при учёте спина электрона. В приближении упругого рассеяния найдены выражения и для диссипативных плотностей потоков тепла и тока проводимости.

Полученные результаты позволяют рассчитать любой термогальваномагнитный эффект в условиях сильного квантующего магнитного поля $\Omega\tau \gg 1$ (Ω циклотронная частота электронов, а τ их время релаксации).

Formulas are obtained for the nondissipative density of thermal and electric currents in the high magnetic field quantum limit, electron-spin being taken into account. Expressions are also obtained for the disspation density of these currents on the basis of the elastic scattering approximation. The results obtained allow any thermogalvanomagnetic effect to be calculated in the high magnetic field quantum limit $\Omega\tau \gg 1$ (Ω cyclotron resonance frequency, τ relaxation-time).

1. Введение

Квантовые эффекты в термогальваномагнитных явлениях представляют большой интерес, поскольку изучение их позволяет получать информацию об энергетическом спектре электронов и о механизмах релаксации в металлах и полупроводниках. Именно поэтому эти явления привлекают внимание многих как экспериментаторов так и теоретиков.

В последнее время было предпринято несколько попыток построения квантовой теории термомагнитных явлений. Однако при внимательном анализе такие попытки оказались противоречащими либо достоверным результатам классической теории либо основным положениям термодинамики. Краткое обсуждение эрих противоречий содержится в работе Образцова [1].[1])

Подчеркнем здесь лишь специфику возникающих трудностей при вычислении плотностей потока тепла и тока проводимости в условиях квантующего магнитного поля.

В классической области магнитных полей $\hbar\Omega \ll T$ (T температура в энергетических единицах) объемная плотность потока заряда $\boldsymbol{j}$, вычисленного с помощью функции распределения (решение ур. Больцмана), как известно, совпадает с током проводимости $\boldsymbol{j}_{np}$, а объемная плотность потока энергии $\boldsymbol{Q}$ сразу же позволяет найти поток тепла $\boldsymbol{Q}_T$

$$\boldsymbol{Q}_T = \boldsymbol{Q} - \frac{\zeta}{e}\,\boldsymbol{j}_{np} \tag{1}$$

(ζ химический потенциал).

[1]) Пользуюсь случаем выразить признательность Образцову за ознакомление с этой работой до её опубликования.

В квантующем магнитном поле ситуация радикальным образом изменяется. Действительно, объемная плотность потока заряда $\boldsymbol{j}$ вычисленная с помощью матрицы плотности, уже не совпадает с плотностью тока проводимости $\boldsymbol{j}_{np}$, поскольку возникает диамагнетизм Ландау, приводящий в пространственно неоднородных системах к появлению „молекулярного" тока $c \operatorname{rot} \boldsymbol{M}$ ($\boldsymbol{M}$ намагниченность электронов проводимости) наряду с током проводимости, поэтому

$$\boldsymbol{j} = \boldsymbol{j}_{np} + c \operatorname{rot} \boldsymbol{M} . \tag{2}$$

Аналогичным образом в объемной плотности потока энергии наряду с потоком тепла $\boldsymbol{Q}_T$ возникает поток магнитной энергии $\boldsymbol{Q}_M$

$$\boldsymbol{Q} - \frac{\zeta}{e} \boldsymbol{j}_{np} = \boldsymbol{Q}_T + \boldsymbol{Q}_M . \tag{3}$$

Именно непонимание этой специфики и отождествление $\boldsymbol{j}$ с $\boldsymbol{j}_{np}$, а $\boldsymbol{Q} - \frac{\zeta}{e} \boldsymbol{j}_{np}$ с $\boldsymbol{Q}_T$ приводило к парадоксальным результатам противоречащим либо классической теории либо основным положениям термодинамики.

В книге Ландау и Лифшица [2] (см. § 25) указано, что для определения гальванотермомагнитных явлений необходимы объемные плотности потока тепла и тока проводимости. В связи с этим и возникают две задачи, а именно: а) вычисление с помощью матрицы плотности объемных плотностей потоков заряда $\boldsymbol{j}$ и энергии $\boldsymbol{Q}_T$, обусловленных пространственными неоднородностями температуры и химического потенциала; б) выделение из них тока проводимости $\boldsymbol{j}_{np}$ и потока тепла $\boldsymbol{Q}_T$, необходимых для построения теории термогальваномагнитных явлений.

В работах Силина и автора [3] были решены эти задачи для системы электронов без учёта их собственного магнитного момента. В условиях квантующего магнитного поля H, когда $\hbar \Omega \gtrapprox \mu H \gg T$ (μ магнитный момент электрона проводимости) зеемановским расщеплением уровней Ландау пренебрегать нельзя.

В условиях сильного магнитного поля, когда частота Лармора значительно превосходит частоту столкновений электронов с рассеивателями возникает малый параметр $(\Omega \tau)^{-1} \ll 1$. Все кинетические коэффициенты можно разложить в ряд по этому параметру. Оказывается, что недиагональные компоненты тензоров кинетических коэффициентов отличны от нуля в нулевом приближении по $(\Omega \tau)^{-1}$ и характеризуют недиссипативные (бесстолкновительные) потоки. Диагональные компоненты этих тензоров пропорциональны $(\Omega \tau)^{-1}$ и характеризуют диссипативные (столкновительные) потоки. Приступим в вычислению этих потоков.

2. Бесстолкновительные потоки

Гамильтонова функция электрона имеет вид

$$\left.\begin{aligned} \hat{H} &= \hat{H}_0 + \hat{H}', \\ \hat{H}_0 &= \left[\hat{p}_x^2 + \left(\hat{p}_y - \frac{e}{c} H x\right)^2 + \hat{p}_z^2\right] \Big/ 2m - e x \mathscr{E}_x + \mu \hat{\sigma} H, \end{aligned}\right\} \tag{4}$$

где $\hat{H}' = V(\boldsymbol{r})$ потенциальная энергия взаимодействия электрона с рассеивателями,

$$\hat{H}_0 \, |n, p_z, x_0, \sigma\rangle = E(n, p_z, x_0, \sigma) | \, n_0 \, p_z, x_0, \sigma\rangle ,$$

$$E(n, p_z, x_0, \sigma) = \hbar \, \Omega \, (n + 1/2) + p_z^2/2 \, m - e \, x_0 \, \mathscr{E}_x + \frac{m \, c^2 \, \mathscr{E}_x^2}{2 \, H^2} + \mu \, \sigma \, H ,$$

$$x_0 = \frac{c \, p_y}{e \, H} + \frac{m \, c^2 \, \mathscr{E}_x}{e \, H^2} ,$$

$$|n, p_z, x_0, \sigma\rangle = (L_y \, L_z \, \alpha)^{-1/2} \exp \, (i \, [p_y \, y + p_z \, z]/\hbar) \, \Phi_n \left(\frac{x - x_0}{\alpha}\right) S(\sigma) ,$$

$$\hat{\sigma} \, |S(\sigma)\rangle = \sigma |s(\sigma)\rangle , \quad \sigma = \pm 1 ,$$

$\Phi_n(x)$ нормированная собственная функция осциллятора, L_y, L_z характеризуют нормировочный объем, $\alpha^2 = \frac{c \, \hbar}{|e| \, H}$. Выписанные формулы позволяют найти квантовомеханические средние плотности потоков заряда и энергии в состоянии $|n, p_z, x_0, \sigma\rangle$ согласно [4] (см. § 114) в следующем виде

$$j_y(n, p_z, x_0, \sigma) = (\alpha \, L_y \, L_z)^{-1} \left\{ e \, \Omega \, (x_0 - x) - \frac{e \, c \, \mathscr{E}_x}{H} - c \, \sigma \, \mu \, \frac{\partial}{\partial x} \right\} \Phi_n^2 \left(\frac{x - x_0}{\alpha}\right), \tag{5}$$

$$Q_y(n, p_z, x_0, \sigma) = \frac{1}{e} \, j_y(n, p_z, x_0, \sigma) \, E(n, p_z, x_0, \sigma) \, . \tag{6}$$

Для вычисления макроскопических объемных плотностей потока заряда и энергии в нулевом приближении по рассеянию электронов (бесстолкновительные потоки!) необходимо усреднить (5), (6) с функцией распределения $f(n, p_z, x_0, \sigma)$ характеризующей заполнение состояний.

В рассматриваемом случае сильных магнитных полей когда ларморовская частота электронов значительно превышает частоту столкновений в нулевом приближении столкновениями можно пренебречь. Тогда очевидно, что в энергетическом представлении матрица плотности диагональна, именно поэтому можно говорить о функции распределения частиц по состояниям, которая может рассматриваться как функция собственных значений оператора Гамильтона и других операторов, коммутирующих с гамильтонианом и образующих полный набор. В частности эту функцию можно записать в виде

$$f_0 \left(\frac{E(n, p_z, x_0, \sigma) - \zeta(x_0)}{T(x_0)}\right), \tag{7}$$

где f_0 функция Ферми.

В этой формуле $\zeta(x_0)$ и $T(x_0)$ приняты зависящими от x_0. Такая зависимость соответствует неоднородному распределению частиц вдоль x поперёк магнитного поля. Покажем, что функция распределения (3), описывающая локально равновесное распределение электронов приближенно обращает в нуль интеграл столкновений. Действительно, в случае упругого рассеяния на беспорядочно распределенных тяжелых рассеивателях интеграл столкновений в представлении $|n, p_z, x_0, \sigma\rangle$ имеет вид

$$J[f(n, p, x_0, \sigma)] = \sum_{n', q} \frac{2\pi}{\hbar} |C_q|^2 F_{n',n}(\alpha^2 q_\perp^2/2) \times$$
$$\times \delta [E(n', p_z + \hbar q_z, x_0 + \alpha^2 q_y, \sigma) - E(n, p_z, x_0, \sigma)] \times$$
$$\times [f(n', p_z + \hbar q_z, x_0 + \alpha^2 q_y, \sigma) - f(n, p_z, x_0, \sigma)], \tag{8}$$

где $q_\perp^2 = q_x^2 + q_y^2$, $\alpha^2 = \frac{c\hbar}{|e| H}$, c_q фурьекомпонентапотенциальной энергии взаимодействия электронов с рассеивателями, а

$$F_{n',n}(\alpha^2 q_\perp^2/2) = \left| \int \frac{dx}{\alpha} \Phi_{n'} \left(\frac{x - x_0 - \alpha^2 q_y}{\alpha}\right) \exp(i q_x x) \Phi_n \left(\frac{x - x_0}{\alpha}\right)\right|^2 .$$

В случае слабых пространственных неоднородностей, когда

$$\gamma = \frac{\alpha^2 q_y}{L} = \left(\frac{\hbar q_y}{\bar{p}}\right)\left(\frac{r_L}{L}\right) \ll 1 ,$$

($L \sim f \left|\frac{\partial f}{\partial x_0}\right|^{-1}$, $\bar{p}$ средний импульс электрона, r_L радиус Лармора) интеграл столкновений можно лазрожить в ряд по γ

$$J[f(n, p_z, x_0, \sigma)] = \sum_{n', q} \frac{2\pi}{\hbar} |C_q|^2 F_{n',n}(\alpha^2 q_\perp^2/2) \times$$
$$\times \Big\{\delta [E(n', p_z + \hbar q_z, \sigma) - E(n, p_z, \sigma)] \times$$
$$\times \left[f(n', p_z + \hbar q_z, x_0, \sigma) - f(n, p_z, x_0, \sigma) + \frac{(\alpha^2 q_y)^2}{2} \frac{\partial^2}{\partial x_0^2} f(n', p_z + \hbar q_z, x_0\, \sigma)\right] +$$
$$+ (\alpha^2 q_y)^2 e \mathcal{E}_x \frac{\partial}{\partial E(n, p_z, \sigma)} \delta [E(n', p_z + \hbar q_z, \sigma) - E(n, p_z, \sigma)] \times$$
$$\times \frac{\partial}{\partial x_0} f(n', p_z + \hbar q_z, x_0, \sigma) + \cdots \Big\}, \tag{9}$$

где $E(n, p_z, \sigma) = \hbar \Omega (n + 1/2) + p^2/2m + \sigma \mu H$.

С точностью до членов (r_L/L) функция распределения (3) действительно обращает в нуль интеграл столкновений. Заметим, что именно благодаря слабым пространственным неоднородностям в системе, когда температура и химический потенциал остаются практически постоянными на радиусе Лармора $r_L \sim \lambda_D$ (λ_D длина волны Деброгля) различием между координатой электрона x и x_0 проекцией центра ларморовой орбиты на ось x можно пренебрегать, поэтому значения T и ζ в точке x могут быть заменены их значениями в точке x_0. Исходя из условия $(r_L/L) \ll 1$ полагаем

$$\zeta(x_0) = \zeta_0 + x_0 \nabla_x \zeta ,$$
$$T(x_0) = T_0 + x_0 \nabla_x T ,$$

где $\nabla_x \zeta, \nabla_x T$ малые величины. Тогда вместо (7) можно записать

$$f_0(n, p_z, x_0, \sigma) = f_0 \left(\frac{E(n, p_z, \sigma) - \zeta_0}{T_0}\right) +$$
$$+ f_0 (1 - f_0) \left[\nabla_x \zeta + \frac{E(n, p_z, \sigma) - \zeta_0}{T_0} \nabla_x T\right] \frac{x_0}{T_0} + \cdots . \tag{10}$$

Усредняя (5), (6) с функцией распределения (10) находим

$$j_y = -\frac{c e N}{H} \mathscr{E}_x + \frac{c}{H T_0} \frac{1}{(2\pi\alpha)^2 \hbar} \sum_{n,\sigma} [E(n,\sigma) - 2\sigma\mu H] \times$$
$$\times \int dp_z f_0 (1 - f_0) \left[\nabla_x \zeta + \frac{E(n, p_z, \sigma) - \zeta_0}{T_0} \nabla_x T\right], \tag{11}$$

$$Q_y = -\frac{c N}{H} \overline{(E + E_\perp)} \mathscr{E}_x + \frac{c}{e H T_0} \frac{1}{(2\pi\alpha)^2 \hbar} \sum_{n,\sigma} [E(n,\sigma) - 2\sigma\mu H] \times$$
$$\times \int dp_z f_0(1 - f_0) E(n, p_z, \sigma) \left[\nabla_x \zeta + \frac{E(n, p_z, \sigma) - \zeta_0}{T_0} \nabla_x T\right], \tag{12}$$
$$E(n,\sigma) = \hbar \Omega (n + 1/2) + \sigma \mu H ,$$
$$N\, \overline{(E + E_\perp)} = \frac{1}{(2\pi\alpha)^2 \hbar} \sum_{n,\sigma} \int dp_z [E(n, p_z, \sigma) + E(n,\sigma)] f_0 . \tag{13}$$

Формулы (11) и (12) решают первую задачу — определяют объемные плотности потоков заряда и энергии.

Для решения второй задачи — нахождения объемной плотности потока тепла и тока проводимости выразим коэффициенты перед $\mathscr{E}_x$, $\nabla_x \zeta$ и $\nabla_x T$ в (11), (12) через термодинамические функции электронного газа энтропию S, намагниченность M и плотность электронов N,

$$S = -\left(\frac{\partial \Omega}{\partial T}\right)_{\zeta, H}, \quad M = -\left(\frac{\partial \Omega}{\partial H}\right)_{\zeta, T}, \quad N = -\left(\frac{\partial \Omega}{\partial \zeta}\right)_{H, T},$$

где термодинамический потенциал

$$\Omega(\zeta, T, H) = -\frac{T}{(2\pi\alpha)^2 \hbar} \sum_{n,\sigma} \int dp_z \ln\left[1 + \exp\left(\frac{\zeta - E(n, p_z, \sigma)}{T}\right)\right]. \tag{14}$$

В результате несложных вычислений находим

$$j_y = (j_{np})_y + c(\mathrm{rot}\, M)_y , \tag{15}$$

$$Q_y = \left\{\frac{c}{e}\left[\zeta_0(\mathrm{rot}\, M)_y + T_0 \left(\frac{\partial}{\partial T_0}\right)_{\zeta_0} \int_{-\infty}^{\zeta_0} d\zeta\, (\mathrm{rot}\, M)_y\right] + c M \mathscr{E}_x\right\} + \frac{\zeta_0}{e} (j_{np})_y + (Q_T)_y, \tag{16}$$

$$(\mathrm{rot}\, M)_y = -\frac{\partial M}{\partial \zeta} \nabla_x \zeta - \frac{\partial M}{\partial T} \nabla_x T ,$$

$$(j_{np})_y = -\frac{c e N}{H}\left(\mathscr{E}_x - \frac{1}{e} \nabla_x \zeta\right) + \frac{c}{H} S \nabla_x T , \tag{17}$$

$$(Q_T)_y = -\frac{c}{H} T_0 S\left(\mathscr{E}_x - \frac{1}{e} \nabla_x \zeta\right) + \frac{c}{e H} \int_{-\infty}^{\zeta_0} d\zeta\, T_0 \left(\frac{\partial S}{\partial T_0}\right)_\zeta \nabla_x T . \tag{18}$$

Как следует из (15) объемная плотность потока заряда складывается из тока проводимости и „молекулярного“ тока $c(\mathrm{rot}\, M)_y$, обусловленного пространственной неоднородностью диа и парамагнитной восприимчивостей. Как известно [2] (см. § 29), вклад в перенос заряда через поперечное сечение проводника вносит лишь ток проводимости. Поток энергии (16) разбивается на несколько слагаемых имеющих различный физический смысл. Слагаемое $c M \mathcal{E}_x$ описывает вклад в вектор Пойнтинга от намагничения $\frac{c}{4\pi}[\mathcal{E} \times (\boldsymbol{H} - \boldsymbol{B})]_y$.

Далее, выражение, содержащее ротор намагничения, так же соответствует плотности потока магнитной энергии, которая, как известно, определяется с точностью до ротора произвольного вектора. Так же как и в случае „молекулярного" тока здесь можно говорить об обращении в нуль вклада от фигурной скобки в (16) в полный перенос энергии через поперечное сечение проводника. Слагаемое $\frac{\zeta_0}{e}(j_{np})_y$ описывает поток энергии, обусловленный переносом частиц (ток проводимости) и, наконец последнее слагаемое является объемной плотностью потока тепла (кинетической энергии хаотического движения).

Определенные формулами (17) и (18) объемные плотности тока проводимости и потока тепла не зависят от намагничения и его градиентов. Наличие таких слагаемых в (17) и (18) привело бы к противоречию с законом возрастания энтропии. В этом можно убедиться с помощью доказательства приведенного в [2] (см. § 25), заменив там градиент давления (плотности) градиентом намагничения (магнитного поля). Такая замена возможна, если вектор намагничения параллелен магнитному полю.

Кинетические коэффициенты в (17) и (18) удовлетворяют как соотношению Эйнштейна так и принципу симметрии Онсагера, в то время как в (11) и (12) или в (15) и (16) соотношение Эйнштейна не выполняется из-за возникновения тока, обусловленного пространственными неоднородностями намагничения электронов проводимости.

3. Диссипативные потоки

Формулы (16), (17) позволяют вычислить лишь дифференциальную термоэлектродвижущую силу в изотермических условиях в нулевом по рассеянию приближении. Для вычисления других термогальваномагнитных коэффициентов необходимы и диссипативные плотности потока тепла и тока проводимости.

В случае медленных процессов в системе, когда характерное время изменения функции распределения ($t^0 \sim f \left|\frac{\partial f}{\partial t}\right|^{-1}$) велико в сравнении с временем релаксации электронов τ

$$\frac{\partial}{\partial t} f(n, p_z, x_0, \sigma) = J[f(n, p_z, x_0, \sigma)] . \tag{19}$$

В системах со слабыми пространственными неоднородностями $(r_L/L) \ll 1$, $f(n, p_z, x_0, \sigma)$ можно искать в виде ряда по (r_L/L). В нулевом по (r_L/L) приближении $f_0(n, p_z, x_0, \sigma)$ согласно (7) удовлетворяет ур. (19). В следующем приближении используя (9), (10) из (19) с точностью до линейных по $\mathscr{E}_x$ и $\nabla_x \zeta$ и $\nabla_x T$ слагаемых получаем

$$\frac{\partial}{\partial t} f(n, p_z, x_0, \sigma) = -\frac{\partial}{\partial x_0}\Big\{-\sum_{n',q} \frac{2\pi}{\hbar} |C_q|^2 F_{n',n}(\alpha^2 q_\perp^2/2) \times$$
$$\times \delta[E(n', p_z + \hbar q_z, \sigma) - E(n, p_z, \sigma)] \times$$
$$\times \frac{1}{2}(\alpha^2 q_y)^2 f_0(1 - f_0)\left[\nabla_x \zeta + \frac{E(n, p_z, \sigma) - \zeta_0}{T_0} \nabla_x T\right] + \cdots\Big\}\,. \quad (20)$$

Этого приближения вполне достаточно для отыскания объемных плотностей потоков различных физических величин.

Действительно, суммируя в (20) по n, p_z, σ найдем уравнение непрерывности

$$\frac{\partial}{\partial t} n(x_0) = -\frac{\partial}{\partial x_0}(j_{np}/e)_y \quad (21)$$

в котором плотность тока проводимости

$$(j_{np})_x = \frac{1}{(2\pi)^2}\left(\frac{\alpha}{2\pi\hbar}\right)^2 \frac{e}{T_0} \sum_{n, n', \sigma} \int dp_z\, d\boldsymbol{q}\, |C_q q_y|^2 F_{n',n}(\alpha^2 q_\perp^2/2) \times$$
$$\times \delta[E(n', p_z + \hbar q_z, \sigma) - E(n, p_z, \sigma)]\, f_0(1 - f_0)\left[\nabla_x \zeta + \frac{E(n, p_z, \sigma) - \zeta_0}{T_0} \nabla_x T\right]. \quad (22)$$

Если умножить (20) на $E(n, p_z, \sigma)$ и так же просуммировать по n, p_z, σ то найдем уравнение непрерывности для энергии

$$\frac{\partial}{\partial t} E(x_0) = -\frac{\partial}{\partial x_0}(Q)_x\,. \quad (23)$$

В этом уравнении плотность потока энергии

$$(Q_T)_x = (Q_x) - \frac{\zeta_0}{e}(j_{np})_x = \frac{1}{(2\pi)^2}\left(\frac{\alpha}{2\pi\hbar}\right)^2 \sum_{n, n', \sigma} \int dp_z\, d\boldsymbol{q}\, |C_q q_y|^2 \times$$
$$\times F_{n',n}(\alpha^2 q_\perp^2/2)\left(\frac{E(n, p_z, \sigma) - \zeta_0}{T_0}\right) \delta[E(n', p_z + \hbar q_z, \sigma) - E(n, p_z, \sigma)] \times$$
$$\times f_0(1 - f_0)\left[\nabla_x \zeta + \frac{E(n, p_z, \sigma) - \zeta_0}{T_0} \nabla_x T\right]. \quad (24)$$

Кинетические коэффициенты в (22), (24) удовлетворяют принципу симметрии Онсагера. Заметим, что коэффициент перед $\left(-\frac{1}{e}\nabla_x \zeta\right)$ в формуле для $(j_{np})_y$ отличается лишь формой записи от найденного в работе Адамса и Гольстейна [5] выражения для электропроводности. Это означает, что соотношение Эйнштейна для диссипативных потоков выполняется.

Таким образом формулы (17), (18) и (22), (24) позволяют вычислить любой термогальваномагнитный коэффициент в приближении упругого рассеяния. Без принципиальных затруднений такая методика вычисления диссипативных потоков легко переносится на случай неупругого рассеяния электронов [6].

В заключение заметим, что с помощью (20) можно без труда найти формулы для коэффициента вязкости электронного газа. Действительно, если имеется гидродинамический поток вдоль магнитного поля со скоростью $U_z(x_0)$, то в (20) в линейном приближении по $U_z(x_0)$ наряду с $\nabla_x \zeta$ и $\nabla_x T$ появится слагаемое $m\, p_z\, \nabla_x U_z(x_0)$. При умножении на p_z такого уравнения в полной аналогии с получением (21), (23) будем иметь уравнение баланса для p_z

$$\frac{\partial}{\partial t}\langle p_z(x_0)\rangle = -\frac{\partial}{\partial x_0}\Pi_{zx},$$

где z, x компонента тензора плотности потока импульса

$$\Pi_{z,x} = \frac{1}{(2\pi)^2}\left(\frac{\alpha}{2\pi\hbar}\right)^2 \frac{1}{T_0} \sum_{n',n,\sigma} \int dp_z\, d\boldsymbol{q}\, |C_q\, q_y|^2\, F_{n',n}(\alpha^2 q_\perp^2/2) \times$$
$$\times\, \delta\,[E(n', p_z + \hbar\, q_z, \sigma) - E(n, p_z, \sigma)] \times (m\, p_z^2)\, f_0(1 - f_0)\, \nabla_x U_z(x_0)\,.$$

Литература

[1] Ю. Н. Образцов, Физ. тверд. тела **6**, 414 (1964).
[2] Л. Д. Ландау и Е. М. Лифшиц, Электродинамика сплошных сред, Гостехиздат, Москва 1957.
[3] П. С. Зырянов и В. П. Силин, Ж. экспер. теор. Физ. **46**, 537 (1964); Физ. металлов и металловедение **17**, 614 (1964).
[4] Л. Д. Ландау и Е. М. Лифшиц, Квантовая механика, изд. 2-е, Физматиздат, Москва 1964 (§ 114).
[5] E. Adams and T. Holstein, J. Phys. Chem. Solids **10**, 254 (1959).
[6] П. С. Зырянов, Физ. тверд. тела **5**, 3011 (1963).

(Received May 11, 1964)

phys. stat. sol. 6, 409 (1964)

Institute of Physics, Polish Academy of Sciences, Warsaw

Thermodynamic Properties of Free Carriers in Semiconductors with Arbitrary Energy Bands

By

W. ZAWADZKI and J. KOŁODZIEJCZAK

Statistical properties, entropy, free energy, internal energy, and specific heat of free carriers in an energy band of arbitrary form are considered. All the properties are expressed in terms of generalized Fermi-Dirac functions. The general formulae are specified for spheroidal and spherical energy bands. It is shown that the nonparabolic conduction bands of III-V compounds may be regarded as a first approximation to general nonparabolic structures. The dependence of Fermi level on temperature, and the internal energy and specific heat are calculated numerically in the nondegenerate region. It is shown that nonparabolicity of the energy band strongly affects the properties discussed, for any degree of degeneracy.

Es werden die statistischen Eigenschaften von freien Ladungsträgern in einem Energieband beliebiger Form untersucht und die entsprechenden Beiträge zur Entropie, freien Energie, inneren Energie und spezifischen Wärme auf der Grundlage verallgemeinerter Fermi-Dirac-Funktionen berechnet. Die allgemeinen Ausdrücke werden auf rotationsellipsoidförmige und kugelförmige Bänder spezialisiert. Es wird gezeigt daß die nichtquadratischen $E(k)$-Abhängigkeiten der III—V-Verbindungen als erste Näherungen allgemeiner nichtquadratischer $E(k)$-Abhängigkeit angesehen werden können. Die Temperaturabhängigkeit der Fermigrenze sowie die innere Energie und spezifische Wärme werden für den nichtentarteten Fall numerisch berechnet. Es ergibt sich, daß die Abweichungen vom quadratischen $E(k)$-Gesetz die physikalischen Eigenschaften bei beliebigem Entartungsgrad stark beeinflussen.

1. General

First we shall consider an energy band of arbitrary form

$$\gamma(\varepsilon) = k^2 \Phi(\vartheta, \varphi) , \tag{1.1}$$

where ε is the energy of a carrier, k the absolute value of wave vector, ϑ, φ are angles in k-space. The arbitrary function $\Phi(\vartheta, \varphi)$ determines the shape of isoenergetic surface. The arbitrary function $\gamma(\varepsilon)$ gives the energy-wave number dependence. The only assumption in (1.1) is that the warping of a band is independent of energy. All the special cases of band structures can be obtained from (1.1).

Let us consider now the integral of an arbitrary function of energy described by (1.1).

Integrating by parts, we find

$$\frac{2}{(2\pi)^3} \int f_0\, \chi(\varepsilon)\, d_3k = I\left(\langle\chi\rangle - \left\langle \int\limits_0^\varepsilon \chi' d\xi \right\rangle\right), \tag{1.2}$$

where

$$I = \frac{1}{3}\,\frac{2}{(2\pi)^3} \int\limits_\Omega \Phi^{-3/2}(\vartheta, \varphi) \sin\vartheta\, d\vartheta\, d\varphi \tag{1.3}$$

and

$$\langle A\rangle = \int_0^\infty \left(-\frac{\partial f_0}{\partial \varepsilon}\right) \hat{A}\, \gamma^{3/2}(\varepsilon)\, d\varepsilon\,, \tag{1.4}$$

f_0 being the Fermi-Dirac distribution function.

The order of factors in the integrand (1.4) should be noted, because $\hat{A}$ may be an operator.

We have, e. g.,

$$\left\langle \int_0^\varepsilon \chi' d\xi \right\rangle = \int_0^\infty \left(-\frac{\partial f_0}{\partial \varepsilon}\right) \left[\int_0^\varepsilon \frac{d\chi(\xi)}{d\xi}\, \gamma^{3/2}(\xi)\, d\xi\right] d\varepsilon\,.$$

The bracket notation (1.4) has been widely applied in the theory of semiconductors with nonstandard energy bands [6, 7, 8, 9, 10, 13, 14, 15, 16, 17].

For systems obeying Maxwell-Boltzmann (M—B) statistics the integral (1.2) can be calculated simply using the relation $f_0 = -k_0 T \dfrac{\partial f_0}{\partial \varepsilon}$

$$\frac{2}{(2\pi)^3} \int f_0\, \chi(\varepsilon)\, d_3 k = I \left\langle \chi \frac{d}{dz} \right\rangle, \tag{1.5}$$

where $z = \varepsilon / k_0 T$ is the reduced energy.

The anisotropy of a band affects the I integral, while the nonparabolicity is related to the fact that $\gamma(\varepsilon)$ is not in general proportional to ε.

For small value of ε, however, we can use the expansion

$$\gamma(\varepsilon) = \gamma_1 \left(\varepsilon + \sum_{l=2}^{\infty} \lambda_l \varepsilon^l\right), \tag{1.6}$$

where

$$\gamma_1 = \left(\frac{d\gamma}{d\varepsilon}\right)_{\varepsilon=0}$$

and

$$\lambda_l = \frac{1}{l!} \frac{1}{\gamma_1} \left(\frac{d^l \gamma}{d\varepsilon_l}\right)_{\varepsilon=0}.$$

γ_1 and λ_l are in general temperature-dependent. If the expansion (1.6) is quickly convergent we may use the first nonparabolic approximation

$$\gamma(\varepsilon) = \gamma_1\, \varepsilon\, (1 + \lambda\, \varepsilon)\,. \tag{1.7}$$

Now we define L functions which will be very useful in following computations:

$$^{n}L_k^m(\eta, \beta) = \int_0^{\infty'} \left(-\frac{\partial f_0}{dz}\right) z^n \left(z + \sum_{l=2}^{\infty} \beta_l z^l\right)^m \left(1 + \sum_{l=2}^{\infty} l\, \beta_l z^{l-1}\right)^k dz\,, \tag{1.8}$$

where $\eta = \zeta / k_0 T$ is the reduced Fermi level and β denotes all the parameters β_l ($\beta_l = \lambda_l (k_0 T)^{l-1}$).

For bands described by (1.7) we have two-parametric functions

$$^{n}L_k^m(\eta, \beta) = \int_0^\infty \left(-\frac{\partial f_0}{\partial z}\right) z^n\, (z + \beta\, z^2)^m\, (1 + 2\,\beta\, z)^k\, dz\,, \tag{1.9}$$

where $\beta = \lambda\, k_0 T$.

These functions have been widely applied in the theory of semiconductors with nonparabolic bands [6, 7, 10, 15, 17].

For M—B statistics

$$ {}^{n}L_{k}^{m}(\eta, \beta) = e^{\eta}\, {}^{n}D_{k}^{m}(\beta) , $$

where

$$ {}^{n}D_{k}^{m}(\beta) = \int_{0}^{\infty} e^{-z} z^{n} (z + \beta z^{2})^{m} (1 + 2 \beta z)^{k} dz . \tag{1.10} $$

In strongly degenerate region ($\eta \gg 0$) the integral (1.4) has the asymptotic expansion [20, see also 21]:

$$ \langle A \rangle = 2 \sum_{l=0}^{\infty} \left(\frac{d^{2l} C}{d\varepsilon^{2l}} \right)_{\varepsilon=\zeta} (k_0 T)^{2l} \frac{\pi^{2l} |(2^{2l-1} - 1) B_{2l}|}{(2 l)!} , \tag{1.11} $$

where $C = \hat{A} \gamma^{3/2}(\varepsilon)$ and B_{2l} are the Bernoulli numbers. We shall note that all the following formulae (with marked exceptions) are valid both in terms of (1.8) and (1.9) L integrals.

Putting $\beta = 0$ we always transform the formulae to relations for a parabolic energy band.

2. Statistics of Free Carriers

The *density of free carriers* in a band can be obtained from (1.2) putting $\chi(\varepsilon) \equiv 1$:

$$ n = I \langle 1 \rangle , \tag{2.1} $$

and using the serial expansion (1.6) we find

$$ n = I(\gamma_1 k_0 T)^{3/2}\, {}^{0}L_{0}^{3/2}(\eta, \beta) . \tag{2.2} $$

It can be shown that in the first approximation to nonparabolic bands given by (1.7) in the nondegenerate region the carrier density is (see [18])

$$ n = \frac{3}{4} I(\gamma_1 k_0 T)^{3/2} \beta^{-1/2} \exp\left(\eta + \frac{1}{2\beta}\right) K_2\left(\frac{1}{2\beta}\right) , \tag{2.3} $$

where K_2 is the Macdonald function (or modified Bessel function [16, 27]).

Using asymptotic relation

$$ K_n(x) \underset{x \to \infty}{\to} \sqrt{\frac{\pi}{2 x}}\, e^{-x} , \tag{2.4} $$

we can obtain for $\beta \to 0$ the classical exponential formula for carrier density in parabolic band in M—B statistics region.

Let us define a *density-of-states mass* by the equation

$$ n = \frac{8}{3 \sqrt{\pi}} \left(\frac{2 \pi m_{ds} k_0 T}{h^2} \right)^{3/2} \int_{0}^{\infty} \left(- \frac{\partial f_0}{\partial z} \right) z^{3/2} dz . \tag{2.5} $$

Hence, by comparison with (2.2)

$$ m_{ds}(\zeta, T) = m_p(T) \left[\frac{{}^{0}L_{0}^{3/2}(\eta, \beta)}{{}^{0}L_{0}^{3/2}(\eta, 0)} \right]^{2/3} , \tag{2.6} $$

where

$$ m_p(T) = \frac{h^2}{8} \left(\frac{3 I}{\pi} \right)^{2/3} \gamma_1(T) . \tag{2.7} $$

Thus in general the density-of-states mass depends on the Fermi level and temperature. $m_p(T)$ is the density-of-states mass for a parabolic energy band. At $T = 0$, β is also zero and both masses are equal.

Using (1.10), (1.11), and (2.3), the density-of-states masses in nondegenerate and strongly degenerate conditions can readily be discussed.

In the general case the *Fermi level* dependence upon carrier density and temperature must be computed numerically on the base of (2.1), (2.2), or (2.3). However, in the case of strong degeneracy we have $-\frac{\partial f_0}{\partial \varepsilon} = \delta(\varepsilon - \zeta)$ (the Dirac delta function) and (2.1) becomes

$$n = I\,\gamma^{3/2}(\zeta, T) \quad \text{with} \quad \gamma^{3/2}(\zeta, T) = \gamma_1^{3/2}\,(\zeta + \lambda\,\zeta^2)^{3/2}\,. \tag{2.8}$$

Hence, using (2.7) the Fermi level is

$$\zeta = \frac{1}{2\,\lambda}\left[\sqrt{\Delta} - 1\right], \tag{2.9}$$

where

$$\Delta = 1 + \frac{1}{2}\left(\frac{3}{\pi}\right)^{2/3} \frac{\lambda\,\hbar^2}{m_p}\, n^{2/3}\,.$$

λ, γ_1 and m_p are all temperature-dependent.

The formula (2.9) is a generalization of relation for isotropic band given in [6b]. In the limit $\lambda \to 0$ the formula for a parabolic band can easily be obtained.

Now we shall discuss two differential relations for $\zeta(n, T)$. (2.1) may be rewritten

$$n - I\,\langle 1 \rangle = F(n, \zeta, T) \equiv 0\,. \tag{2.10}$$

If excess concentrations are present in a material, we have to deal with the quasi Fermi level defined by the relation (2.10).

We are going to give interrelation between the quasi Fermi level and excess carrier density at constant temperature:

$$\left(\frac{\partial n}{\partial \eta}\right)_T = -\frac{\partial F}{\partial \eta}\Big/\frac{\partial F}{\partial n} = I\,\frac{\partial}{\partial \eta}\,\langle 1 \rangle = n\,\left(\overline{\frac{d}{dz}}\right). \tag{2.11}$$

The relation $\frac{\partial f_0}{\partial \eta} = -\frac{\partial f_0}{\partial z}$ has been used here. The average value is defined

$$\bar{A} = \frac{\langle A \rangle}{\langle 1 \rangle}\,.$$

For small amounts of the additional concentrations

$$\eta - \eta_0 = \left(\overline{\frac{d}{dz}}\right)_{\eta_0}^{-1} \ln \frac{n}{n_0}\,.$$

For a system obeing M—B statistics $\frac{\partial}{\partial \eta}\langle 1 \rangle \equiv \langle 1 \rangle$, hence $\left(\overline{\frac{d}{dz}}\right)_{\text{M}-\text{B}} = 1$ regardless of the band structure.

By (2.11), (1.4), (1.6), and (1.7) we find for any degree of degeneracy

$$n\left(\frac{\partial \eta}{\partial n}\right)_T = \frac{2}{3}\,\frac{{}^0L_0^{3/2}(\eta, \beta)}{{}^0L_1^{1/2}(\eta, \beta)|}\,, \tag{2.12}$$

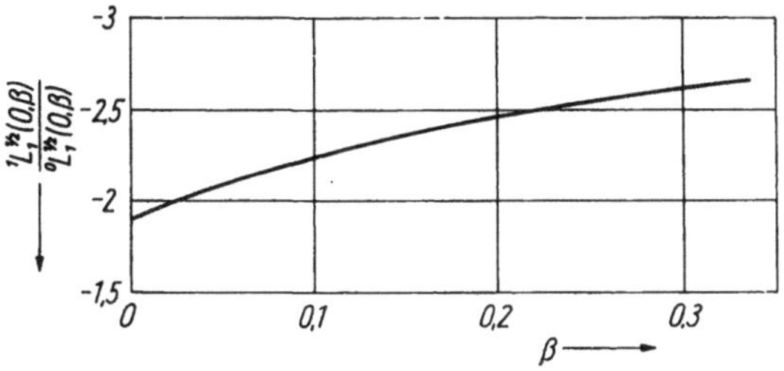

Fig. 1. The values of $\left(\frac{\partial \zeta}{\partial k_0 T}\right)_n$ *versus* β for the Fermi level passing the bottom of a band. For $\beta = 0$ the well known value -1.898 for a parabolic band is obtained

which is identical with the relation for a spherical band. Numerical values of (2.12) can be found in [10b]. Similarly we find

$$\left(\frac{\partial \zeta}{\partial T}\right)_n = -\frac{\partial F}{\partial T}\Big/\frac{\partial F}{\partial \zeta} = k_0\left(\eta - \frac{\left\langle z\frac{d}{dz}\right\rangle}{\left\langle\frac{d}{dz}\right\rangle}\right), \tag{2.13}$$

the relation $\frac{\partial f_0}{\partial T} = -\frac{\partial f_0}{\partial \varepsilon}\frac{\varepsilon - \zeta}{T}$ has been used here. The formula (2.13) is useful in highly degenerate region.

In terms of L integrals we have

$$\frac{\partial \zeta}{\partial k_0 T} = \eta - \frac{{}^1L_1^{1/2}(\eta, \beta)}{{}^0L_1^{1/2}(\eta, \beta)}\,. \tag{2.14}$$

For the Fermi level passing the bottom of conduction band

$$\left(\frac{\partial \zeta}{\partial k_0 T}\right)_{\zeta=0} = -\frac{{}^1L_1^{1/2}(0, \beta)}{{}^0L_1^{1/2}(0, \beta)}\,. \tag{2.15}$$

For a parabolic band ($\beta = 0$) this is a universal constant independent of the carrier density and temperature and is equal to -1.898 (i. e. for different carrier densities the curves $\zeta(T)$ have the same slope as they pass through the bottom of a band [22]). For a nonparabolic band, however, (2.15) is a function of β, thus it depends on temperature.

For M—B statistics

$$\left(\frac{\partial \zeta}{\partial k_0 T}\right)_{\text{M-B}} = \eta - \frac{{}^1D_1^{1/2}(\beta)}{{}^0D_1^{1/2}(\beta)} \tag{2.16}$$

which also may significantly differ from the relation for parabolic band:

$$\left(\frac{\partial \zeta}{\partial k_0 T}\right)_{\text{M-B}} = \eta - \frac{3}{2}\,.$$

The expression (2.13) for highly degenerate conditions will be discussed later.

3. Thermodynamic Functions and Specific Heat

In order to calculate the total internal energy due to free carriers in a band, we have to put to the formula (1.2) $\chi(\varepsilon) \equiv \varepsilon$.

Hence

$$U = I\,k_0\,T\left(\langle z\rangle - \left\langle\int_0^z d\xi\right\rangle\right). \tag{3.1}$$

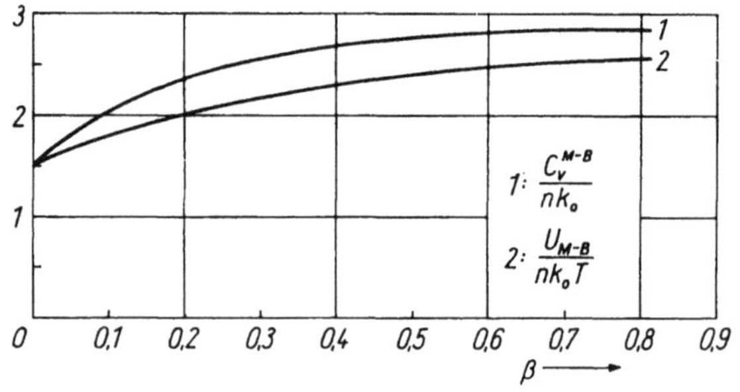

Fig. 2. The mean internal energy U and specific heat C_v *versus* $\beta = \lambda k_0 T$ for a carrier gas obeying Maxwell-Boltzmann statistics. For $\beta = 0$ the classical values 3/2 are obtained

In M—B statistics case we find from (1.5)

$$U_{\text{M-B}} = n\,k_0\,T \left(\overline{z \frac{d}{dz}}\right)_{\text{M-B}} \tag{3.2}$$

and using (1.4), (1.7), (1.10)

$$\frac{U_{\text{M-B}}}{n} = \frac{3}{2}\,k_0\,T\,\frac{{}^1D_1^{1/2}(\beta)}{{}^0D_0^{3/2}(\beta)}\,. \tag{3.3}$$

This can also be expressed by the Macdonald functions in the form

$$\frac{U_{\text{M-B}}}{n} = \frac{3}{2}\,k_0\,T + \frac{k_0\,T}{2\,\beta}\left[\frac{K_1\left(\frac{1}{2\,\beta}\right)}{K_2\left(\frac{1}{2\,\beta}\right)} + 3\,\beta - 1\right]. \tag{3.4}$$

It can be easily seen from (3.3) and (3.4) that for the general nonparabolic band we do not obtain classical value $U_{\text{M-B}} = \frac{3}{2}\,n\,k_0\,T$. The classical value can be found from (3.3) and (3.4) by putting $\beta \to 0$.[1])

Similarly (see e. g. [23]) we find the free energy of a system at constant volume

$$F = n\,k_0\,T\left[\eta - \frac{2}{3}\frac{\langle z\rangle}{\langle 1\rangle} + \frac{2}{3}\frac{\left\langle \int_0^z d\xi \right\rangle}{\langle 1\rangle}\right] \tag{3.5}$$

and

$$F_{\text{M-B}} = n\,k_0\,T\left[\eta - \frac{2}{3}\left(\overline{z\frac{d}{dz}}\right)_{\text{M-B}}\right]. \tag{3.6}$$

---

[1])

$$\frac{K_1\left(\frac{1}{2\,\beta}\right)}{K_2\left(\frac{1}{2\,\beta}\right)} \approx \begin{cases} 1 - 3\,\beta & \text{for small values of } \beta \\ \frac{1}{4\,\beta} & \text{for large values of } \beta \end{cases}$$

The entropy is

$$S = n\,k_0 \left[\frac{5}{3}\frac{\langle z\rangle}{\langle 1\rangle} - \frac{5}{3}\frac{\left\langle \int\limits_0^z d\xi \right\rangle}{\langle 1\rangle} - \eta\right]; \tag{3.7}$$

$$S_{\mathrm{M-B}} = n\,k_0 \left[\frac{5}{3}\overline{\left(z\frac{d}{dz}\right)}_{\mathrm{M-B}} - \eta\right]. \tag{3.8}$$

Now we shall derive the expression for the specific heat at constant volume and carrier density independent of temperature:

$$C_v = \frac{2}{(2\,\pi)^3}\frac{d}{dT}\int f_0\,\varepsilon\,d_3k = k_0\,n\overline{\left(\frac{d}{dz}\right)}\,\overline{[\delta(z)]^2}\,. \tag{3.9}$$

The relation (2.13) was utilized here.
$\overline{\left(\frac{d}{dz}\right)} = \left\langle\frac{d}{dz}\right\rangle \Big/ \langle 1\rangle$ is a usual mean value, and $\overline{[\delta(z)]^2}$ is a mean quadratic deviation of energy from its mean value.

$$\overline{[\delta(z)]^2} = \frac{\left\langle z^2\frac{d}{dz}\right\rangle}{\left\langle\frac{d}{dz}\right\rangle} - \frac{\left\langle z\frac{d}{dz}\right\rangle^2}{\left\langle\frac{d}{dz}\right\rangle^2} = \overline{z^2} - \bar{z}^2\,. \tag{3.10}$$

Here the mean values are taken by the use of function $\left(-\frac{\partial f_0}{\partial\varepsilon}\right)\frac{d}{d\varepsilon}\gamma^{3/2}(\varepsilon)$. For M—B statistics

$$C_v^{\mathrm{M-B}} = k_0\,n\overline{[\delta(z)]^2}\,. \tag{3.11}$$

In terms of L and D integrals

$$\frac{C_v}{n\,k_0} = \frac{3}{2}\frac{{}^0L_1^{1/2}}{{}^0L_0^{3/2}}\left[\frac{{}^2L_1^{1/2}}{{}^0L_1^{1/2}} - \left(\frac{{}^1L_1^{1/2}}{{}^0L_1^{1/2}}\right)^2\right]; \tag{3.12}$$

$$\frac{C_v^{\mathrm{M-B}}}{n\,k_0} = \frac{3}{2}\left[\frac{{}^2D_1^{1/2}}{{}^0D_1^{1/2}} - \left(\frac{{}^1D_1^{1/2}}{{}^0D_1^{1/2}}\right)^2\right]. \tag{3.13}$$

Thus $C_v^{\mathrm{M-B}}/n\,k_0$ does not depend on η, but it depends through β on the temperature.
The specific heat due to free carriers with a carrier density temperature-dependent can be calculated by use of standard procedure (see e. g. [25]).

4. Spheroidal and Spherical Energy Bands

A spheroidal energy band structure can be described by relation

$$\gamma(\varepsilon) = a_{\alpha\beta}\,k_\alpha\,k_\beta\,, \tag{4.1}$$

where k_i are components of the wave vector and $a_{\alpha\beta}$ are arbitrary constants. (The sum convention is used in (4.1).) $\hat{a}$ tensor is obviously symmetric.
The band described by (4.1) presents a special case of general band structure given by (1.1). Thus all the formulae derived hold also in this case. The I integral may be now calculated directly to give $I = (3\,\pi^2)^{-1}A^{-1/2}$, where $A = \det\,(a_{\alpha\beta})$.

For a band given by (4.1) it is possible to introduce a tensor of reciprocal momentum effective mass [13]:

$$m_{ij}^{-1} = \frac{2\,a_{ij}}{\hbar^2}\left(\frac{d\gamma}{d\varepsilon}\right)^{-1}. \tag{4.2}$$

In general the effective masses depend on the energy.

Now we shall discuss the properties of a system in highly degenerate conditions. The asymptotic expansion (1.11) is to be used in order to evaluate the expressions obtained before.

The carrier density in the case of absolute degeneracy is given by (2.8).

The internal energy (3.1) becomes

$$U = n\,k_0\,T\left(\eta - \gamma^{-3/2}(\eta)\int_0^{\eta}\gamma^{3/2}(\xi)\,d\xi\right). \tag{4.3}$$

The absolute degeneracy can occur only at $T = 0$. Thus we have $\beta = 0$ and both parabolic and nonparabolic cases become identical. However, at very low temperatures $\beta \neq 0$ and on the other hand the approximation of absolute degeneracy may still be useful.

The first nonparabolic approximation to (4.3) ($\beta\,\eta < 1$) is

$$U = n\,k_0\,T\left(\frac{3}{5}\eta + \frac{6}{35}\beta\,\eta^2\right). \tag{4.4}$$

The first term presents a well known result for metals. The second one gives the first contribution caused by the nonparabolicity of a band.

Similarly the free energy (3.5) and the entropy (3.7) of highly degenerate electron gas can be evaluated. $\frac{\partial\zeta}{\partial T}$ and C_v are of course equal to zero in absolutely degenerate conditions.

Taking into account two first terms in the expansion (1.11), we get from (2.13)

$$\frac{\partial\zeta}{\partial T} = -\frac{\pi^2}{6}\,k_0^2\,T\,\gamma^{-1}(\zeta)\left(\frac{d\gamma}{d\varepsilon}\right)_{\zeta}\left[1 + 2\,\gamma\left(\frac{d\gamma}{d\varepsilon}\right)^{-2}\frac{d^2\gamma}{d\varepsilon^2}\right]_{\zeta}. \tag{4.5}$$

Hence, using (2.8) and (4.2)

$$\frac{\partial\zeta}{\partial T} = -\frac{4\,\pi^3}{9}\left(\frac{3}{\pi}\right)^{1/3}\frac{M_\zeta}{h^2\,n^{2/3}}\,(1 + 3\,\varkappa_\zeta)\,, \tag{4.6}$$

where

$$\varkappa_\zeta = \frac{2}{3}\,\gamma\left(\frac{d\gamma}{d\varepsilon}\right)^{-2}\frac{d^2\gamma}{d\varepsilon^2} = \frac{n}{M_\zeta}\,\frac{dM_\zeta}{dn}\,, \tag{4.7}$$

with $M = (m_1\,m_2\,m_3)^{1/3}$ being a geometric mean value of momentum effective mass. m_1, m_2, m_3 denote the principal momentum effective masses of diagonalized tensor (4.2). Subscript ζ means that we should take the values for the Fermi level corresponding to the carrier density n (from the relation (2.9)).

Using (4.9), (4.2), (2.9), (1.7), we find after a little manipulation

$$\varkappa_\zeta = \frac{1}{3}\,\frac{\Delta - 1}{\Delta}\,, \tag{4.8}$$

where Δ is defined by (2.9).

The parameter $\varkappa$ is a very useful quantity in the theory of semiconductors with nonparabolic bands. For an isotropic nonparabolic band it was introduced in reference [9] (see also [10a, d; 15, 17].

For the parabolic case $\gamma(\varepsilon)$ is proportional to ε, thus $m_i = \text{const}$, and we have $\varkappa \equiv 0$.

As follows from (2.9), for $\lambda > 0$ (positive nonparabolicity) we have always $\varkappa \geqq 0$ and it increases from 0 to 1/3 with increasing electron concentration [7].

Quite similarly the expansion (1.11) may be utilized to evaluate the specific heat C_v (3.9) in degenerate conditions: (cf. [1])

$$C_v = \frac{4\,\pi^3}{3}\left(\frac{3}{\pi}\right)^{1/3}\frac{k_0^2}{\hbar^2}\,T\,M_\zeta\,n^{1/3}\,. \tag{4.9}$$

It must be born in mind that for a nonparabolic band M_ζ may strongly depend on the electron density n (see e. g. [6b]). PINES [24] has considered the influence of electron interaction on the one-electron energy curves $\varepsilon(k)$ in metals and applied the results to evaluation of specific heats.

Spherical energy band structure may be obtained from relations (1.1) and (4.1) putting $\Phi(\vartheta, \varphi) \equiv 1$ and $a_{\alpha\beta} = a\,\delta_{\alpha\beta}$, respectively. Hence we have $\gamma_1 = 2\,m_n/\hbar^2$ where m_n means the effective mass at the bottom of a band. The notation (1.4) is to be replaced now by

$$\langle A \rangle = \int\limits_0^\infty \left(-\frac{\partial f_0}{\partial \varepsilon}\right)\hat{A}\,k^3(\varepsilon)\,d\varepsilon\,. \tag{4.10}$$

As follows from considerations of EHRENREICH [3, 5] and KOŁODZIEJCZAK [6b] who discussed the KANES model [19], for conduction bands of III—V compounds the relation between wave number and energy is of the form

$$k^2 = \frac{2\,m_n}{\hbar^2}\,\varepsilon\left(1 + \frac{\varepsilon}{\varepsilon_g}\right), \tag{4.11}$$

where ε_g is the energy gap. The relation (4.11) is a good approximation to the $\varepsilon(k)$ relation for small values of energy.

Thus (4.11) presents the special case of general expansion (1.6) (cf. (1.7)). We have $\lambda = 1/\varepsilon_g$, and $\beta = k_0 T/\varepsilon_g$.

5. Conclusions

Statistical properties, entropy, free energy, internal energy, and specific heat of free carriers in an energy band of arbitrary form are considered. All the properties are strongly affected by the nonparabolic form of a band. In particular the internal energy and specific heat in nondegenerate region are temperature-dependent and they may significantly differ from the classical values. It is shown that the $\varepsilon(k)$ dependence for conduction bands in III—V compounds based on the KANES band model may be regarded as a first nonparabolic approximation to general nonparabolic structures. The theory includes the parabolic structure as a special simplified case.

It should be noted that for spheroidal bands the quantities discussed may be expressed in terms of momentum effective masses. This is also the case in the theory of transport and magnetic phenomena [2, 6, 7, 9, 10, 13, 15, 17]. Thus it is the momentum effective mass rather than the inertial mass that should be regarded as a basic quantity. For a parabolic band both masses are identical [2,7]. Thus the common use of the inertial effective mass in the theory of metals and semiconductors with the parabolic band structure seems to be rather accidental.

The authors would like to thank Prof. L. SOSNOWSKI for kind attention and critical remarks on the subject.

References

[1] A. H. WILSON, The Theory of Metals, Cambridge 1954.
[2] R. BARRIE and J. T. EDMOND, J. Electronics and Control **1**, 161 (1955).
[3] H. EHRENREICH, J. Phys. Chem. Solids **2**, 131 (1957).
[4] M. RODOT, Ann. Phys. (France) **5**, 1085 (1960).
[5] H. EHRENREICH, J. appl. Phys. **32**, 2155 (1961).
[6] J. KOŁODZIEJCZAK, Acta phys. Polon. a) **20**, 379 (1961); b) **20**, 289 (1961); c) **21**, 637 (1962).
[7] J. KOŁODZIEJCZAK and L. SOSNOWSKI, Acta phys. Polon. **21**, 399 (1962).
[8] L. SOSNOWSKI, Summer School of Physics, Varenna 1961, Nuovo Cimento.
[9] J. KOŁODZIEJCZAK, L. SOSNOWSKI, and W. ZAWADZKI, Rep. Int. Conf. Semicond. Phys. Exeter 1962.
[10] W. ZAWADZKI, phys. stat. sol. a) **2**, 385 (1962); b) **3**, 692 (1963); c) **3**, 1006 (1963); d) **3**, 1421 (1963); e) Phys. Letters **4**, 190 (1963).
[11] T. C. HARMAN and J. H. HONIG, J. Phys. Chem. Solids **23**, 913 (1962).
[12] G. I. GUSEVA and I. M. TSIDIL'KOVSKII, Fiz. tverd. Tela **4**, 2490 (1962).
[13] S. ŻUKOTYŃSKI and J. KOŁODZIEJCZAK, phys. stat. sol. **3**, 990 (1963).
[14] L. SOSNOWSKI, Summer School of Physics, Hercegnovii 1963.
[15] J. KOŁODZIEJCZAK and S. ŻUKOTYŃSKI, phys. stat. sol. **5**, 145 (1964).
[16] W. ZAWADZKI and J. KOŁODZIEJCZAK, phys. stat. sol. **6**, 419 (1964).
[17] W. ZAWADZKI, Magnetic Susceptibility of Free Carriers in Semiconductors Rep. Int. Conf. Semicond. Phys., Paris 1964.
[18] W. ZAWADZKI, Properties of Generalized Fermi-Dirac Integrals (to be published).
[19] E. O. KANE, J. Phys. Chem. Solids **1**, 249 (1957).
[20] R. BLANKENBECLER, Amer. J. Phys. **25**, 279 (1957).
[21] C. KITTEL, Elementary Statistical Physics, New York/London (1958); Russian transl.: Moscow 1960.
[22] J. BLAKEMORE, Semiconductor Statistics, Pergamon Press 1962.
[23] A. H. WILSON, Thermodynamics and Statistical Mechanics, Cambridge 1957.
[24] D. PINES, Solid State Physics **1**, 367 (1955).
[25] A. I. ANSELM, Introduction to Theory of Semiconductors, Moscow 1962.
[26] H. M. MACDONALD, Proc. London Math. Soc. **30**, 167 (1899).
[27] E. T. WHITTAKER and G. N. WATSON, A Course of Modern Analysis, Cambridge 1927.

(Received May 12, 1964)

phys. stat. sol. **6**, 419 (1964)

Institute of Physics, Polish Academy of Sciences, Warsaw

Transport Properties of Cubic Semiconductors with Nonparabolic Energy Bands

By

W. ZAWADZKI and J. KOŁODZIEJCZAK

Transport phenomena are considered in a semiconductor crystal having cubic symmetry and an arbitrary relation between energy and wave number. The theory is formulated in terms of three basic transport tensors. Expressions for galvanomagnetic and thermomagnetic effects in the limiting cases of weak and strong magnetic fields are obtained. The results apply both to the many-ellipsoid cubic energy bands of Si and Ge and to the nonparabolic band structure of III-V compounds.

Es werden die Transporterscheinungen in einem kubischen Halbleiterkristall mit beliebigem $E(k)$-Gesetz auf der Grundlage von drei Transporttensoren behandelt. Die galvanomagnetischen und thermomagnetischen Koeffizienten werden in den Grenzfällen schwacher und starker Magnetfelder berechnet. Es wird die Anwendung der Ergebnisse auf Halbleiter mit mehreren Ellipsoiden (Ge, Si) und einfachen nichtquadratischen $E(k)$-Abhängigkeiten (III–V-Verbindungen) diskutiert.

Introduction

A theory of transport phenomena in anisotropic materials with a quadratic dependence of energy on the absolute value of the wave vector has been given by a number of authors [1 to 20]. On the other hand a theory of transport properties of metals and semiconductors with spherical but nonparabolic energy bands is being quickly developed [21 to 43]. It has been found that the nonparabolic structure of a band can greatly affect some of the transport effects. Recently KOŁODZIEJCZAK and ŻUKOTYŃSKI [38, 39] have developed an approach which enables to consider the two models simultaneously (see also [40, 41, 42]). The authors have given the theory of transport phenomena due to free carriers in a single ellipsoidal energy shell with an arbitrary $\varepsilon(k)$ dependence. The present paper is a continuation of this work dealing with many-ellipsoid cubic energy structures. The calculations are made as general as possible stressing arbitrariness of cubic structure. The work is based upon the paper [39] and the fundamental contribution of HERRING [9, quoted in 20].

At the beginning we shall summarize the starting points as given in [39]. The current density J_ν and the heat flux Q_ν in the presence of not very strong electric field and temperature gradient are given by

$$J_\nu = \sigma_{\nu\mu} E_\mu - \theta_{\nu\mu} \frac{\partial T}{\partial x_\mu}, \qquad Q_\nu = \chi_{\nu\mu} E_\mu - \xi_{\nu\mu} \frac{\partial T}{\partial x_\mu}, \tag{0.1}$$

where E_μ and $\partial T/\partial x_\mu$ are the electric field and the temperature gradient respectively. $\sigma_{\nu\mu}, \theta_{\nu\mu}, \chi_{\nu\mu}, \xi_{\nu\mu}$ are the transport tensors depending in general on a magnetic field. From the Onsager relations

$$\chi_{\nu\mu}(H) = T\,\theta_{\mu\nu}(-H)\,. \tag{0.2}$$

Thus we have three independent basic transport tensors.

The equations (0.1) can be rewritten in a more traditional form

$$E_\nu = \varrho_{\nu\mu} J_\mu + \alpha_{\nu\mu} \frac{\partial T}{\partial x_\mu}, \qquad Q_\nu = \pi_{\nu\mu} J_\mu - \varkappa_{\nu\mu} \frac{\partial T}{\partial x_\mu} \tag{0.3}$$

with

$$\left.\begin{array}{ll} \varrho_{\nu\mu} = \sigma^{-1}_{\nu\mu} & \text{the resistivity tensor,} \\ \alpha_{\nu\mu} = \sigma^{-1}_{\nu\beta}\,\theta_{\beta\mu} & \text{the thermoelectric tensor,} \\ \pi_{\nu\mu} = \chi_{\nu\beta}\,\sigma^{-1}_{\beta\mu} & \text{the Peltier tensor,} \\ \varkappa_{\nu\mu} = \xi_{\nu\mu} - \chi_{\nu\beta}\,\alpha_{\beta\mu} & \text{the thermal conductivity tensor.} \end{array}\right\} \tag{0.4}$$

And again we have

$$\pi_{\nu\mu}(H) = T\,\alpha_{\mu\nu}(-H)\,. \tag{0.5}$$

The band structure considered can be described by the relation

$$\gamma(\varepsilon) = a_{\alpha\beta}\, k_\alpha\, k_\beta\,. \tag{0.6}$$

The components of symmetric tensor $a_{\alpha\beta}$ are assumed to be constant (independent of energy). An energy surface described by (0.6) is an ellipsoid in $\boldsymbol{k}$-space. For $a_{\alpha\beta} = a\,\delta_{\alpha\beta}$ it becomes a sphere.

The function $\gamma(\varepsilon)$ determines the non-parabolicity of the band and in the particular case of $\gamma(\varepsilon) \equiv \varepsilon$ the energy band becomes parabolic.

A mobility tensor is given by

$$u_{\beta\mu} = q\,\tau\, m^{-1}_{\beta\mu}, \tag{0.7}$$

where q means electric charge ($q = \mp e$ with the upper sign for electrons, the lower for holes) and τ the relaxation time dependent on energy. $m^{-1}_{\beta\mu}$ denotes a momentum effective mass-tensor

$$m^{-1}_{\beta\mu} = \frac{2}{\hbar^2}\left(\frac{d\gamma}{d\varepsilon}\right)^{-1} a_{\beta\mu}\,. \tag{0.8}$$

The components $m^{-1}_{\beta\mu}$ are energy-dependent since in general $\gamma(\varepsilon) \not\equiv \varepsilon$. The solution of Boltzmann transport equation gives finally following expressions for the three basic tensors:

$$\sigma_{\nu\mu} = q\,n\frac{\langle G_{\nu\mu}\rangle}{\langle 1\rangle}, \qquad \theta_{\nu\mu} = k_0\,n\frac{\langle(\epsilon-\eta)\,G_{\nu\mu}\rangle}{\langle 1\rangle}, \qquad \xi_{\nu\mu} = \frac{k_0^2}{q}\,T\,n\frac{\langle(\epsilon-\eta)^2\,G_{\nu\mu}\rangle}{\langle 1\rangle}, \tag{0.9}$$

where $\epsilon = \frac{\varepsilon}{k_0 T}$ is the reduced energy, $\eta = \frac{\zeta}{k_0 T}$ is the reduced Fermi level and n is a density of carriers in the ellipsoid.

The bracket $\langle X\rangle$ is defined

$$\langle X\rangle = \int_0^\infty \left(-\frac{\partial f_0}{\partial \varepsilon}\right) X(\varepsilon)\,\gamma^{3/2}(\varepsilon)\,d\varepsilon\,, \tag{0.10}$$

f_0 being the Fermi-Dirac distribution function. We have for example

$$n = \frac{1}{3\,\pi^2\,|A|^{1/2}}\langle 1\rangle, \tag{0.11}$$

where $|A|$ is the determinant of $a_{\alpha\beta}$ matrix.

The tensor $G_{\nu\mu}$ is

$$G_{\nu\mu} = \frac{1}{\Delta}\left\{u_{\nu\mu} - \frac{1}{c}\,\varepsilon_{\alpha\beta\gamma}\,H_\alpha\,u_{\nu\gamma}\,u_{\beta\mu} + \frac{|u|}{c^2}\,H_\nu\,H_\mu\right\} \tag{0.12}$$

with

$$\Delta = 1 + \frac{|u|}{c^2}\,a_{\gamma\alpha}^{-1}\,H_\gamma\,H_\alpha\,,$$

where $|u|$ is the determinant of $u_{\beta\mu}$ matrix (c velocity of light).

On the base of (0.7) and (0.8) we can write

$$u_{\beta\mu} = 2\,\mu\,a_{\beta\mu}\,, \tag{0.13}$$

where μ depends on energy and mechanism of scattering. Taking this into account and transforming the second term in (0.12), $G_{\nu\mu}$ can be found in the form

$$G_{\nu\mu} = \frac{1}{\Delta}\left\{2\,\mu\,a_{\nu\mu} + \frac{4\,\mu^2}{c}\,|A|\,\varepsilon_{\nu\mu\alpha}\,a_{\zeta\alpha}^{-1}\,H_\zeta + \frac{8\,\mu^3}{c^2}\,|A|\,H_\nu\,H_\mu\right\} \tag{0.14}$$

with

$$\Delta = 1 + \frac{4\,\mu^2}{c^2}\,|A|\,a_{\gamma\alpha}^{-1}\,H_\gamma\,H_\alpha\,,$$

where $\varepsilon_{\nu\mu\zeta}$ is the antisymmetric tensor of Levi-Civita.

As it can be easily seen, there is no restriction made concerning the configuration of a magnetic field, electric current, temperature gradient, and the orientation of the ellipsoid. The expression (0.14) is not entirely new (see, e. g., [1, 2, 8, 17, 18]. However, by the use of (0.6), (0.8), (0.9), and (0.10) it includes also the nonparabolic case.

1. Transport Phenomena in the Absence of a Magnetic Field

The above given expressions describe the transport phenomena in a single ellipsoid. In the case when intervalley scattering can be neglected, separate valleys may be considered as different bands and the basic transport tensors (0.9) are to be replaced by sums of tensors corresponding to different valleys.

We shall assume further that the different ellipsoids have the same shape and they occupy equivalent positions in a Brillouin zone. Then the $\varepsilon(k)$ dependence will also be identical for every ellipsoid. Because the determinent $|A|$ in (0.11) is an invariant for revolutions in $\boldsymbol{k}$-space, the carrier concentration is the same in every valley.

In the absence of a magnetic field $G_{\nu\mu}^{(i)} = 2\,\mu\,a_{\nu\mu}^{(i)}$ and taking into account the cubic symmetry of a structure, we have

$$\sigma_{\nu\mu} = \sum_{i=1}^{l} \sigma_{\nu\mu}^{(i)} = q\,N\,\bar{u}\,\delta_{\nu\mu}\,, \qquad \varrho_{\nu\mu} = \sigma_{\nu\mu}^{-1} = \frac{1}{q\,N\,\bar{u}}\,\delta_{\nu\mu}\,, \tag{1.1}$$

where $N = n\,l$ is the total density of carriers in l ellipsoids, and

$$\bar{u} = \frac{2}{3}\,\frac{\langle\mu\rangle}{\langle 1\rangle}\,Sp\,A \tag{1.2}$$

is the so-called ohmic mobility. $Sp\,A$ denotes the trace of $a_{\alpha\beta}$ matrix and is of course independent of a coordinate system orientation. Similarly

$$\theta_{\nu\mu} = k_0\,N\,\frac{\langle(\epsilon-\eta)\,\mu\rangle}{\langle\mu\rangle}\,\bar{u}\,\delta_{\nu\mu}\,, \qquad \xi_{\nu\mu} = \frac{k_0^2\,T}{q}\,N\,\frac{\langle(\epsilon-\eta)^2\,\mu\rangle}{\langle\mu\rangle}\,\bar{u}\,\delta_{\nu\mu}\,. \tag{1.3}$$

The relations (1.1) and (1.3) can be easily proved in the special cases of six ellipsoids of revolution situated along ⟨100⟩ directions (Si) or eight ellipsoids in ⟨111⟩ directions (Ge). They are also valid for any cubic structure consisting of ellipsoids (see e. g. [5] or [16]).

Hence the thermoelectric power is

$$\alpha_{\nu\mu} = \sigma^{-1}_{\nu\beta}\,\theta_{\beta\mu} = \frac{k_0}{q}\left(\frac{\langle\epsilon\,\mu\rangle}{\langle\mu\rangle} - \eta\right)\delta_{\nu\mu} \tag{1.4}$$

and the thermal conductivity

$$\varkappa_{\nu\mu} = \frac{k_0^2}{q}\,T\,N\left[\frac{\langle\epsilon^2\,\mu\rangle}{\langle\mu\rangle} - \frac{\langle\epsilon\,\mu\rangle^2}{\langle\mu\rangle^2}\right]\bar{u}\,\delta_{\nu\mu}\,. \tag{1.5}$$

The Wiedemann-Franz law becomes

$$\varrho_{\nu\alpha}\,\varkappa_{\alpha\mu} = \left(\frac{k_0}{q}\right)^2 T\left[\frac{\langle\epsilon^2\,\mu\rangle}{\langle\mu\rangle} - \frac{\langle\epsilon\,\mu\rangle^2}{\langle\mu\rangle^2}\right]\delta_{\nu\mu}\,. \tag{1.6}$$

Thus all the quantities have scalar properties. In the one ellipsoid case only the thermoelectric power and the Lorenz number were found to be isotropic [39].

2. Case of Weak Magnetic Fields

To consider the transport properties in the presence of a magnetic field we have to sum up the transport tensors in the general case given by (0.14). It is possible in principle, but the task is rather tedious since Δ is not the same for different ellipsoids. For this reason we shall consider the case of weak magnetic fields determined by the inequality $\frac{4\,\mu^2}{c^2}\,|A|\,a^{-1}_{\gamma\alpha}\,H_\gamma\,H_\alpha \ll 1$. Then

$$\frac{1}{\Delta} \approx 1 - \frac{4\,\mu^2}{c^2}\,|A|\,a^{-1}_{\gamma\alpha}\,H_\gamma\,H_\alpha$$

and by making use of (0.9) and (0.14) we get for (i)-th ellipsoid

$$\sigma^{(i)}_{\nu\mu} = q\,n\left\{2\frac{\langle\mu\rangle}{\langle 1\rangle}\,a^{(i)}_{\nu\mu} + \frac{4}{c}\,\frac{\langle\mu^2\rangle}{\langle 1\rangle}\,|A|\,\varepsilon_{\nu\mu\alpha}\,a^{-1(i)}_{\zeta\alpha}\,H_\zeta + \right.$$

$$\left. + \frac{8}{c^2}\,\frac{\langle\mu^3\rangle}{\langle 1\rangle}\,|A|\left(\delta_{\nu m}\,\delta_{\mu n} - a^{(i)}_{\nu\mu}\,a^{-1(i)}_{mn}\right)H_m\,H_n\right\}. \tag{2.1}$$

For a cubic crystal the relation of the Hall current to $\boldsymbol{E}$ and $\boldsymbol{H}$ must be isotropic, i. e. the second term in (2.1) after summation over the different ellipsoids must be proportional to $\varepsilon_{\nu\mu\zeta}\,H_\zeta$. Making use of this assumption and the relation (1.1), we have finally

$$\sigma_{\nu\mu} = q\,N\,(K_1\,\delta_{\nu\mu} + K_2\,\varepsilon_{\nu\mu\zeta}\,H_\zeta + K_3\,\beta_{\nu\mu mn}\,H_m\,H_n)\,, \tag{2.2}$$

where

$$K_1 = \frac{2}{3}\,\frac{\langle\mu\rangle}{\langle 1\rangle}\,Sp\,A\,,\qquad K_2 = \frac{4}{3c}\,\frac{\langle\mu^2\rangle}{\langle 1\rangle}\,A_{\alpha\alpha}\,,\qquad K_3 = \frac{8}{c^2}\,\frac{\langle\mu^3\rangle}{\langle 1\rangle}\,|A|\,, \tag{2.3}$$

and

$$\beta_{\nu\mu mn} = \frac{1}{2}\,(\delta_{\nu m}\,\delta_{\mu n} + \delta_{\nu n}\,\delta_{\mu m}) - \frac{1}{l}\sum_{i=1}^{l} a^{(i)}_{\nu\mu}\,a^{-1(i)}_{mn}\,. \tag{2.4}$$

$A_{\alpha\alpha}$ is a sum of leading second order minors of the $a_{\alpha\beta}$ matrix. K_1, K_2, and K_3 are proportional to three invariants of $a_{\alpha\beta}$ matrix with respect to changes in the orientation of the coordinate system. Moreover, they are the same for any many-ellipsoidal cubic structure.

On the other hand the tensor $\beta_{\nu\mu mn}$ may differ for various cubic materials. We have used here the symmetrized $\beta_{\nu\mu mn}$ tensor (2.4) (in regard to $m\,n$) because the term $H_m H_n$ in (2.2) is also symmetric. It is easy to see from the definition (2.4) that the relations

$$\beta_{\nu\mu mn} = \beta_{\nu\mu nm} = \beta_{\mu\nu mn} \tag{2.5}$$

are fulfilled.

Moreover, it is well known [4, 5] that for any cubic symmetry we have only three different components of $\beta_{\nu\mu mn}$ (there is no summation over the subscripts in the parentheses):

$$\beta_{(\lambda\lambda\lambda\lambda)}, \quad \beta_{(\lambda\zeta\lambda\zeta)}, \quad \beta_{(\lambda\lambda\zeta\zeta)}, \qquad (\lambda, \zeta = 1, 2, 3) \tag{2.6}$$

where the relations (2.5) hold, all the other components being zero.

Hence in the coordinate system directed along principal cubic axes we have

$$\beta_{\nu\mu mn} H_m H_n = \beta_{(\lambda\lambda\zeta\zeta)} H^2 \delta_{\nu\mu} + 2\,\beta_{(\lambda\zeta\lambda\zeta)} H_\nu H_\mu + \\ + [\beta_{(\lambda\lambda\lambda\lambda)} - \beta_{(\lambda\lambda\zeta\zeta)} - 2\,\beta_{(\lambda\zeta\lambda\zeta)}]\, H^2_{(\nu)}\, \delta_{(\nu)\mu}\,,$$

and finally (2.2) becomes

$$\sigma_{\nu\mu} = \sigma\,\delta_{\nu\mu} + \eta\,\varepsilon_{\nu\mu\zeta} H_\zeta + \varphi_1 H^2 \delta_{\nu\mu} + \varphi_2 H_\nu H_\mu + \varphi_3 H^2_{(\nu)}\,\delta_{(\nu)\mu}\,, \tag{2.7}$$

where

$$\sigma = q N K_1, \quad \eta = q N K_2, \quad \varphi_1 = q N K_3 \varphi_1', \quad \varphi_2 = q N K_3 \varphi_2', \quad \varphi_3 = q N K_3 \varphi_3' \tag{2.8}$$

with

$$\varphi_1' = \beta_{(\lambda\lambda\zeta\zeta)}, \quad \varphi_2' = 2\,\beta_{(\lambda\zeta\lambda\zeta)}, \quad \varphi_3' = \beta_{(\lambda\lambda\lambda\lambda)} - \varphi_1' - \varphi_2'\,. \tag{2.9}$$

The inverse tensor $\overset{-1}{\sigma_{\nu\mu}}$ may be found to give

$$\varrho_{\nu\mu} = \varrho\,\delta_{\nu\mu} + h\,\varepsilon_{\nu\mu\zeta} H_\zeta + f_1 H^2 \delta_{\nu\mu} + f_2 H_\nu H_\mu + f_3 H^2_{(\nu)}\,\delta_{(\nu)\mu} \tag{2.10}$$

(no summation over (ν) in the last term)
where

$$\left.\begin{aligned} &\varrho = \frac{1}{\sigma}\,, && h = -\frac{\eta}{\sigma^2}\,, \\ &f_1 = -\frac{1}{\sigma^2}\left(\varphi_1 + \frac{\eta^2}{\sigma}\right), && f_2 = -\frac{1}{\sigma^2}\left(\varphi_2 - \frac{\eta^2}{\sigma}\right), \qquad f_3 = -\frac{\varphi_3}{\sigma^2}\,. \end{aligned}\right\} \tag{2.11}$$

By the use of (1.2), (2.3), (2.8), and (2.11) we find finally the antisymmetric and symmetric part of magnetoresistivity tensor in the form

$$\varrho^a_{\nu\mu} = \frac{a_r}{q N c}\,\varepsilon_{\nu\mu\zeta} H_\zeta\,, \tag{2.12}$$

$$\varrho^s_{\nu\mu} = \frac{1}{q N \bar{u}}\left[\delta_{\nu\mu} - (\varphi_1' b_r + a_r^2)\frac{\overline{u^2}}{c^2} H^2 \delta_{\nu\mu} - (\varphi_2' b_r - a_r^2)\frac{\overline{u^2}}{c^2} H_\nu H_\mu - \varphi_3' b_r \frac{\overline{u^2}}{c^2} H^2_{(\nu)}\,\delta_{(\nu)\mu}\right], \tag{2.13}$$

where

$$a_r = \frac{3\,A_{\alpha\alpha}}{(Sp\,A)^2}\,\frac{\langle\mu^2\rangle\langle 1\rangle}{\langle\mu\rangle^2}\,, \qquad b_r = \frac{27\,|A|}{(Sp\,A)^3}\,\frac{\langle\mu^3\rangle\langle 1\rangle^2}{\langle\mu\rangle^3}\,, \tag{2.14}$$

are scattering factors. They are expressed by the invariants of $a_{\alpha\beta}$ matrix and they depend both on the nonspherical and nonparabolic structure of a band. For a spherical energy band the coefficients depending on the anisotropy become equal to 1 (cf. for example [29b, 35]).

The antisymmetric part of the magnetoresistivity (2.12) gives the Hall constant in weak magnetic fields. It is isotropic to this approximation. However, in higher magnetic fields, in the region where the Hall coefficient depends on a magnetic field (in the approximation taking into account terms $\sim H^3$), it is anisotropic. A measurement of this anisotropy provides useful informations in band structure investigations [13].

For a spherical band structure $a_{\alpha\beta} = a\,\delta_{\alpha\beta}$ and by (2.4), (2.6), and (2.9) it is easy to find

$$\varphi_1' = -1\,; \qquad \varphi_2' = 1\,; \qquad \varphi_3' = 0\,. \tag{2.15}$$

Thus in this case the last term in (2.13) is equal to zero and the longitudinal magnetoresistance vanishes identically.

For conduction band of Si we have $\beta_{(\lambda\lambda\lambda\lambda)} = 0$ and if a magnetic field is situated along $\langle 100\rangle$ direction the longitudinal magnetoresistance vanishes. In the case of n-Ge $\beta_{(\lambda\lambda\lambda\lambda)} \neq 0$ and the longitudinal effect is not ever equal to zero.

In the general case of cubic symmetry by (2.13) the longitudinal magnetoresistivity does not vanish. It is of interest to note here that for a single ellipsoid the longitudinal effect is always equal to zero [39]. (In the previous literature this fact was mentioned only for a magnetic field directed along the principal axes of the ellipsoid.) It is a rather surprising conclusion if one takes into account that the cubic structure possesses higher symmetry than the single ellipsoid.

The planar Hall effect suggested and measured by GOLDBERG and DAVIS [12, 19] can also give useful informations about a band structure. The effect is anisotropic both in the general case given by (2.13) and for a single ellipsoidal band. In materials with spherical band structure the effect is isotropic.

Using (0.9) and (2.7) we find the second basic transport tensor

$$\theta_{\nu\mu} = \theta\,\delta_{\nu\mu} + \lambda\,\varepsilon_{\nu\mu\zeta}\,H_\zeta + \gamma_1\,H^2\,\delta_{\nu\mu} + \gamma_2\,H_\nu\,H_\mu + \gamma_3\,H_{(\nu)}^2\,\delta_{(\nu)\mu}\,, \tag{2.16}$$

where

$$\left.\begin{aligned} \theta &= k_0\,N\,P_1\,, \qquad \lambda = k_0\,N\,P_2\,, \\ \gamma_1 &= k_0\,N\,P_3\,\varphi_1'\,, \quad \gamma_2 = k_0\,N\,P_3\,\varphi_2'\,, \qquad \gamma_3 = k_0\,N\,P_3\,\varphi_3'\,, \end{aligned}\right\} \tag{2.17}$$

with

$$\left.\begin{aligned} P_1 &= \frac{2}{3}\,\frac{\langle(\epsilon-\eta)\,\mu\rangle}{\langle 1\rangle}\,Sp\,A\,, \qquad P_2 = \frac{4}{3c}\,\frac{\langle(\epsilon-\eta)\,\mu^2\rangle}{\langle 1\rangle}\,A_{\alpha\alpha}\,, \\ P_3 &= \frac{8}{c^2}\,\frac{\langle(\epsilon-\eta)\,\mu^3\rangle}{\langle 1\rangle}\,|A|\,. \end{aligned}\right\} \tag{2.18}$$

Hence by (0.4), (2.8), (2.10), (2.11), (2.16), (2.17), and (2.18) the thermoelectric tensor is

$$\alpha^a_{\nu\mu} = a_r\,\alpha_1\,\frac{\bar u}{c}\,\varepsilon_{\nu\mu\zeta}\,H_\zeta\,, \tag{2.19}$$

$$\alpha^s_{\nu\mu} = \alpha\,\delta_{\nu\mu} + (\varphi_1'\,b_r\,\alpha_2 + a_r^2\,\alpha_1)\,\frac{\overline{u^2}}{c^2}\,H^2\,\delta_{\nu\mu} + (\varphi_2'\,b_r\,\alpha_2 - a_r^2\,\alpha_1)\,\frac{\overline{u^2}}{c^2}\,H_\nu\,H_\mu + \\ + \varphi_3'\,b_r\,\alpha_2\,\frac{\overline{u^2}}{c^2}\,H_{(\nu)}^2\,\delta_{(\nu)\mu}\,, \tag{2.20}$$

where

$$\alpha = \frac{k_0}{q}\left(\frac{\langle\epsilon\,\mu\rangle}{\langle\mu\rangle} - \eta\right), \quad \alpha_1 = \frac{k_0}{q}\left(\frac{\langle\epsilon\,\mu^2\rangle}{\langle\mu^2\rangle} - \frac{\langle\epsilon\,\mu\rangle}{\langle\mu\rangle}\right), \quad \alpha_2 = \frac{k_0}{q}\left(\frac{\langle\epsilon\,\mu^3\rangle}{\langle\mu^3\rangle} - \frac{\langle\epsilon\,\mu\rangle}{\langle\mu\rangle}\right). \tag{2.21}$$

The antisymmetric part (2.19) determines the Nernst-Ettingshausen effect in weak magnetic fields $E_{\perp} = P_{N-E} \cdot \boldsymbol{H} \cdot \text{grad } T$. Using phenomenological parameters: The Hall constant R and conductivity σ, we have

$$P_{N-E} = R \sigma \alpha_1 . \tag{2.22}$$

The form of (2.20) is very similar to the symmetric part of magnetoresistivity tensor (2.13). This fact can also be deduced on the base of general considerations of symmetry properties (cf. [10a]).

Now we shall consider briefly the planar Nernst-Ettingshausen effect found by Bass and Tsidil'kovskii [14, 18] (the authors called it "longitudinal-transversal thermomagnetic effect"). If we set a gradient of temperature in $\boldsymbol{x}_1$ direction and a magnetic field in $\boldsymbol{x}_1$, $\boldsymbol{x}_2$ plane, the electric field arises which has nonvanishing component also along $\boldsymbol{x}_2$ direction. Thus we can detect the electric field in grad T, $\boldsymbol{H}$ plane. In the simplest case, when the coordinate system is directed along principal cubic axes, we get from (2.20)

$$\alpha^s_{21} = \frac{1}{2} (\varphi_2' \, b_r \, \alpha_2 - a_r^2 \, \alpha_1) \frac{H^2 \, \overline{u}^2}{c^2} \sin 2\psi , \tag{2.23}$$

where ψ is the angle between grad T and $\boldsymbol{H}$.

The effect is anisotropic in the general case of cubic structure and for a single ellipsoidal band. In materials with spherical band structure the effect is isotropic.

3. Case of Strong Magnetic Fields

The region of strong magnetic fields is determined by the condition

$$1 \ll \frac{4\,\mu^2}{c^2} |A| \, a_{\gamma\alpha}^{-1} \, H_\gamma \, H_\alpha ,$$

and $G_{\nu\mu}$ (0.14) becomes:

$$G_{\nu\mu} = \frac{c^2}{4\,\mu^2 \, |A| \, a_{\gamma\alpha}^{-1} \, H_\gamma \, H_\alpha} \left\{ 2\,\mu \, a_{\nu\mu} + \frac{4\,\mu^2}{c} |A| \, \varepsilon_{\nu\mu\eta} \, a_{\zeta\eta}^{-1} \, H_\zeta + \frac{8\,\mu^3}{c^2} |A| \, H_\nu \, H_\mu \right\} . \tag{3.1}$$

Like in the general case given by (0.14) it is hard to sum up this expression over ellipsoids. For this reason we are going to carry the calculations only for special directions of a magnetic field (cf. [7, 8, 19]) to emphasize the main points.

Let us consider the cubic conduction band structure of Si with six ellipsoids of revolution. Let the magnetic field be directed along $\boldsymbol{x}_3$ in the coordinate system attached to the principal cubic axes. In this case it is easy to sum up (3.1) over the ellipsoids and to find the inverse tensor. Finally the magnetoresistivity tensor in a strong magnetic field is

$$\left.\begin{aligned} &\varrho_{\nu\mu} = \begin{pmatrix} \varrho_{11} & \varrho_{12} & \varrho \\ -\varrho_{12} & \varrho_{11} & \varrho \\ \varrho & 0 & \varrho_{33} \end{pmatrix}, \\ &\varrho_{11} = \frac{1}{q\,N} \left\langle \frac{1}{\mu} \right\rangle \frac{a}{\langle 1 \rangle}, \qquad \varrho_{12} = -\frac{H}{q\,N\,c}, \qquad \varrho_{33} = \frac{1}{q\,N} \frac{\langle 1 \rangle}{\langle \mu \rangle} \frac{1}{b}, \end{aligned}\right\} \tag{3.2}$$

where

$$a = \frac{1}{6} \frac{2\,a_1\,a_2 + a_2^2}{a_1\,a_2^2}, \qquad b = \frac{2}{3}(a_1 + 2\,a_2)$$

with a_1 and a_2 being associated with the long and short ellipsoid axes respectively.

On the other hand the $\theta_{\nu\mu}$ tensor becomes

$$\left.\begin{aligned}
&\theta_{\nu\mu} = \begin{pmatrix} \theta_{11} & \theta_{12} & 0 \\ -\theta_{12} & \theta_{11} & 0 \\ 0 & 0 & \theta_{33} \end{pmatrix}, \qquad \theta_{11} = k_0\, N\, \frac{c^2}{H^2}\, a \left\langle \frac{\epsilon - \eta}{\mu} \right\rangle \frac{1}{\langle 1 \rangle}, \\
&\theta_{12} = k_0\, N\, \frac{c}{H}\, \frac{\langle \epsilon - \eta \rangle}{\langle 1 \rangle}, \qquad \theta_{33} = k_0\, N\, b\, \frac{\langle (\epsilon - \eta)\, \mu \rangle}{\langle 1 \rangle}\,.
\end{aligned}\right\} \tag{3.3}$$

From (3.2) we find the Hall constant to be $R(\infty) = \frac{1}{q\,N\,c}$. It has been found (cf. [7, 8, 9]) that this result is independent of the magnetic field orientation. The same result has been obtained for one ellipsoid band [39]. Thus the Hall coefficient is constant and isotropic in the limiting cases of weak and strong magnetic fields. In the intermediate region the coefficient is anisotropic and it depends upon a magnetic field strength. Using (3.2) and (3.3) the thermoelectric power in a transversal magnetic field is obtained:

$$\alpha_{\perp}(\infty) = \frac{k_0}{q} \left(\frac{\langle \epsilon \rangle}{\langle 1 \rangle} - \eta \right). \tag{3.4}$$

This result also holds for any orientation of temperature gradient and transversal magnetic field (cf. [10a, 18]). Thus it should be isotropic. HERRING et al. [10a] have found the thermoelectric power in n-Ge to be strongly anisotropic at liquid air temperatures. This fact is, however, related to the large phonon drag contribution to the observed phenomenon.

The Nernst-Ettingshausen field can also be found:

$$E_{\perp} = \frac{k_0}{q} \left(\frac{H\,\bar{u}}{e} \right)^{-1} q_r \left(\frac{\langle \epsilon \rangle}{\langle 1 \rangle} - \frac{\langle \epsilon/\mu \rangle}{\langle 1/\mu \rangle} \right) \frac{\partial T}{\partial x}\,. \tag{3.5}$$

This is quite a general result with q_r being, however, dependent on the temperature gradient and magnetic field orientations.

The scattering factor q_r can be written in the form

$$q_r = \lambda\, \frac{\langle \mu \rangle \langle 1/\mu \rangle}{\langle 1 \rangle^2} = \lambda\, \bar{\mu}\, \overline{\left(\frac{1}{\mu} \right)}\,. \tag{3.6}$$

For the spherical band structure $\lambda = 1$ [29b, 35]. The anisotropy coefficient λ for conduction bands of Si and Ge for several orientations of temperature gradient and magnetic field can be found in [10a, 18, 19].

4. Conclusions

The model has been worked out which enables to treat cubic many valley structures and nonparabolic energy bands from one point of view. The results apply directly to the many valley conduction bands of Si and Ge and the nonparabolic spherical conduction bands of III—V compounds. The theory is valid also for structures that are both nonspherical and nonparabolic.

In the quoted papers the influence of deviations of a band structure from the simplest case given by $\varepsilon = \hbar^2 k^2/2\, m^*$ on the transport properties are discussed in details. For this reason we are going to stress only the main points concerning deviations from the parabolic shape of a band.

In general case of nonparabolic band the effective mass (0.8) depends upon energy. The relaxation time appearing in (0.7) cannot be expressed in this case by a simple power function of energy [24, 31, 34, 39]. Thus the resulting mobility differs greatly in parabolic and nonparabolic cases [24, 25, 29a, 31]. This must also be taken into account in evaluation of the scattering factors a_r, b_r, q_r [27, 29a, 36].

The expressions for α, α_1, α_2 ((2.21) and (3.4), (3.5)) are of particular interest. They depend upon a parameter which is equal to zero for a parabolic band which can reverse the sign of the expressions if a nonparabolicity is strong enough [31, 35, 39, see also 29]. In various regions of carrier concentration α, α_1, α_2 can have different signs which strongly affect the thermoelectric power in a magnetic field (2.20), and normal and planar Nernst effects ((2.19) and (2.23)).

Thus the mentioned effects depend both on a mechanism of scattering and the shape of a band. For this reason it is of particular interest that the thermoelectric power in a strong transversal magnetic field (3.4) is isotropic and independent of the mechanism of scattering [10a, 18, 39]. (It was Rodot [15] who noticed this fact for the simple structure.) From this point of view the thermoelectric power is similar to the Hall constant in strong magnetic fields. These facts have been used in band structure investigations [45, 46, 47]. As we have already mentioned, one must consider here carefully the phonon drag contribution [10, 11].

The authors would like to thank Prof. L. Sosnowski for critical remarks on the subject.

References

[1] M. Bronstein, Phys. Z. Sowjetunion **2**, 28 (1932).
[2] D. Blochinzev and L. Nordheim, Z. Phys. **84**, 168 (1933).
[3] L. Davis, Phys. Rev. **56**, 93 (1939).
[4] F. Seitz, Phys Rev. **79**, 376 (1950).
[5] G. L. Pearson and H. Suhl, Phys. Rev. **83**, 768 (1951).
[6] G. L. Pearson and C. Herring, Physica's Grav. **20**, 975 (1954).
[7] B. Abeles and S. Meiboom, Phys. Rev. **93**, 1121 (1954); **95**, 31 (1954).
[8] M. Shibuya, Phys. Rev. **95**, 1385 (1954).
[9] C. Herring, Bell Syst. tech. J. **24**, 237 (1955).
[10] C. Herring, T. H. Geballe, and J. E. Kunzler, a) Phys. Rev. **111**, 36 (1958); b) Bell Syst. tech. J. **38**, 657 (1959).
[11] I. V. Motchan, Yu. N. Obraztsow, and T. V. Krylova, J. tech. Phys. (U.S.S.R.) **27**, 242 (1957).
[12] C. Goldberg and R. E. Davis, Phys. Rev. **94**, 1121 (1954).
[13] P. I. Baranskii and P. M. Kurilo, Fiz. tverd. Tela **6**, 54 (1964).
[14] F. T. Bass and I. M. Tsidil'kovskii, J. tech. Phys. (U.S.S.R.) **24**, 1834 (1954).
[15] M. Rodot, C. R. Acad. Sci. (France) **243**, 129 (1956).
[16] J. F. Nye, Physical Properties of Crystals, Clarendon Press, Oxford 1957; Russian transl.: Moscow 1960.
[17] F. J. Blatt, Theory of Mobility of Electrons in Solids, Solid State Physics **4** (1957); Russian transl.: Moscow 1963.
[18] I. M. Tsidil'kovskii, Thermomagnetic Phenomena in Semiconductors, Fizmatgiz, Moscow 1960.
[19] E. H. Putley, The Hall Effect and Related Phenomena, Butterworths, London 1960.
[20] A. I. Anselm, Introduction to Theory of Semiconductors, Fizmatgiz, Moscow 1962.
[21] A. H. Wilson, The Theory of Metals, Cambridge U. P., 1954.
[22] D. Pines, Electron Interaction in Metals, Solid State Physics **1** (1955) (some values in this paper are quoted erroneously, see [25, 17]).

[23] R. BARRIE and J. T. EDMOND, J. Electronics and Control **1**, 164 (1955).
[24] R. BARRIE, Proc. Phys. Soc. **B 69**, 553 (1956).
[25] R. BARRIE, Phys. Rev. **103**, 1581 (1956).
[26] E. O. KANE, J. Phys. Chem. Solids **1**, 249 (1957).
[27] H. EHRENREICH, J. Phys. Chem. Solids **2**, 131 (1957).
[28] M. RODOT, Ann. Phys. (France) **5**, 1085 (1690).
[29] J. KOŁODZIEJCZAK, Acta phys. Polon., a) **20**, 289 (1961); b) **20**, 379 (1961).
[30] T. C. HARMAN and J. M. HONIG, J. Phys. Chem. Solids **23**, 913 (1962).
[31] J. KOŁODZIEJCZAK and L. SOSNOWSKI, Acta phys. Polon. **21**, 399 (1962).
[32] J. KOŁODZIEJCZAK, Acta phys. Polon. **21**, 637 (1962).
[33] L. SOSNOWSKI, Summer School of Physics, Varenna 1961, Nuovo Cimento.
[34] J. KOŁODZIEJCZAK, L. SOSNOWSKI, and W. ZAWADZKI, Rep. Int. Conf. Semicond. Phys., Exeter 1962.
[35] W. ZAWADZKI, phys. stat. sol. **2**, 385 (1962).
[36] W. ZAWADZKI, phys. stat. sol. **3**, 1006 (1963).
[37] W. ZAWADZKI, phys. stat. sol. **3**, 1421 (1963); Phys. Letters **4**, 190 (1963); phys. stat. sol. **3**, 692 (1963).
[38] S. ZUKOTYŃSKI and J. KOŁODZIEJCZAK, phys. stat. sol. **3**, 990 (1963).
[39] J. KOŁODZIEJCZAK and S. ŻUKOTYŃSKI, phys. stat. sol. **5**, 145 (1964).
[40] L. SOSNOWSKI, Summer School of Physics, Hercegnovii 1963.
[41] W. ZAWADZKI and J. KOŁODZIEJCZAK, phys. stat. sol. **6**, 409 (1964).
[42] W. ZAWADZKI, Magnetic Susceptibility of Semiconductors with Nonstandard Energy Bands, Rep. Int. Conf. Semicond. Phys., Paris 1964.
[43] G. I. GUSEVA and I. M. TSIDIL'KOVSKII, Fiz. tverd. Tela **4**, 2490 (1962).
[44] L. L. KORENBLIT, D. V. MASHOVETS, and S. S. SHALYT, Fiz. tverd. Tela **6**, 559 (1964).
[45] I. V. MOTCHAN, Y. N. OBRAZTSOV, and T. V. SMIRNOVA, Fiz. tverd. Tela **4**, 1021 (1962).
[46] V. V. GALAVANOV, F. P. KASAMANLY, and O. V. YEMELYANENKO, Fiz. tverd. Tela **5**, 616 (1963).
[47] P. BYSZEWSKI, J. KOŁODZIEJCZAK, and S. ŻUKOTYŃSKI, phys. stat. sol. **3**, 1880 (1963).

(Received May 12, 1964)

phys. stat. sol. 6, 429 (1964)

Institute of Physics, Polish Academy of Sciences, Warsaw

Theory of Carrier Recombination at Dislocations in Germanium

By

T. Figielski

On the basis of Read's dislocation model, a theory of the recombination of carriers in n-type germanium is developed, including both the transient and steady state. The starting point for the present considerations is the assumption that the carrier capture cross-sections are dependent on the charge density of the dislocation line, an idea similar to that given previously by Morrison. The theory deals mainly with temporary trapping of holes (a phenomenon which was observed recently in plastically deformed germanium) and yields good agreement with experiment.

Auf der Grundlage des Readschen Versetzungsmodells wird eine Theorie der Rekombination von Ladungsträgern an Versetzungen in n-leitendem Ge entwickelt, die den stationären und nichtstationären Fall umfaßt. In Übereinstimmung mit den Vorstellungen von Morrison beruhen die Rechnungen auf der Annahme, daß die Einfangquerschnitte von der Zahl der Ladungen auf der Versetzungslinie abhängen. In erster Linie wird das Anhaften von Löchern behandelt (das in plastisch deformiertem Ge experimentell beobachtet wurde). Die Ergebnisse stimmen gut mit experimentellen Daten überein.

1. Introduction

It is well known that edge dislocations act as effective centres of recombination for excess current carriers in germanium [1, 2, 3, 4, 5, 6]. The simple inverse proportional relation between minority carrier lifetime and dislocation density has been confirmed in both n- and p-type materials, when recombination occurred through dislocation levels. The lifetime increases with the temperature indicating an activation energy ranged usually between 0.15 eV and 0.35 eV (as was observed by the present author, these values depend also on the injection level). Below 200 °K the lifetime varies as T^{-n}, where n lies between 2,5 and 4. The recent experiments by Figielski and Belyaev [7, 8] have shown that at low temperature and low injection levels the rate of electron capture is reduced by dislocations in n-Ge; thus the dislocations act rather as temporary traps for holes. The trapping at dislocations is a very effective process and it may be wholly responsible for the photoconductive characteristics of n-type germanium at low temperatures.

In germanium a dislocation behaves as a row of acceptor centres. The electrostatic energy of such system increases, when electrons are captured in it, limiting the occupation of dislocation levels. As a result, the Fermi-Dirac distribution function is not useful in this case. Equally the usual Read-Shockley theory of recombination via impurity levels [9] cannot be applied here. Statistics of occupation of dislocation centres have been considered by Read [10, 11]. He has found, using several approximate methods, that the fraction of occupied sites at a dislocation decreases from about 0.1 at absolute zero to about 0.03 at room temperature. This change is too small to explain directly the observed temperature dependence of the lifetime.

Yet in 1956 MORRISON [12] in his theory of recombination called attention to the role of the electrostatic potential around a dislocation in the electron capturing process. However, his approximate considerations could not fully explain the experimental characteristics.

Recently [13], an attempt was made to describe the recombination properties of dislocations on the basis of BONCH-BRUEVICH's theory of dislocation bands [14]. According to this theory the whole observed temperature variation of the lifetime has to be attributed to the change of the capture cross sections [6]. The low temperature trapping phenomena are beyond the scope of theory.

A number of new experimental observations [7, 8] induced us to find a theoretical model which describes, at least qualitatively, the behaviour of a dislocation as a trap in n-type germanium. The theory starts from Read's model of the dislocation, and it is in a certain manner a development of MORRISON's idea. The basic for the present considerations is the concept of a variable effective probability of electron capture, depending on the charge in the dislocation line (or on the potential barrier high). Though such an assumption may be questionable in the light of recent papers of BONCH-BRUEVICH and his collaborators [15, 16], we shall accept it. We assume also that a single acceptor level is responsible for all observed properties of dislocations.

2. Model and Notations

We consider n-type germanium with the density of dislocations N_d(cm^{-2}). The site at a dislocation, which is able to accept an electron, corresponds to a dangling bond at the edge of an extra lattice plane. Each site represents a single acceptor level. Let a be the spacing between the sites. Thus $1/a$ is the number of sites per unit length of dislocation and N_d/a — the number of sites per cm³. f is the fraction of sites that are occupied by electrons.

A cylinder of positive space charge is formed around the dislocation. Inside the space charge region the electrostatic potential decreases toward the dislocation. Assuming a constant spacing between the charges at the sites, the potential Φ at one accepted electron due to all others and the positive space charge is according to READ [8]

$$e\,\Phi = -\,\varepsilon_0\, f\,[3 \ln (f/f_c) - 1.232]\,, \tag{1}$$

where $\varepsilon_0 = 0.9\,a$ (eV/Å), and f_c is a parameter determined by the surplus concentration of donors over acceptors $N_d - N_a$, as $f_c = a\,\pi^{1/3}\,(N_d - N_a)^{1/3}$. In deriving of (1) the contribution of current carriers to the space charge was neglected. We shall identify Φ with the potential barrier height at the dislocation. The distorted band about the dislocation is shown schematically in Fig. 1.

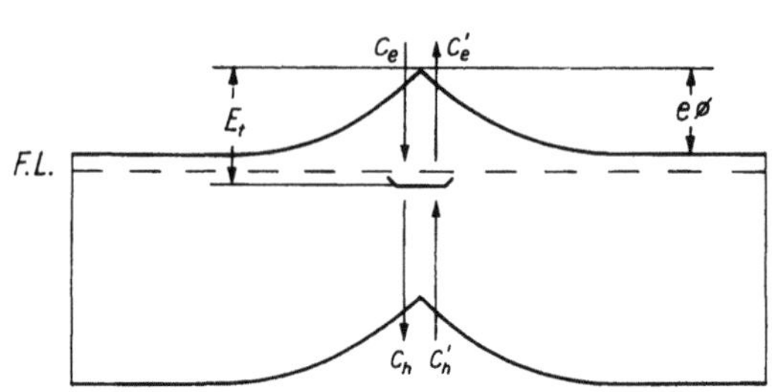

Fig. 1. Energy band distortion at dislocation

By n, p will be denoted the concentrations of electrons and holes outside the space charge region, and by n_b, p_b the respective concentrations at the dislocation (in the centre of space charge region). To mark that a quantity refers to the equilibrium state, the index 0 will be added; for instance n_0, n_{b0}, f_0. We shall use the relations

$$n_b = n \exp(-\beta \Phi) \,, \quad p_b = p \exp(\beta \Phi)$$

and

$$n_{b0} = n_0 \exp(-\beta \Phi_0) \,, \quad p_{b0} = p_0 \exp(\beta \Phi_0) \,, \tag{2}$$

where $\beta = e/k\,T$.

We define the following probabilities: the probability per unit time that an electron will be captured by any of available sites — c_e; the probability that an electron from a full site will be emitted into conduction band — c_e'. Similar probabilities for the exchange of a hole between the sites and valence band are c_h, c_h'. All probabilities have the dimension (cm^3 s^{-1}).

Effects of lattice deformation around the dislocation as well as possible Cottrell atmospheres will be neglected in this paper.

3. Steady State

Neglecting the electron tunnelling through the potential barrier, one can express the net rate of electron capture into the dislocation level, R_e (cm^{-3} s^{-1}), as

$$R_e = \frac{N_d}{a} \left[c_e\, n_b\, (1 - f) - c_e'\, N_c\, f\right] \tag{3}$$

and similarly the net rate of hole capture

$$R_h = \frac{N_d}{a} \left[c_h\, p_b\, f - c_h'\, N_v\, (1 - f)\right], \tag{4}$$

where N_c, N_v are the numbers of states (per cm^3) in the conduction and valence bands.

At thermodynamical equilibrium $R_e = R_h = 0$ and hence

$$\left.\begin{aligned} c_e' &= c_e\, n_{b0}\, (1 - f_0)/N_c\, f_0 \,; \\ c_h' &= c_h\, p_{b0}\, f_0/N_v\, (1 - f_0) \,. \end{aligned}\right\} \tag{5}$$

According to Read statistics f_0 is small as compared with unity; therefore the bracket forms in (5) can be omitted. From (3), (4), and (5) one gets

$$R_e = \frac{c_e\, N_d}{a} \left[n_b\, (1 - f) - n_{b0}\, f/f_0\right] ; \tag{6}$$

$$R_h = \frac{c_h\, N_d}{a} \left[p_b\, f - p_{b0}\, (1 - f)\, f_0\right] . \tag{7}$$

Let us design $n = n_0 + \Delta n$, $p = p_0 + \Delta p$, $f = f_0 - \Delta f$, $\Phi = \Phi_0 - \Delta\Phi$; moreover Δn can be neglected as compared with n_0 in the interesting range of generation.

With these notations and using (2) equation (6) can be rewritten in the following form:[1])

$$R_e = \frac{c_e\, n_0\, N_d}{a} \exp(-\beta \Phi) \left[1 - \exp(-\beta \Delta\Phi) + \frac{\Delta f}{f_0} \left(1 + \exp(-\beta\, \Delta\Phi)\right)\right] . \tag{8}$$

---

[1]) Alternatively, we could consider all n_0 electrons per cm^3 as contributing to the capture but with the effective probability $c_e \exp(-\beta \Phi)$.

It is useful to introduce the quantity $\eta = \frac{\Delta f}{f_0}$ which can be called the trapping level (by analogy to the term injection level).

As the potential Φ at the dislocation is proportional to f (omitting the insensitive logarithmic term in (1)), one can write

$$\Phi = \Phi_0 (1 - \eta) , \; \Delta\Phi = \Phi_0 \eta .$$

Introducing these notations into (8), one obtains

$$R_e = \frac{c_e n_0 N_d}{a} \exp[-\beta \Phi_0 (1 - \eta)] [1 - (1 - \eta) \exp(-\beta \Phi_0 \eta)] . \tag{9}$$

Similarly, (7) can be written as

$$R_h = \frac{c_h p_0 f_0 N_d}{a} \exp[\beta \Phi_0 (1 - \eta)] \times$$
$$\times \left[\left(1 + \frac{\Delta p}{p_0}\right)(1-\eta) - \exp(\beta \Phi_0 \eta)\right] . \tag{10}$$

At steady state $R_e = R_h = G$, where G is the rate of generation of electron-hole pairs in the unit volume. Comparing (9) with (10) one obtains

$$(\eta - 1) \Delta p/p_0 = 1 - \eta - \exp(\beta \Phi_0 \eta) -$$
$$- \frac{c_e n_0}{c_h p_0 f_0} \exp(-2 \beta \Phi_0 (1 - \eta)) [1 - (1 - \eta) \exp(-\beta \Phi_0 \eta)] . \tag{11}$$

It can be shown that in all cases, considered in this paper, the term $1 - \eta - \exp(\beta \Phi_0 \eta)$ is negligible compared with the second one on the right hand of equation (11). Thus (11) simplifies to

$$\Delta p = \frac{c_e n_0}{c_h f_0 (1 - \eta)} \exp(-2 \beta \Phi_0 (1 - \eta)) [1 - (1 - \eta) \exp(-\beta \Phi_0 \eta)] . \tag{12}$$

From (9) and (12) we obtain the relation between Δp and G

$$\Delta p = G \tau_h , \tag{13}$$

where the quantity $\tau_h = a \exp(-\beta \Phi_0 (1 - \eta))/c_h f_0 N_d (1 - \eta)$ can be defined as the stationary lifetime of holes.

(9) together with (13) are the basic equations for the steady state. We rewrite them in the form

$$G = \frac{c_e n_0 N_d}{a} \exp(-\beta \Phi_0 (1 - \eta)) [1 - (1 - \eta) \exp(-\beta \Phi_0 \eta)] ; \tag{14}$$

$$\tau_h = a \exp(-\beta \Phi_0 (1 - \eta))/c_h f_0 N_d (1 - \eta) . \tag{15}$$

If $\Phi_0(T)$ is a known function, η can be obtained using (14) as a function of generation at different temperatures, and hence η versus T at constant G is obtainable. Subsequently, making use of the $\eta(T)$ relation, we can find the temperature dependence of hole lifetime τ_h from (15).

To obtain standard graphs it is useful to introduce the reduced (dimensionless) values of G and τ_h defined as follows

$$G^* = G a/c_e n_0 N_d , \quad \tau_h^* = \tau_h c_h N_d/a .$$

Fig. 2. Trapping level $\eta = \Delta f/f_0$ in dependence on the reduced generation rate at different temperatures

Fig. 3. Trapping level versus reduced lifetime of holes at different temperatures

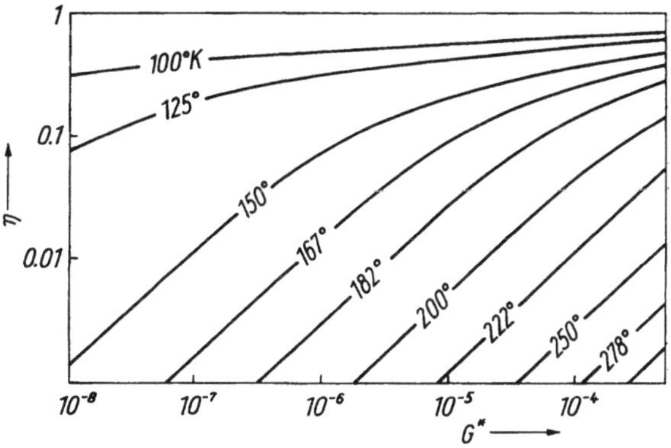

Fig. 2

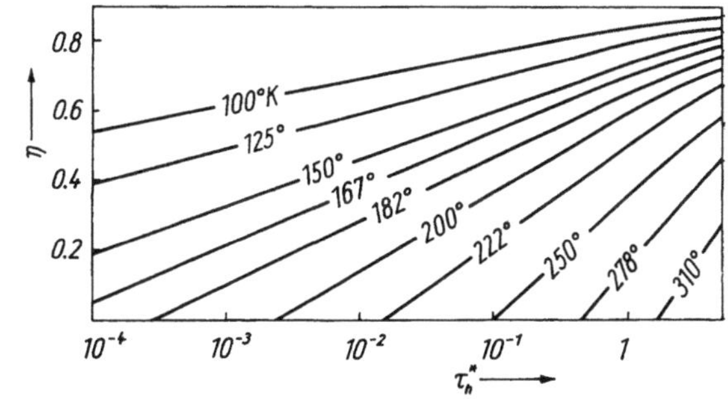

Fig. 3

4. Stationary Characteristics

We shall now apply the equations (13) and (14) to find photoelectric characteristics of a real germanium crystal. We must choose at first a proper function of population $f_0(T)$. So as the exact position of dislocation level is still uncertain one could introduce it as a parameter of the theory. However, the recent experiments [8] performed on plastically deformed Ge samples have shown the longwavelength treshold of photoconductivity at a photon energy of about 0.34 eV. It suggests that the level associated with dislocations lies at the middle of the energy gap. Although this poistion differs from that assumed by READ (0.225 eV below the conduction band), we shall accept it as obtained from direct measurement.

The READ theory has been applied to find the function $f_0(T)$. According to READ's considerations, his minimum energy approximation probably is better at low temperatures and the Fermi statistics at higher. We used a compromise function lying between the two mentioned above. Calculations were done for germanium with 10^{14} electrons per cm^3 ($\varrho \cong 16\ \Omega$cm). In the interesting range of temperature (100 to 300 °K), the function of population can be approximated by

$$f_0 = 0.135 - 3 \times 10^{-4}\, T\,(^\circ\mathrm{K})\,. \tag{16}$$

On the basis of equations (14) and (15), making use of (1) and (16), the curves of $\eta(G^*)$ and $\tau^*(\eta)$ were calculated. They are plotted in Fig. 2 and 3.

The increase in the conductivity of the crystal due to homogeneous illumination is of course

$$\Delta\sigma = e\left[\Delta p\,(\mu_e + \mu_h) + f_0 \frac{N_d}{a}\,\eta\,\mu_e\right] = e\,\mu_e\left(1.5\,G\,\tau_h + f_0\,\frac{N_d}{a}\,\eta\right). \tag{17}$$

We have assumed $\mu_e = 2\,\mu_h$ here. The first term in the bracket describes the photoconductivity of electron-hole pairs and the second one the photoconductivity of free electrons which compensate the excess charge of holes trapped at dislocations. The relative change of conductivity is from (17)

$$\frac{\Delta\sigma}{\sigma} = 1.5\,G^*\,\tau_h^*\,c_e/c_h\,f_0 + f_0\,\frac{N_d}{a\,n_0}\,\eta\,. \tag{18}$$

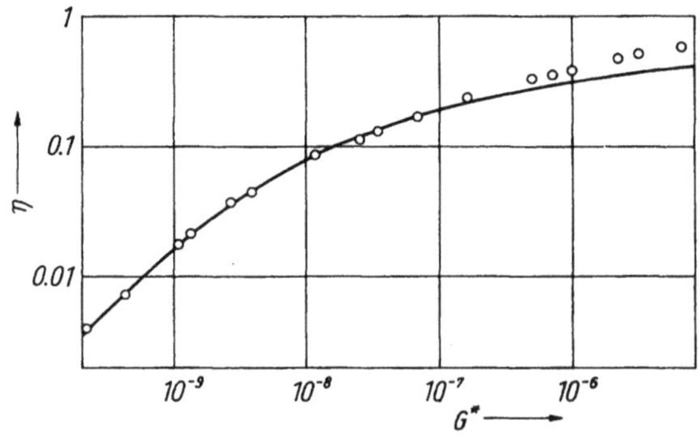

Fig. 4. Comparison of the monopolar electron photocurrent in plastically deformed Ge as a function of light intensity (circles) with the theoretical curve of $\eta(G^*)$ (continuous line). $T = 125\,°\mathrm{K}$

In order to compare the experimental characteristics with those given by (18), the temperature behaviour of the c_e and c_h coefficients ought to be known. So, as the adaptation of LAX's cascade theory [17] in the case of dislocation may be questionable, we shall treat the coefficients c_e, c_h as independent on the temperature at first approximation. This corresponds to the assumption of constant G^* at a given rate of generation.

The graphs in Fig. 2 show in fact the relative number of holes trapped at the dislocation as a function of light intensity. We can compare the shape of this relation with the experimental one obtained at 125 °K (Fig. 4). The agreement is satisfactory, excluding a few final points at high intensities, where appears difficulty in correct separation of the monopolar electron photoconductivity from the total one. If there are also other recombination centres in the sample, the observed characteristics may differ from that in Fig. 2.

Going along the vertical lines ($G^* = \text{const}$) on the diagram in Fig. 2 we can find the temperature dependence of η. Making use of this relation and of the graphs in Fig. 3, one can plot the lifetime of holes τ_h^* versus $1/T$ (Fig. 5). The general shape of this relation reminds that given by the usual Read-Shockly statistics. In our case, however, the values of τ_h^* at low temperature ("plateau" region) depend strongly on the injection level. It is worth noting that this relation cannot be simply proofed experimentally. For instance, applying a strong background illumination to eliminate the trapping phenomena, the value of η and at the same time τ_h will change. On the other hand one cannot simply use the photomagnetoelectric effect to determine τ_h at low temperature for the reason which will be given in the following section. We shall compare rather $\Delta\sigma/\sigma$ as given by the formula (18) with corresponding experimental characteristics. For exact comparison the ratio c_e/c_h has to be known. This ratio can be expected to be

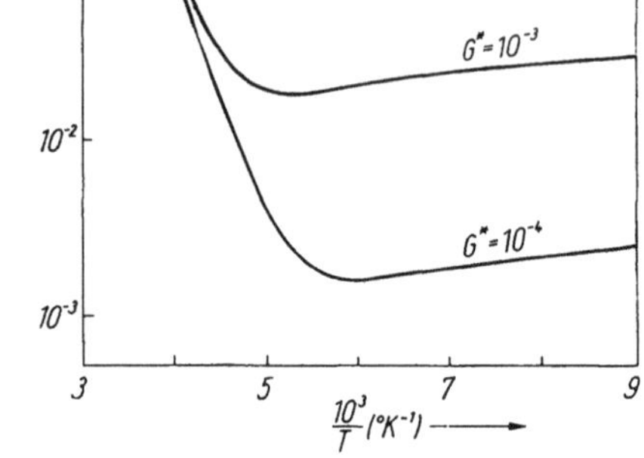

Fig. 5. Temperature dependence of the reduced lifetime of holes at different generation rates

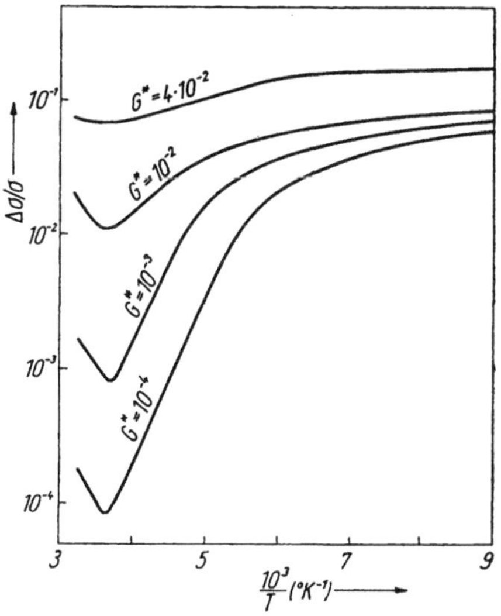

Fig. 6

Fig. 6. Relative photoconductivity as a function of reciprocal temperature at different rates of generation ($N_d = a\,n_0$)

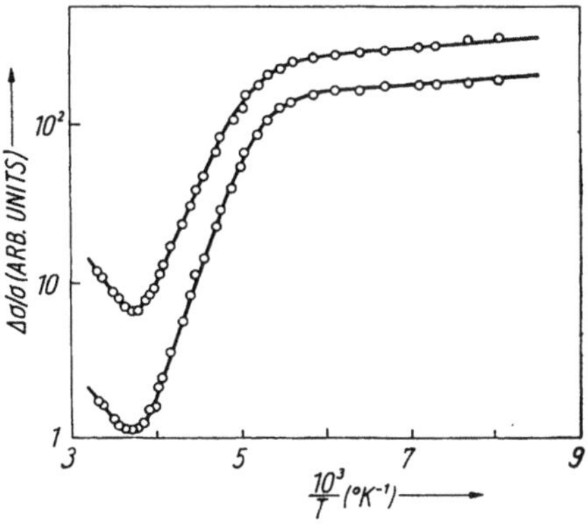

Fig. 7

Fig. 7. Experimental curves of $\Delta\sigma/\sigma(1/T)$ for plastically deformed Ge with $N_d = 10^7$ cm^{-2}, at two illuminations

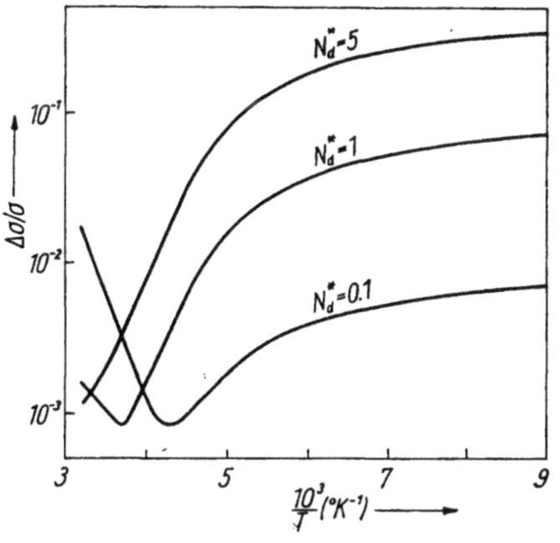

Fig. 8

Fig. 8. Relative photoconductivity as a function of reciprocal temperature at constant generation ($G^* = 10^{-3}$) for different numbers of dislocations ($N_d^* = N_d/a\,n_0$)

close to unity on account of the middle position of the dislocation level relatively to the band edges.

With this assumption the runs of $\Delta\sigma/\sigma(1/T)$ are plotted in Fig. 6 for some concentration of dislocations and different illuminations. They may be directly compared with the experimental data (Fig. 7). The lower curve in Fig. 7 was taken at a light intensity as low as possible. About the same slopes of the increasing parts of the bottom curves in Fig. 6 and 7 give evidence that the assumption about the position of the dislocation level was essentially correct. At higher intensities these slopes become light sensitive. The falling parts in the room-temperature range (connected with the change of hole lifetime) are less significant because the assumed population according to (16) may be incorrect in this range. If the coefficient c_e decreased with the temperature, which is probably true at low temperatures, then the saturation parts of the curves would be flatter as compared with those shown in Fig. 7.

Fig. 8 shows the curves of $\Delta\sigma/\sigma(1/T)$ at a constant illumination and different densities of dislocations. The displacement of the rising part of the characteristics towards the higher temperatures with increasing number of dislocations is noticeable, in agreement with observations (7).

5. Inhomogeneous Generation

Let us consider the surface generation, e.g. the situation when strongly absorbed light falls on one side of the sample. We want to find the distribution of holes (or electrons) inside the sample along the axis perpendicular to the illuminated surface plane. At a low trapping level ($\eta \ll 1$) the stationary lifetime of holes τ_h has a constant value. In such case the hole distribution has the usual exponential form and can be characterized by the mean diffusion length of holes $L_h = (D_h \tau_h)^{1/2}$, where D_h is the diffusion coefficient. When the condition $\eta \ll 1$ is not fulfilled (low temperatures, high injection levels), the situation becomes more complicated; τ_h is now a function of η.

In the one-dimensional case we have to solve the continuity equation

$$D_h \frac{d^2(\Delta p)}{dx^2} = \frac{\Delta p}{\tau_h}, \tag{19}$$

where Δp and τ_h are mutually connected through the parameter η (equations (13) and (15)). Limiting the consideration to the range of high but not too much values of η, one can neglect the variation of $1 - \eta$ as compared with that of $\exp[-\beta \Phi_0 (1 - \eta)]$ in (13). Then, using (13), one can approximately write

$$D_h \frac{d^2(\Delta p)}{dx^2} = A(\Delta p)^{1/2}, \tag{20}$$

where $A = \frac{N_d}{a D_h} (c_e c_h f_0 n_0 (\overline{1 - \eta}))^{1/2}$, and $\overline{1 - \eta}$ is some mean value of $1 - \eta$.

The solution of (20) which satisfies the boundary conditions

$$\Delta p|_{x=\xi} = \Delta p(\xi), \qquad \left.\frac{d(\Delta p)}{dx}\right|_{x=\xi} = \Delta p'(\xi)$$

is of the form

$$x - \xi = \int_{\Delta p(\xi)}^{\Delta p} \{4/3\, A\, [(\Delta p)^{3/2} - (\Delta p(\xi))^{3/2}] + [\Delta p'(\xi)]^2\}^{-1/2}\, d(\Delta p). \tag{21}$$

We assume that the plane of generation lies at $x = 0$ and the crystal extends to $x = \infty$. Then the concentration as well as diffusion current of holes has to fall with increasing x. Thus one can assume that at some distance ξ $\Delta p(\xi) = \Delta p'(\xi) = 0$. Within these conditions the excess hole distribution is from (21)

$$\left.\begin{aligned} \Delta p &= (A/12)^2 (x - \xi)^4 &&\text{for } x < \xi; \\ \Delta p &= 0 &&\text{for } x \geqq \xi. \end{aligned}\right\} \tag{22}$$

From (22) we can write the following relations:

$$\left.\begin{aligned} -\left.\frac{d(\Delta p)}{dx}\right|_{x=0} &= \frac{4\,\Delta p(0)}{\xi}, \\ \Delta P = \int_0^{\xi} \Delta p\, dx &= \frac{\Delta p(0)}{5}\,\xi, \end{aligned}\right\} \tag{23}$$

where $\Delta p(0)$ is the value of Δp at $x = 0$.

Let us compare the formulæ (23) with those obtained for the usual exponential distribution of carriers,

$$\left.\begin{aligned} -\left.\frac{d(\Delta p)}{dx}\right|_{x=0} &= \frac{\Delta p(0)}{L_h}; \\ \Delta P = \int_0^{\xi} \Delta p\, dx &= \Delta p(0) \cdot L_h. \end{aligned}\right\} \tag{24}$$

It can be seen that the quantity $\xi/4$ (or $\xi/5$) plays the role of the diffusion length in (23). It may be called the effective diffusion length of holes for the case of high trapping level, L_h^*.

Within the boundary condition at the surface $D_h \left. \frac{d(\Delta p)}{dx} \right|_{x=0} = -\Gamma$, where Γ is the rate of surface generation, we obtain for L_h^*

$$L_h^* = \xi/4 = \left[\frac{9}{16} \cdot \frac{D_h\, a^2\, \Gamma}{N_d^2\, c_e\, c_h\, f_0\, n_0\, (1-\eta)}\right]^{1/3}. \tag{25}$$

It must be remembered that beyond some distance from the surface (where $\eta \ll 1$) the distribution given by (22) is followed by the usual exponential tail.

There is also an other reason for which one cannot simply use the idea of mean diffusion length. The apparent carrier diffusion may occur preferably along the dislocation lines. It is due to the fact that the dislocation in n-type material acts as a transmission line for a hole which entered into a p-type cylinder. Owing to that a hole can be transmitted along the dislocation line to a considerable distance before recombining. Such behaviour was observed for diluted arrays of dislocations [18] as well as for grain boundaries [19, 20]. This effect may be dominant at low temperature.

So the discussion of experimental results based on the concept of diffusion length (as for instance PME-effect) has to be done with great caution.

6. Transient Process

Let us consider the decay of excess current cariers at low temperature in the presence of dislocations. We are interested in the relaxation of photocurrent after illumination of the sample with a rectangular light pulse, the duration of which was sufficiently long to attain the steady state. If there are no other recombination centres in the sample, the decay of excess carriers is only limited by the net rate of electron capture into dislocation sites. Hence on can write, using (9),

$$-\frac{d\eta}{dt} = \frac{1}{f_0}\frac{df}{dt} = c_e\, n_0\, f_0^{-1} \exp\left(-\beta\, \Phi_0\, (1-\eta)\right) \left[1-(1-\eta) \exp\left(-\beta\, \Phi_0\, \eta\right)\right]. \tag{26}$$

This is the general form of the kinetic equation forthe relaxation process. Let us consider the solution of (15) in two extreme cases for which integrating can be performed analytically.

When the deviation of the number of charges in dislocations from its equilibrium value is small ($\eta \ll 1$), thus the equation (15) reduces to

$$-\frac{d\eta}{dt} = \frac{\eta}{\tau_e}, \tag{27}$$

where $\tau_e = \exp(\beta\, \Phi_0)/c_e\, n_0\, f_0^{-1}\, \beta\, \Phi$ has the meaning of electron lifetime at low trapping level.

The solution of (27) is of course

$$\eta = \eta(0) \exp(-t/\tau_e), \tag{28}$$

where $\eta(0)$ is the value of η at $t = 0$.

Hence a simple exponential decay occurs, characterized by the relaxation time τ_e which does not depend on the dislocation density and is a decreasing function of the temperature.

In the opposite case, when the number of charges on dislocations deviates strongly from its equilibrium value $|1 > \eta \gtrsim (\beta\,\Phi_0)^{-1}|$, the equation (16) becomes

$$\frac{d\eta}{dt} = c_e\, n_0\, f_0^{-1} \exp\left[-\beta\,\Phi_0\,(1-\eta)\right]. \tag{29}$$

The solution of (29) is of the form

$$\eta = \eta(0) - (\beta\,\Phi_0)^{-1} \ln\,(1 + t/\tau_0)\,, \tag{30}$$

where

$$\tau_0^{-1} = c_e\, n_0\, f_0^{-1}\, \beta\,\Phi_0 \exp\left[-\beta\,\Phi_0\,(1-\eta)\right] = \beta\,\Phi_0 \frac{G\,a}{f_0\, N_d}$$

in order to satisfy the boundary condition

$$\left.\frac{d\eta}{dt}\right|_{t=0} = -\,G\,a/f_0\, N_d\,.$$

According to (30) the excess electron current (or η) decreases logarithmically with time.[2])

We return once more to (26). For a fixed temperature, the rate of decay at any moment is a function of η only. So we can draw a standard decay curve for each temperature. An example of that is shown in Fig. 9.

Let us assume, the initial state of the crystal at the moment of interruption of illumination corresponds to the value of η close to unity: for instance $\eta = 0.9$. It means that one tenth of the equilibrium electron charges remains only in the sites at dislocations. The potential barrier is then low and the rate of electron capture is high. As time elapses, electrons are being captured on dislocations and the potential barrier is being restored. The decay rate is steadily decreasing. When η becomes small compared with unity, which means that almost all of the possible sites are occupied, the further decay occurs at a constant potential barrier, e. g. at a steady rate. Then the process becomes an exponential one and can be characterized by the relaxation time given by (28).

In fact one observes both fast and the following slow decay in germanium at low temperature. However, they were usually attributed to different processes connected with several sets of levels in the forbidden gap.

According to (28) the realaxation time of the exponential tail would increase steadily with lowering temperature as $f_0 \exp(\beta\,\Phi_0)/\beta\,\Phi_0$. Experimentally a ten-

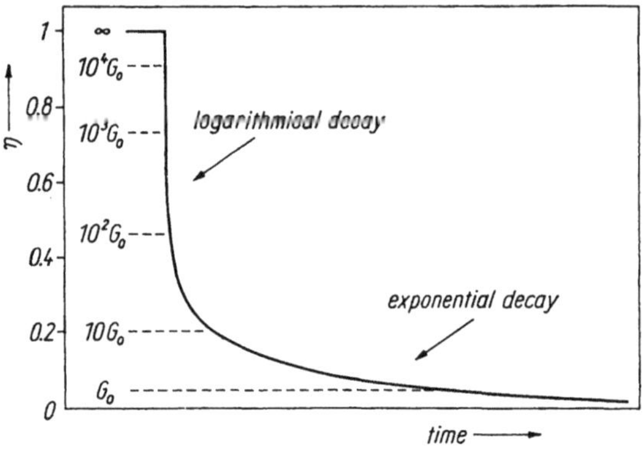

Fig. 9. Decay curve of excess carriers at low temperature. The dashed lines correspond to the different initial light intensities

---

[2]) Such a law of photocurrent relaxation has been recently observed in boron at room temperature [21].

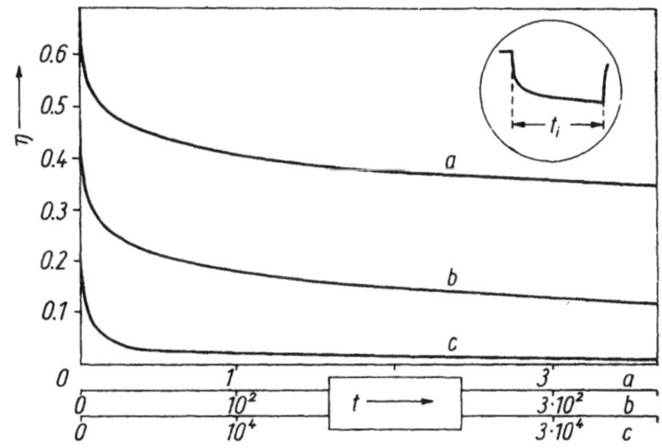

Fig. 10. Decay curve of excess carriers at low temperature drawn at three different time bases

dency to reach saturation is observed below 200 °K. It will be shown beneath that the apparent saturation of τ_e can be explained as a simple apparatus effect.

An experimentator usually observes the decay of photocurrent on the oscilloscope screen at the repetition rate of light pulses of order of 100 Hz. Let us look at Fig. 10 in which the theoretical curve of decay is shown, drawn on the basis of equation (26). Though the run of the two upper curves satisfies the logarithmic law of decay, the observer will confirm that the decay consists of at least two processes, the fast nonexponential following by the slower and exponential one.

In fact, one can draw the logarithm of photoresponse of the "slow process" as a function of time. Assuming as the zero level the photoresponse just before incoming of the subsequent light pulse, one obtains a straight line (excluding few final points, which can be easily attributed to the experimental error). It may be noted that the relaxation time determined from the slope of this line is equal to the time interval between two successive light pulses t_i divided by 2.7.

We can now explain the apparent temperature dependence of the decay time. Going from room to lower temperature one observes an increasing value of the relaxation time of the exponential process. When this value exceeds the inverse of the repetition rate, the apparent time of the logarithmic decay is rather measured (equal to $t_i/2.7$). Thus one finds the maximum value of τ_e of the order of 10 ms at usual repetition rates in agreement with observations. In Fig. 11 there are shown experimental values of τ_e taken from [7], and a theoretical curve drawn on the basis of equation (27).

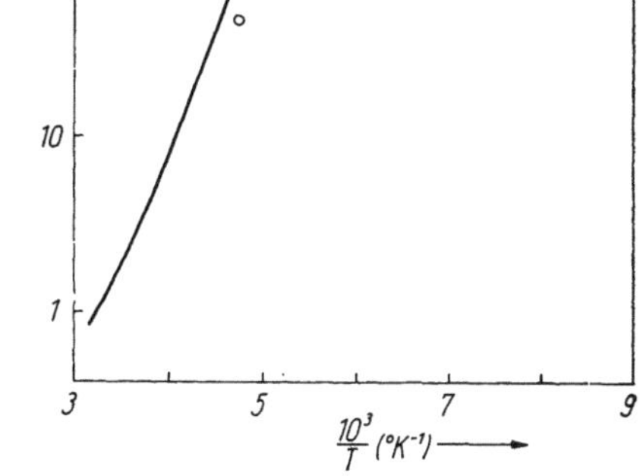

Fig. 11. Experimental dependence of $\tau_e(1/T)$ (circles) in comparison with a theoretical curve drawn on the basis of equation (27) (continuous line). Both curves are fitted at 200 °K. The dashed line corresponds to the apparent limit of τ_e connected with the light pulse repetition rate 100 Hz

7. Final Remarks

A dislocation represents a giant crystal defect extending over a great number of interatomic distances. The whole defect behaves (at least at low temperatures) as a single system which can be in different charge states. The number of electrons per unit length of dislocation may continuously change from zero to a value of order of one million. The rates of the capture of excess electrons or holes strongly depend on the electrostatic potential connected with the charge on the dislocation line.

One can estimate the values of probability coefficients for the capture of an electron and a hole c_e, c_h. Using the experimental value of τ_e (for instance in Ge with $n = 10^{14}$ cm^{-3} τ_e is equal to about 200 μs at 200 °K) we find from (27) $c_e \simeq 4 \times 10^{-8}$ s^{-1} cm^3. On the other hand it is known that the factor $N_d \tau_h$ in n-type Ge is equal to about unity at room temperature. Because the value of τ_h^* according to the plot in Fig. 3 is also close to unity at this temperature, we obtain $c_h/a = 1$ and hence $c_h \simeq 4 \times 10^{-8}$ s^{-1} cm^3. Thus one has $c_e \simeq c_h$ in agreement with our previous assumption.

The author is greatly indebted to Prof. L. SOSNOWSKI for his useful comments and criticism. Thanks are also due to Mrs. J. WAŚNIEWSKA and Mr. R. KUŚNIERZ who made numerical calculations.

References

[1] A. KURTZ, S. KULIN, and B. AVERBACH, Phys. Rev. **101**, 1285 (1956).
[2] J. OKADA, J. Phys. Soc. Japan **10**, 1110 (1955).
[3] G. WERTHEIM and G. PEARSON, Phys. Rev. **107**, 694 (1957).
[4] A. D. BELYAEV, V. N. VASILEVSKAYA, and E. G. MISELYUK, Fiz. tverd. Tela **2** 227 (1960).
[5] M. I. IGLITSYN and L. I. KOLESNIK, Fiz. tverd. Tela **2**, 1542 (1960).
[6] L. I. KOLESNIK, Fiz. tverd. Tela **4**, 1449 (1962).
[7] T. FIGIELSKI and A. D. BELYAEV, Fiz. tverd. Tela (in press).
[8] A. D. BELYAEV and T. FIGIELSKI, Ukr. fiz. Zh. **7**, 1179 (1963).
[9] W. SHOCKLEY and W. T. READ, Phys. Rev. **87**, 835 (1952).
[10] W. T. READ, Phil. Mag. **45**, 775 (1954).
[11] W. T. READ, Phil. Mag. **45**, 1119 (1954).
[12] S. R. MORRISON, Phys. Rev. **104**, 619 (1956).
[13] YU. V. GULYAEV, Fiz. tverd. Tela, **3**, 1094 (1961).
[14] V. L. BONCH-BRUEVICH, Fiz. tverd. Tela **2** (Suppl.), 182 (1959).
[15] V. L. BONCH-BRUEVICH and V. B. GLASKO, Fiz. tverd. Tela **3**, 36 (1961).
[16] YU. V. GULYAEV, Fiz. tverd. Tela **4**, 1285 (1962).
[17] M. LAX, Phys. Rev. **119**, 1502 (1960).
[18] A. F. GIBSON and E. G. S. PAIGE, Phil. Mag. **3**, 950 (1958).
[19] T. FIGIELSKI and L. SOSNOWSKI, Solid State Electronics **1**, Part 1, 109 (1960).
[20] T. FIGIELSKI, Acta phys. Polon. **19**, 607 (1960).
[21] A. NADOLNY and J. OSTROWSKI, private communication.

(Received May 13, 1964)

phys. stat. sol. 6, 441 (1964)

Institut für angewandte Physik der Technischen Hochschule Karlsruhe

Über den elektrischen Leitungsmechanismus von hexagonalen Selen-Einkristallen

Von

J. Stuke

An hexagonalen Selen-Einkristallen wird die Änderung der Temperatur- und Spannungsabhängigkeit der Leitfähigkeit durch plastische Verformung untersucht. Im ohmschen Gebiet bei hohen Temperaturen nimmt dadurch die Leitfähigkeit zu, im spannungsabhängigen Gebiet bei tiefen Temperaturen dagegen ab. Aus Thermokraftmessungen ergibt sich, daß die Löcherkonzentration in erster Näherung bis −180 °C nicht von der Temperatur und auch nicht von der Beleuchtung abhängt. Bei einer Verformung nimmt die Konzentration dagegen proportional zur Leitfähigkeit bei 20 °C zu.

Die verschiedenen Meßergebnisse lassen sich einheitlich deuten, wenn man in den Selen-Kristallen Potentialschwellen annimmt, die für die Temperatur- und Spannungsabhängigkeit der Leitfähigkeit verantwortlich sind, und deren Höhen durch Lichteinstrahlung stetig verringert werden können.

Changes in the temperature and voltage dependence of conductivity, induced by plastic deformation, are investigated for hexagonal Se single crystals. The conductivity increases in the ohmic region at high temperatures and decreases in the voltage dependent region at low temperatures. From thermal e.m.f. measurements the hole concentration is found to be to a first approximation, independent of light and temperature down to −180 °C. By deformation the concentration increases proportionally to the conductivity at 20 °C.

The various results can be explained if potential walls are assumed to occur in the Se single crystals. These walls are responsible for the temperature and voltage dependence of the conductivity and their height can be decreased by illumination.

1. Experimentelles

1.1 Einleitung

Der elektrische Leitungsmechanismus von hexagonalem Selen ist noch weitgehend ungeklärt. Neben den zahlreichen Untersuchungen [1] an polykristallinem Material liegen auch eine Reihe von Arbeiten über Einkristalle vor. Sie haben im wesentlichen folgendes Ergebnis: Die spezifische Leitfähigkeit von Selen-Kristallen, welche bei Zimmertemperatur etwa 10^{-5} bis $10^{-6}\ \Omega^{-1}\,\mathrm{cm}^{-1}$ beträgt, zeigt eine Temperaturabhängigkeit gemäß der Beziehung $\sigma = \sigma_0 \cdot e^{-\frac{\Delta\varepsilon}{kT}}$. Die Aktivierungsenergie $\Delta\varepsilon$ hat dabei Werte zwischen 0,1 und 0,3 eV [2, 3, 4, 5]. Aus Thermokraftmessungen [3, 4] ergibt sich, daß Selen-Kristalle stets p-leitend sind mit einer Löcherkonzentration von etwa 10^{14} bis $10^{15}\ \mathrm{cm}^{-3}$, die von der Temperatur nur wenig abhängt. Halleffekts-Messungen, die Plessner [4] an einem gutleitenden Kristall durchführte, zeigen dasselbe Ergebnis. Die exponentielle Temperaturabhängigkeit der Leitfähigkeit ist deswegen — wenigstens formal — der Beweglichkeit zuzuschreiben, welche bei Zimmertemperatur etwa 0,1 $\mathrm{cm}^2/\mathrm{Vs}$ beträgt.

Die Leitfähigkeit hängt von der Kristall-Orientierung ab, sie ist in Richtung der c-Achse, d. h. in Kettenrichtung größer als senkrecht dazu [3, 4]. Das Verhältnis $\sigma_{\parallel}/\sigma_{\perp}$ liegt zwischen den Werten 3 und 10. Dabei ist die Aktivierungsenergie $\Delta\varepsilon$ bei einem Kristall in beiden Richtungen annähernd gleich.

Bei Temperaturen unterhalb Zimmertemperatur treten häufig Abweichungen vom Ohmschen Gesetz auf [2, 3, 4, 5]. Der Strom nimmt hier schon bei relativ kleinen Feldstärken stärker als proportional mit der Spannung zu. In einem Falle ergibt sich über mehrere Zehnerpotenzen eine quadratische Kennlinie [5]. Die Temperaturabhängigkeit ist im quadratischen Bereich relativ gering mit einer Aktivierungsenergie von 0,035 eV.

Selen-Kristalle zeigen bei Wechselspannung ähnlich wie polykristallines Selen eine Zunahme der Leitfähigkeit mit der Frequenz [6], die mit einer Abnahme von $\Delta\varepsilon$ gekoppelt ist.

In der Photoleitung tritt ein Maximum an der Kante der Eigenabsorption bei etwa 800 nm auf [7]. Der Photostrom in diesem Maximum ist — besonders bei tiefen Temperaturen — mit einer großen Trägheit behaftet.

Eine Deutung dieser von dem normalen Verhalten von Halbleitern teilweise stark abweichenden Eigenschaften ist in geschlossener Form bisher nicht gelungen. In den meisten Fällen wurden Modell-Vorstellungen an Hand von Einzelergebnissen mit mehr oder weniger Erfolg entwickelt. Die exponentielle Temperaturabhängigkeit der Beweglichkeit legt den Schluß nahe, daß auch in Einkristallen Potentialschwellen vorhanden sind, welche die Löcher durch thermische Aktivierung überwinden müssen. Billig [8] nimmt an, daß diese Schwellen sich an den Kettenenden befinden. Es wurden aber auch besondere Beweglichkeitsmechanismen zur Deutung herangezogen. Schottky [9] schlägt zur Erklärung des kleinen Wertes der Beweglichkeit eine Störbandleitung vor, Gobrecht und Tausend [10] diskutieren eine Art hopping-Prozeß von Atom zu Atom. Ludwig und Eckart [11] schließen aus der Frequenzabhängigkeit der dielektrischen Verluste von polykristallinem Selen auf einen Mechanismus, bei dem die Leitfähigkeit durch Platzwechselvorgänge der Löcher zwischen Potentialmulden zustande kommt. Mit diesen Vorstellungen sind aber andere Eigenschaften, besonders die Spannungsabhängigkeit der Leitfähigkeit, nur schwer zu verstehen. Letzten Endes läuft das vieldiskutierte Problem des Leitungsmechanismus von hexagonalem Selen auf die Frage hinaus, ob die verschiedenen anomalen Eigenschaften auf einen besonderen gittereigenen Mechanismus zurückzuführen sind, der mit der starken Anisotropie der Bindung eng verknüpft ist, oder ob der Leitungsmechanismus im Prinzip ähnlich wie beim isomorphen Tellur ist, wo eine richtungsabhängige Gitter-Beweglichkeit normaler Größe und normaler Temperaturabhängigkeit vorliegt [12]. Im ersten Fall ist Selen als ein Übergang zu den organischen Halbleitern mit Molekülstruktur anzusehen. Im zweiten Fall ist die Ursache der Besonderheiten in einem fehlerhaften Gitterbau oder in einer inhomogenen Störstellenverteilung zu suchen.

Die vorliegende Arbeit beschäftigt sich hauptsächlich mit dieser wichtigen Frage. Es wird dazu die Änderung der elektrischen Eigenschaften von Selen-Kristallen durch plastische Verformung untersucht. Hier stehen besonders die Temperatur- und Spannungsabhängigkeit der Leitfähigkeit im Vordergrund, wobei Thermokraftmessungen zur Ermittlung der auftretenden Konzentrationsänderungen herangezogen werden. Außerdem wird der Einfluß der Verformung auf die Photoleitung, speziell auf ihre spektrale Verteilung, untersucht.

1.2 Meßmethode

Bei den Messungen wurden neben aus der Dampfphase [13] hergestellten Kristallen auch solche benutzt, die in der Schmelze [14] gewachsen waren. Drehkristallaufnahmen zeigen, daß es sich in beiden Fällen um Einkristalle ohne gröbere Fehler handelt. Während die Dampf-Kristalle in ihrer ursprünglichen Form, nämlich als dünne, häufig hohle Nadeln oder dünne Plättchen Verwendung fanden, wurden aus den unregelmäßigen Schmelzkristallen durch Spalten längs der c-Achse die bei der Messung benutzten Plättchen oder Stäbchen hergestellt. Zur Vermeidung von Verformung wurde das in einigen Fällen bei -180 °C durchgeführt.

Als Kontakte bei den Leitfähigkeitsmessungen dienten Gold-Elektroden, die im Vakuum (10^{-5} Torr) aufgedampft wurden. Die in einer früheren Arbeit [5] benutzte Kontaktform, bei der sich die Elektroden auf zwei gegenüberliegenden Flächen eines Einkristallplättchens befinden, erwies sich für die Verformungsversuche als nicht zweckmäßig. Für die Messung der Spannungsabhängigkeit benötigt man einen kleinen Elektrodenabstand, d. h. relativ dünne Kristallplättchen, deren gleichmäßige Verformung schwierig ist. Deshalb wurden die beiden Elektroden nebeneinander auf eine Kristalloberfläche aufgedampft, wobei der Elektrodenabstand durch Abschatten mit schmalen Metallbändern bis 50 μm verringert werden konnte. Für die Messung der Richtungsabhängigkeit wurden zwei Spalte kreuzförmig angeordnet, der eine senkrecht und der andere parallel zur c-Achse. Durch einfaches Umschalten ließ sich dann die Orientierung des Stroms ändern.

Bei den Thermokraftmessungen war eine andere Kontaktanordnung zweckmäßiger. Zwei 50 μm dicke Thermoelemente aus NiCr-Konstantan wurden durch Heizen mit Joulscher Wärme in Kristallstäbchen eingeschmolzen, die eine Dicke von 1 bis 2 mm und eine Länge von 5 bis 10 mm hatten. Die hierbei entstandene dünne Zone aus amorphem Selen konnte durch anschließendes Tempern bei etwa 180 °C in polykristallines Selen umgewandelt werden. Da selbst eine dünne amorphe Zwischenschicht den Wert der Thermokraft nicht verändert, entsteht dadurch kein Meßfehler. Die NiCr-Schenkel der Thermoelemente dienten zur Messung der Thermospannung und als Potentialsonden für die Leitfähigkeitsmessung.

Die elektrischen Messungen wurden mit hochohmigen Röhrenvoltmetern und mit X-Y-Koordinatenschreibern durchgeführt. Die Proben befanden sich in einem Kühltopf üblicher Bauart im Vakuum, dessen Güte keinen nennenswerten Einfluß auf die Meßergebnisse hatte. Zur Messung der Zeitabhängigkeit des Photostroms konnten wegen der großen Trägheit Kompensations-Schreiber verwendet werden.

Die plastische Verformung wurde durch Stauchen mit einem kleinen Stempel oder durch Pressen in einem kleinen Schraubstock durchgeführt. Selen-Kristalle lassen sich — ähnlich wie Tellur [15] — senkrecht zur Kettenrichtung relativ leicht verformen. Man kann bei Zimmertemperatur ohne weiteres Längenänderungen von 50 vH und mehr erreichen. In Kettenrichtung ist dagegen eine Verformung sehr viel schwieriger; die Kristalle reißen bei Zug- oder Biegeversuchen, bevor eine nennenswerte plastische Verformung eintritt, und beim Stauchen splittern sie leicht in viele kleine Kristallbänder auf. Bei den hier vorliegenden Untersuchungen wurde deshalb die Verformung nur senkrecht zur c-Achse durchgeführt, und zwar wegen der zum Teil sehr geringen Größe der Kristalle in qualitativer Weise.

1.3 Meßergebnisse

1.31 Dunkel-Leitfähigkeit

Die spezifische Leitfähigkeit bei Zimmertemperatur nimmt durch plastische Verformung beträchtlich zu. An Kristallen, die vorher eine relativ niedrige Leitfähigkeit von etwa 10^{-6} bis 10^{-7} $\Omega\,cm^{-1}$ zeigen, kann man durch starke Verformung eine Zunahme um zwei Zehnerpotenzen erreichen. Die spannungsabhängige Leitfähigkeit bei tiefen Temperaturen nimmt dagegen ab. Diese Abnahme ist meistens sogar wesentlich größer als die Zunahme bei Zimmertemperatur. Geht man z. B. von einem Kristall aus, der vor der Verformung oberhalb Zimmertemperatur eine ohmsche Strom-Spannungs-Beziehung und bei — 180 °C eine annähernd quadratische Kennlinie hat (Fig. 1a, ausgezogene Kurven)[1]), so ergibt sich nach der Verformung ein anderes Bild (gestrichelte Kurven).

Die Überschneidung der Kennlinien zeigt, daß die Abweichung vom Ohmschen Gesetz nach der Verformung erst bei größerer Spannung einsetzt. Bei — 180 °C ist die Kurve aber steiler als vor der Verformung. Besonders deutlich kommt dieser Zusammenhang in der Temperaturabhängigkeit zum Ausdruck (Fig. 1b). Im ohmschen Bereich bei hohen Temperaturen wird die Kurve durch die Verformung annähernd parallel zu höherer Leitfähigkeit verschoben. Mit abnehmender Temperatur setzt die Spannungsabhängigkeit später ein, wird dann aber sehr ausgeprägt. Der Charakter der Temperatur- und Spannungsabhängigkeit wird offenbar etwas geändert. Vor der Verformung tritt beim Übergang vom ohmschen zum spannungsabhängigen Bereich ein deutlicher Knick auf, der durch die verschiedene Aktivierungsenergie im ohmschen und im quadratischen Bereich bedingt ist. Nach der Verformung zeigen die Kurven dagegen eine stetige Krümmung, die mit der Spannung stark wächst. An sehr stark verformten Kristallen kann man für kleine Spannungen ($\approx$ 1 V) eine Temperaturabhängigkeit erhalten, die bis an-

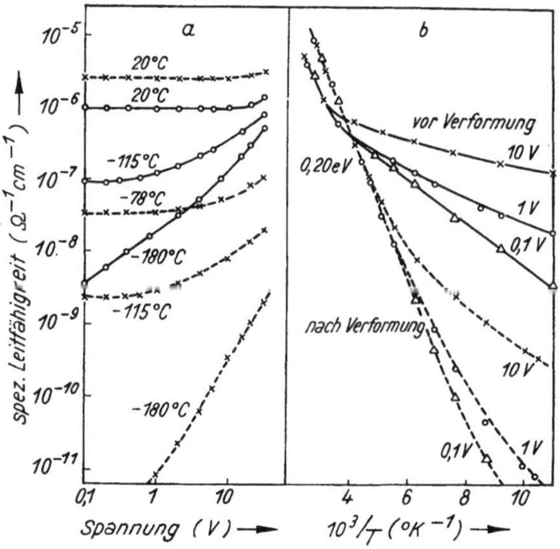

Fig. 1. Temperatur- und Spannungsabhängigkeit der Leitfähigkeit eines hexagonalen Selen-Kristalls vor und nach plastischer Verformung. Elektrodenabstand $L = 1 \times 10^{-2}$ cm; Stromrichtung ‖ zur c-Achse

[1]) Im strengen Sinn ist eine spezifische Leitfähigkeit nur in dem ohmschen Bereich definiert. Damit die Messungen an verschiedenen Kristallen miteinander verglichen werden können, wird diese Darstellung auch im spannungsabhängigen Gebiet beibehalten.

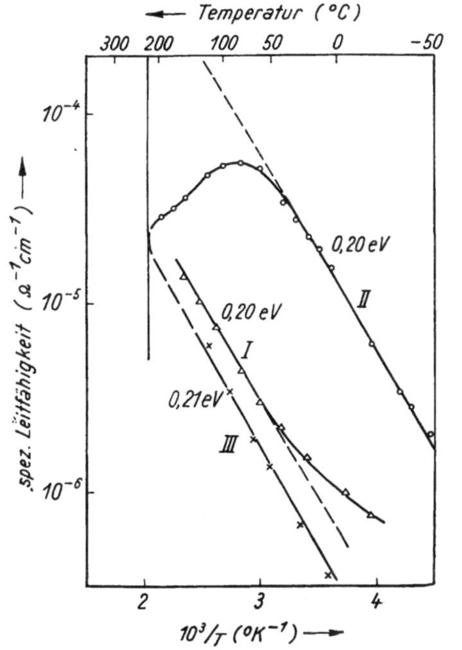

Fig. 2. Isochrone Ausheilung der durch Verformung erzeugten Leitfähigkeit.
I: Messung vor Verformung;
II: nach Verformung;
III: nach Ausheilung
$dT/dt \approx 10°/\text{min}$

nähernd — 180 °C über vier Zehnerpotenzen der Leitfähigkeit eine einheitliche Aktivierungsenergie zeigt (Fig. 7a und Fig. 9b).

Die Zunahme der Leitfähigkeit im ohmschen Bereich läßt sich durch Tempern meistens vollständig wieder rückgängig machen. Dabei bleibt die Aktivierungsenergie in vielen Fällen konstant. Im spannungsabhängigen Bereich bei tiefen Temperaturen hat Tempern umgekehrt eine Zunahme der Leitfähigkeit zur Folge, die aber nicht wieder zu den Werten vor der Verformung führt. Nach starker Verformung ist die Änderung hier relativ gering. Dieses Ergebnis zeigt, daß bei der Verformung zwei Arten von Gitterfehlern entstehen, von denen die eine Sorte die Leitfähigkeit erhöht und leicht ausheilbar ist, die andere dagegen die Temperatur- und Spannungsabhängigkeit beeinflußt und nur schwer ausheilt. Zur Klärung der Natur dieser Gitterfehler ist die Kenntnis der Versetzungsdichte und deren Verhalten bei der Verformung und Ausheilung erforderlich. Solche Untersuchungen sind im Gange. Leider sind die Ätzverfahren noch nicht so weit entwickelt, daß eine verläßliche Bestimmung der Versetzungsdichte möglich ist [16]. Eine wichtige Aussage über die leicht ausheilenden Gitterfehler, welche wegen ihrer Leitfähigkeitssteigerung offenbar Akzeptoren sind, erhält man durch eine isochrone Ausheilung (Fig. 2). Dabei wird der Verlauf der Leitfähigkeit bei einer mit konstanter Geschwindigkeit wachsenden Temperatur — hier mit etwa 10°/min — gemessen. Aus der Abweichung vom Exponentialgesetz läßt sich dann die zur Ausheilung der Gitterfehler notwendige Energie bestimmen. Mit einem von Balarin und Zetzsche [17] angegebenen Auswertungsverfahren ergibt sich dafür ein Wert von 0,5 bis 0,6 eV.

Nach diesen Ergebnissen sollten Selen-Kristalle mit sehr wenig Gitterfehlern auch bei tiefen Temperaturen keine Spannungsabhängigkeit der Leitfähigkeit zeigen. Dieser Fall ist nun offenbar möglich. Einige aus dem Dampf gewachsenen Kristalle haben, wenn man Verformungen bei der Kontaktierung vermeidet, bei — 180 °C bis zu relativ hohen Feldstärken eine ohmsche Charakteristik (Fig. 3a). Die Temperaturabhängigkeit der Leitfähigkeit ist dabei gering (Fig. 3b), ebenso ihre Zunahme bei Belichtung.

Nach einer Verformung ergibt sich auch hier ein ähnliches Bild wie in Fig. 1, nur liegt im ganzen Temperaturbereich die Leitfähigkeit unterhalb der Kurve vor der Verformung. Dieses Ergebnis zeigt in aller Deutlichkeit, daß für die Temperatur- und Spannungsabhängigkeit der Leitfähigkeit von Selen-Kristallen Gitterfehler von ausschlaggebender Bedeutung sind.

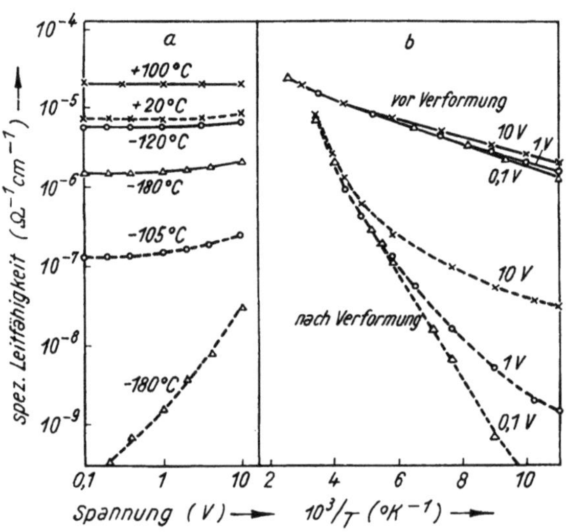

Fig. 3. Temperatur- und Spannungsabhängigkeit der Leitfähigkeit eines hexagonalen Selen-Kristalls vor und nach plastischer Verformung. Elektrodenabstand $L = 1 \times 10^{-2}$ cm, Stromrichtung ∥ zur c-Achse

1.32 Thermokraft

Zur Klärung der Frage, ob die Leitfähigkeitszunahme bei der Verformung durch eine Konzentrationszunahme bedingt ist, wurden Thermokraftmessungen durchgeführt. Die Temperaturabhängigkeit der Thermokraft entspricht einer annähernd temperaturunabhängigen Konzentration. Diese bei hohen Temperaturen bekannte Tatsache konnte bis − 180 °C bestätigt werden.[2])

Die eingezeichneten Geraden in Fig. 4 wurden nach der Beziehung $\Theta = \frac{k}{e}\left(\ln\frac{N_v}{p} + 2\right)$ für konstantes p unter der Annahme berechnet, daß die effektive Masse gleich der Elektronenmasse ist. Der Kristall hat zunächst eine Löcherkonzentration von $p = 4{,}4 \cdot 10^{14}$ cm^{-3} (Kurve 1), die durch Tempern auf $p = 1{,}5 \times 10^{14}$ cm^{-3} absinkt (Kurve 2). Durch Verformen steigt die Konzentration auf den Wert $p = 2{,}5 \times 10^{15}$ cm^{-3} (Kurve 3) an, der durch anschließendes Tempern wieder auf $8{,}0 \times 10^{14}$ cm^{-3} vermindert wird (Kurve 4). Die Beziehung zwischen der Leitfähigkeit bei Zimmertemperatur und der zugehörigen Thermokraft entspricht der obigen einfachen Beziehung mit $p \sim \sigma$. In Fig. 5 sind neben den Werten aus

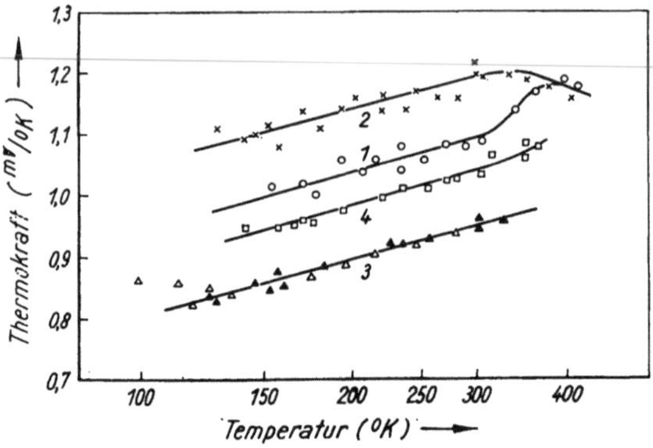

Fig. 4. Temperaturabhängigkeit der Thermokraft eines hexagonalen Selen-Kristalls; 1) unbehandelt; 2) nach 1. Temperung; 3) nach Verformung; 4) nach 2. Temperung. Die schwarzen Dreiecke in Kurve 3 wurden im Dunkeln gemessen, die offenen Dreiecke nach Belichtung

[2]) Vergleiche auch J. Stuke und K. Wendt, Physikalische Verhandlungen 7, 226 (1963).

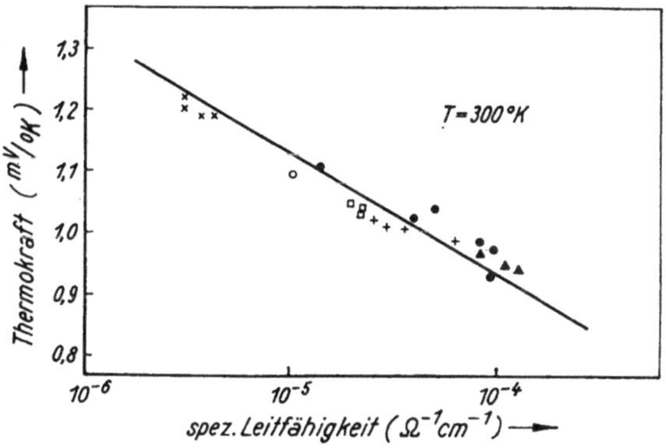

Fig. 5. Zusammenhang zwischen der Thermokraft und der spezifischen Leitfähigkeit parallel zur c-Achse verschiedener Selen-Kristalle bei 300 °K

Fig. 4 weitere Messungen eingetragen, die durch Verformung und Ausheilung erhalten wurden. Sie lassen sich alle angenähert durch eine Gerade mit der Neigung $\frac{k}{e} = 86\ \mu V/°K$ darstellen.

Es entstehen also bei der Verformung Akzeptoren, deren Aktivierungsenergie E_A so klein ist, daß bei — 180 °C noch praktisch alle Löcher angeregt sind. E_A muß demnach in der Größenordnung von 0,01 eV oder niedriger liegen. Durch diese Messungen wird die häufig geäußerte Vermutung bewiesen, daß Gitterfehler als Akzeptoren in Selen-Kristallen eine wichtige Rolle spielen.

Neben der im Dunkeln gemessenen Thermokraft sind in Fig. 4 Kurve 3 noch einige Werte eingetragen, die nach Belichtung erhalten wurden (offene Dreiecke). Das etwas überraschende Ergebnis ist, daß eine Änderung der Thermokraft durch die Belichtung nicht auftritt. Im folgenden Kapitel wird hierauf nun ausführlicher eingegangen.

1.33 Photo-Leitfähigkeit

Bei den lichtelektrischen Untersuchungen ist eine Beschränkung auf die Ergebnisse notwendig, welche über die Kernfrage nach dem Leitungsmechanismus Auskunft geben. Deshalb steht der Einfluß von Licht auf die Temperaturabhängigkeit der Leitfähigkeit und die Ladungsträger-Konzentration hier im Vordergrund. Dabei läßt sich die große Trägheit des Photostroms ausnutzen. Bei tiefer Temperatur bleibt der Photostrom nach einer Bestrahlung im Maximum der spektralen Empfindlichkeit, das bei — 180° nahe 700 nm liegt, weitgehend eingefroren. Er fällt bei logarithmisch aufgetragener Zeit linear ab (Fig. 6). Der Abfall wird also mit zunehmender Zeit immer schwächer, nach einigen Minuten ist er so gering, daß man den Strom für viele Zwecke als stationär ansehen kann. Dieses langsame Abklingen ist die Grundlage einer Methode, die im folgenden kurz als Stufen-Glow-Messung bezeichnet werden soll. Dabei wird zunächst die Temperaturabhängigkeit der Dunkelleitfähigkeit bis herab zu — 180 °C gemessen (Kurve D in Fig. 7a). Anschließend wird bei der tiefen Temperatur mit Licht der Wellenlänge $\lambda = 700$ nm bestrahlt und das Licht wieder abgeschaltet. Nach etwa 15 Minuten Wartezeit wird dann die Temperatur mit konstanter Geschwindigkeit von etwa 10 Grad/min erhöht. Im Gegensatz zu einer normalen Glow-Kurve (Kurve G) steigert man die Temperatur aber nur ein Stück, kühlt dann relativ schnell auf — 180 °C ab und mißt anschließend wieder die Leitfähigkeit mit wachsender Temperatur, die diesmal etwas weiter erhöht wird. Auf diese Weise wurden in verschiedenen Stufen die Kurven nacheinander gewonnen. Sie haben gegenüber

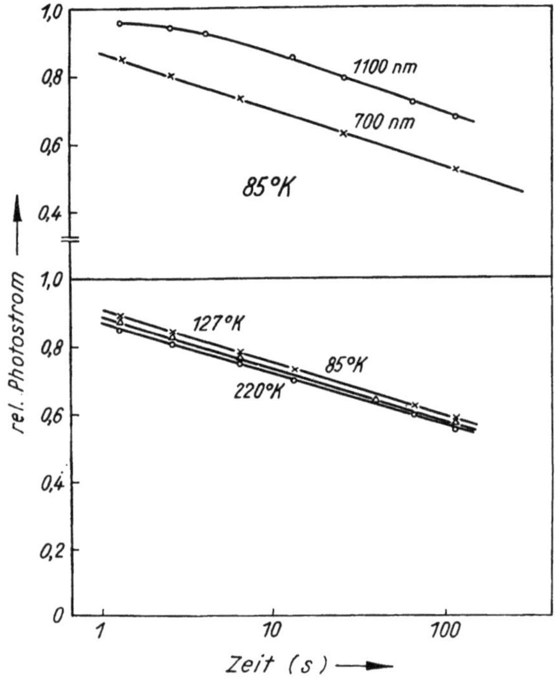

Fig. 6. Abklingen des Photostromes bei verschiedenen Temperaturen und Wellenlängen des eingestrahlten Lichtes

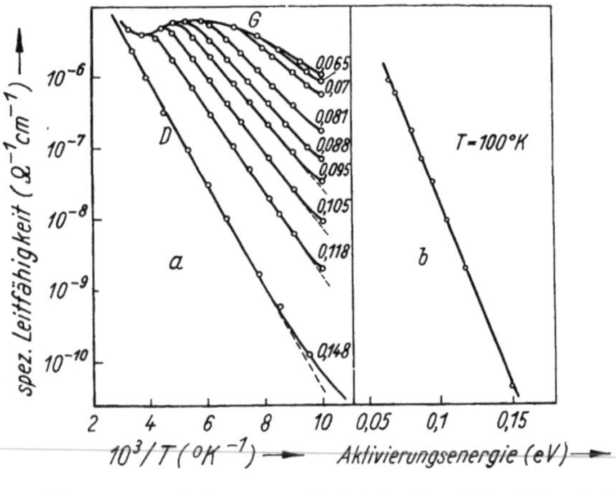

Fig. 7. Stufen-Glow-Messung nach starker Verformung.
D Dunkelleitfähigkeit;
G normale Glow-Kurve.
$dT/dt \approx 10°/\mathrm{min}$

der normalen Glow-Kurve, die ihrer Natur nach irreversibel ist, den Vorteil, daß sie annähernd reversibel durchlaufen werden.

Besonders aufschlußreich ist eine solche Stufen-Glow-Messung, wenn man sie an einem Kristall durchführt, der im Dunkeln — wie in Fig. 7a — nach starker Verformung eine einheitliche Temperaturabhängigkeit über mehrere Zehnerpotenzen zeigt. Man erhält dann ebenfalls Geraden, deren Neigung mit zunehmender Leitfähigkeit stetig abnehmen. Die Aktivierungsenergie nimmt von etwa 0,15 eV im Dunkeln auf 0,065 eV nach Belichtung ab. Extrapoliert man diese Geraden zu höheren Temperaturen, so schneiden sie sich alle mit guter Genauigkeit bei $1/T = 0$. Sie gehorchen demnach einer Beziehung $\sigma = \sigma_0 \cdot e^{-\frac{\Delta\varepsilon}{kT}}$, wo $\sigma_0 = \mathrm{const} =$

Fig. 8. Abhängigkeit der Thermokraft von der Leitfähigkeit nach Bestrahlung mit Licht

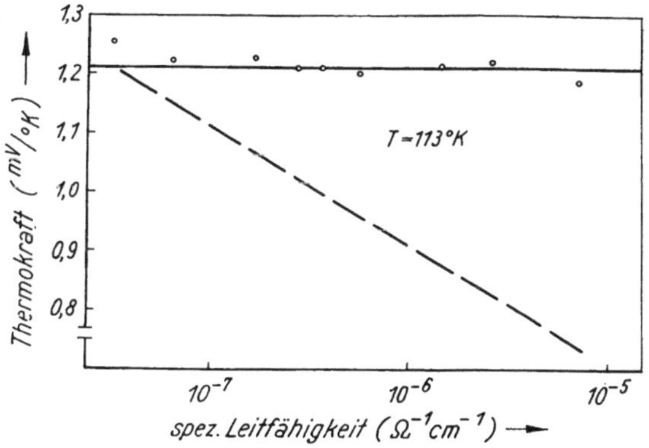

$= 1{,}6 \times 10^{-3}\ \Omega^{-1}\ \mathrm{cm}^{-1}$ und $\Delta\varepsilon$ Parameter ist. Zur Veranschaulichung dieses Zusammenhangs sind in Fig. 7b die Aktivierungsenergien der einzelnen Kurven und die zugehörigen (extrapolierten) Leitfähigkeitswerte bei 100 °K aufgetragen. Die Neigung dieser Kurve entspricht genau der Beziehung $\Delta\varepsilon = k\,T \ln \frac{\sigma_0}{\sigma}$ für $T =$ 100 °K. Die Zunahme der Leitfähigkeit bei Belichtung kommt also durch eine stetige Verminderung der Aktivierungsenergie zustande.

Aus Thermokraftmessungen geht hervor, daß dabei die Änderung der Ladungsträgerkonzentration nur eine untergeordnete Rolle spielt. Zur Vermeidung jeglicher Photospannung wurde nicht dauernd eingestrahlt, sondern nach Belichtung bei konstanter tiefer Temperatur im Dunkeln gemessen. Die eingefrorene Leitfähigkeit wurde durch stufenweises Tempern und Wiederabkühlen — ähnlich wie bei einer Stufen-Glow-Messung — schrittweise verringert. Die zugehörige Thermokraft bleibt trotz einer Abnahme der Leitfähigkeit um $2^1/_2$ Zehnerpotenzen dabei nahezu konstant (Fig. 8). Die gestrichelte Gerade wäre zu erwarten, wenn diese Leitfähigkeitsänderung auf eine Änderung der Löcherkonzentration zurückzuführen wäre. Aus der starken Abweichung ist zu schließen, daß auch die Photoleitung hauptsächlich durch eine Erhöhung der Beweglichkeit zustande kommt. Die Kurven in Fig. 7a geben demnach die Temperaturabhängigkeit der Beweglichkeit für verschiedene temperaturunabhängige Anregungsgrade wieder. Der wesentliche Einfluß einer Belichtung bei 700 nm besteht also darin, daß sie die Aktivierungsenergie der Beweglichkeit stetig verringert.

Durch Belichtung nimmt neben der Aktivierungsenergie $\Delta\varepsilon$ auch die Spannungsabhängigkeit der Leitfähigkeit deutlich ab. Deren Uursache ist also eng mit der Ursache der Temperaturabhängigkeit verknüpft. Bei tiefer Temperatur ist aber die Abhängigkeit von der Spannung bei einer bestimmten durch Licht erzeugten Leitfähigkeit noch deutlich größer als bei dem gleichen im Dunkeln durch Temperaturerhöhung erreichten Wert [5]. Eine einfache Beziehung zwischen Temperaturabhängigkeit und Spannungsabhängigkeit besteht demnach nicht.

1.34 Richtungsabhängigkeit der Leitfähigkeit

Die im vorhergehenden geschilderten Ergebnisse wurden in Richtung der *c*-Achse, also in Kettenrichtung gemessen, sie gelten aber im Prinzip auch in der Richtung senkrecht dazu. Der wesentliche Unterschied zwischen den beiden Orientierungen besteht nur darin, daß $\sigma_{\parallel}$ um einen konstanten Faktor größer ist als $\sigma_{\perp}$. Die Kurven laufen vor der Verformung parallel (Fig. 9a), und zwar im

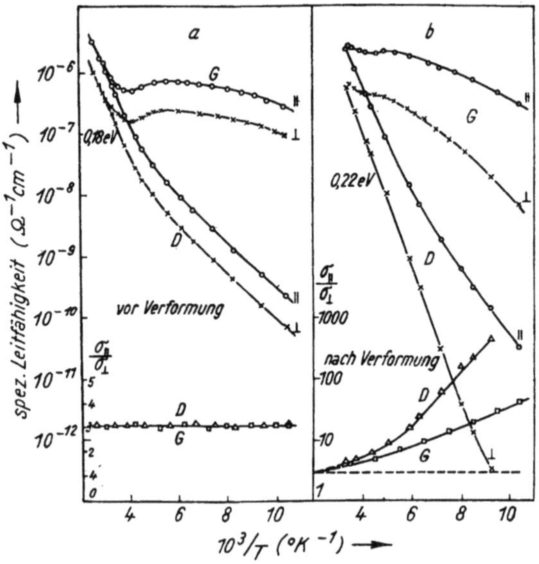

Fig. 9. Richtungsabhängigkeit der Leitfähigkeit eines Selen-Kristalles vor und nach Verformung. *D* Dunkelleitfähigkeit; *G* Glow-Kurven. Das Verhältnis $\sigma_{\parallel}/\sigma_{\perp}$ ist in a) linear und b) logarithmisch aufgetragen

ohmschen Bereich bei hohen Temperaturen und auch im spannungsabhängigen Gebiet bei tiefen Temperaturen. Das gilt sowohl für die im Dunkeln gemessenen Kurven D als auch für die nach Belichtung gemessenen Glow-Kurven G. Das Verhältnis $\sigma_{\parallel}/\sigma_{\perp}$ hat unter diesen verschiedenen Bedingungen den Wert 3. Das bedeutet also, daß das Beweglichkeitsverhältnis der Löcher in hexagonalem Selen $b = \mu_{\parallel}/\mu_{\perp} = 3$ ist und nicht von der Temperatur abhängt.

Durch plastische Verformung ändert sich nun b erheblich (Fig. 9b). Sie hat offensichtlich senkrecht zu den Ketten einen größeren Einfluß auf die Temperatur- und Spannungsabhängigkeit der Leitfähigkeit als in Kettenrichtung. Das ist verständlich, da die Gleitung senkrecht zur c-Achse erfolgt und richtungsabhängige Gitterstörungen wie Versetzungen hier bevorzugt auftreten werden. Dadurch wächst b sehr stark an, bei $-180\,°C$ hat es im Dunkeln einen Wert weit über 100. Bei den Glow-Kurven ist er kleiner, liegt aber immer noch größenordnungsmäßig über dem gestrichelt eingezeichneten Wert $b = 3$. Mit zunehmender Temperatur nimmt diese Abweichung stark ab, bei hohen Temperaturen münden beide Kurven in den Wert $b = 3$ ein. Diese Kurven geben also nicht das Beweglichkeitsverhältnis des ungestörten Gitters wieder, sondern sind durch richtungsabhängige Gitterstörungen stark verfälscht.

1.35 Spektrale Verteilung der Photoleitung

Neben der Änderung der elektrischen Eigenschaften tritt durch plastische Verformung auch eine beträchtliche Veränderung der spektralen Verteilung des Photostromes auf. Vor der Verformung zeigen die Kristalle das bekannte Maximum an der Kante der Eigenabsorption, das sich beim Abkühlen auf $-180\,°C$ entsprechend der Kantenverschiebung nach etwa 700 nm verschiebt. Außerdem wurde im langwelligen Gebiet vor diesem Maximum eine deutliche Tilgung der Photoleitung gefunden. Bei der Tilgungsmessung wird hier nicht, wie sonst üblich, die Abnahme des Stroms gegenüber seinem Wert bei schwacher Grundgitteranregung gemessen sondern die Änderung gegenüber dem Abfall des Photostromes

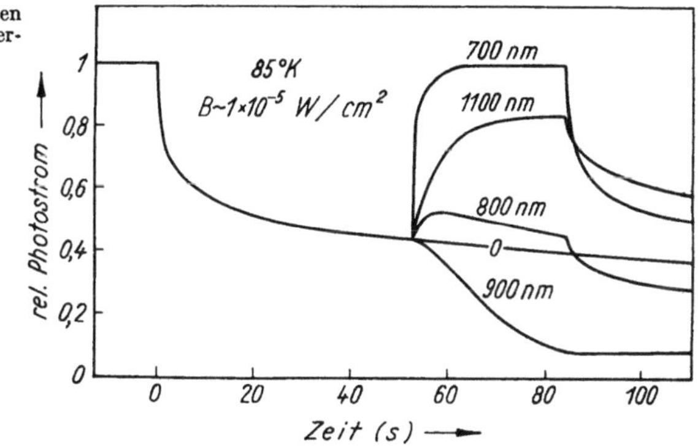

Fig. 10. Verlauf des Photostromes beim Abklingen und anschließender Bestrahlung mit Licht verschiedener Wellenlängen

nach Abschalten des Lichts (Fig. 10, Kurve 0). Dieses Verfahren hat den Vorteil größerer Empfindlichkeit, da die Anregung fehlt, setzt aber einen sehr langsamen Abklingvorgang voraus. Das Maximum der Tilgung liegt bei etwa 900 nm. An verformten Kristallen tritt außerdem im langwelligeren Gebiet bei 1100 nm ein starker positiver Effekt auf. Die genauere Abhängigkeit der beiden Prozesse von der Wellenlänge zeigt Fig. 11. Hier ist eine positive Änderung des Stroms gegenüber dem Abfall ohne Licht nach oben und die negativ wirkende Tilgung nach unten aufgetragen. Alle Kurven sind so normiert, daß das Maximum bei 700 nm, d. h. in dieser Darstellung bei etwa 1,8 eV, den Wert 1 hat.

Nach starker Verformung tritt vor diesem Maximum eine ausgeprägte Störstellen-Photoleitung mit einem Maximum bei 1,2 eV auf (Kurve I); die Tilgung ist relativ gering. Mit der bei Temperung durch Ausheilen der Gitterfehler abnehmenden Leitfähigkeit verringert sich auch die relative Höhe des zweiten Maximums. Die Kurven II, III und IV wurden nach Tempern bei stufenweiser erhöhter Temperatur gemessen. Dabei nimmt die Tilgung zu. Bei Kristallen, die solange getempert wurden, daß sie nicht mehr weiter zunimmt, setzt sie bei 0,9 eV ein und hat ein Maximum bei 1,5 eV. Als charakteristischer Energiewert ist aber der Einsatzpunkt bei 0,9 eV anzusehen, da das Maximum nur durch den starken Anstieg der Anregung bedingt ist. Dadurch wird die Tilgung überkompensiert, so daß sie nach außen hin stark abgeschwächt oder schließlich gar nicht mehr in Erscheinung tritt. Das ist offenbar auch der Fall, wenn durch Verformung das zweite Maximum sehr stark wird. Es ist deshalb anzunehmen, daß diese Abnahme der Tilgung nicht durch eine Verringerung der wirksamen

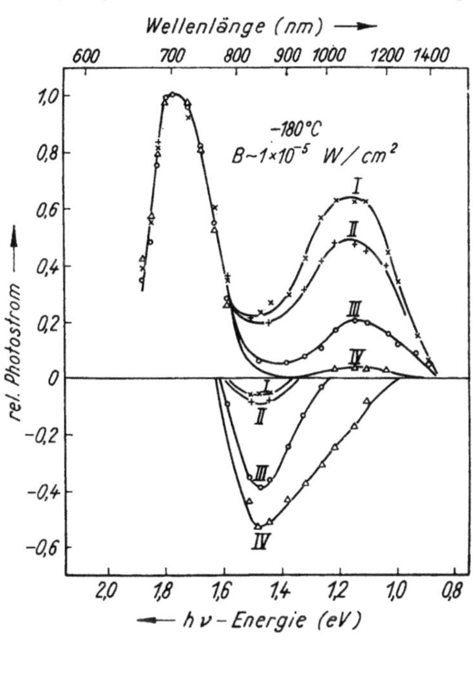

Fig. 11. Spektrale Verteilung des Photostromes und der Tilgung bei −180 °C.
I ist die Messung nach Verformung; II, III und IV wurden nach Temperungen gemessen

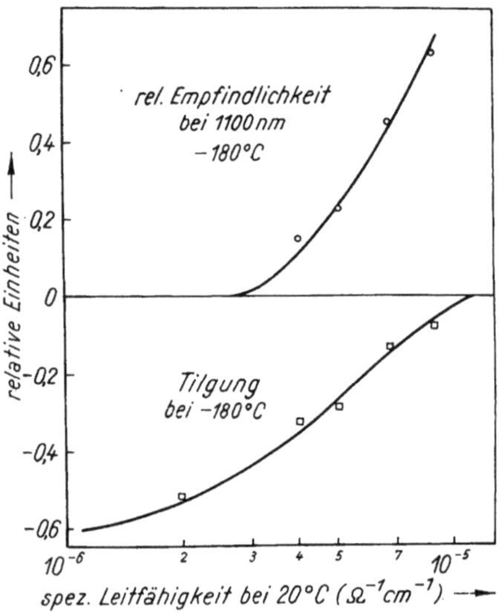

Fig. 12. Beziehung zwischen der spezifischen Leitfähigkeit im Dunkeln bei 20 °C und der Höhe des Maximums bei 1,2 eV sowie der Tilgung bei 1,5 eV, gemessen bei −180 °C

Haftstellen-Konzentration sondern durch die Zunahme der Störstellen-Photoleitung zustande kommt.

Bei der Ausheilung ist die relative Abnahme des Störstellen-Maximums größer als die zugehörige relative Leitfähigkeitsabnahme (Fig. 12). Die Beziehung ist annähernd quadratisch, wahrscheinlich ist sie aber durch die im gleichen Spektralgebiet vorhandeneTilgung stark verfälscht. Als charakteristischen Energiewert der Störstellen-Photoleitung pflegt man ebenfalls den Einsatzpunkt zu nehmen, der hier bei 0,8 eV liegt.

2. Deutung

2.1 Beschreibung des Modells

Das Verhalten der Selen-Kristalle läßt sich qualitativ durch die Annahme einer stark inhomogenen Verteilung der Löcherkonzentration erklären. Die wichtigsten Fälle solcher Inhomogenitäten sind in Fig. 13 zusammengestellt: Im größten Teil des Gitters wird eine Löcherkonzentration p_0 angenommen; die hierfür verantwortlichen Akzeptoren und möglicherweise noch vorhandene kompensierende Donatoren sind nicht eingezeichnet. Sie sollen sich im ganzen Halbleiter, d. h. auch im Gebiet der Inhomogenitäten befinden. Dann kommen die Anreicherungszonen I und II durch zusätzliche Akzeptoren, die Verarmungszonen III und IV durch zusätzliche kompensierende Donatoren zustande. Der Fall „amorpher“ Zwischenschichten bleibt außer Betracht, da er in Selen-Einkristallen sehr unwahrscheinlich ist.

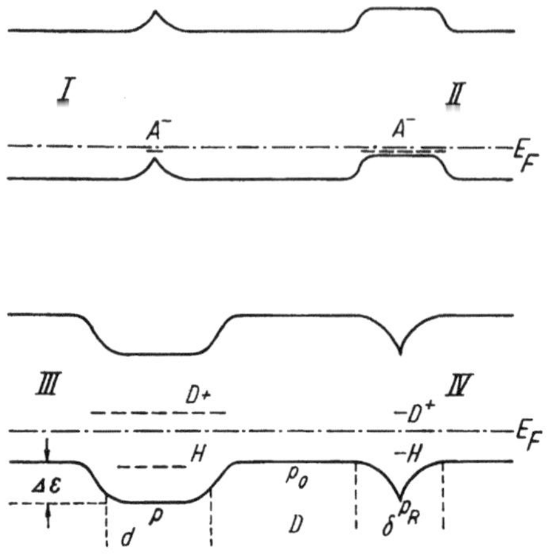

Fig. 13. Schematische Darstellung von Inhomogenitäten; I und II sind Anreicherungszonen, III und IV Verarmungsgebiete

Die Anreicherungszone I kann z. B. an einer inneren Oberfläche entstehen, da die hier vorhandenen freien Kettenenden als Akzeptoren wirken. Der mit ihr verbundene Potentialverlauf hat für die Löcher in dem ungestörten Kristallbereich aber nicht den Charakter einer Potentialschwelle, sondern vielmehr den einer anziehenden Haftstelle. Auch einzelne freie Kettenenden z. B. in Leerstellen führen nicht zu abstoßenden Potentialschwellen, da im ionisierten Zustand — nur dieser Fall interessiert hier — eine Coulombsche Anziehung zwischen Loch und negativ geladenem Akzeptor besteht, die durch Zufuhr thermischer Energie überwunden werden muß.

Zur Entstehung von Potentialschwellen sind Verarmungszonen notwendig, wie sie für die beiden Grenzfälle III und IV in Fig. 13 dargestellt sind: bei III ein relativ breites Verarmungsgebiet, das in der Mitte raumdladungsfrei ist, bei IV dagegen eine Doppel-Randschicht mit positiver Flächenladung und negativer Raumladung. Der Grundgedanke des Modells ist nun, daß sich aus Thermokraft oder Hall-Effekt-Messungen die Löcherkonzentration p_0 ergibt, während die Leitfähigkeit durch die in den Verarmungsgebieten vorhandene wesentlich geringere Konzentration p bzw. p_R bestimmt wird. Dazu muß einmal deren Dicke d bzw. δ klein gegen die Dicke der gutleitenden Kristallbereiche D sein. Zum anderen muß gelten: $\frac{d}{D} \gg \frac{p}{p_0}$, da die Verarmungszonen trotz ihrer geringen Dicke für die Leitfähigkeit maßgebend sind. Außerdem müssen diese den Kristall in einem geschlossenen Netzwerk durchziehen.

Die Annahme von Donatoren steht nicht in Widerspruch mit der Tatsache, daß Selen nur p-leitend ist. Das bedeutet ja nicht, daß Donatorterme überhaupt auszuschließen sind, sondern daß diese energetisch so weit vom Valenzband entfernt sein müssen, daß eine n-Leitung normalerweise nicht auftritt. Wenn ein Akzeptor sehr nahe am Valenzband liegt, der Donator aber z. B. in Bandmitte oder noch darunter liegt, so kommt ein Umschlag zur n-Leitung erst zustande, wenn die Donatorkonzentration D so groß gegenüber der Akzeptorkonzentration A ist, daß trotz der sehr geringen Übergangswahrscheinlichkeit mehr Elektronen ins Leitungsband gelangen als Löcher ins Valenzband. Es kommt dann nicht für $A = D$ zur n-Leitung sondern es besteht für $D > A$ ein sehr weiter Konzentrationsbereich von D mit p-Leitung geringer Löcher-Konzentrationen. Bekannte Beispiele dieser Art sind — allerdings in reziprokem Sinn für n-Leitung — der hochliegende Gold-Akzeptor in n-leitendem Silizium [18] und Kupfer als Akzeptor in CdS [19].

Bei Selen können also tiefliegende Donatoren mit inhomogener Verteilung im Gitter die Ursache der Verarmungszonen sein. Diese Vorstellung wird durch die Ergebnisse der Photoleitung gestützt. Aus den Tilgungsmessungen geht hervor, daß Haftstellen für Elektronen 0,9 eV unterhalb des Leitungsbandes vorhanden sind. Da der Bandabstand 1,76 eV [20] beträgt, liegen sie etwas unterhalb der Bandmitte. Der Charakter von Haftstellen erfordert einen großen Einfang-Querschnitt, was bei ein- oder besser zweifach positiver Ladung am einfachsten zu erklären ist. Bei den Haftstellen in vielen Photoleitern handelt es sich bekanntlich um kompensierte Donatoren und Akzeptoren [21].

Leider lassen sich über die genaueren Ursachen der Verarmungszonen sehr wenig Angaben machen. Da in dem Tilgungsspektrum bei Kristallen, die aus verschiedenen Selen-Sorten hergestellt sind, gewisse Unterschiede auftreten, wird vermutet, daß bei den Elektronenhaftstellen Fremdatome eine Rolle spielen. Es wäre z. B. denkbar, daß es sich dabei um Metallatome auf Zwischengitterplätzen handelt, die wie Kupfer in Germanium und Silizium Donator-Charakter haben. Wenn sich diese bevorzugt an Versetzungen abscheiden, und zwar sowohl an den in Kettenrichtung liegenden Stufenversetzungen als auch an den senkrecht dazu liegenden Schraubenversetzungen, so erhält man bei genügender Dichte der Versetzungen ein geschlossenes Netzwerk von Verarmungszonen. Die enge Verknüpfung von Potentialschwellen und mechanischer Behandlung ist dann verständlich. Man kann auch die Möglichkeit nicht ausschließen, daß die meistens im Selen vorhandenen Halogen-Atome diese Donatoren sind, wenn sie auf Gitterplätzen sitzen. Hier ist allerdings die inhomogene Verteilung schwieriger zu deuten. Zur Klärung dieser Zusammenhänge sind noch umfangreiche Untersuchungen über den Einfluß von Fremdatomen notwendig.

Zunächst sollen nun die Eigenschaften der Selen-Kristalle an Hand dieses Modells diskutiert werden. Anschließend folgen noch einige allgemeinere Bemerkungen über das Modell selbst und über die Beweglichkeit der Löcher.

2.2 Diskussion der experimentellen Ergebnisse

2.21 Strom-Spannungs-Abhängigkeit

Der durch Verarmungszonen fließende Strom hängt stark von der Spannung ab. Bei kleinen Spannungen ist im Potentialtopf III die Konzentration p im Verarmungsgebiet für den Strom maßgebend, und es gilt das Ohmsche Gesetz. Bei höheren Spannungen werden aus den benachbarten Bereichen auf der Seite der Anode Löcher in das Gebiet hineingeschwemmt. Dann gelten die Gesetzmäßigkeiten für raumladungsbegrenzte Ströme, nur daß die an dem Topf anliegende Teilspannung ΔU und die Dicke d anstelle der an dem Elektrodenabstand L anliegenden Gesamtspannung U einzusetzen ist. Man erhält — bei schwacher Besetzung der Haftstellen — eine ohmsche Kennlinie bis $\Delta U_0 = \frac{e}{2\,\varepsilon\,\varepsilon_0} \cdot p \cdot d^2 \cdot \frac{H}{N_v} \cdot e^{\frac{E_H}{kT}}$, wo H die Konzentration der Haftstellen für Löcher im Bereich d, E_H deren Tiefe und N_v die effektive Zustandsdichte im Valenzband ist [5, 22]. Bei diesen Haftstellen kann es sich z. B. um die unterhalb des Fermi-Niveaus liegenden kompensierten Akzeptoren handeln. Befinden sich N solcher Verarmungszonen innerhalb des Elektrodenabstands L, so ist $L = N\,(D + d) \approx N\,D$. Die quadratische Kennlinie setzt dann bei der äußeren Spannung

$$U_0 = N \cdot \frac{e}{2\,\varepsilon\,\varepsilon_0}\, p\, d^2 \cdot \frac{H}{N_v} \cdot e^{\frac{E_H}{kT}} \tag{1}$$

ein. Die Haftstellenkonzentration H ist von großem Einfluß, da diese um viele Zehnerpotenzen verschieden sein kann. Für das Einsetzen des raumladungsbegrenzten Stromes ist gewissermaßen die Dicke

$$d_{\text{eff}} = d \cdot \sqrt{\frac{N \cdot H}{N_v}\, e^{\frac{E_H}{kT}}} \tag{2}$$

maßgebend, die unter Umständen wesentlich größer ist als die geometrische Dicke $N \cdot d$. Man kann also bei dem Potentialverlauf III den Formalismus für raumladungsbegrenzte Ströme in etwas modifizierter Form anwenden, so daß die gefundenen annähernd quadratischen Kennlinien damit zu erklären sind. Ein Verarmungsgebiet mit stärker besetzten Haftstellen kann eine stärker als quadratisch ansteigende Kennlinie haben, die bei Selen-Kristallen auch häufig gefunden wird.

Auch die Doppelrandschicht IV zeigt eine starke Spannungsabhängigkeit. Fließen die Löcher von links nach rechts, so ist die linke Seite in Durchlaß-Richtung, die rechte in Sperrichtung gepolt. Mit zunehmender Spannung nimmt die Randkonzentration p_R von der linken Sperrschicht her in exponentieller Form zu. Wegen der starken Zunahme von p_R mit der Spannung stellt die rechte Randschicht bei kleinen Spannungen keinen Sperrkontakt dar. Zu einer Sperrwirkung kommt es dagegen, wenn p_R von der Spannung weitgehend unabhängig wird. Dieser Fall trifft z. B. für Korngrenzen in n-Germanium zu, er wurde von TAYLOR, ODELL und FAN [23] und auch von MUELLER [24] ausführlich behandelt. Der Umschlag von der überproportionalen zu einer unterproportionalen Kennlinie erfolgt bei einer kritischen Spannung, deren Größe von den Abmessungen der Verarmungsrandschicht sowie von den Konzentrationen der freien Ladungsträger und der Haftstellen abhängt.

Eine numerische Abschätzung ist schwierig, da über die Haftstellen-Konzentration und die geometrischen Parameter nichts bekannt ist. Für die Dicke δ der Doppel-Randschicht ergibt sich bei $\Delta\varepsilon = 0{,}2$ eV und $p_0 = 10^{14}$ cm^{-3} ein Wert von etwa 10^{-4} cm. Bei dem experimentellen Elektrodenabstand $L = 1 \times 10^{-2}$ cm und einem angenommenen Wert von $d/D = 0{,}1$ ist die Zahl der Verarmungszonen zwischen den Elektroden $N \approx 10$. Auch bei der relativ kleinen Löcherkonzentration von 10^{14} cm^{-2} ist also unter diesen Bedingungen eine ausreichende Zahl von Verarmungsgebieten möglich.

2.22 Photoleitung

Auch die wichtigsten lichtelektrischen Eigenschaften von Selen-Kristallen lassen sich qualitativ mit diesem Modell verstehen. Dazu wollen wir der Einfachheit halber den Fall IV in Fig. 13 betrachten. Strahlt man in diese Anordnung Licht ein, so werden die ins Leitungsband gehobenen Elektronen in das Raumladungsgebiet gezogen und von den ionisierten Donatoren eingefangen. Mit zunehmender Besetzung der Donatoren durch Elektronen nimmt die positive Flächenladung und damit $\Delta\varepsilon$ stetig ab. Auch die große Trägheit des Photostromes läßt sich so ohne weiteres erklären. Der in der Verarmungszone vorhandene Potentialverlauf trennt nämlich die Elektronen und Löcher voneinander und wirkt deshalb der Rekombination entgegen. Sieht man den Donator als Rekombinationszentrum an, so führt das, wie MORRISON [25] gezeigt hat, zu einem relativ langsamen nichtexponentiellen Abfall des Photostroms nach Belichtung. Aus dem Rekombinationszentrum wird durch den ihn umgebenden Potentialwall eine Haftstelle. MORRISON konnte damit die an Germanium und Silizium gemessenen Zeitkonstanten von etwa 20 ms erklären. Da beim hexagonalen Selen das Abklingen um mehrere Größenanordnungen langsamer verläuft, wird angenommen, daß hier der Donator schon bei homogener Verteilung im Gitter den Charakter einer Haftstelle hat. Die Kombination von Haftstellen und Potentialwall führt dann zu einem sehr langsamen Abklingen, bei dem die Zeitkonstante im Laufe des Abklingvorgangs stark wächst, da nicht nur die Konzentration der zur Rekombination verfügbaren Ladungsträger abnimmt, sondern außerdem noch der rekombinationshemmende Potentialwall höher wird. Im Gegensatz zu den Überlegungen von MORRISON wird die Leitfähigkeit bei Selen aber nicht durch die Konzentration der Löcher, sondern durch die Höhe der Potentialschwellen $\Delta\varepsilon$ bestimmt. Die Thermokraftmessungen (Fig. 8) zeigen ja, daß die Konzentration p_0 durch Licht in erster Näherung nicht geändert wird. Wir haben es also mit einer Sperrschicht-Photoleitung zu tun, deren Theorie von RITTNER [26], SLATER [27] und PETRITZ [28] entwickelt worden ist. Sie wurde bisher im wesentlichen zur

Deutung der Photoleitung von polykristallinen PbS- und PbSe-Schichten herangezogen, ist aber etwas umstritten [29], da hier nach Hall-Effekt-Messungen ebenfalls eine deutliche Konzentrationsänderung in den einzelnen Kristalliten stattfindet. Die von Slater gefundene Bedingung, daß das Produkt der zur Zeit t vorhandenen Leitfähigkeit σ mit der zugehörigen Lebensdauer τ konstant ist: $\sigma \cdot \tau = \text{const}$, reicht zur Erklärung des an Selen gefundenen Abklingverlaufs noch nicht aus. Hier muß die Lebensdauer bei abnehmender Leitfähigkeit stärker ansteigen, etwa nach der Beziehung $e^{\sigma/\sigma_0} \cdot \tau = \text{const}$. Einfache Überlegungen führen jedoch ebenfalls zu einem hyperbolischen Abklinggesetz, das der Slater'schen Beziehung entspricht. Es ist deshalb anzunehmen, daß die aus mathematischen Gründen notwendigen Vereinfachungen noch zu stark sind, um das annähernd logarithmische Abklinggesetz zu erhalten. Wahrscheinlich nimmt $\Delta\varepsilon$ nicht linear mit der Zahl der Elektronen in den Haftstellen ab, wie angenommen wurde, sondern schwächer. Das führt unter sonst gleichbleibenden Bedingungen zu einem langsameren Abklingen.

2.23 Einfluß von plastischer Verformung

Den Einfluß plastischer Verformung auf die Temperatur und Spannungsabhängigkeit der Leitfähigkeit kann man wie folgt verstehen: Einmal nimmt die von der Temperatur unabhängige Löcherkonzentration p_0 bei der Verformung zu. Das hat eine Abnahme der Thermokraft und, da $p = p_0 \cdot e^{-\frac{\Delta\varepsilon}{kT}}$ ist, eine Zunahme der Leitfähigkeit im ohmschen Bereich bei hohen Temperaturen zur Folge, wenn $\Delta\varepsilon$ sich bei der Verformung nicht ändert. In vielen Fällen trifft das in erster Näherung zu. Bei dem Kristall in Fig. 3 tritt außerdem eine starke Zunahme von $\Delta\varepsilon$ bei der Verformung auf. Dadurch wird die Zunahme der Löcherkonzentration p_0 überkompensiert, so daß hier im ganzen Temperaturbereich eine Abnahme der Leitfähigkeit zu verzeichnen ist. Die bei der Verformung auftretende Leitfähigkeitsabnahme im spannungsabhängigen Gebiet bei tiefen Temperaturen läßt sich im Rahmen der raumladungsbegrenzten Ströme durch eine Zunahme der effektiven Dicke d_{eff} (Gleichung 2) der Verarmungszonen beschreiben, da die Abweichung vom Ohmschen Gesetz dann erst bei höherer äußerer Spannung U einsetzt. Die Ursache kann einmal die Zunahme der Zahl der Verarmungszonen N sein. Nimmt man an, daß diese mit Versetzungen zusammenhängen, so kann man das leicht erklären und auch die Tatsache, daß sich bei Temperung N nur wenig ändert. Zum anderen kann aber auch die Zunahme der Haftstellenkonzentration H bzw. die Einführung von zusätzlichen Haftstellen-Sorten größerer Tiefe E_H der Grund für das Wachsen von d_{eff} bei der Verformung sein.

Die Zunahme der Löcherkonzentration p_0 bei der Verformung ist auf die Einführung von Akzeptoren zurückzuführen, deren Abstand vom Valenzband in der Größenordnung 0,01 eV liegt. Es handelt sich dabei wahrscheinlich um Leerstellen die beim Wandern von Versetzungen gebildet werden und sich beim Tempern wieder an Versetzungen oder an anderen Leerstellen ausscheiden. Die Versetzungen selbst kommen als Ursache der Konzentrationsänderung wohl nicht in Frage, da sie nicht so leicht ausheilen und es außerdem unwahrscheinlich ist, daß sie im Selen-Gitter einen Akzeptor-Charakter haben. Reine Stufenversetzungen führen nämlich nicht zu freien Valenzen, da sie genau in Kettenrichtung liegen. Diese treten nur an Sprüngen von Stufenversetzungen auf, welche auch die Leerstellen bei der Verformung ausscheiden. Das Verhalten von Schraubenversetzungen ist nicht einfach zu übersehen. Es ist aber anzunehmen, daß hier ebenfalls keine freienValenzen auftreten, falls nicht bei der vomValenzwinkel etwas abweichenden Atom-Anordnung die Bindung zwischen zwei Atomen reißt. Neben dem flachen Akzeptor tritt in der Photoleitung nach der Verformung ein Term bei etwa 0,8 eV auf; beide sind offenbar eng miteinander verknüpft. Sie können von zwei verschiedenen Gitterfehlern stammen, die bei der Verformung gemeinsam entstehen und auch zusammen wieder ausheilen. Dafür kommen z. B. Leerstellen und Zwischengitter-Atome oder Einfach- und Doppel-Leerstellen in Frage. Es ist aber auch möglich, daß es sich dabei um einen einzigen Gitterfehler mit zwei Niveaus handelt, z. B. eine Leerstelle mit zwei Akzeptortermen, von denen nur der untere bis zum Schmelzpunkt bei 220 °C thermisch angeregt ist. Es spricht manches für diese letztere Vorstellung, eine Entscheidung ist aber im Augenblick nicht möglich.

2.24 Weitere experimentelle Beobachtungen

Das Modell von den relativ großen Kristallbereichen mit einer Löcherkonzentration p_0 von etwa 10^{14} bis 10^{15} cm^{-3} und dünnen Verarmungszonen wird durch eine optische Konzentrationsmessung mittels p-Banden gestützt, die von Kessler und Sutter [30] durchgeführt wurde. Während die ultraroten Absorptionsbanden der thermisch vorhandenen Konzentration sich bei einer einfachen Absorptionsmessung [10, 20, 30] ergeben, läßt sich die Konzentrationszunahme bei Belichtung nur mit einer sehr empfindlichen Differenz-Methode [31] messen. Die durch Licht zusätzlich erzeugte Löcherkonzentration liegt weit unter der thermischen Konzentration, eine Abschätzung [32] führt in die Größenordnung von 10^{11} bis 10^{12} cm^{-3}.

Auch die Zunahme der Leitfähigkeit mit der Frequenz weist auf dünne Verarmungsgebiete innerhalb der Kristalle hin. Die Frequenzabhängigkeit ist aber, wie Jaumann und Neckenbürger [33] gezeigt haben, mit einem einfachen Zweischichten-Modell nicht zu erklären, jedenfalls nicht bei polykristallinem Selen. Daraus ist zu schließen, daß die Inhomogenitäten ein relativ breites Spektrum umfassen. Wegen ihrer engen Kopplung mit Gitterstörungen ist das auch besser verständlich als nur eine einzige Art von Verarmungsgebieten. Da bei polykristallinem Material an den Kristallit-Oberflächen außerdem Anreicherungs-Randschichten auftreten können [34], muß man hier wahrscheinlich zur Deutung des Frequenzverhaltens auch die beiden Fälle I und II in Fig. 13 hinzunehmen.

Messungen der Frequenz- und Temperaturabhängigkeit der dielektrischen Verluste an polykristallinem Se deuten Ludwig und Eckart [11] als Debeye-Relaxation, bei der die Löcher zwischen Potentialmulden ausgetauscht werden, die durch Schwellen unterschiedlicher Höhe voneinander getrennt sind. Diese Potentialmulden sollen keine Eigenschaft des ungestörten Gitters sein, sondern sich in Gitterstörungen befinden. Mit dem hier geschilderten Modell kann diese Vorstellung in Einklang gebracht werden, wenn es sich dabei um Haftstellen für Löcher mit verschiedener Tiefe handelt, die in den Verarmungszonen lokalisiert sind.

Eine weitere qualitative Bestätigung für die Annahme einer stark inhomogenen Löcherkonzentration innerhalb von Selen-Einkristallen sind lokale Leitfähigkeitsmessungen mit einer feinen Potential-Doppelsonde von Gobrecht, Tausend und Schmidt [35]. An dünnen Kristall-Plättchen ergaben sich dabei Unterschiede bis zu vier Zehnerpotenzen. Diese Größenordnung ist nur durch Verarmungsgebiete etwa der hier geschilderten Art zu verstehen. Lichtsonden-Messungen [36] zeigen, daß im Innern von optisch und kristallographisch einwandfreien Selen-

Einkristallen Photospannungen auftreten, was ebenfalls auf starke Dichteschwankungen der Löcher hinweist. Die spaltförmige Sonde von etwa 10 μm Breite war aber noch zu groß, um eine genauere Auflösung zu erzielen. Versuche, die geometrische Anordnung der Verarmungsgebiete durch Aufschwemmung von sehr feinen TiO_2-Kristallen direkt sichtbar zu machen, blieben ohne klares Ergebnis. Aussicht auf Erfolg besteht hier allerdings auch nur, wenn die Inhomogenitäten gröberer Natur sind.

2.3 Diskussion der Beweglichkeit

Die Vorstellung von relativ starken Inhomogenitäten in Einkristallen mag etwas befremden. Man ist eher geneigt , hier eine weitgehend homogene Störstellenverteilung vorauszusetzen. Nun zeigt aber eine Untersuchung von WEISSBERG [37], daß dies in vielen Fällen nicht zutrifft. Neben den Schichtungen der Störstellenkonzentration in Richtung des Kristallwachstums (striations) bei gezogenen Kristallen [38] findet man auch häufig Gitterstörungen, die den Betrag der Beweglichkeit erheblich vermindern. Die Ursache sind Raumladungsgebiete, die um inhomogen verteilte Störstellen entstehen. Je nach der Geometrie der Inhomogenitäten haben sie verschiedene Formen, z. B. Raumladungskugeln bei punktförmigen Störungen oder Schläuche bei linearen Störungen wie Versetzungen. Bei kleiner Trägerkonzentration erstreckt sich das Raumladungsgebiet viel weiter in den Kristall hinein als die Störung selbst. Dadurch wird der raumladungsfreie Bereich u. U. erheblich verkleinert. Die Träger werden gewissermaßen an den Raumladungen gestreut, wobei Lichteinstrahlung den Durchmesser dieser riesigen Streuzentren vermindert. An solchen Halbleitern muß also neben der Konzentrationszunahme bei Belichtung ein Anstieg der Beweglichkeit auftreten. Das ist von BUBE und McDONALD [39] an einigen GaAs-Kristallen auch gemessen worden. Hall-Effekt-Messungen zeigen, daß hier die Zunahme der Leitfähigkeit durch Licht zu einem großen Teil durch eine Zunahme der Beweglichkeit zustande kommt. An verschiedenen anderen Stoffen sind ebenfalls solche riesigen Streuzentren gefunden worden, welche die Beweglichkeit größenordnungsmäßig verringern. Dazu gehört auch Tellur, das sogar eine besonders große Schwankungsbreite der gemessenen Beweglichkeitswerte zeigt, die offenbar eng mit der mechanischen Vorgeschichte zusammenhängt [40], da eine Ausheilung stattfindet.

Die Temperaturabhängigkeit der Beweglichkeit wird bei den von WEISSBERG diskutierten unzusammenhängenden Raumladungszonen nicht grundsätzlich geändert, sie gehorcht annähernd einem T^{-1}-Gesetz. Eine wesentliche Änderung ist aber zu erwarten, wenn sich z. B. eine Reihe von Raumladungsschläuchen zu einer geschlossenen Potentialschwelle zusammenschließen. Der für die Hall-Beweglichkeit maßgebende Prozeß ist jetzt die Überwindung dieser Schwelle. Das führt zu einem kleinen Absolutwert und weiterhin zu einer exponentiellen Zunahme mit der Temperatur. Es ist möglich, daß Fig. 3 einen solchen Übergang zu einer geschlossenen Potentialschwelle zeigt. Vor der Verformung sind in diesem Kristall wahrscheinlich noch schmale Bereiche vorhanden, in denen keine Spannungsabhängigkeit der Leitfähigkeit auftritt. Durch die Verformung entsteht dann die geschlossene Schwelle mit ihrer wesentlich stärkeren Temperatur- und Spannungsabhängigkeit.

In dem oben diskutierten Modell ist die experimentell bestimmte exponentielle Temperaturabhängigkeit der Beweglichkeit nur formaler Natur, da die aus Thermokraftmessungen bestimmte Konzentration p_0 mit der für die Leitfähigkeit verantwortlichen geringen Konzentration p in den Verarmungszonen kombiniert wird. Die Gitterbeweglichkeit sollte also wesentlich größer als die mit Gleichstrom gemessenen Werte von etwa 0,1 cm²/Vs sein. Bei Wechselstrom der Frequenz 200 MHz ergibt sich an Einkristallen nun auch eine Beweglichkeit von etwa 10 cm²/Vs [6]. Die schwache Zunahme mit der Temperatur zeigt, daß dieser Wert wohl noch zu niedrig ist. Daraus ist zu schließen, daß die Gitterbeweglichkeit wahrscheinlich in einer Größenanordnung liegt, die durch Streuung an Gitterschwingungen oder durch Störstellenstreuung zustande kommt. Eine grobe Abschätzung erhält man ebenfalls, wenn man die gemessene Temperaturabhängigkeit der Beweglichkeit entsprechend der Beziehung: $\mu = \mu_0 \cdot e^{-\frac{\Delta\varepsilon}{kT}}$ nach sehr hohen Temperaturen ($1/T \to 0$) extrapoliert. Das führt zu Werten von μ_0, die zwischen 10 und 50 cm²/Vs liegen und demnach mit dem bei 200 MHz gemessenen Werten in der Größenordnung übereinstimmen.

Es sprechen auch noch andere Gründe dafür, daß Selen sich in seiner Gitterbeweglichkeit nicht grundsätzlich von Tellur unterscheidet. Die Anisotropie der Bindung ist bei Selen zwar stärker als bei Tellur, aber nicht so viel, daß es in den beiden Kristall-Richtungen zu ganz verschiedenen Beweglichkeitsmechanismen kommt. Bei beiden Stoffen äußert sich die Anisotropie in gleicher Weise, nämlich durch ein temperaturunabhängiges Beweglichkeitsverhältnis $b = \mu_{\parallel}/\mu_{\perp}$, das bei Tellur den Wert 2 [41] hat. Bei Selen ist $b = 3$; hier werden durch den starken Einfluß von Gitterstörungen häufig größere Werte und auch eine Temperatur- und Spannungsabhängigkeit von b vorgetäuscht. Verglichen mit anderen anisotropen Elementen, wie z. B. Antimon, dessen Beweglichkeitsverhältnis etwa 1,3 beträgt, deutet der Wert 3 auf eine relativ starke Anisotropie ohne Änderung des Beweglichkeitsmechanismus hin. Nur bei extrem anisotropen Stoffen wie Graphit kommt es zu wesentlich größeren Werten. Die Brandstrukturen von Selen und Tellur [42] sind offenbar ähnlich; sie unterscheiden sich, wie Reflexionsmessungen [43] zeigen, im wesentlichen nur durch die energetische Lage der verschiedenen Bänder. In anderen isomorphen Reihen ist kein Fall bekannt, wo sich durch eine quantitative Änderung der Bandstruktur-Parameter der Beweglichkeitsmechanismus völlig ändert. Die Gitterbeweglichkeit von hexagonalem Selen ist deshalb aller Wahrscheinlichkeit nach ähnlich wie bei Tellur von normaler Größenordnung.

Zusammenfassend ist zu sagen, daß man die wesentlichen elektrischen und lichtelektrischen Eigenschaften von hexagonalen Selen-Kristallen mit dem vorgeschlagenen Modell viel besser erklären kann als mit der Annahme eines besonderen Beweglichkeitsmechanismus. Vor allem läßt sich damit die Schwierigkeit beheben, daß sich aus Thermokraft- und auch aus Hall-Effekt-Messungen eine von der Temperatur *nicht* abhängige Löcherkonzentration ergibt, während sich andererseits die Temperatur- und Spannungsabhängigkeit der Leitfähigkeit durch raumladungsbegrenzte Ströme qualitativ richtig beschreiben läßt, jedoch nur mit der Annahme, daß die für die Leitfähigkeit maßgebende Konzentration *stark* von der Temperatur *abhängt*. Mit einer exponentiellen Temperaturabhängigkeit der Gitterbeweglichkeit läßt sich der Einfluß der Temperatur auf die $I(U)$-Kennlinien nicht erklären. Das vorgeschlagene Modell bietet hier eine Lösung an, welche außerdem auch die wichtigsten lichtelektrischen Eigenschaften der Kristalle beschreibt.

Herrn Prof. Dr. F. Stöckmann danke ich für eine Reihe wertvoller Diskussionen. Herrn cand. phys. W. Fuhs und Herrn cand. phys. K. Wendt danke ich für die Durchführung von Messungen.

Literatur

[1] Zusammenfassende Darstellung: GMELINS Handbuch anorg. Chemie, 8. Aufl. Nr. 10, Teil A, Verlag Chemie, Berlin 1953.
[2] T. MÜLLER, Sitzungsbericht phys. med. Soz. Erlangen **70**, 7 (1938).
[3] H. W. HENCKELS, J. appl. Phys. **22**, 916 (1951).
[4] K. W. PLESSNER, Proc. Phys. Soc. **B 64**, 671 (1951).
[5] M. POLKE, J. STUKE und E. VINARICKY, phys. stat. sol. **3**, 1885 (1963).
[6] H. W. HENCKELS und J. MACZUK, Phys. Rev. **91**, 1562 (1953).
[7] L. P. SIEG und F. C. BROWN, Phys. Rev. **4**, 507 (1914).
[8] E. BILLIG, Proc. Phys. Soc. **B 65**, 216 (1952).
[9] W. SCHOTTKY, Z. Naturf. **8a**, 457 (1953).
[10] H. GOBRECHT und A. TAUSEND, Z. Phys. **161**, 205 (1961).
[11] W. LUDWIG und F. ECKART, Z. Phys. **162**, 127 (1961).
[12] H. KRONMÜLLER, J. JAUMANN und K. SEILER, Z. Naturf. **11a**, 243 (1956).
[13] F. C. BROWN, Phys. Rev. **4**, 85 (1914).
[14] H. W. HENCKELS, Phys. Rev. **76**, 1737 (1949).
[15] L. C. LOVELL, J. H. WERNICK und K. E. BENSON, Acta metall. **6**, 716 (1958).
[16] F. ECKART und W. HENRION, phys. stat. sol. **3**, 1247 (1963).
[17] M. BALARIN und A. ZETZSCHE, phys. stat. sol. **2**, 1670 (1962).
[18] C. B. COLLINS, R. O. CARLSON und G. J. GALLAGHER, Phys. Rev. **105**, 1168 (1957).
[19] R. H. BUBE, J. Phys. Chem. Solids **1**, 234 (1957).
[20] F. ECKART und W. HENRION, phys. stat. sol. **2**, 841 (1962).
[21] R. H. BUBE, Photoconductivity of Solids, John Wiley and Sons, New York—London 1960.
[22] F. STÖCKMANN, Halbleiterprobleme VI, Vieweg, Braunschweig 1961.
[23] W. E. TAYLOR, N. H. ODELL und H. Y. FAN, Phys. Rev. **88**, 867 (1952).
[24] R. K. MUELLER, J. appl. Phys. **32**, 635 (1961).
[25] S. R. MORRISON, Phys. Rev. **104**, 619 (1956).
[26] E. S. RITTNER, Science, **111**, 685 (1950).
[27] J. C. SLATER, Phys. Rev. **103**, 1631 (1956).
[28] R. L. PETRITZ, Phys. Rev. **104**, 1508 (1956).
[29] T. S. MOSS, Proc. IRE **43**, 1869 (1955).
[30] F. R. KESSLER und E. SUTTER, Z. Phys. **173**, 54 (1962).
[31] F. R. KESSLER, Z. Naturf. **13a**, 295 (1958).
[32] F. R. KESSLER, persönliche Mitteilung.
[33] J. JAUMANN und E. NECKENBÜRGER, Z. Phys. **151**, 72 (1958).
[34] H. GOBRECHT, A. TAUSEND und M. PLÜMECKE, Z. angew. Phys. **15**, 496 (1963).
[35] H. GOBRECHT, A. TAUSEND und H. SCHMIDT, Z. angew. Phys. **14**, 453 (1962).
[36] M. POLKE, Dissertation, Karlsruhe 1963.
[37] L. R. WEISSBERG, J. appl. Phys. **33**, 1817 (1962).
[38] F. R. KESSLER und H. J. METZGER, Phys. kondens. Materie **1**, 263 (1963).
[39] R. H. BUBE und H. E. MCDONALD, Phys. Rev. **128**, 2062 (1962).
[40] R. V. PARFEN'EV, J. J. FARBSHTEIN und S. S. SHALYT, Soviet Phys. — Solid State Physics **2**, 2599 (1961).
[41] I. L. DRICKKO, I. V. MOCHOM und YU. N. OBRAZTSOV, Sov. Phys. — Solid State Physics **4**, 1842 (1963).
[42] R. GASPAR, Acta phys. Acad. Sci. Hung. **27**, 289 (1957).
J. R. REITZ, Phys. Rev. **105**, 1233 (1957).
[43] H. KELLER und J. STUKE, noch unveröffentlicht.

(Received May 15, 1964)

phys. stat. sol. 6, 461 (1964)

Commonwealth Scientific and Industrial Research Organization,
Division of Tribophysics, University of Melbourne

The [111] Dislocation in a Cubic Crystal

By

A. K. HEAD

Exact expressions are obtained for the energy of a dislocation lying in a [111] direction in a cubic crystal and for three components of the stress field. Sufficient information is given to enable the remaining stress components, displacements, etc., to be obtained.

Es werden exakte Formeln für die Energie einer Versetzung in [111]-Richtung in einem kubischen Kristall und für drei Komponenten des Spannungsfeldes abgeleitet. Auf der Grundlage der vorliegenden Untersuchungen können die übrigen Spannungskomponenten, Verrückungen usw. berechnet werden.

1. Introduction

The theory of the elastic properties of dislocations in an anisotropic solid has been given by ESHELBY et al. [1] and in an alternate but equivalent form by STROH [2]. In general the analysis can only be carried through numerically. However, for a cubic crystal, if the direction of the dislocation line is either [100] or [110] then an explicit solution has been obtained [1]. This is because for these orientations (and no other) the elastic displacements produced by the dislocation separate, the screw component producing only displacements parallel to the dislocation line and the edge component producing only displacements perpendicular to the dislocation line. It will be shown that there is a third direction of dislocation line in a cubic crystal, namely [111], for which an explicit solution can be obtained, even though the displacements do not separate. A solution is possible in this case essentially because of the three fold symmetry of a cubic crystal about [111].

2. Analysis of a [111] Dislocation

The method described by STROH [2] will be used and equations in that paper will be referred to by the prefix S.

Consider a cubic crystal with elastic coefficients s_{11}, s_{12}, s_{44} with respect to the cube axes. It is necessary to choose another set of axes, which we will call the dislocation axes, with the x_3 axis in the direction of the dislocation line [111]. We choose $Ox_1 = [11\bar{2}]$ and $Ox_2 = [\bar{1}10]$ to complete a right handed set. This choice is convenient but arbitrary as any other pair of perpendicular directions in the (111) plane would have been equally valid. Relative to the dislocation axes, the elastic coefficients are

$$\left.\begin{array}{cccccc}
s_{11}-\frac{1}{2}\delta & & & & & \\
s_{12}+\frac{1}{6}\delta & s_{11}-\frac{1}{2}\delta & & & & \\
s_{12}+\frac{1}{3}\delta & s_{12}+\frac{1}{3}\delta & s_{11}-\frac{2}{3}\delta & & & \\
0 & 0 & 0 & s_{44}+\frac{4}{3}\delta & & \\
-\frac{\sqrt{2}}{3}\delta & \frac{\sqrt{2}}{3}\delta & 0 & 0 & s_{44}+\frac{4}{3}\delta & \\
0 & 0 & 0 & 2\frac{\sqrt{2}}{3}\delta & 0 & s_{44}+\frac{2}{3}\delta
\end{array}\right\} \tag{1}$$

where $\delta = s_{11} - s_{12} - \frac{1}{2} s_{44}$. If the crystal happens to be isotropic, then $\delta = 0$.

STROH defines in (S 66) a set of reduced elastic coefficients S_{ij}. For the array (1) the non-zero elements of S_{ij} are

$$\left.\begin{aligned}
S_{11} = S_{22} &= \left[\left(s_{11}-\frac{1}{2}\delta\right)\left(s_{11}-\frac{2}{3}\delta\right)-\left(s_{12}+\frac{1}{3}\delta\right)^2\right]\Big/\left(s_{11}-\frac{2}{3}\delta\right),\\
S_{12} &= \left[\left(s_{12}+\frac{1}{6}\delta\right)\left(s_{11}-\frac{2}{3}\delta\right)-\left(s_{12}+\frac{1}{3}\delta\right)^2\right]\Big/\left(s_{11}-\frac{2}{3}\delta\right),\\
S_{44} = S_{55} &= s_{44}+\frac{4}{3}\delta\,,\\
S_{66} &= s_{44}+\frac{2}{3}\delta\,,\\
S_{25} = -S_{15} &= \frac{1}{2}S_{46} = \frac{\sqrt{2}}{3}\delta,
\end{aligned}\right\} \tag{2}$$

and the sextic equation (S 74) for p becomes

$$S_{11}\,S_{44}\,(p^2+1)^3 - S_{15}^2\,p^2\,(p^2-3)^2 = 0 \tag{3}$$

on using the equalities in (2) and that $2\,S_{12} + S_{66} = 2\,S_{11}$. It is known that the roots of (3) will all be complex. They may be found by the substitution

$$p = \tan\varepsilon\,,$$

where ε will be complex. With this substitution (3) becomes

$$\tan^2 3\,\varepsilon = \frac{S_{11}\,S_{44}}{S_{15}^2 - S_{11}\,S_{44}}\,. \tag{4}$$

The right hand side of (4) is always negative for allowable values of the elastic coefficients s_{11}, s_{12}, s_{44} so that writing

$$R = \left|\sqrt{\frac{S_{11}\,S_{44}}{S_{11}\,S_{44} - S_{15}^2}}\right| \tag{5}$$

equation (4) becomes

$$\tan 3\,\varepsilon = i\,R \qquad \text{or} \qquad -\,i\,R\,. \tag{6a, b}$$

From the properties of the complex tangent [3] it will be seen that if ε is a root of (6a) then ε has positive imaginary part and so does the corresponding $p = \tan\varepsilon$. Corresponding to such an ε, its complex conjugate $\bar{\varepsilon}$ is a root of (6b) and the corresponding $\bar{p} = \tan\bar{\varepsilon}$ is the conjugate of p. Further, if ε is a root of (6a) then so also is $\varepsilon + 2\,n\,\pi/3$. Thus if $p_1 = \tan\varepsilon_1$ is a root of (6a) then the six roots of the sextic equation (3) are

$$\left.\begin{aligned} p_1 &= \tan\varepsilon_1\,,\\ p_2 &= \tan\varepsilon_2 = \tan\left(\varepsilon_1 + \frac{2\,\pi}{3}\right) = (p_1 - \sqrt{3})/(1 + p_1\sqrt{3})\,,\\ p_3 &= \tan\varepsilon_3 = \tan\left(\varepsilon_1 + \frac{4\,\pi}{3}\right) = (p_1 + \sqrt{3})/(1 - p_1\sqrt{3})\,, \end{aligned}\right\} \tag{7}$$

which have positive imaginary parts and $\bar{p}_1$, $\bar{p}_2$, $\bar{p}_3$ which have negative imaginary parts. From [3] it will be seen that there is one root of (6a) which is pure imaginary and it is convenient to take this as p_1. This is

$$p_1 = i\,\frac{(R+1)^{1/3} - (R-1)^{1/3}}{(R+1)^{1/3} + (R-1)^{1/3}}\,. \tag{8}$$

in which the real cube roots are to be taken.

Following Stroh's analysis we have

$$[L_{i\alpha}] = \begin{bmatrix} -\tan\varepsilon_1 & -\tan\varepsilon_2 & -\tan\varepsilon_3 \\ 1 & 1 & 1 \\ -\dfrac{S_{15}\sin 3\,\varepsilon_1}{S_{44}\cos\varepsilon_1} & -\dfrac{S_{15}\sin 3\,\varepsilon_2}{S_{44}\cos\varepsilon_2} & -\dfrac{S_{15}\sin 3\,\varepsilon_3}{S_{44}\cos\varepsilon_3} \end{bmatrix} \tag{S 77}$$

$$[M_{\alpha i}] = \begin{bmatrix} -\dfrac{1}{3}\sin 2\,\varepsilon_1 & \dfrac{2}{3}\cos^2\varepsilon_1 & -\dfrac{1}{3}\dfrac{S_{44}\cos\varepsilon_1}{S_{15}\sin 3\,\varepsilon_1} \\ -\dfrac{1}{3}\sin 2\,\varepsilon_2 & \dfrac{2}{3}\cos^2\varepsilon_2 & -\dfrac{1}{3}\dfrac{S_{44}\cos\varepsilon_2}{S_{15}\sin 3\,\varepsilon_2} \\ -\dfrac{1}{3}\sin 2\,\varepsilon_3 & \dfrac{2}{3}\cos^2\varepsilon_3 & -\dfrac{1}{3}\dfrac{S_{44}\cos\varepsilon_3}{S_{15}\sin 3\,\varepsilon_3} \end{bmatrix} \tag{S 78}$$

and on solving the set of linear equations (S 41—44, 67—68)

$$\left.\begin{aligned} B_{11} &= B_{22} = 2\,[S_{11}\,(S_{11}\,S_{44} - S_{15}^2)/S_{44}]^{1/2}\,,\\ B_{33} &= [S_{44}\,(S_{11}\,S_{44} - S_{15}^2)/S_{11}]^{1/2}\,,\\ B_{ij} &= 0 \text{ for } i \neq j\,. \end{aligned}\right\} \tag{9}$$

Since $[B_{ij}]$ is a diagonal matrix it follows from (S 50) that

$$d_i = b_i/B_{ii} \qquad \text{(no summation)}\,. \tag{10}$$

Thus from (S 55) the energy of a dislocation with direction [111] and components of Burgers vector b_1, b_2, b_3 relative to the dislocation axes defined above is given by

$$E = [K_{11}(b_1^2 + b_2^2) + K_{33} b_3^2] \cdot \frac{1}{4\pi} \ln\left(\frac{r_1}{r_0}\right), \tag{11}$$

where r_0 and r_1 are the conventional inner and outer cutoff radii and

$$\left.\begin{aligned} K_{11} &= \frac{1}{2}[S_{11}(S_{11} S_{44} - S_{15}^2)/S_{44}]^{-1/2}, \\ K_{33} &= [S_{44}(S_{11} S_{44} - S_{15}^2)/S_{11}]^{-1/2}. \end{aligned}\right\} \tag{12}$$

Up to this point it has not been necessary to use the actual value of p_1 from (8) for in following STROHs analysis only symmetrical combinations of ε_1, ε_2 and ε_3 occur which can all be evaluated by (6a) as it stands or in terms of the coefficients of the expanded form of (6a)

$$\tan^3 \varepsilon - 3\, i\, R \tan^2 \varepsilon - 3 \tan \varepsilon + i\, R = 0\,.$$

The evaluation of the stresses by (S 51—52) is straightforward knowing $[L_{i\alpha}]$, $[M_{\alpha i}]$ and $[B_{ij}]$ but as the actual values of the p_i are involved the result is rather cumbersome and will not be given here. However, as shown in the following section, three components of stress $\sigma_{r\theta}$, $\sigma_{\theta\theta}$ and $\sigma_{z\theta}$ have a simpler form.

3. The Relation between Energy and Stress in General Anisotropy

FOREMAN [4] has shown that the energy of a dislocation in the general anisotropic solid is

$$E = [K_{ij}\, b_i\, b_j] \cdot \frac{1}{4\pi} \ln\left(\frac{r_1}{r_0}\right), \tag{13}$$

where the b_i are the components of the Burgers vector relative to dislocation axes, i.e. the x_3 axis along the dislocation line and the x_1 and x_2 axes completing a right handed set, and the K_{ij} depend only on the elastic constants (also measured relative to the dislocation axes).

It is now convenient to introduce cylindrical coordinates r, θ, z with z along the dislocation line Ox_3 and θ measured from Ox_1. The dislocation can be made from the unstrained solid in the classical way by cutting along any plane radially from the dislocation line and giving the two sides of the cut the necessary relative displacement. In doing this, work will be done against three of the components of stress $\sigma_{r\theta}$, $\sigma_{\theta\theta}$ and $\sigma_{z\theta}$. Since along any radial plane all components of stress vary as r^{-1} [1] and since the total work done in a sequence of partial displacements must be independent of the order of the sequence, it follows, on equating the work done with the energy of the dislocation (13), that when the dislocation has been formed these three stress components are given by

$$\left.\begin{aligned} \sigma_{r\theta} &= [(K_{11} b_1 + K_{12} b_2 + K_{13} b_3) \cos\theta + (K_{21} b_1 + K_{22} b_2 + K_{23} b_3) \sin\theta]/2\pi r, \\ \sigma_{\theta\theta} &= [(K_{21} b_1 + K_{22} b_2 + K_{23} b_3) \cos\theta - (K_{11} b_1 + K_{12} b_2 + K_{13} b_3) \sin\theta]/2\pi r, \\ \sigma_{z\theta} &= [K_{31} b_1 + K_{32} b_2 + K_{33} b_3]/2\pi r\,. \end{aligned}\right\} \tag{14}$$

Applying this to the [111] dislocation in a cubic crystal, it will be seen that

$$\left.\begin{aligned} \sigma_{r\theta} &= K_{11}\,(b_1 \cos\theta + b_2 \sin\theta)/2\,\pi\, r\,, \\ \sigma_{\theta\theta} &= K_{11}\,(b_2 \cos\theta - b_1 \sin\theta)/2\,\pi\, r\,, \\ \sigma_{z\theta} &= K_{33}\, b_3/2\,\pi\, r\,, \end{aligned}\right\} \tag{15}$$

where K_{11} and K_{33} are given by (12).

In [5] an approximation was given for the energy of a screw dislocation with an arbitrary direction in a cubic crystal. This interpolates between the energy in the three directions [100], [110], and [111]. It is now seen that H_{111} in Table 2 of [5] is actually $s_{44}\, K_{33}$ where K_{33} is given by (12). If this interpolation is used to estimate the energy of a screw dislocation as

$$E = \frac{b^2\, c_{44}\, H}{4\,\pi} \ln\left(\frac{r_1}{r_0}\right),$$

then from (14), the stress component $\sigma_{z\theta}$ will be

$$\sigma_{z\theta} = b\, c_{41}\, H/2\,\pi\, r\,.$$

References

[1] J. D. Eshelby, W. T. Read, and W. Shockley, Acta metall. **1**, 251 (1953).
[2] A. N. Stroh, Phil. Mag. **3**, 625 (1958).
[3] E. Jahnke and F. Emde, Tables of Functions, Dover, New York 1945 (Addenda p. 71).
[4] A. J. E. Foreman, Acta metall. **3**, 322 (1955).
[5] A. K. Head, phys. stat. sol. **5**, 51 (1964).

(Received May 19, 1964)

phys. stat. sol. 6, 467 (1964)

Institut für Physikalische Chemie der Friedrich-Schiller-Universität Jena

Über die Bestimmung von Molekülschwingungen im Kristallgitter aus Röntgenaufnahmen bei verschiedenen Temperaturen

Von

J. Krausse und H. Dunken

Es wird ein kritischer Überblick über röntgenographische Methoden zur Bestimmung von Molekülschwingungen im Gitter gegeben, ohne daß dabei auf die bekannten Verfahren der röntgenographischen Strukturanalyse näher eingegangen wird. Insbesondere wird auf eine neuartige Methode zur Bestimmung der Translationsschwingungen hingewiesen, welche auf der Ermittlung von Elektronendichteverteilungen bei etwa 100 °K und 300 °K beruht. Derartige Diagramme werden von Benzoesäure und Salizylsäure angegeben und ausführlich diskutiert. Die genannte Methode scheint besonders wichtig für die Trennung von Torsions- und Translationsamplituden und für Probleme der zwischenmolekularen Bindung zu sein.

A critical survey is given of the determination of molecular vibrations in a lattice by means of X-ray diffraction. A new method is used for the determination of translational vibrations, which is based on the calculation of the electron density distributions at about 100 °K and 300 °K. Appropriate diagrams are given for benzoic and salicylic acid and these are critically discussed. This method seems to be particularly important for the separation of torsional and translational amplitudes and for intermolecular bonding problems.

Bei der röntgenographischen Strukturaufklärung werden zur genauen Lagebestimmung der Atome individuelle anisotrope Temperaturfaktoren verwendet. Diese Faktoren sollen sämtliche thermische Schwingungen einschließlich der Nullpunktsschwingungen der Atome bzw. Moleküle im Festkörper berücksichtigen. Ihre Ermittlung geschieht durch Angleichen der aus der gestreuten Intensität erhaltenen Strukturamplituden F_{ob} an die berechneten Strukturamplituden F_c, wobei bekanntlich

$$F_c = \sum_{n=1}^{N} f_n \cdot e^{2\pi i (h x_n + k y_n + l z_n)} \cdot e^{-({}_nB_{11} h^2 + {}_nB_{22} k^2 + {}_nB_{33} l^2 + {}_nB_{12} h k + {}_nB_{13} h l + {}_nB_{23} k l) \frac{\sin^2 \vartheta}{\lambda^2}} \tag{1}$$

ist und der zweite Exponentialfaktor als anisotroper Temperaturfaktor des Atoms n bezeichnet wird. Die ${}_nB_{ij}$ können transformiert werden zu Komponenten eines symmetrischen Schwingungstensors ${}_nU$, so daß der Temperaturfaktor die Form

$$e^{-8\pi^2 \left(\sum\limits_{i,j}^{3} U_{ij} l_i l_j\right)_n \frac{\sin^2 \vartheta}{\lambda^2}} \tag{2}$$

annimmt, wobei die Schwingungsamplitude des Atoms n in der Richtung l[1]) gegeben ist [1] durch

$$\overline{u^2} = \sum_{i,j}^{3} U_{ij} l_i l_j \,. \tag{3a}$$

[1]) Komponenten l_1, l_2, l_3 entlang x, y, z.

Die optimale Anpassung zwischen den F_c- und F_{ob}-Werten wird in der Regel durch Variation der Ortskoordinaten x_n, y_n, z_n und der Temperaturfaktorparameter $_nB_{ij}$ nach der Methode der kleinsten Quadrate erreicht. Die Berechnung erfolgt auf Rechenautomaten, wobei häufig anstelle der Basis e die Zahl 2 benutzt wird. Es muß aber bei diesem Verfahren immer daran gedacht werden, daß die F_{ob}- und f_n-Werte nicht fehlerfrei sind. Das hat zur Folge, daß die anisotropen Parameter besonders durch systematische Fehler und anisotrope Ladungsverteilung fehlerhaft werden können. In hohem Maße gilt dies für die Strukturbestimmungen aromatischer Verbindungen [2]. Lonsdale und Milledge [3] veröffentlichten unter der Überschrift „Analysis of Thermal Vibrations in Crystals: a Warning" Berechnungen der $_nB_{ij}$-Werte für Anthracen aus zwei unabhängigen Meßserien verschiedener Autoren. Sie erhielten Unterschiede in den Schwingungsamplituden und in den Schwingungsrichtungen und konnten dadurch die Fehlerhaftigkeit solcher Berechnungen klar zeigen. Es erscheint daher im Sinne von Londsdale und Milledge, wenn aus den röntgenographisch ermittelten Atomschwingungen solche Schwingungen festgelegt werden, die auf Grund von physikalischen Überlegungen bei der Meßtemperatur wirklich stark angeregt sein müssen und die nach Möglichkeit aus vielen Atomschwingungen gemittelt werden können. Wie die weiteren Betrachtungen zeigen werden, handelt es sich bei den Molekülschwingungen organischer Strukturen um solche Schwingungen. Aus der mittleren quadratischen Schwingungsamplitude erhält man diese Molekülschwingungen nach Cruickshank [4] aus der folgenden Gleichung

$$\overline{u^2} = \sum_{i,j}^{3} [T_{ij}\, l_i\, l_j + \omega_{ij}\, m_i\, m_j]\ ^{2)}, \tag{3b}$$

wobei der erste Summand den Translationsschwingungen und der zweite den Torsionsschwingungen entspricht. Einige auf diese Weise bestimmte Molekülschwingungen verschiedener Moleküle sind in Tabelle 1 zusammengestellt.

Die angegebenen Amplituden wurden aus den Diagonalgliedern der Schwingungstensoren berechnet, wobei

$$\bar{a}_i = \sqrt{T_{ii}} \qquad \text{(Translation entlang der Achse } i\text{)}$$

und

$$\bar{\alpha}_i = \frac{360^\circ \sqrt{\omega_{ii}}}{2\,\pi} \qquad \text{(Drehung um die Achse } i\text{)}.$$

Der Tabelle 1 entnimmt man eine relativ kleine Translationsamplitude in Richtung der durch Wasserstoffbrückenbindung verbundenen Molekülreihen des Bernsteinsäurekristalls. Dies steht nach Gleichung (4b) in Übereinstimmung mit den kleineren Frequenzen bei der Wasserstoffbrückenschwingung gegenüber den Schwingungen der durch van der Waals-Wechselwirkungen gebundenen Molekülgitter, worauf noch näher eingegangen wird. Auch die Torsionsamplituden stehen im sinnvollen Zusammenhang mit den entsprechenden Trägheitsmomenten. Es sei jedoch darauf hingewiesen, daß einige in der Tabelle 1 aufgeführte Autoren

---

[2]) Es bedeuten: T_{ij} — Komponenten des Translationstensors, ω_{ij} — Komponenten des Torsionstensors, $m_i \,||\, (\mathfrak{e} \cdot \mathfrak{r})_i$, wobei $\mathfrak{e}$ der Einheitsvektor in Bewegungsrichtung ist und $\mathfrak{r}$ die Lage des Atoms angibt.

Tabelle 1

Substanz	Bezeichnungsweise der Achsen in der Zeichenebene*)	Translations-amplituden (Å)			Torsions-amplituden			Unter-suchungs-tempera-tur	Lit.
		$\bar{a}_1$	$\bar{a}_2$	$\bar{a}_3$	$\bar{\alpha}_1$	$\bar{\alpha}_2$	$\bar{\alpha}_3$		
Anthracen		0,16	0,20	0,16	2,2°	3,8°	3,1°	300 °K	[1]
Diphenyl-tetrafluor-äthan	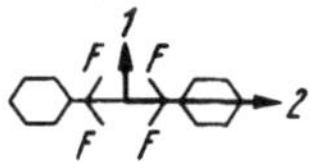	0,21	0,22	0,19	1,6°	6,9°	1,8°	300 °K	[6]
Bernstein-säure	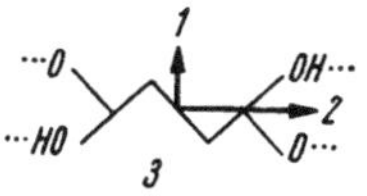	0,18	0,10	0,16	0–2°	9°	0–2°	300 °K	[1]
Harnstoff	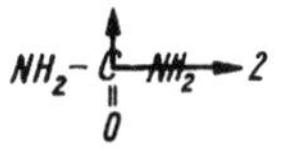	0,20	0,20	0,12	3,1°	3,1°	11,6°	295 °K	[7]
		0,13	0,13		2,3°	2,3°	8,3°	112 °K	
Benzol		0,25	0,23	0,21	5,1°	5,0°	7,9°	270 °K	[5]

*) Die fehlende Achse steht senkrecht auf den angegebenen Richtungen.

große Fehler für die genannten Amplituden angeben. Die Gründe liegen einmal in der Schwierigkeit beim Trennen von Translations- und Torsionsanteil, zum anderen tritt aber auch hierbei wieder die Fehlerhaftigkeit der einzelnen ${}_nB_{ij}$-Werte zutage.

Eine neuartige Methode [7] zur röntgenographischen Bestimmung der Schwingungsrichtung im Kristall basiert auf der Temperaturabhängigkeit der Anregung von Schwingungen. Die Informationen werden aus Beugungsmessungen bei verschiedenen Temperaturen am gleichen Kristall unter den gleichen apparativen Bedingungen auf dem gleichen Film ermittelt und sind somit von systematischen Fehlern frei. Auch Verfälschungen durch anisotrope Ladungsverteilungen als Folge chemischer Bindung sind eliminiert. Die Anwendung dieser Methode betrifft besonders die Strukturuntersuchungen organischer Verbindungen. Diese Moleküle sind, abgesehen von wenigen Ausnahmen, durch schwache van der Waals-Wechselwirkungen im Gitter gebunden, so daß deren Schwingungen relativ niederfrequent ($\tilde{\nu} < 200$ cm^{-1}) und daher bei Zimmertemperatur stark angeregt sind. Beispielsweise ergibt sich nach der Theorie der spezifischen Wärmen von Einstein, daß bei Zimmertemperatur mehr als 95% der möglichen Gitterschwingungen zur spezifischen Wärme beitragen, während die meisten typischen Valenzschwingungen nur zu einem kleinen Prozentsatz angeregt sind. Weiterhin zeigen die niedrigen Schmelztemperaturen (in der Regel $F_p < 600$ °K) und die relativ großen und anisotropen Temperaturfaktoren (B_{isotrop} etwa zwischen 3,5 und 4,5 Å^2, 10 grd unterhalb F_p zwischen 6 und 9 Å^2) einen hohen Anregungsgrad an und weisen andererseits auf einen Zusammenhang zwischen den Temperaturfaktoren und den Molekülschwingungen hin. Zur Abschätzung der Temperaturabhängigkeit der Schwingungsamplitude von Molekülen kann man nach CRUICKSHANK [8] die Debye-Waller-Formel

$$\overline{u^2_{\text{trans}}} = \frac{3\,\boldsymbol{h}^2}{4\,\pi^2\,M\,\boldsymbol{k}\,\Theta_{\max}}\left\{\frac{\Phi(x)}{x} + \frac{1}{4}\right\}, \qquad \Phi(x) = \frac{1}{x}\int\limits_0^x \frac{\xi\,d\xi}{e^{\xi}-1}\,, \qquad x = \frac{\Theta_{\max}}{T} \tag{4a}$$

M Masse, $\Theta_{\max}$ charakteristische Temperatur

benutzen, obwohl diese Formel speziell nur für ein kubisches Gitter aufgestellt wurde. Weiterhin trägt die Verteilungsfunktion der Schwingungsfrequenzen nach DEBYE nicht ganz den wahren Verhältnissen Rechnung. Einmal zeigt das kurzwellige Ende dieses Spektrums Abweichungen gegenüber der Frequenzverteilung nach der Gittertheorie, zum anderen werden die meßbaren Reflexe von den langwelligen Gitterschwingungen nicht entsprechend dem Anregungsgrad der Verteilungsfunktion beeinflußt. Die Annahme einer einzelnen Schwingung wird jedoch schlechter als die Annahme nach DEBYE sein, da der Kristall ein dreidimensionales System gekoppelter Oszillatoren darstellt. Dennoch müssen zur genauen Berechnung des Temperaturverlaufes unabhängige Oszillatoren für die hohen Frequenzen angenommen werden. Das gilt sowohl für die spezifischen Wärmen wie für die röntgenographisch bestimmten Schwingungsamplituden. Die Anwendung dieses Gedankens scheitert meist an den großen Schwierigkeiten bei der Ermittlung der Schwingungsfrequenzen und deren Entartungsgrades. Ein eindrucksvolles Beispiel für die Anwendung der Debye- und Einstein-Funktion hat RAMAN [9] bei der Berechnung der spezifischen Wärme von Diamant im Bereich zwischen 15 und 1100 °K gegeben. Für die Temperaturabhängigkeit der Schwingungsamplitude eines harmonischen Oszillators gilt nach JAMES [10]

$$\overline{u^2_{\text{trans}}} = \frac{\boldsymbol{h}}{8\,\pi^2\,M\,\nu}\coth\left(\frac{\boldsymbol{h}\,\nu}{2\,\boldsymbol{k}\,T}\right); \qquad M = \frac{m_1 \cdot m_2}{m_1 + m_2}\,. \tag{4b}$$

Ein Vergleich der mittleren Schwingungsamplitude $\bar{a} = \sqrt{\overline{u^2}}$ nach (4a) und (4b) ist in Fig. 1 für eine Molekülmasse $M = 100 \cdot 1{,}67 \cdot 10^{-24}$ g und den Bereich der Molekül- und niederfrequenten Valenzschwingungen dargestellt. Bei zwei Molekülen in der Elementarzelle und bei geringer Symmetrie sind unter Berücksichtigung der Phasenbeziehungen akustische und optische Schwingungen möglich;

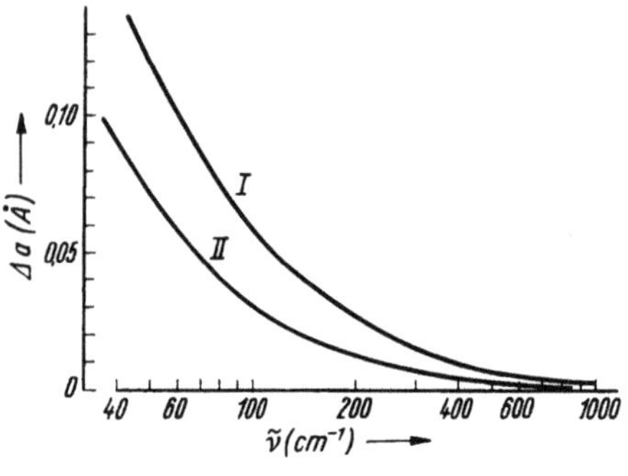

Fig. 1. Mittlere Schwingungsamplitude bei 300 °K. Frequenzverteilung: Kurve I nach DEBYE mit $\nu = \boldsymbol{k}\,\Theta_{\max}/\boldsymbol{h}$, Kurve II für eine Oszillatorfrequenz

nach CRUICKSHANK [8] kann man dann anstelle der Gleichung (4a) setzen

$$\overline{u^2_{\text{trans}}} = \frac{1}{2}\left\{\frac{3\,h^2}{4\,\pi^2\,M\,k\,\Theta_{\max}}\left(\frac{\Phi(x)}{x}+\frac{1}{4}\right)+\frac{h}{8\,\pi^2\,M\,\nu}\coth\left(\frac{h\,\nu}{2\,k\,T}\right)\right\}, \qquad (4\text{c})$$

wobei die Frequenz ν zum optischen Zweig gehört. Nimmt man auf Grund der Anisotropie der zwischenmolekularen Bindung r charakteristische Oszillatorfrequenzen an, so erhält man in Analogie zu dieser Gleichung für das örtlich und zeitlich gemittelte Amplitudenquadrat (ohne Berücksichtigung des Entartungsgrades)

$$\overline{u^2_{\text{trans}}} = \frac{1}{r+1}\left\{\frac{3\,h^2}{4\,\pi^2\,M\,k\,\Theta'_{\max}}\left(\frac{\Phi(x')}{x'}+\frac{1}{4}\right)+\frac{h}{8\,\pi^2\,M}\sum_{i=1}^{r}\frac{1}{\nu_i}\coth\left(\frac{h\,\nu_i}{2\,k\,T}\right)\right\}, \qquad (4\text{d})$$

$$\nu_{\Theta'_{\max}} < \nu_{i\,\min},\ \Theta'_{\max} < \Theta_{\max}\,.$$

In Fig. 2 sind zwei typische Fälle für die Annahme charakteristischer Frequenzen bei hoher Symmetrie wiedergegeben. Es soll damit gezeigt werden, daß bei Annahme dreier charakteristischer Oszillatorfrequenzen entsprechend den drei Translationsfreiheitsgraden eines Moleküls diese im allgemeinen niederfrequenter als die zugehörige maximale Grenzfrequenz nach DEBYE sind. Das bedeutet, daß in den Schwingungsrichtungen entsprechend Fig. 1 die Amplituden der Molekülschwingungen bei 300 °K etwa 0,1 bis 0,25 Å betragen. Die Temperaturabhängigkeit der Torsionsamplitude ist in Analogie zur Gleichung (4b) gegeben durch

$$\overline{\varphi^2} = \frac{h}{8\,\pi^2\,I\,\nu}\coth\left(\frac{h\,\nu}{2\,k\,T}\right), \qquad (4\text{e})$$

$$I = \text{Trägheitsmoment}\,,$$

für den Fall zweier Moleküle in der Elementarzelle nach [8] durch

$$\overline{\varphi^2} = \frac{h}{8\,\pi^2\,I}\left\{\frac{1}{2\,\nu_1}\coth\left(\frac{h\,\nu_1}{2\,k\,T}\right)+\frac{1}{2\,\nu_2}\coth\left(\frac{h\,\nu_2}{2\,k\,T}\right)\right\}, \qquad (4\text{f})$$

$$\nu_1 \text{ — in-phase},\ \nu_2 \text{ — out-of-phase}\,.$$

Die Torsionsamplituden zeigen also die gleiche Temperaturabhängigkeit wie die charakteristischen Translationsamplituden.

Die erwähnte neuartige Methode besteht in der Bestimmung der Veränderung der temperaturabhängigen Schwingungsamplitude mittels eines Differenzdiagramms der Elektronendichten bei verschiedenen Temperaturen; gemeint ist das Diagramm der Funktion

$$\Delta_T\varrho = \varrho_{T_2} - \varrho_{T_1}\,, \qquad T_2 < T_1\,. \qquad (5\text{a})$$

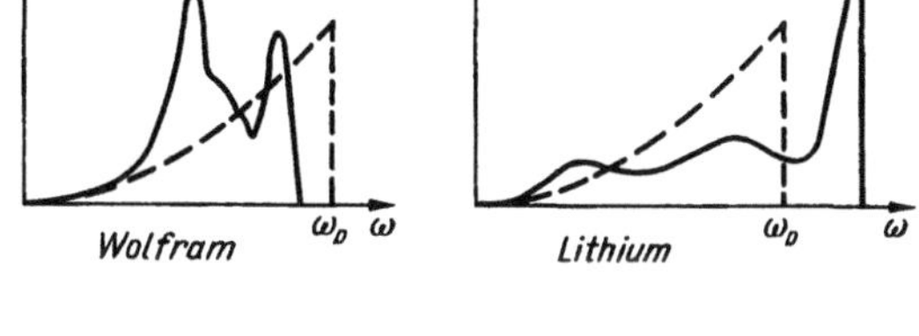

Fig. 2. Vergleich von Gitterspektrum und Debye-Spektrum (nach [11])

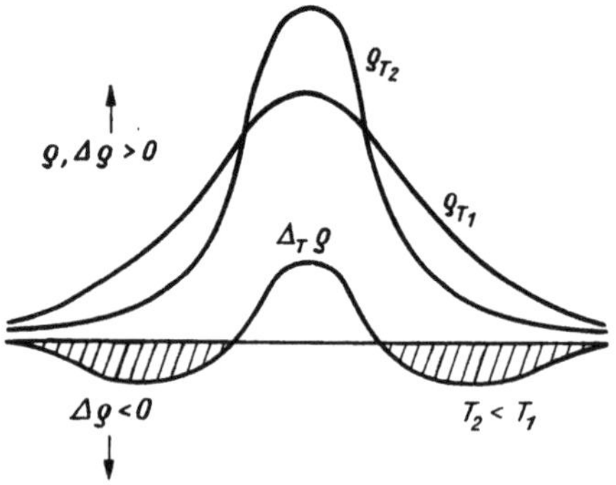

Fig. 3. Elektronendichteverteilung bei verschiedenen Temperaturen im Schnitt durch die Atomlage (schematische Darstellung)

Entsprechend der Fig. 3 erscheinen an jenen Stellen Minima, wo bei tiefer Temperatur die Elektronendichte kleiner als bei Zimmertemperatur ist. Selbstverständlich ist bei der Bestimmung von Molekülschwingungen über eine Vielzahl scheinbar einzelner Atomschwingungen eines Moleküls zu mitteln, worauf bereits hingewiesen wurde. Experimentell können solche Messungen meist mit siedendem Stickstoff durchgeführt werden, so daß T_2 bestenfalls 77 °K sein kann, häufig aber größer ist. Eine genaue Bestimmung der Meßtemperatur ist für die folgenden Betrachtungen nicht erforderlich; im Mittel kann sie 100 °K sein. Die Abnahme der mittleren Translationsamplitude bei Abkühlung auf diese Temperatur ist hinsichtlich der Frequenzabhängigkeit in Fig. 4 dargestellt und ergibt einen sehr ähnlichen Verlauf wie die Amplituden bei 300 °K in Fig. 1. Zur Überprüfung dieser Feststellungen sind in den Figuren 5, 6 und 7 solche Elektronendichtedifferenzdiagramme als Projektionen wiedergegeben. Bei der Benzoesäure (Fig. 5) und der Salicylsäure (Fig. 6) sind in dieser Projektion Translationsschwingungen dimerer Molekülkomplexe zu erwarten. Die beiden Säuremoleküle sind durch Wasserstoffbrücken zwischen den Karboxylgruppen gebunden, deren Assoziatschwingung bei 200 cm^{-1} liegt (12), so daß die Abnahme der Schwingungsamplitude nach Fig. 4 nicht merklich sein wird. Dasselbe gilt in noch höherem Maße für die Valenz- und anderen Atomschwingungen. Zur gleichen Erkenntnis kommt ROBERTSON [1] bei der Bestimmung der Wärmebewegung von Anthracen und Bernsteinsäure. Im $\Delta_T\varrho$-Diagramm treten also die dimeren Moleküle als nahezu starre Körper in Erscheinung.

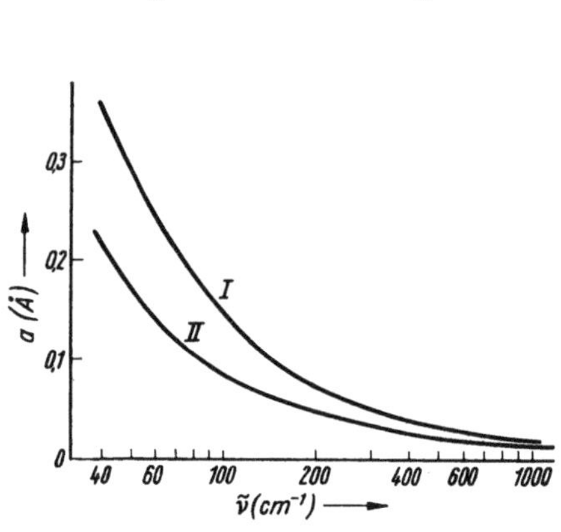

Fig. 4. Differenz der mittleren Schwingungsamplituden bei 300 °K und 100 °K
Frequenzverteilung: Kurve I nach DEBYE mit $\nu = k\,\Theta_{max}/h$, Kurve II für eine Oszillatorfrequenz

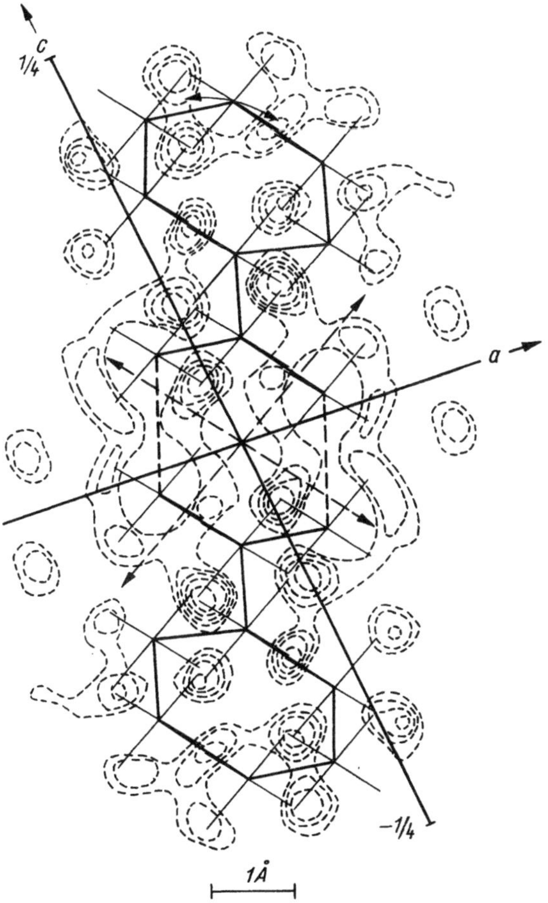

Fig. 5. $\Delta_{T}\varrho$-Diagramm der Benzoesäure als Projektion auf (010). Elektronendichteangaben: −0,25; −0,50; −0,75; ... Å^{-2}. Meßtemperaturen: $T_1 \approx 300$ °K und $T_2 \approx 120$ °K

Zur Deutung der $\Delta_{T}\varrho$-Diagramme von Fig. 5 und 6 werden entsprechend der Lage und Größe der Minima Translationsschwingungen der dimeren Molekülkomplexe angenommen. Der Übersichtlichkeit halber sind die Maxima an den Stellen der Atome weggelassen; sie sind nahezu kreisförmig und daher bedeutungslos. Verbindet man im Diagramm jeweils zwei Minima, die in der Nähe eines Atoms gegenüberliegen, durch Geraden und mittelt über alle nahezu parallelen Geraden durch die Atome, so erhält man bei Translationsschwingungen die Richtungen größter Aufenthaltswahrscheinlichkeit in der Projektion. Entsprechend den drei Freiheitsgraden werden die Minima im allgemeinen Fall die Eckpunkte eines Parallelepipedes darstellen. Dies gilt wohl auch noch für ein Teilchen eines gekoppelten Systems, jedoch wird der erwähnte Körper größer sein. Aus der Fig. 5 (Benzoesäure) muß man aus der klaren Anordnung von jeweils vier Minima um ein Atom entnehmen, daß eine Achse des erwähnten Parallelepipedes nahezu senkrecht auf der Projektionsebene steht. Dies steht in Übereinstimmung mit der zwischenmolekularen Bindung im Gitter [2]. Die Torsionsamplitude $\bar{\alpha} = \sqrt{\overline{\varphi^2}}$ um die Drehachse senkrecht zur Molekülebene kann nach (4f) berechnet werden und ergibt mit den Frequenzen $\tilde{\nu}_1 = 28$ cm^{-1} und $\tilde{\nu}_2 = 45$ cm^{-1} nach Maier und Schifferdecker [13] und einem Trägheitsmoment $I = 6500 \cdot 10^{-40}$ g cm^2 einen Torsionswinkel von −2,3 bei 300 °K. Die Abnahme der Torsionsamplitude beträgt beim Abkühlen auf 120 °K etwa 0,9°, so daß für das Kohlenstoffatom, welches von der Drehachse am weitesten entfernt ist, in der betrachteten Projektion eine Änderung der Auslenkung um 0,08 Å eintreten würde. Im $\Delta_{T}\varrho$-Diagramm (Fig. 5) sind die Elektronendichteminima der Torsionsschwingung dieses Kohlenstoffatoms durch Pfeile gekennzeichnet. Für alle anderen Atome ist diese Änderung kleiner. Die Abschätzung der Translationsamplituden aus den gemessenen isotropen Temperaturfaktoren $B_{T_1} = 4{,}6$ Å^2 und $B_{T_2} = 2{,}0$ Å^2 ergibt unter der Annahme charakteristischer Oszillatorfrequenzen eine Änderung von

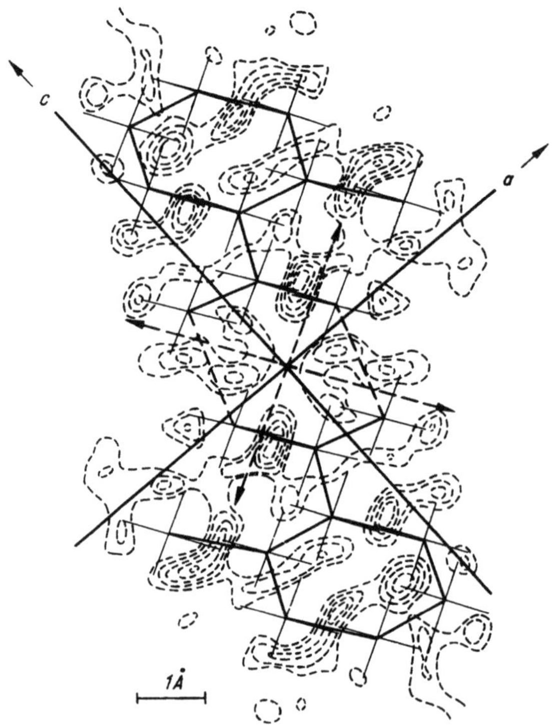

Fig. 6. $\Delta_T\varrho$-Diagramm der Salizylsäure als Projektion auf (001). Elektronendichteangaben: −0,25; −0,50; −0,75; ... Å^{-3}. Meßtemperaturen: $T_1 \approx 300$ °K und $T_2 \approx 100$°K

etwa 0,08 Å, wobei im Gegensatz zur Torsion die Auslenkung für alle Atome gleich sein muß. Bei der Salizylsäure (Fig. 6) liegen die Verhältnisse im Gitterspektrum vermutlich nicht so günstig. Durch allseitig stärkere zwischenmolekulare Bindung im Gitter [2] werden die charakteristischen Oszillatorfrequenzen größer als bei der Benzoesäure sein, so daß die entsprechenden Schwingungsamplituden kleiner sind. Dafür sprechen auch die etwas kleineren Temperaturfaktoren ($B_{T_1} = 4{,}0$ Å^2 und $B_{T_2} = 1{,}9$ Å^2). Demzufolge sind die Änderungen der Schwingungsamplituden relativ klein und in diesem Falle durch Reihenabbruchseffekte bei der Berechnung von $\Delta_T\varrho$ stark verfälscht. Betrachtet man jedoch die Abstände vom Atommittelpunkt bis zur Stelle minimaler Elektronendichte in den Richtungen größter Aufenthaltswahrscheinlichkeit (Tabelle 2), so ergibt sich durch eine befriedigende Konstanz ein Hinweis für die Realität der Translationsschwingungen.

Weiterhin zeigt ein Vergleich von Fig. 5 und 6, daß die Richtung der zwischenmolekularen Wasserstoffbrückenbindungen etwa die Winkelhalbierende der eingezeichneten Geraden ist. Sollte die Brückenschwingung trotz der erwähnten relativ hohen Frequenz dennoch merklichen Einfluß auf diese $\Delta_T\varrho$-Diagramme haben, so kann dies in Übereinstimmung mit den erwähnten Vorstellungen über die Molekülschwingungen als weitere Stütze für die Richtigkeit der Diagrammdeutung angesehen werden.

Ein Beispiel für den Einfluß von Torsionsschwingungen auf das $\Delta_T\varrho$-Diagramm zeigt Fig. 7 für Harnstoff, wie es von Toman und Tichý [7] angegeben wurde. Erkennbar ist die Torsionsschwingung an der Anisotropie des Stickstoffmaximums und den beiden relativ schwachen Minima. Die Schwingungsanalyse wurde von den genannten Autoren außerdem nach dem anfangs erwähnten Verfahren (Bestimmung der ${}_nB_{ij}$-Werte und Anwendung der Gleichung (3b)) durchgeführt; ihre Ergebnisse sind in Tabelle 1 aufgeführt und weisen ebenfalls auf eine relativ große Torsionsamplitude hin. Die ermittelten Torsionsamplituden ergaben nach (4e) eine Torsionsfrequenz von 54 cm^{-1}, welche in guter Übereinstimmung mit der Ramanfrequenz von 58 cm^{-1} steht [7].

Tabelle 2

Abweichungen vom Mittelwert der Abstände Atommittelpunkt — Minimum im $\Delta_T\varrho$-Diagramm in den beiden Schwingungsrichtungen in Å-Einheiten. Eingeklammerte Werte sind fehlerhaft oder nicht bestimmt worden

	Benzoesäure		Salizylsäure	
Abweichungen vom Mittelwert	0,00	0,03	0,03	0,04
	0,08	0,03	0,03	0,12
	0,02	0,03	0,00	0,05
	(0,22)	0,02	0,03	(0,17)
	0,03	0,08	(?)	0,01
	0,00	0,15	0,00	0,02
	0,03	0,10	0,03	0,01
	0,02	(0,15)	0,03	0,04
	0,00	0,05	0,03	0,10
	(0,20)	0,00	0,03	0,12
	0,05	0,00	0,06	0,10
	(?)	0,00	0,00	(0,21)
	0,12	0,07	0,06	0,04
	0,03	(0,22)	0,06	0,05
	0,07	0,05	0,03	(0,17)
	0,05	0,10	0,03	0,02
	(0,10)	0,00	0,06	0,04
	0,03	0,00	0,06	0,01
			0,03	0,08
			(0,12)	0,10
Mittelwert	$0,67_5$	$0,78_4$	$0,69_0$	$0,71_2$

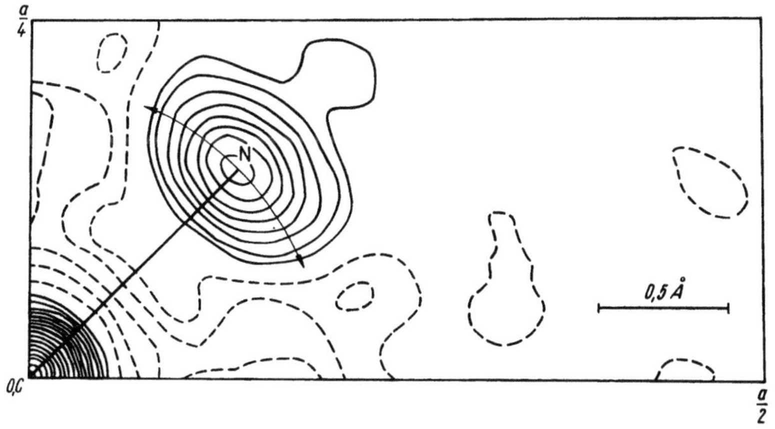

Fig. 7. $\Delta_T\varrho$-Diagramm von Harnstoff als Projektion auf (001). Elektronendichteangaben in relativen Einheiten: . . . 7; 6; 5; 4; 3; 2; 1; 0,5; −0,5; −1; −2; . . . Meßtemperaturen: $T_1 = 295\,°K$ und $T_2 = 112\,°K$

Die $\Delta_T\varrho$-Diagramme von Harnstoff, Benzoesäure und Salizylsäure wurden berechnet mittels folgender Fourierreihe:

$$\Delta_T\varrho = \frac{1}{V_{T_2}} \sum_{h,k,l} \left(F_{ob,T_2}(h,k,l) - \frac{V_{T_2}}{V_{T_1}} F_{ob,T_1}(h,k,l) \right) e^{-2\pi i (hx+ky+lz)} e^{-B' \frac{\sin^2\vartheta}{\lambda^2}}, \quad (5\text{b})$$

wobei V das Volumen der Einheitszelle, bei Projektionen die Fläche, auf die projiziert wird, und B' den Parameter einer sogenannten künstlichen Rechentemperatur bedeutet. Obwohl, wie bereits erwähnt, die gemessenen Strukturamplituden F_{ob} keine systematischen Fehler aufweisen, ist die Fourierreihenberechnung nicht fehlerfrei. Hauptsächlich bestehen die Fehler bei der Berechnung mehrdimensionaler Fourierreihen im unkontrollierbaren Reihenabbruch. Diese Fehler sind besonders zu beachten, da die Belichtungszeiten wegen Mangels an flüssigem Stichstoff meistens nur 8 bis 10 Stunden betrugen, so daß auch bei großen Kristallen zu wenige Tieftemperaturreflexe gemessen wurden. Weiterhin bewirken ein ungenauer Normierungsfaktor zwischen den experimentellen Strukturamplituden $F_{ob,300\,°\mathrm{K}}$ und $F_{ob,100\,°\mathrm{K}}$ und eventuelle Lageveränderungen in der Einheitszelle Verfälschungen der Extrema im $\Delta_T\varrho$-Diagramm. Die $\Delta_T\varrho$-Diagramme von Salizylsäure und Harnstoff zeigten merkliche Verbesserungen bei Benutzung eines „isotropen Temperaturfaktors" (sogenannte Rechentemperatur, $B' \approx 1{,}0$ Å^2), vor allem kam es zum Verschwinden unrealer Extrema. Allerdings verflachen dann die Extrema. Die Lageveränderungen der Atome in der Einheitszelle bei der Temperaturänderung wurden sorgfältig kontrolliert. Sie waren relativ klein; bei der Salizylsäure betrug die Veränderung der Lageparameter maximal 0,065 Å, während die Karboxylgruppe praktisch unverändert blieb. Im Mittel war den Parameteränderungen bei Benzoe- und Salizylsäure hauptsächlich eine kleine Drehung des dimeren Molekülkomplexes bei der Abkühlung zu entnehmen, die in den betrachteten Projektionen die Parameteränderungen durch Verkürzen der zwischenmolekularen Abstände weitgehend kompensierte. Diese Parameteränderungen wirken sich zwar systematisch aus, jedoch sind die beobachteten vier charakteristischen Minima um jedes Atom damit nicht zu erklären. Weiterhin zeigte ein $\Delta_T\varrho$-Diagramm mit berechneten Tieftemperaturreflexen (isotroper Temperaturparameter $B = 2$ Å^2 bei gleichen Atomlagen wie bei 300 °K) ebenfalls an jenen Stellen diese charakteristischen Minima, jedoch nicht sehr klar ausgeprägt. Bekanntlich treten in diesem Falle neben den Reihenabbruchseffekten die bereits am Anfang dieser Veröffentlichung erwähnten Fehler auf. Ein Versuch, Reihenabbruchseffekte und Koordinatenveränderungen im $\Delta_T\varrho$-Diagramm durch Verwendung von Koeffizienten der Gestalt

$$F_{ob,100\,°\mathrm{K}} - F_{ob,300\,°\mathrm{K}} + F_c(\mathfrak{r}, B_{\mathrm{isotrop}})_{300\,°\mathrm{K}} - F_c(\mathfrak{r}, B_{\mathrm{isotrop}})_{100\,°\mathrm{K}}$$

zu vermeiden, führte zu keiner Verbesserung der erwähnten charakteristischen Minima. Fehler durch ungenaue Normierung der F_{ob,T_2}-Werte bezüglich der F_{ob,T_1}-Werte verursachten keine charakteristischen Verfälschungen. Dies gilt auch für den Faktor $\frac{V_{T_2}}{V_{T_1}}$, der in (5b) vorkommt; er ist nahezu gleich 1 und beträgt für die angegebenen Projektionen bei Salizylsäure 0,970, bei Harnstoff 0,973 und bei Benzoesäure 0,978.

Neben der direkten Ermittlung von Schwingungsrichtungen durch Elektronendichtedifferenzdiagramme wird ihre Ermittlung aus physikalisch sinnvollen experimentellen Meßdaten für die Trennung von Torsions- und Translationsanteil der ${}_nB_{ij}$-Werte sehr wichtig sein.

Schlußbemerkung

Die im vorliegenden Artikel beschriebenen röntgenographisch bestimmbaren Größen sind in verschiedener Hinsicht nützlich. Insbesondere gilt dies für die Ermittlung akustischer Schwingungen im Gitter. So gibt CRUICKSHANK [8] die charakteristischen Gitterfrequenzen der Translations- und Torsionsschwingungen für Benzol, Naphthalin und Anthracen aus Röntgen- und Raman-Untersuchungen an. Andererseits ergeben sich wertvolle Hinweise für die zwischenmolekularen Wechselwirkungen in Molekülgittern. Derartige Betrachtungen bei den orthosubstituierten Benzoesäuren werden an anderer Stelle veröffentlicht. Weiterhin werden die temperaturabhängigen Schwingungsamplituden für Benzoesäure und Salizylsäure berechnet, sobald die Möglichkeit besteht, nach einem anisotropen „Least-squares-Programm" zu rechnen. In diesem Zusammenhang wird für die Röntgenstrukturanalyse auf die Möglichkeiten hingewiesen, Aussagen aus dem Elektronendichtedifferenzdiagramm über eine etwaige statistische Atomverteilung oder anisotrope Temperaturfaktoren zu erhalten.

Herrn Prof. Dr. K. TOMAN und Herrn K. TICHÝ, Prag, gilt besonderer Dank für förderliche Diskussionen und die Überlassung von Daten der Harnstoffstrukturanalyse. Weiterhin danken die Verfasser Herrn G. PÄSELT für die Tieftemperaturuntersuchungen an der Benzoesäure im Rahmen seiner Diplomarbeit [14].

Literatur

[1] J. M. ROBERTSON, Z. Krist. **112**, 68 (1959).
[2] J. KRAUSSE,Vortrag, gehalten in Hamonia bei Bratislava, September 1963
[3] K. LONSDALE und J. MILLEDGE, Acta cryst. **14**, 59 (1961).
[4] D. W. J. CRUICKSHANK, Acta cryst. **9**, 754 (1956).
[5] E. G. COX, D. W. J. CRUICKSHANK und J. A. S. SMITH, Proc. Roy. Soc. **247**, 1 (1958).
[6] D. W. J. CRUICKSHANK, G. A. JEFFREY und S. C. NYBURG, Z. Krist. **112**, 385 (1959).
[7] K. TOMAN und K. TICHÝ, Czech. J. Phys. **B 12**, 119 (1962).
[8] D. W. J. CRUICKSHANK, Acta. cryst. **9**, 1005 (1956).
[9] C. V. RAMAN, Proc. Indian Acad. Sci., Sect. A, **46**, 323 (1958).
[10] R. W. JAMES, Phys. Z. **33**, 737 (1932).
[11] R. BECKER, Theorie der Wärme, Springer-Verlag, Berlin/Göttingen/Heidelberg 1955 (S. 226).
[12] H. J. SPANGENBERG, Dissertation, Jena 1960.
[13] M. MAIER und O. SCHIFFERDECKER, Spectrochim. Acta **18**, 709 (1962).
[14] G. PÄSELT, Diplomarbeit, Jena 1961.

(Received May 19, 1964)

phys. stat. sol. 6, 479 (1964)

Abteilung Technische Physik des Physikalischen Institutes der Karl-Marx-Universität, Leipzig

Über die Niederfrequenzdispersion der Dielektrizitätskonstanten und der elektrischen Leitfähigkeit polykristalliner Ferrite[1])

Von

M. KRÖTZSCH

Das Ferritmodell von KOOPS wird durch Annahme einer Rechteckverteilung der Zeitkonstanten erweitert. Die Dispersionsformeln für die Dielektrizitätskonstante und die Leitfähigkeit sind geeignet, die experimentellen Ergebnisse mit einiger Genauigkeit wiederzugeben. Messungen unter hohem hydrostatischem Druck zeigen, daß der spezifische Widerstand der Ferritkristallite mit dem Druck abnimmt. Eine Abhängigkeit des Niederfrequenzniveaus der DK von Temperatur und Druck wird nicht gefunden.

The ferrite model due to KOOPS is extended by assuming a rectangular distribution of relaxation times. The dispersion formulas for the dielectric constant and conductivity agree well with experimental results. Measurements at high hydrostatic pressure show that the resistivity of ferrite grains decreases with pressure. The low-frequency limit of the dielectric constant is found to be independent of temperature and pressure.

1. Einleitung

Im Jahre 1938 hat BLECHSCHMIDT [1] festgestellt, daß Mn-Ferrite bei niedrigen Frequenzen sehr große Dielektrizitätskonstanten haben. Autoren, die in der Folgezeit andere Ferrite untersuchten, kamen zu ähnlichen Ergebnissen [2, 3, 4, 5, 6, 7, 8, 9]. Alle Autoren fanden, daß die DK von einem hohen Niederfrequenzwert mit wachsender Frequenz abnimmt und sich bei hohen Frequenzen einem Wert zwischen 10 und 20 nähert.

Schon BLECHSCHMIDT hat versucht, die Dispersion der DK zu erklären. Zu diesem Zweck stellte er eine Probe aus Graphitpulver und Paraffinwachs her. Da diese Probe ähnliche dielektrische Eigenschaften wie die untersuchten Ferrite hatte, schloß er, daß Ferrite aus Kristalliten guter Leitfähigkeit bestehen, die durch Schichten mit hohem spezifischem Widerstand voneinander getrennt sind. Diese Überlegungen wurden von KOOPS [9] als Grundlage für den Aufbau eines rechnerisch zu erfassenden, einfachen Ferritmodells gewählt. Er nahm an, daß alle Kristallite gleiche Leitfähigkeit σ_2 und gleiche DK ε_2 haben und von kubischer Gestalt sind. Die einzelnen Kristallite werden durch Verunreinigungsschichten, die untereinander auch gleiche elektrische und dielektrische Eigenschaften (σ_1, ε_1) haben sollen, getrennt. Wenn man nun die Zwischenschichten parallel zur Stromrichtung vernachlässigt, erhält man für eine Ferritplatte ein Modell aus einander abwechselnden Kristallschichten (Index 2) und Zwischenschichten (Index 1). Auf Grund der gemachten Annahmen kann man alle Schichten mit gleichem Index zusammenfassen. Die Dicke der Zwischenschichten sei d_1, die der Kristallite d_2. Das Verhältnis der beiden Dicken nennen wir x.

$$x = d_1/d_2 .$$

[1]) Teilauszug aus der Dissertation des Verfassers, Leipzig 1961.

Das so erhaltene Zweischichtenmodell für Ferrite läßt sich mit der Theorie inhomogener Dielektrika von Maxwell [10] bzw. Wagner [11] beschreiben.

$$\sigma = \sigma^{\infty} + \frac{\sigma^0 - \sigma^{\infty}}{1 + \omega^2 \tau^2}, \tag{1}$$ [2]

$$\varepsilon = \varepsilon^{\infty} + \frac{\varepsilon^0 - \varepsilon^{\infty}}{1 + \omega^2 \tau^2}, \tag{2}$$

ω Kreisfrequenz, τ Zeitkonstante.
σ^{∞} bzw. ε^{∞} sind die Leitfähigkeit bzw. DK bei sehr hohen, σ^0 bzw. ε^0 die Leitfähigkeit bzw. DK bei sehr tiefen Frequenzen.

σ^{∞}, σ^0, ε^{∞}, ε^0 und τ sind Funktionen von σ_1, σ_2, ε_1, ε_2 und x. Zwischen diesen Funktionen besteht der allgemeine Zusammenhang

$$\tau = \varepsilon_0 \frac{\varepsilon^0 - \varepsilon^{\infty}}{\sigma^{\infty} - \sigma^0}, \tag{3}$$

ε_0 Influenzkonstante.
Die übrigen vier Funktionen lassen sich mit den Annahmen

$$1.\ x \ll 1, \quad 2.\ \sigma_1 \ll \sigma_2, \quad 3.\ \varepsilon_1 = \varepsilon_2$$

leicht nach den interessierenden Größen auflösen:

$$\sigma_1 = \frac{\varepsilon^{\infty} \sigma^0 \sigma^{\infty} [\sigma^{\infty} - \sigma^0]}{\varepsilon^0 [\sigma^{\infty}]^2 - \varepsilon^{\infty} [\sigma^0]^2}, \tag{4}$$

$$\sigma_2 = \sigma^{\infty}, \tag{5}$$

$$\varepsilon_1 = \varepsilon_2 = \varepsilon^{\infty}, \tag{6}$$

$$x = \frac{\varepsilon^{\infty} [\sigma^{\infty} - \sigma^0]^2}{\varepsilon^0 [\sigma^{\infty}]^2 - \varepsilon^{\infty} [\sigma^0]^2}. \tag{7}$$

Infolge der Dispersion von σ und ε kann der Verlustfaktor $\operatorname{tg} \delta$ von Ferriten Extremwerte $\operatorname{tg} \delta_E$ haben. Das Modell liefert

$$\tau^2 [\omega(\operatorname{tg} \delta_E)]^2 = \frac{1}{2} \left\{ \left[\frac{\varepsilon^0}{\varepsilon^{\infty}} - 3 \frac{\sigma^0}{\sigma^{\infty}} \right] \pm \sqrt{\left[\frac{\varepsilon^0}{\varepsilon^{\infty}} - \frac{\sigma^0}{\sigma^{\infty}} \right] \left[\frac{\varepsilon^0}{\varepsilon^{\infty}} - 9 \frac{\sigma^0}{\sigma^{\infty}} \right]} \right\}. \tag{8}$$

Das positive Vorzeichen vor der Wurzel gilt für die Frequenz des Maximums, das negative für die des Minimums. Setzt man die Gültigkeit des Modells voraus, so hat der Verlustfaktor Extremwerte, wenn der Radikand größer als Null ist, d. h.

$$\frac{\varepsilon^0}{\varepsilon^{\infty}} > 9 \frac{\sigma^0}{\sigma^{\infty}}. \tag{9}$$

Bedingung (9) ist für Ferrite wahrscheinlich stets erfüllt.

Alle Autoren mußten feststellen, daß die Dispersionskurven nach (1) und (2) zwar den experimentellen Kurven ähnlich sind, von einer quantitativen Übereinstimmung aber im allgemeinen keine Rede sein kann. Der Dispersionsmittelpunkt der experimentellen ε-Kurve liegt bei einer Frequenz, die kleiner als die

[2]) Koops [9] hat die Dispersionsformeln für den spezifischen Widerstand und für die Dielektrizitätskonstante aufgeschrieben. Dann erhält man zwei Zeitkonstanten τ_ε und τ_ϱ, zwischen denen aber eine bestimmte Abhängigkeit besteht.

des Dispersionsmittelpunktes der σ-Kurve ist. Nach dem Modell stimmen beide Frequenzen überein. Nach dem Experiment ist also das Frequenzband, in dem eine Dispersion von ε und σ zu finden ist, breiter als für die Modellkurven (1) und (2).

Parker und Smith [12] haben 1958 das Modell von Koops erweitert, indem sie eine beliebige Verteilung der Schichtdicken und Schichtorientierungen betrachteten. Diese Verallgemeinerung führt aber zu den gleichen Dispersionsformeln, die Koops aufgestellt hat. Sie versuchten, die Abweichungen zwischen Experiment und Modell dadurch zu erklären, daß der Stromdichtevektor bei niedrigen Frequenzen im Innern des Materials von Punkt zu Punkt sowohl in seiner Größe als auch in seiner Richtung variiert.

Koops [9] hat die Vermutung ausgesprochen, daß die Ursache für die Differenzen zwischen experimentellen Kurven und Modellkurven in der Existenz von mehr als einer Zeitkonstante zu suchen ist. Dieser Gedanke wird in der vorliegenden Arbeit zu einer Erweiterung des Modells von Koops herangezogen.

2. Eine Erweiterung des Modells von Koops

Eine Ferritplatte vom Querschnitt F und von der Dicke d bestehe aus einer sehr großen Anzahl von Quadern mit Querschnitten ΔF_n und der Dicke d. Die einzelnen Quader seien durch dünne Zwischenschichten, die als Isolatoren behandelt werden, voneinander getrennt. Für jeden Quader gelte das Modell von Koops. Innerhalb eines Quaders sollen σ^0 und σ^∞ die gleiche Aktivierungsenergie U haben, d. h.

$$\tau \sim e^{\frac{U}{kT}} .$$

Dagegen sollen die Aktivierungsenergien von Quader zu Quader variieren. Der komplexe Leitwert der Ferritplatte ist dann

$$\mathfrak{G} = G + j\,\omega\, C$$

mit

$$G = \int_0^\infty f(U)\, G(U)\, dU\,, \qquad C = \int_0^\infty f(U)\, C(U)\, dU\,.$$

Man führt nun zweckmäßigerweise τ als Integrationsvariable ein und geht vom Leitwert zur Leitfähigkeit und von der Kapazität zur Dielektrizitätskonstanten über:

$$\sigma_P = \int_0^\infty g(\tau)\, \sigma(\tau)\, d\ln\tau\,,$$

$$\varepsilon_P = \int_0^\infty g(\tau)\, \varepsilon(\tau)\, d\ln\tau\,.$$

Nimmt man für die Verteilungsfunktion $g(\tau)$ eine Gaußfunktion um einen wahrscheinlichsten Wert der Zeitkonstanten an, so können die Integrale analytisch nicht ausgewertet werden. Zu übersichtlichen Formeln kommt man, wenn man nach Richter [13] die Gaußfunktion durch ein Rechteck annähert, d. h.

$$g(\tau) = 0 \quad \text{für} \quad \tau < \tau\,(E - \Delta E) \quad \text{und} \quad \tau > \tau\,(E + \Delta E)\,,$$

$$g(\tau) = c \quad \text{für} \quad \tau\,(E - \Delta E) \leqq \tau \leqq \tau\,(E + \Delta E)\,.$$

Wegen

$$\int g(\tau)\, d \ln \tau = 1$$

ist

$$c = \frac{1}{\ln \dfrac{\tau\,(E + \varDelta E)}{\tau\,(E - \varDelta E)}}\,.$$

Die Integration liefert für die Leitfähigkeit

$$\sigma_P = \frac{1}{2 \ln b} \left\{ \sigma^0(E) \left[b - \frac{1}{b} \right] + [\sigma^\infty(E) - \sigma^0(E)]\; \Omega \operatorname{arc\,tg} \frac{\Omega \left[b - \dfrac{1}{b} \right]}{1 + \Omega^2} \right\}, \tag{10}$$

σ_P Leitfähigkeit der Ferritplatte, $\sigma^0(E)$ bzw. $\sigma^\infty(E)$ Leitfähigkeit des Bereichs mit der Aktivierungsenergie E bei sehr niedrigen bzw. sehr hohen Frequenzen.

$$\tau_M = \sqrt{\tau\,(E - \varDelta E) \cdot \tau\,(E + \varDelta E)} = \tau(E)\,, \tag{11}$$

$$b = \sqrt{\frac{\tau\,(E + \varDelta E)}{\tau\,(E - \varDelta E)}} = e^{\frac{\varDelta E}{k\,T}}\,, \tag{12}$$

$$\Omega = \omega\, \tau_M\,. \tag{13}$$

Für $\omega = \Omega = 0$ erhält man

$$\sigma_P^0 = \sigma^0(E) \frac{b - \dfrac{1}{b}}{2 \ln b}\,. \tag{14}$$

Für $\Omega \gg 1$, also für sehr hohe Frequenzen, ist

$$\Omega \operatorname{arc\,tg} \frac{\Omega \left[b - \dfrac{1}{b} \right]}{1 + \Omega^2} = b - \frac{1}{b} \qquad \sigma_P^\infty = \sigma^\infty(E) \frac{b - \dfrac{1}{b}}{2 \ln b}\,. \tag{15}$$

σ_P^0 und σ_P^∞ haben die gleiche Temperaturabhängigkeit. Mit (14) und (15) kann man (10) folgendermaßen schreiben:

$$\sigma_P = \sigma_P^0 + [\sigma_P^\infty - \sigma_P^0] \frac{\Omega}{b - \dfrac{1}{b}} \operatorname{arc\,tg} \frac{\Omega \left[b - \dfrac{1}{b} \right]}{1 + \Omega^2} \tag{16}$$

Die Integration liefert für die Dielektrizitätskonstante

$$\varepsilon_P = \varepsilon_P^0 - [\varepsilon_P^0 - \varepsilon_P^\infty] \frac{\ln \left(\dfrac{1 + \Omega^2 b^2}{1 + \dfrac{\Omega^2}{b^2}} \right)}{4 \ln b}\,. \tag{17}$$

Für $\Omega = 0$ ist $\varepsilon_P = \varepsilon_P^0$, für $\Omega \gg 1$ $\varepsilon_P = \varepsilon_P^\infty$. Für $\Omega = \omega\, \tau_M = 1$ erhält man $\varepsilon_P = \dfrac{1}{2} [\varepsilon_P^0 + \varepsilon_P^\infty]$.

Damit ist klar, wie man τ_M zu bestimmen hat:

$$\tau_M = \frac{1}{\omega\left(\frac{\varepsilon_P^0 + \varepsilon_P^\infty}{2}\right)}. \tag{18}$$

Zur Ermittlung von b formt man (17) um in

$$b^{4\left(1 - \frac{\varepsilon_P - \varepsilon_P^\infty}{\varepsilon_P^0 - \varepsilon_P^\infty}\right)} = \frac{1 + \Omega^2 b^2}{1 + \frac{\Omega^2}{b^2}}.$$

Diese Gleichung läßt sich für zwei Werte von Ω leicht nach b lösen:

$$\frac{\varepsilon_P - \varepsilon_P^\infty}{\varepsilon_P^0 - \varepsilon_P^\infty} = \frac{3}{4} \quad \text{für} \quad \Omega = \Omega_1.$$

Dann ist

$$b = \frac{1}{2}\left[\frac{1}{\Omega_1^2} - 1\right] + \sqrt{\frac{1}{4}\left[\frac{1}{\Omega_1^2} - 1\right]^2 - 1}. \tag{19}$$

$$\frac{\varepsilon_P - \varepsilon_P^\infty}{\varepsilon_P^0 - \varepsilon_P^\infty} = \frac{1}{4} \quad \text{für} \quad \Omega = \Omega_2.$$

Dann ist

$$b = \frac{1}{2}[\Omega_2^2 - 1] + \sqrt{\frac{1}{4}[\Omega_2^2 - 1]^2 - 1}. \tag{20}$$

Stimmen die beiden Zahlenwerte b nicht überein, rechnet man mit dem Mittelwert. Die Wahl von τ_M und b hat zur Folge, daß die durch (17) dargestellte Kurve mit der experimentellen ε-Kurve im wesentlichen übereinstimmt. Da die Beziehungen (16) und (17) wie auch die Beziehungen (1) und (2) den Kronig-Kramers-Relationen [14] genügen, muß die Gleichung (16) die experimentelle σ-Kurve mit genügender Genauigkeit wiedergeben.

Das Verhältnis x der Schichtdicken kann durch eine Zeitkonstantenverteilung nicht beeinflußt werden. Es gilt also wie im einfachen Koops-Modell Formel (7).

Die Werte für die Leitfähigkeit von Ferritkristallit und Zwischenschicht sind natürlich von Bereich zu Bereich verschieden. Die unteren Grenzen der Intervalle erhält man, indem man die σ_2- bzw. σ_1-Werte des einfachen Modells von Koops mit dem Faktor

$$\frac{2 \ln b}{b^2 - 1}$$

multipliziert. Dieser Faktor ist kleiner als 1. Die oberen Grenzen der Intervalle findet man durch Multiplikation der σ_2- bzw. σ_1-Werte des einfachen Koops-Modells mit dem Faktor

$$b^2 \frac{2 \ln b}{b^2 - 1}.$$

Dieser Faktor ist größer als 1.

Die Relation des einfachen Modells $\varepsilon_1 = \varepsilon_2 = \varepsilon^\infty$ gilt auch in dem erweiterten Modell.

3. Experimentelle Untersuchungen

Für die Messungen stand eine Scheringbrücke zur Verfügung, die Posselt gebaut und in seiner Diplomarbeit [15] ausführlich beschrieben hat. Die Brücke gestattet, Messungen an dielektrischen Proben im Frequenzgebiet von 0,2 bis 300 kHz auszuführen. Dabei können die zu messenden Kapazitäten zwischen 10^{-12} und 10^{-6} F, die Verlustfaktoren zwischen 10^{-3} und 20 (in einigen Fällen noch höher) liegen. Der Abgleich der Brücke wird bei fest gewähltem Widerstandsverhältnis mit einem verlustarmen Normaldrehkondensator C_N (C-Abgleich) und einer C-Dekade C_A (Verlustabgleich) vorgenommen. Zunächst wurden Leermessungen durchgeführt, d. h., es wurden Kapazität und Verlustfaktor des im Innern einer Hochdruckkammer befindlichen Plattenkondensators (ohne Ferritprobe) und der Leitung von diesem Kondensator zur Brücke gemessen. Danach wurden die Messungen mit der Ferritprobe wiederholt. Die Abgleichempfindlichkeit der Brücke hängt sehr von den elektrischen und dielektrischen Eigenschaften der zu untersuchenden Probe und von der Meßfrequenz ab. Bei den Messungen an den Proben 1, 2 und 89c waren die Abgleichfehler im unteren und mittleren Frequenzbereich (bis etwa 10 kHz) sowohl für die Probenkapazität als auch für die Leitfähigkeit $< 2\%$. Mit wachsender Frequenz nehmen die Abgleichfehler für die Leitfähigkeit von Probe 1 auf etwa 4% bei 100 kHz zu, während sie bei Probe 89c auch bei 300 kHz 2% nicht übersteigen. Die größten Fehler treten bei der Bestimmung der Probenkapazität bei den höchsten Meßfrequenzen auf. Mit $\Delta C_N = \pm 0{,}5$ pF und $\Delta C_A = \pm 10$ pF lassen sich sowohl die Leerkapazität (95...97 pF) als auch die Kapazität der Leitung und des Plattenkondensators mit Probe auf je $\pm 0{,}5$ pF genau angeben. Die Probenkapazität ist daher bei Frequenzen von 100 kHz und mehr nur auf 1 pF sicher. Probe 1 hat bei 50,0 °C und 100 kHz eine Kapazität von 5,1 pF, Probe 89c bei 30,0 °C und 100 kHz eine solche von 4,9 pF und bei 300 kHz von 2,5 pF. In Fig. 1 bis 7 sind Mittelwerte eingezeichnet, die aus je vier Einzelmessungen stammen. Die Abweichungen der einzelnen Meßwerte vom Mittelwert sind in keinem Falle größer als die maximalen Abgleichfehler.

3.1 Meßergebnisse von Probe 1

Die Probe hat die Zusammensetzung

$$(NiO)_{0,4}(ZnO)_{0,6}(Fe_2O_3)_{1,01} \, .$$

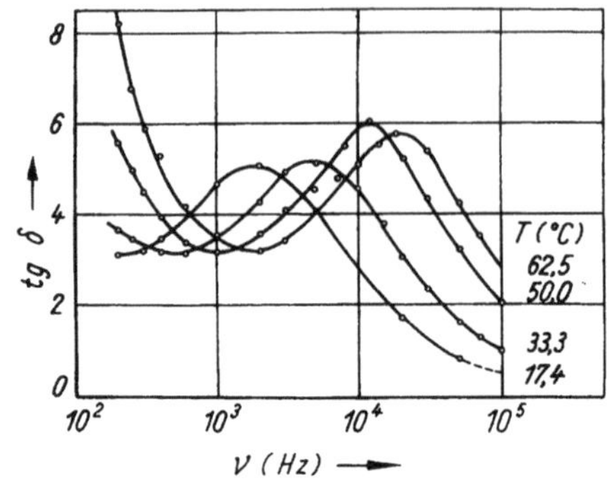

Fig. 1. Frequenzabhängigkeit des Verlustfaktors von Probe 1

Die Herstellungsbedingungen sind in [16] beschrieben. Nach [17] läßt sich der spezifische Gleichstromwiderstand

$$\varrho^0 = A_1 [1 + \alpha p] e^{\frac{E}{kT}} \tag{21}$$

schreiben ($A_1 = 0{,}94\ \Omega\text{cm}$, $\alpha = -16{,}8 \cdot 10^{-6}\ \text{cm}^2 \cdot \text{kp}^{-1}$, $E = 0{,}386\ \text{eV}$).

Fig. 1 zeigt die Frequenzabhängigkeit des Verlustfaktors bei vier verschiedenen Temperaturen und Luftdruck. Die Frequenzen, bei denen tg δ ein Maximum hat, genügen der Beziehung

$$\nu(\text{tg}\,\delta_{\max}) = C\, e^{-\frac{E}{kT}},$$

wobei der Zahlenwert von E sehr gut mit dem Wert übereinstimmt, der sich aus der Temperaturabhängigkeit des Gleichstromwiderstandes ergibt.

Fig. 2 zeigt, daß bei keiner der vier Temperaturen das Niederfrequenzniveau bzw. das Hochfrequenzniveau der DK im Frequenzbereich von 0,2 bis 100 kHz liegt. Die extrapolierten Kurven führen unabhängig von der Temperatur zu

$$\varepsilon^0 = 2500\,, \qquad \varepsilon^\infty = 15\,.$$

Man kann die Werte der Niveaus auch anders abschätzen. Trägt man z. B. die ε-Werte bei 200 Hz über der reziproken Temperatur auf und extrapoliert die Kurve nach höheren Temperaturen hin, so findet man, daß die DK oberhalb von 120 °C den konstanten Wert 2500 hat.

Fig. 3 zeigt die Frequenzabhängigkeit der Leitfähigkeit bei verschiedenen Versuchsbedingungen. Bei 10 Hz sind die Gleichstromwerte eingetragen. Die Werte der Hochfrequenzniveaus lassen sich mit einiger Sicherheit ablesen. Die Leitfähigkeitswerte, die unter hohem hydrostatischem Druck gemessen wurden, sind größer als die zugehörigen Werte bei Luftdruck.

Aus Tabelle 1 ist zu erkennen, daß das Verhältnis σ^0/σ^∞ weder von der Temperatur noch vom Druck abhängt. Die Zahlenwerte streuen mit einem Fehler von maximal 5% um

$$\frac{\varrho^\infty}{\varrho^0} = \frac{\sigma^0}{\sigma^\infty} = 0{,}43\,.$$

Daraus folgt

$$\varrho^\infty = A_2 [1 + \alpha p] e^{\frac{E}{kT}},$$

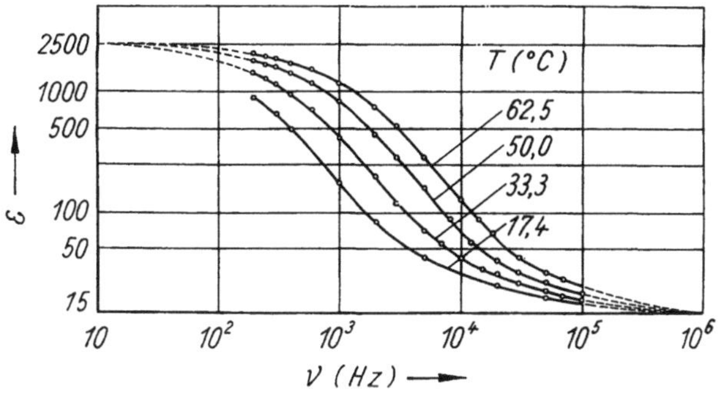

Fig. 2. Frequenzabhängigkeit der DK von Probe 1

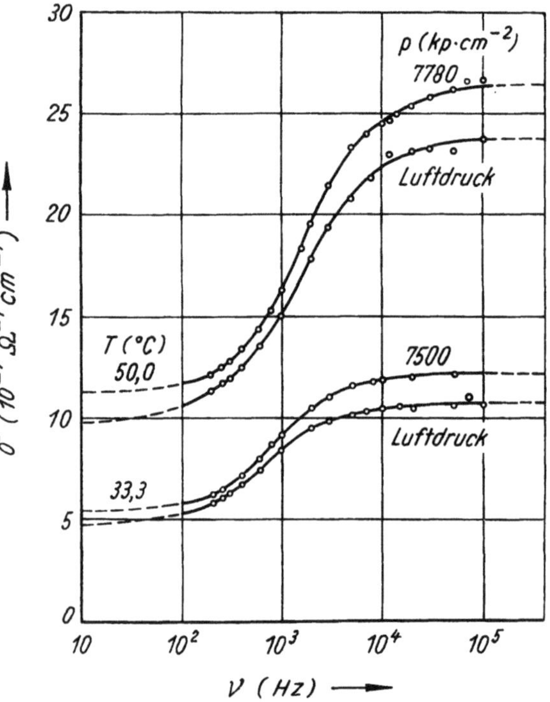

Fig. 3. Frequenzabhängigkeit der Leitfähigkeit von Probe 1

wobei der Druckkoeffizient α und die Aktivierungsenergie E mit den entsprechen den Gleichstromwerten übereinstimmen.

Tabelle 1

T (°C)	p ($kp \cdot cm^{-2}$)	$\frac{\sigma^0}{\sigma^\infty}$
17,4	1	0,439
33,3	1 7500	0,444 0,446
50,0	1 7780	0,412 0,426
62,5	1 7910	0,428 0,451

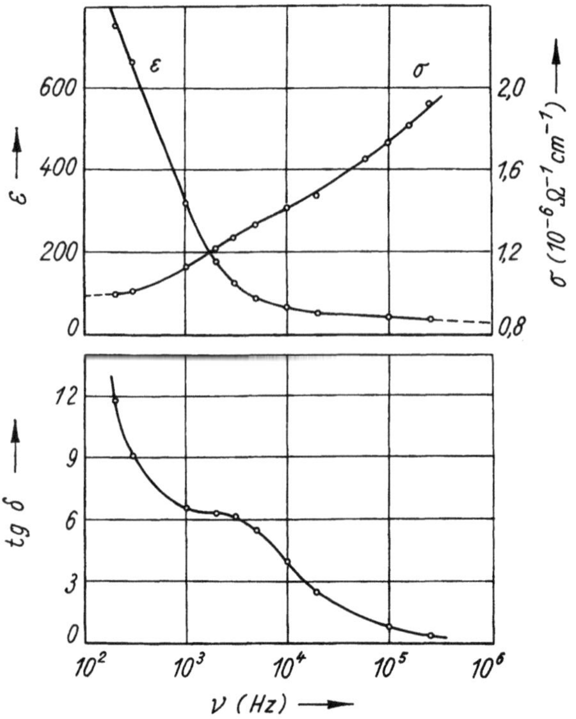

Fig. 4. Frequenzabhängigkeit der DK, der Leitfähigkeit und des Verlustfaktors von Probe 2

3.2 Meßergebnisse von Probe 2

Diese Probe stammt aus dem in [16] beschriebenen Brand f. Fig. 4 zeigt die Frequenzabhängigkeit der DK, der Leitfähigkeit und des Verlustfaktors bei 18,9 °C und Luftdruck. Wie bei allen anderen untersuchten Proben aus den in [16] beschriebenen Bränden konnten ε und σ_∞ im Frequenzbereich von 0,2 bis 300 kHz nicht bestimmt werden.

3.3 Meßergebnisse von Probe 89c

Die Probe hat die Zusammensetzung

$$(NiO)_{0,4}(ZnO)_{0,6}Fe_2O_3 .$$

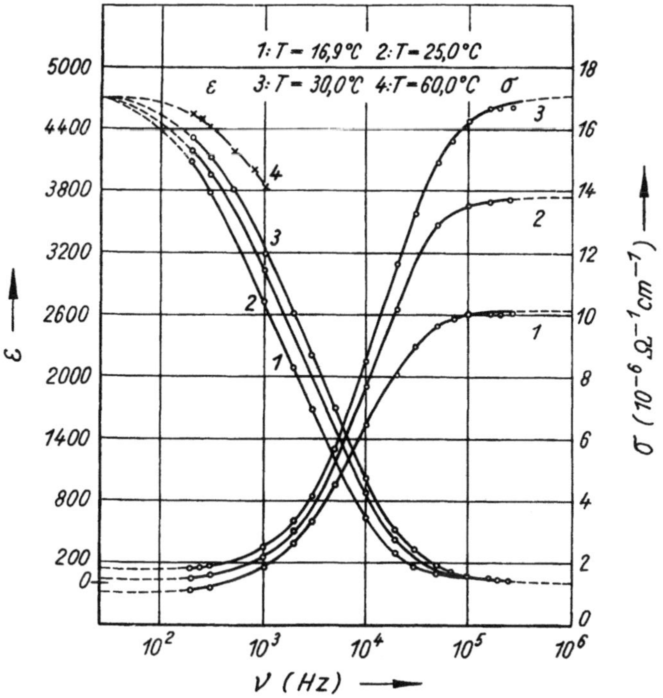

Fig. 5. Frequenzabhängigkeit der DK und der Leitfähigkeit von Probe 89 c

Sie wurde freundlicherweise vom VEB Werk für Bauelemente der Nachrichtentechnik „Carl von Ossietzky" zur Verfügung gestellt. Nach [17] genügt der spezifische Gleichstromwiderstand der Gleichung (21) mit $A_1 = 3{,}20\ \Omega\mathrm{cm}$, $E = 0{,}316$ eV und $\alpha = +\,10{,}0 \cdot 10^{-6}\ \mathrm{cm}^2 \cdot \mathrm{kp}^{-1}$.

Fig. 5 zeigt die Frequenzabhängigkeit der DK und der Leitfähigkeit bei drei verschiedenen Temperaturen und Luftdruck. Auch bei dieser Probe konnten die Niveaus der DK nur durch Extrapolation gewonnen werden. Zur Bestimmung von ε^0 sind einige ε-Werte eingezeichnet, die POSSELT [15] bei 60 °C gemessen hat.

$$\varepsilon^0 = 4700\,, \qquad \varepsilon^\infty = 15\,.$$

Tabelle 2

T (°C)	16,9	17,7	25,0	30,0
$\frac{\sigma^0}{\sigma^\infty}$	0,099	0,100	0,103	0,102

Aus Tabelle 2 folgt, daß das Hochfrequenzniveau σ^∞ die gleiche Temperaturabhängigkeit wie die Gleichstromleitfähigkeit hat.

In Fig. 6 ist die Leitfähigkeit im oberen Frequenzbereich bei 30 °C und Luftdruck bzw. hohem hydrostatischem Druck dargestellt. Bei niedrigen Frequenzen sind die σ-Werte unter Druck kleiner als die entsprechenden Luftdruckwerte.

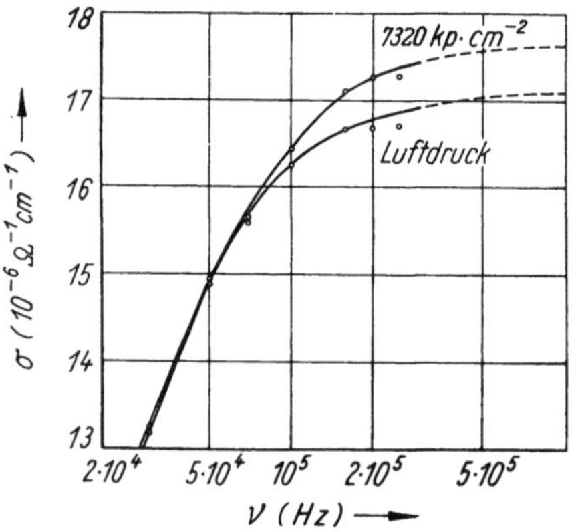

Fig. 6. Frequenzabhängigkeit der Leitfähigkeit von Probe 89c bei 30,0 °C und zwei verschiedenen Drücken

Zwischen 30 und 70 kHz stimmen die beiden Kurven im Rahmen der Meßgenauigkeit überein. Bei Frequenzen oberhalb von 70 kHz sind die σ-Werte unter Druck größer als die zugehörigen Luftdruckwerte. Unter Berücksichtigung aller Fehlermöglichkeiten läßt sich sagen, daß das Hochfrequenzniveau bei 7300 kp · cm^{-2} oberhalb des Niveaus bei Luftdruck liegt.

3.4 Meßergebnisse von Probe 3

Die Probe hat die Zusammensetzung

$$(NiO)_{0,5}(ZnO)_{0,5}Fe_2O_3 \,.$$

Der spezifische Widerstand ist 430 Ωcm bei 20 °C. Während bei allen anderen untersuchten Proben das Niederfrequenzniveau der DK nur durch Extrapolation gefunden werden konnte, hat diese Probe unterhalb von 70 kHz eine konstante DK. Es wurden Messungen bei 19,8; 38,0 und 50,0 °C und Luftdruck sowie bei 19,8 °C und 7100 kp · cm^{-2} durchgeführt. Aus den Meßreihen folgt

$$\varepsilon^0 = 21\,800 \pm 2\% \,,$$

unabhängig von der Temperatur und dem Druck.

4. Vergleich zwischen Experiment und Modell

4.1 Probe 1

Die quantitative Übereinstimmung zwischen dem Experiment und den Modellkurven nach Koops (1) und (2) ist bei allen Temperaturen und Drücken schlecht.

Die nach (19) und (20) berechneten Werte für die Intervallbreite sind im Mittel

$$b = 6$$

für Temperaturen zwischen 17 und 65 °C und Drücke zwischen Luftdruck und 8000 kp · cm^{-2}. Daraus folgt nach (12)

$$\Delta E = 1{,}54 \cdot 10^{-4}\, T \,,$$

T in °K, ΔE in eV.

Für den Fall, daß $b = f(T)$, kann weder die Temperaturabhängigkeit der Gleichstromleitfähigkeit noch die der Leitfähigkeit bei hohen Frequenzen durch eine einfache Exponentialfunktion beschrieben werden (vgl. (14) und (15)).

Nach (7) ist

$$x = 1{,}9 \cdot 10^{-3} .$$

Für die Leitfähigkeit der Kristallite bzw. der Zwischenschichten ergeben sich die Intervalle

$$0{,}1 \leqq \frac{\sigma_2}{\sigma_P^\infty} \leqq 3{,}7 ,$$

$$3{,}5 \cdot 10^{-4} \leqq \frac{\sigma_1}{\sigma_P^0} \leqq 1{,}3 \cdot 10^{-2} .$$

Da diese Zahlenwerte das Verhältnis $\varepsilon^\infty/\varepsilon^0$ enthalten, können sie nur die richtige Größenordnung angeben.

4.2 Probe 2

Der Verlauf der in Fig. 4 dargestellten Leitfähigkeitskurve kann weder nach dem einfachen Modell von Koops noch nach dem erweiterten Modell erklärt werden. In diesem Zusammenhang sei erwähnt, daß Habel [18] im Anschluß an die hier beschriebenen Messungen bei einigen Proben aus dem Brand g [16] ein ausgeprägtes Zwischenniveau der Leitfähigkeit gefunden hat. Dieses Verhalten konnte Habel durch Annahme einer Doppelrechteckverteilung von τ erklären.

4.3 Probe 89c

Für 16,9, 25,0 und 30,0 °C und Luftdruck ist in guter Näherung

$$b = 11{,}5 ,$$

$$\Delta E = 2{,}10 \cdot 10^{-4}\, T ,$$

T in °K, ΔE in eV.

In Fig. 7 sind die Modellkurven für Probe 89c bei 30 °C und Luftdruck dargestellt. Die eingezeichneten Punkte sind die experimentell ermittelten Werte. Kurve 1 wurde nach Gleichung (2), Kurve 2 nach Gleichung (1) mit τ nach (3) berechnet. In der Nähe von 2 kHz sind die Dielektrizitätskonstanten nach (2) um etwa 70% zu groß, oberhalb von 50 kHz die um über 100% zu klein. Auch

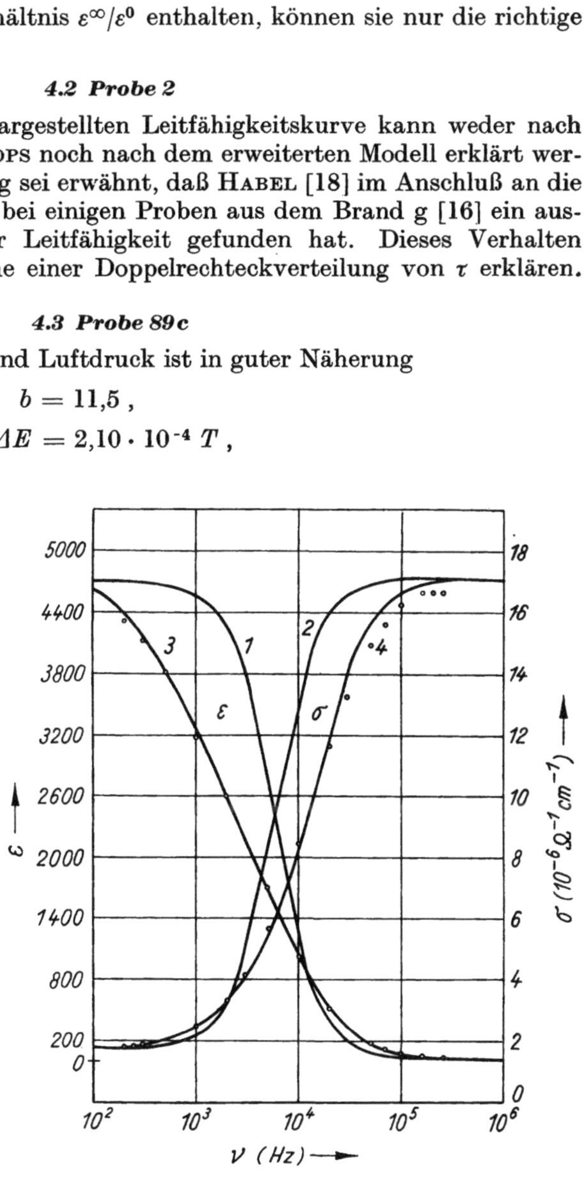

Fig. 7. Modellkurven für Probe 89 c bei 30,0 °C und Luftdruck

die Übereinstimmung zwischen den experimentellen Leitfähigkeitswerten und Kurve 2 ist unbefriedigend. Kurve 3 wurde nach Gleichung (17), Kurve 4 nach Gleichung (16) mit $\tau_M = 6{,}32 \cdot 10^{-5}$ s, $b = 11{,}5$ berechnet. Merkliche Abweichungen zwischen den experimentellen Dielektrizitätskonstanten und Kurve 3 sind nur beim Übergang in das Hochfrequenzniveau zu finden. Die experimentellen Leitfähigkeitswerte unterscheiden sich im gesamten Frequenzbereich um weniger als 5% von Kurve 4. Wenn man bedenkt, daß die der Berechnung zugrunde liegenden Größen ε^0, ε^∞, σ^∞, τ_M und b mit Fehlern behaftet sind, ist die Übereinstimmung der Modellkurven 3 und 4 mit dem Experiment als gut zu bezeichnen.

Auch nach dem erweiterten Modell ist

$$\sigma_2 \sim \sigma_P^\infty \quad \text{oder} \quad \varrho_2 \sim \varrho_P^\infty \,.$$

Aus Fig. 6 kann man danach schließen, daß der spezifische Widerstand der Kristallite dieses Ferrites mit wachsendem Druck abnimmt. Der positive Druckkoeffizient des Gleichstromwiderstandes wird verständlich, wenn man dem spezifischen Widerstand der Zwischenschichten einen positiven Druckkoeffizienten zuschreibt.

Herrn Prof. Holzmüller möchte ich für die Förderung danken, die er mir bei der Durchführung dieser Arbeit gewährt hat. Herr Dr. Dietzmann hat mir in zahlreichen Diskussionen wertvolle Ratschläge gegeben. Dafür spreche ich ihm meinen besonderen Dank aus.

Literatur

[1] E. Blechschmidt, Phys. Z. **39**, 212 (1938).
[2] F. W. Brockman, P. H. Dowling und W. G. Steneck, Phys. Rev. **77**, 85 (1950).
[3] D. Polder, Proc. Inst. Electr. Engrs. **97**, II, 246 (1950).
[4] G. T. Rado, R. W. Wright und W. H. Emerson, Phys. Rev. **80**, 273 (1950).
[5] K. Kamiyoshi, Sci. Rep. Res. Inst. Tohoku Univ. **A 3**, 716 (1951).
[6] T. Okamura, T. Fujimura und M. Date, Sci. Rep. Res. Inst. Tohoku Univ. **73**, 191 (1952).
[7] G. Möltgen, Z. angew. Phys. **4**, 216 (1952).
[8] A. Fairweather und D. J. Frost, Proc. Inst. Electr. Engrs. **100**, II A, 15 (1953).
[9] C. G. Koops, Phys. Rev. **83**, 121 (1951).
[10] J. C. Maxwell, Electricity and Magnetism, Bd. 1, London 1873.
[11] K. W. Wagner, Ann. Phys. **40**, 817 (1913).
[12] R. Parker und M. S. Smith, J. Electronics and Control, 1. Serie, **5**, 354 (1958).
[13] G. Richter, Probleme der technischen Magnetisierungskurve, Berlin 1938.
[14] Neben anderen B. Gross und R. H. Cole, Phys. Rev. **59**, 748 (1941); Phys. Rev. **60**, 172 A (1941).
[15] K. Posselt, Diplomarbeit, Leipzig 1959.
[16] G. Dietzmann, M. Krötzsch und S. Wolf, phys. stat. sol. **2**, 1762 (1962).
[17] M. Krötzsch, phys. stat. sol. **2**, 1768 (1962).
[18] H. Habel, Diplomarbeit, Leipzig 1961.

(Received May 25, 1964)

phys. stat. sol. 6, 491 (1964)

Max-Planck-Institut für Metallforschung, Stuttgart

Das Verfestigungsverhalten von vielkristallinem Nickel

Von

E. Macherauch und O. Vöhringer

An Nickel wurde der Einfluß der Verformungstemperatur, des Reinheitsgrades und der Korngröße auf das Verfestigungsverhalten untersucht. Die Verfestigungskurve hat je nach Reinheitsgrad unterschiedliche Form, jedoch stets einen Bereich linearer Verfestigung, der bei kleinen Deformationen auftritt. Die Streckgrenze der reinsten Nickelsorte (99,98% Ni) zeigt die gleiche Temperaturabhängigkeit wie sie die Seegersche Fließspannungstheorie fordert. Zusätze an Fremdatomen ändern Absolutbetrag und Temperaturabhängigkeit der Streckgrenze. Bei dem unreinen Nickel (99,8% Ni) tritt ein Fließbereich auf. Zwischen seiner Länge, der Streckgrenze und der Verformungstemperatur besteht ein eindeutiger Zusammenhang. Der Verfestigungsanstieg im linearen Bereich ist bei den verschiedenen Nickelqualitäten temperaturunabhängig, der Korngrößeneinfluß auf den Verfestigungsanstieg verschieden. Der Logarithmus der Fließspannung am Ende des Bereichs des konstanten Verfestigungsanstiegs ist bei allen Nickelsorten linear von der Temperatur abhängig.

The influence of deformation temperature, impurities, and grain size on the work-hardening properties of nickel is investigated. The shape of the work-hardening curve depends on the degree of impurity, but in all cases a stage of linear work-hardening occurs for small degrees of deformation. The yield point of the purest material (99.98% Ni) shows the temperature dependence given by Seeger's theory of work-hardening. The absolute value and temperature dependence of the yield point are changed by adding impurity atoms. The least pure material (99.8% Ni) shows a yield elongation. There exists a simple relation between the yield elongation, the yield point and the temperature of deformation. The work-hardening coefficient of the linear stage is independent of the deformation temperature, its dependence on grain size being different in the various kinds of nickel investigated. The logarithm of the yield stress at the end of the linear work-hardening stage varies linearily with the temperature of deformation.

1. Einleitung

Die im letzten Jahrzehnt erzielten Fortschritte auf dem Gebiete der Einkristallplastizität [1 bis 4] haben neuerdings mehrfach zu genaueren Untersuchungen über das plastische Verhalten von Vielkristallen kubisch flächenzentrierter, reiner Metalle angeregt. Von besonderem Interesse war dabei die Festlegung der Einflüsse von Korngröße, Verformungsgeschwindigkeit und Verformungstemperatur auf die Kenngrößen der Vielkristallverfestigungskurven. Ausführlicher untersucht wurden reines Aluminium [5 bis 7] und Kupfer [7 bis 9]. Auch an Nickel sind neuere Beobachtungsergebnisse bekannt geworden. So lieferte Christian [10] Angaben über den Temperatur- und Korngrößeneinfluß auf die Verfestigung, Zankl [11] berichtete über kombinierte magnetische und elektronenmikroskopische Untersuchungen und Macherauch, Kolb und Christian [12] beobachteten eine Streckgrenzen- und Alterungserscheinung bei Nickelvielkristallen. Besonders der letzte Befund legte nahe, die Streckgrenzen und das Verfestigungsverhalten von reinem Nickel mit verschiedenen Gehalten an Zusatzelementen genauer zu untersuchen. Über die an drei Nickelqualitäten durchgeführten Versuche und die dabei erhaltenen Ergebnisse wird nachfolgend berichtet.

2. Versuchsmaterial und Versuchsdurchführung

Das Untersuchungsmaterial — im folgenden mit Ni1, Ni2 und Ni3 bezeichnet — hatte die in Tabelle 1 angegebene Zusammensetzung. Aus den vorliegenden Nickelstangen wurden zylindrische Rundproben mit 4 mm Durchmesser und einer Meßlänge von 60 mm hergestellt. Die Einspannenden waren 25 mm lang und hatten einen Durchmesser von 6 mm. Nach der Probenherstellung wurden die aus Tabelle 2 ersichtlichen Rekristallisationsglühungen im Hochvakuum vorgenommen, die zu den angegebenen mittleren Korngrößen führten.

Tabelle 1

Charge	Ni	C	N_2	O_2 in Gew.%	S	Fe	Co	Spuren
Ni1	99,98	0,003	$\leqq$0,001	0,002	0,0007	Spur	Spur	Mn, H_2
Ni2	99,8	0,040	$\leqq$0,001	0,001	0,0025	~0,04	~0,015	Cu, Mg, Mn, H_2
Ni3	99,8	0,048	$\leqq$0,001	<0,001	0,0024	~0,04	~0,015	Cu, Mg, Mn, H_2

Tabelle 2

Material	Glühtemperatur (°K)	Glühzeit (h)	Korndurchmesser (μm)
Ni1	973	2	50
	1173	2	140
	1473	2	440
Ni2	973	2	95
	1223	2	190
	1273	2	225
	1473	2	420
Ni3	973	2	50

Die Zugversuche erfolgten in einer Polanyiapparatur mit elektrischer Kraftmessung durch Dehnungsmeßstreifen. Tieftemperaturversuche wurden in einem von KNÖLL [7] gebauten Kryostaten, Verformungsversuche bei höheren Temperaturen als Raumtemperatur mit Hilfe eines Wobser-Thermostaten durchgeführt. Die Last-Verlängerungsdiagramme der Registrierstreifen der Verformungsanlage wurden unter Berücksichtigung der Eigendehnung der Apparatur, der Nichtlinearitäten der Registrierung und der Querdehnung der Versuchsproben in Spannungs-Dehnungsdiagramme (σ, ε-Diagramme) umgerechnet.

3. Übersicht über den Korngrößen- und Temperatureinfluß auf die Verfestigungskurven

Bei Ni1 wurde bei allen Verformungstemperaturen der in Fig. 1a schematisch gezeichnete Typ von Verfestigungskurven beobachtet, der in drei Bereiche unterteilt werden kann. An Bereich I, der ein Übergangsintervall vom elastischen zum

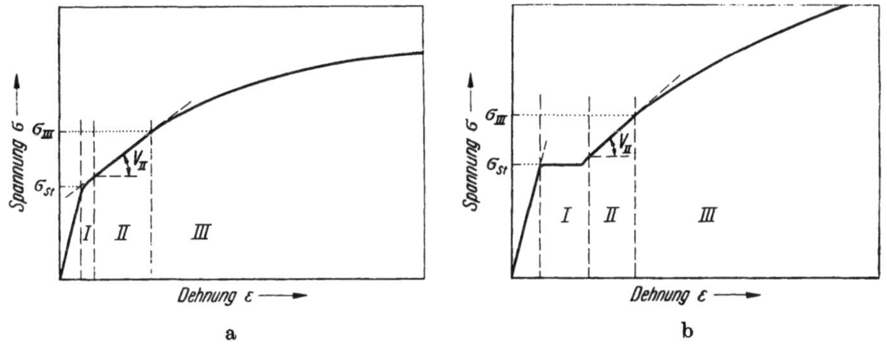

Fig. 1. Schematische Verfestigungskurve a) von Ni1 und b) von Ni2 und Ni3

plastischen Teil der Verfestigungskurve darstellt, schließt sich Bereich II mit konstantem Verfestigungsanstieg $V_{II} = \left(\frac{d\sigma}{d\varepsilon}\right)_{II}$ an. Oberhalb einer bestimmten Spannung σ_{III} beginnt Bereich III, in dem der Verfestigungsanstieg mit zunehmender Deformation abnimmt.

Fig. 2a gibt die bei Verformungstemperaturen zwischen 90° und 370 °K gewonnenen Verfestigungskurven von Ni1 mit einer mittleren Korngröße von $d = 50\ \mu m$ wieder. Das Ende des Bereichs II ist jeweils durch einen kleinen senkrechten Strich markiert. Im Bereich III nimmt bei gegebener Dehnung die Verfestigung mit abnehmender Temperatur zu, was zum Auffächern der Verfestigungskurven mit sinkender Temperatur führt. Der Korngrößeneinfluß auf die Raumtemperaturverfestigungskurven von Ni1 geht aus Fig. 3a hervor. Die dort dargestellten Kurven sind Mittelwerte aus mehreren Messungen. Die Schwankungsbreite der Fließspannungen gleich behandelter Nickelproben ist relativ klein und überschreitet beispielsweise bei einer Korngröße von 440 μm nicht 5%.

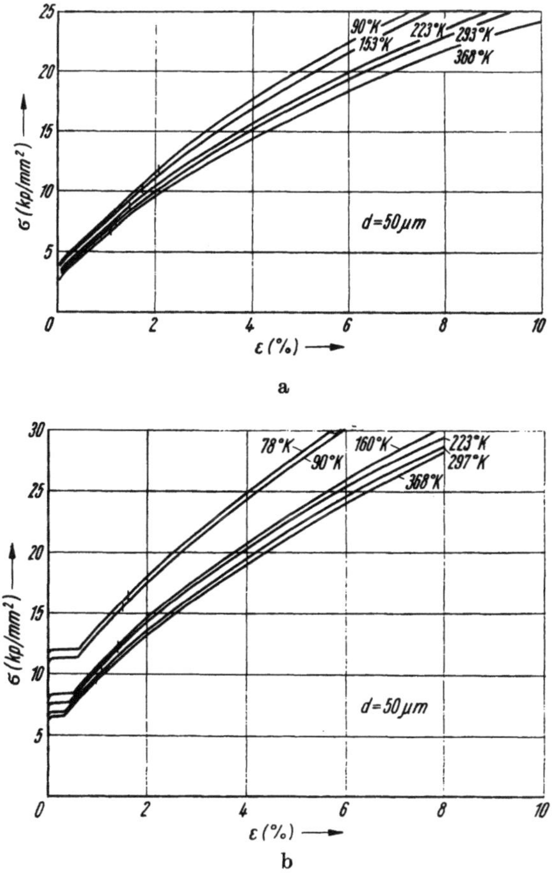

Fig. 2. Temperaturabhängigkeit der Verfestigungskurven a) von Ni1 und b) von Ni3

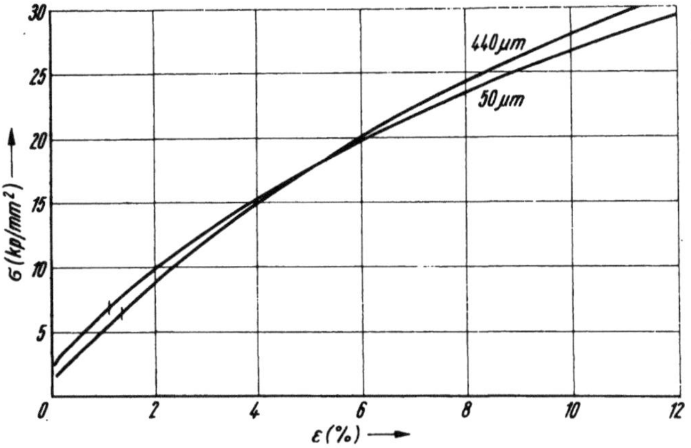

a

Fig. 3. Der Korngrößeneinfluß auf den Verlauf der Raumtemperaturverfestigungskurven bei a) Ni 1 und b) Ni 2

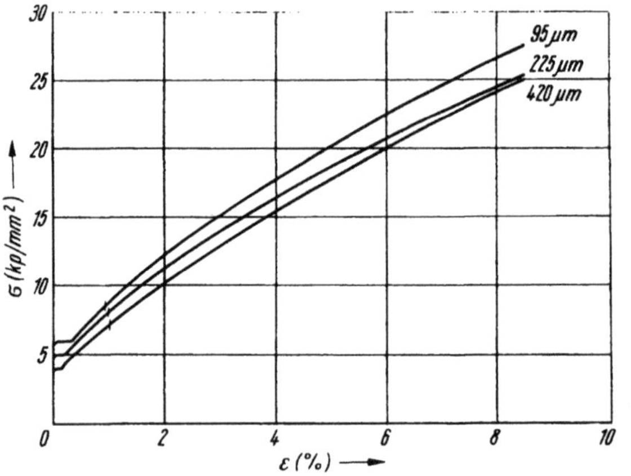

b

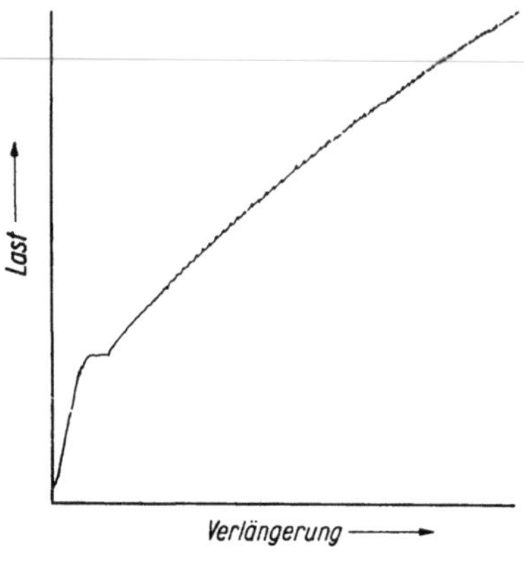

Fig. 4. Last-Verlängerungsdiagramm einer bei Raumtemperatur verformten Ni 3-Probe

Fig. 5. Temperaturabhängigkeit der Streckgrenze a) von Ni 1 mit zwei verschiedenen Korngrößen, b) von Ni 2 bei drei verschiedenen Korngrößen und c) von Ni 3

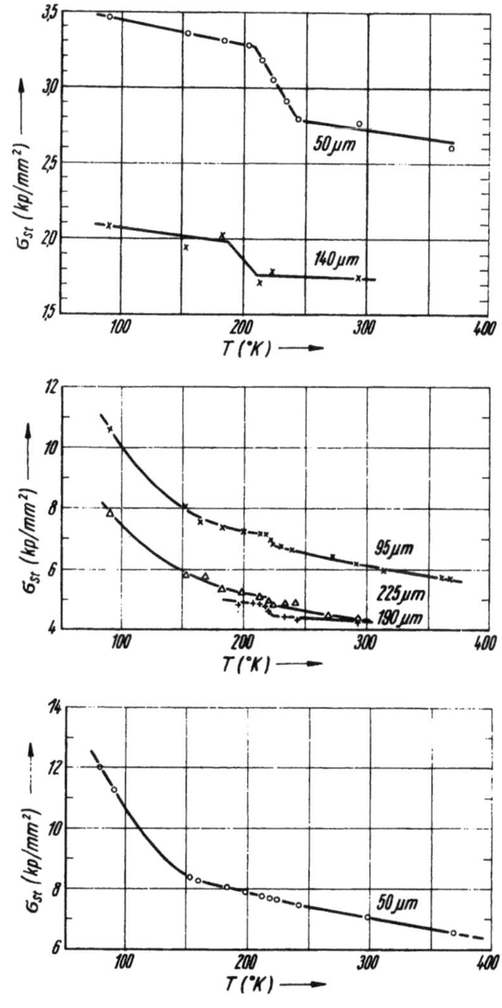

a

b

c

Die Streckgrenze ist größer beim kleineren Korn, dagegen ist bei höheren Verformungsgraden der Verfestigungsanstieg für das größere Korn geringer als für das kleinere. Als Folge davon überschneiden sich die Verfestigungskurven. Qualitative Beobachtungen bis zu 40% Gesamtdehnung ergaben, daß oberhalb 12% Dehnung die Verfestigungskurven keineswegs parallel zueinander verlaufen, sondern die grobkristallineren Proben ihren größeren Verfestigungsanstieg bis zum Beginn der Brucheinschnürung beibehalten.

Die Verfestigungskurven der unreineren Nickelsorten Ni2 und Ni3 zeigten bei Verformungstemperaturen zwischen 78° und 370 °K den in Fig. 1b schematisch dargestellten Verlauf. Auch hier bietet sich eine dreifache Unterteilung des Anfangsteils der Kurve an. Bereich I umfaßt den im Anschluß an die Hookesche Gerade auftretenden Fließbereich und ein angrenzendes kleines Übergangsintervall zu Bereich II. In Bereich II besitzt die Verfestigungskurve praktisch konstanten Anstieg, in Bereich III, oberhalb σ_{III}, nimmt der Verfestigungsanstieg mit zunehmender Deformation ab. Bei Verformungstemperaturen größer als 273 °K wurde ein sägezahnartiger Verlauf der Verfestigungskurven beobachtet. Dieser Portevin-Le Chatelier-Effekt [13, 14] war umso ausgeprägter, je höher die Verformungstemperatur lag. Er war ferner bei Ni3 stärker vorhanden als bei Ni2.

Fig. 2b zeigt die zwischen 78° und 370 °K erhaltenen Verfestigungskurven von Ni3 mit einer Korngröße von 50 µm. Ein Vergleich dieser Abbildung mit Fig. 2a zeigt, daß bei Ni3 — ebenso auch bei Ni2 — ein stärkerer Einfluß der Verformungstemperatur auf die Streckgrenze und Verfestigung besteht als bei Ni1. Da die zeichnerische Übertragung der durch den Portevin-Le Chatelier-Effekt hervorgerufenen Unstetigkeiten im Verlauf der Verfestigungskurve nicht möglich ist, wurden für Temperaturen oberhalb von 273 °K in Fig. 2b ausgleichende

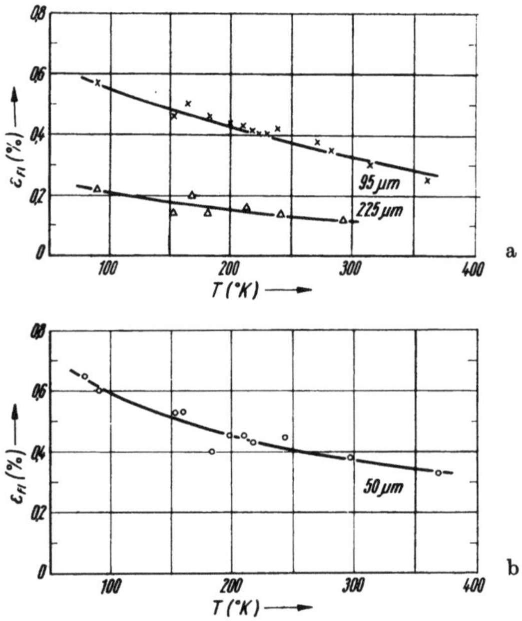

Fig. 6. Temperatureinfluß auf die Länge des Fließbereiches a) bei Ni 2 und b) bei Ni 3

Kurvenzüge aufgezeichnet. Das Original einer sägezahnförmigen Registrierkurve ist in Fig. 4 wiedergegeben. Der Korngrößeneinfluß bei den weniger reinen Nikkelqualitäten auf die Raumtemperaturverfestigungskurve geht aus Fig. 3b hervor. Die Streckgrenzenerscheinung wird umso ausgeprägter, je kleiner die Korngröße ist. Überschneidungseffekte wie in Fig. 3a treten in dem dargestellten Deformationsbereich nicht auf.

4. Korngrößen- und Temperaturabhängigkeit der Verfestigungskenngrößen

4.1 Streckgrenze

Temperatur- und Korngrößeneinfluß auf die Streckgrenze (vgl. Fig. 1a) von Ni1 geht aus Fig. 5a hervor. Bei den Proben mit 50 μm mittlerer Kristallitgröße ist die Streckgrenze im Bereich $90° \leqq T \leqq 203$ °K und für $T > 243$ °K wenig, im Temperaturintervall zwischen 203° und 243 °K jedoch relativ stark von der Temperatur abhängig. Beim größeren Korn (140 μm) deutet sich ein Bereich stärkerer Temperaturabhängigkeit der Streckgrenze zwischen 183° und 213 °K an. Die 0,05%-Dehngrenzen führen in beiden Fällen mit anderen Absolutwerten auf völlig vergleichbare Temperaturabhängigkeiten.

Gegenüber Ni1 zeigen die beiden anderen Nickelsorten bei vergleichbarer Kristallitgröße wesentlich höhere Streckgrenzen (vgl. Fig. 1b) und eine in charakteristischer Weise veränderte Temperaturabhängigkeit. Sowohl bei Ni2 in Fig. 5b als auch bei Ni3 in Fig. 5c ändert sich die Streckgrenze bei tieferen Temperaturen wesentlich stärker als bei Ni1 (vgl. Fig. 5a). Wie aus Fig. 5b zu ersehen ist, besitzt Ni2 bei den vermessenen Korngrößen noch einen Sprung in der Temperaturabhängigkeit der Streckgrenze zwischen 213° und 223 °K. Dieser Sprung liegt bei $d = 95$ μm und bei $d = 190$ μm deutlich außerhalb der Fehlerbreite der Messungen, ist dagegen im Rahmen der Schwankungsbreite der Meßpunkte bei $d = 225$ μm nicht mehr zu erkennen. Nur bei Ni3 ist die Temperaturabhängigkeit der Streckgrenze durch einen stetigen Kurvenzug beschreibbar (vgl. Fig. 5c), der durch keinen Sprung unterbrochen wird.

4.2 Fließbereich

Im Gegensatz zu Ni1 wird bei Ni2 und Ni3 nach Erreichen der Streckgrenze ein verfestigungsfreier Fließbereich beobachtet. Die Längen der Fließbereiche

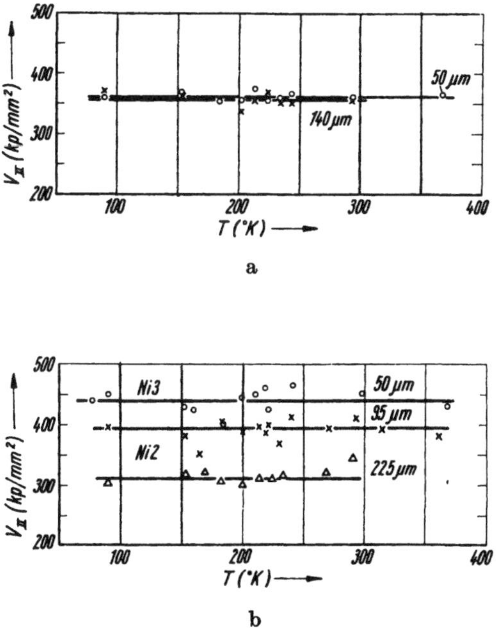

Fig. 7. Temperatureinfluß auf den Verfestigungsanstieg im Bereich II a) von Ni 1 und b) von Ni2 und Ni 3

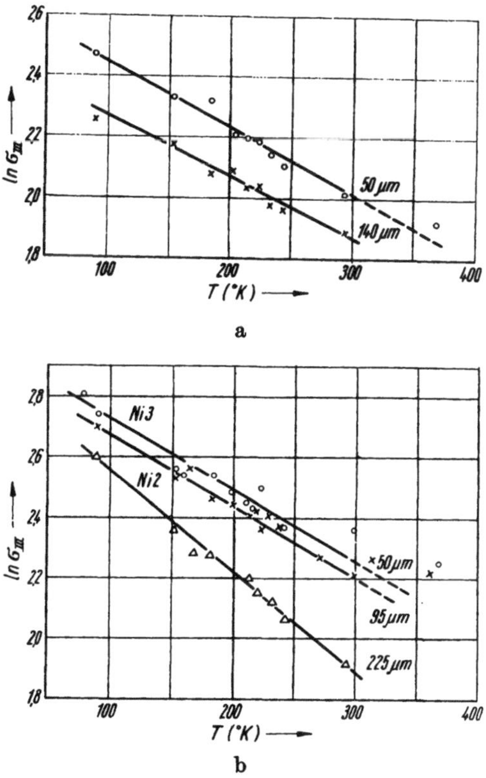

Fig. 8. ln σ_{III}-T-Diagramme a) von N 1 mit zwei Korngrößen und b) von Ni 2 und Ni 3

sind, wie aus Fig. 6a und 6b hervorgeht, relativ stark von der Verformungstemperatur abhängig. Bei 90 °K sind sie fast doppelt so lang wie bei Raumtemperatur. Die Proben mit großem Korn haben einen kürzeren Fließbereich als die feinkristallinen Proben.

4.3 Lineare Verfestigung

Im Bereich II der Verfestigungskurve wird der Verfestigungsanstieg $V_{II} = \left(\frac{d\sigma}{d\varepsilon}\right)_{II}$ praktisch nicht durch die Verformungstemperatur beeinflußt. Wie aus den in den Fig. 7a und 7b wiedergegebenen Meßwerten hervorgeht, ist kein die Meßgenauigkeit überschreitender Temperaturgang festzustellen, obwohl man einen solchen im Ausmaße der Temperaturabhängigkeit der elastischen Moduln erwarten würde. Bei Ni1 wird nach Fig. 7a der Verfestigungsanstieg in Bereich II auch nicht durch die Korngröße beeinflußt. Bei den unreineren Nickelqualitäten ist dagegen V_{II} nach Fig. 7b korngrößenabhängig und nimmt mit zunehmenden Kornabmessungen ab.

Die Länge des Verfestigungsbereichs II beträgt bei Nickelvielkristallen maximal $\varepsilon = 2\%$ und wird mit zunehmender Verformungstemperatur kleiner.

4.4 Übergang zu Bereich III

Als σ_{III} wurde die Spannung am Ende von Bereich II bestimmt, bei der die Verfestigungskurve die ersten deutlich erfaßbaren Abweichungen vom konstanten Verfestigungsanstieg zeigt. In den Fig. 8a und 8b sind die natürlichen Logarithmen der beobachteten σ_{III}-Werte als Funktion der Verformungstemperatur aufgetragen. Bei allen Nickelarten und Korngrößen ist für $T \leqq 293$ °K der Zusammenhang

$$\ln \sigma_{III} = A - B\,T \tag{1}$$

erfüllt. Als Ordinatenabschnitte A und Anstiege B ergeben sich die in Tabelle 3 zusammengestellten Werte. Die ln σ_{III}-Werte sind bei gleicher Nickelsorte umso

Tabelle 3

Nickelsorte	Korndurchmesser (μm)	V_{II} (kp/mm²)	A	$B \cdot 10^3$ (°K⁻¹)	$\sigma_{III}(0)$ (kp/mm²)	$\tau_{III}(0)$ (kp/mm²)
Ni1	50	362	2,68	2,2	14,5	4,73
Ni1	140	356	2,47	2,0	11,4	3,40
Ni2	95	396	2,91	2,4	18,4	6,02
Ni2	225	312	2,90	3,4	18,3	5,98
Ni3	50	440	2,95	2,4	19,7	6,43

größer, je kleiner die Korngröße ist. Bei den Proben mit größten Kristallitabmessungen (Ni2) wird der höchste Anstieg beobachtet.

5. Diskussion

5.1 Form der Verfestigungskurven

Die Verfestigungskurven der untersuchten Nickelqualitäten weisen charakteristische Unterschiede auf. Diese rühren von den verschiedenen Gehalten an Fremdatomen her, die maßgebend die Grundstruktur der Materialien beeinflussen [15, 16]. Beim reinsten vermessenen Nickel ist der Einfluß der Verformungstemperatur auf die Streckgrenze und die Fließspannung bei kleinen Verformungsgraden gering, bei den unreineren Nickelsorten stärker ausgeprägt. Diese Unterschiede zeigen, daß in beiden Fällen nicht dieselben Prozesse im gleichen Ausmaß für die Temperaturabhängigkeit verantwortlich sind. Neben verschiedenen Schneidprozessen ist die elastische Wechselwirkung zwischen Fremdatomen — im vorliegenden Falle vor allem wohl die des Kohlenstoffs — und Versetzungen für die Temperaturabhängigkeit bestimmend.

Das unreinere Nickel, Ni2, zeigt hinsichtlich der Korngrößenabhängigkeit eine Auffächerung der Verfestigungskurven wie man sie als Normalfall erwartet und z. B. auch bei Aluminium [7] beobachtet. Dagegen überschneiden sich bei Ni1 die Verfestigungskurven der verschiedenen Korngrößen. Bei reinem Kupfer [7] wurde ähnliches beobachtet. Auch bei älteren Messungen an Kupfer [17] sowie Kupfer und weniger reinem Aluminium [5, 9] deuten sich solche Überschneidungseffekte an. Der dieser Erscheinung im einzelnen zugrundeliegende Mechanismus ist unklar und wird z. Z. noch näher untersucht. Zweifellos muß es sich um einen Prozeß handeln, bei dem die Dichte oder Stärke der Deformationshindernisse mit zunehmender Verformung bei den Proben großer Korngröße stärker anwächst als bei feinkristallinen Proben. Wir kommen später nochmals darauf zurück.

5.2 Streckgrenze

Der Korngrößeneinfluß auf die Streckgrenze (vgl. Fig. 5a und 5b) besteht darin, die Absolutwerte beim kleineren Korn nach höheren Werten zu schieben. Folgende Ursachen können hierfür verantwortlich sein:

Fig. 9. Temperaturabhängigkeit des Streckgrenzenverhältnisses aller Nickelsorten

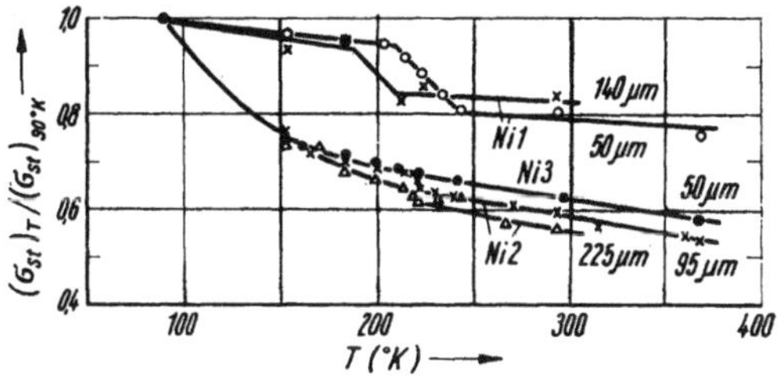

Von Quellen emittierte Versetzungen werden an Korngrenzen aufgestaut und erzeugen dort eine Aufstauungsspannung [18]

$$\hat{\tau} = \alpha\, n\, \tau\,, \tag{2}$$

wobei α ein numerischer Faktor, n die Zahl der an der Korngrenze aufgestauten Versetzungen und τ die in der Gleitebene wirkende Schubspannung ist. Bei einer kritischen Schubspannung $\tau_{\text{krit}} = \alpha\, n\, \tau_0$ werden Gleitprozesse in Nachbarkörnern induziert [19, 20] oder die Korngrenzen von den aufgestauten Versetzungen durchbrochen [21, 22]. Sieht man τ_{krit} als korngrößenunabhängig an, so ist τ_0 der Zahl der aufgestauten Versetzungen umgekehrt proportional. Da die größtmögliche Zahl der aufgestauten Versetzungen n im Korn geringer Abmessungen kleiner ist als im größeren Korn, ist bei kleinem Korn eine entsprechend höhere Schubspannung τ_0 erforderlich, um τ_{krit} zu erreichen. Da τ_0 der Streckgrenze σ_{St} proportional ist, wächst diese mit kleiner werdendem Korn.

Bei korngrößenabhängiger Grundstruktur ist ein etwa mit $\sqrt{N}$ (wobei N die mittlere Versetzungsdichte ist) ansteigende Streckgrenze zu erwarten. Eine anteilmäßige Bewertung beider Einflüsse ist zur Zeit nicht möglich.

Die charakteristischen Unterschiede im Temperaturverhalten der Streckgrenze der verschiedenen Nickelarten sind in Fig. 9 zusammengefaßt. Dort sind alle Meßwerte auf die bei 90 °K gemessenen Streckgrenzen bezogen. Die Temperaturabhängigkeit dieser Streckgrenzenverhältnisse ist für Ni1 deutlich verschieden von den bei Ni2 und Ni3 erhaltenen. Von 90 °K bis zu den jeweiligen Sprungtemperaturen zeigen sowohl Ni1 als auch Ni2 und Ni3 unabhängig von der Korngröße nahezu das gleiche Streckgrenzenverhältnis.

Bei Ni1 tritt bei den untersuchten Korngrößen eine Dreiteilung der Kurven auf, wie es die Seegersche Fließspannungstheorie für Einkristalle mit größerer Aufspaltungsweite der Halbversetzungen fordert [23, 24]. Diese Theorie kann für die experimentellen Ergebnisse übernommen werden. Nimmt man an, daß die temperaturabhängigen Prozesse zu Beginn der Verformung in den einzelnen Kristalliten des Vielkristallverbandes wie beim Einkristall verlaufen und die Korngrenzen keinen zusätzlichen temperaturabhängigen Anteil zu den Schneidprozessen liefern, so bleibt sie auch bei Mittlung über alle Orientierungen gültig. Die Dreiteilung ist demnach auf die Auswirkung unterschiedlicher Schneidprozesse zurückzuführen.

Der praktisch nicht vorhandene Korngrößeneinfluß auf die Temperaturabhängigkeit des Streckgrenzenverhältnisses in Fig. 9 läßt eine weitere Folgerung zu. Wir nehmen an, daß die Versetzungsstruktur korngrößenabhängig ist und die Streckgrenze sich gemäß

$$\sigma_{\text{St}}(T, d) = \sigma_{\text{S}}(T, d) + \sigma_{\text{G}}(d) \tag{3}$$

zusammensetzt. Der von Schneidprozessen herrührende Anteil σ_{S} sei von der Temperatur und dem Korngrößenparameter d abhängig, der von den inneren Spannungsfeldern gelieferte Anteil σ_{G} werde als temperaturunabhängig angenommen. Experimentell ergibt sich näherungsweise für das Streckgrenzenverhältnis

$$\sigma_{St}(T, d)/\sigma_{St}(T_1, d) = f(T)\,, \tag{4}$$

also eine Funktion von T allein, unabhängig von d. Aus den Gleichungen (3) und (4) folgt dann aber

$$\frac{[\sigma_S(T, d)/\sigma_G(d)] + 1}{[\sigma_S(T_1, d)/\sigma_G(d)] + 1} = f(T)\,, \tag{5}$$

was mit

$$\sigma_S(T, d)/\sigma_G(d) = a(T) \tag{6}$$

identisch erfüllt ist. Physikalisch besagt (6), daß sich die Spannung σ_S proportional σ_G ändert, wenn bei konstanter Verformungstemperatur T die Korngröße variiert wird. Dieser Zusammenhang gilt für jede beliebige Temperatur im untersuchten Temperaturintervall. Ob die Proportionalität zwischen σ_S und σ_G auch für die Fließspannungen und ein anderes als hier überstrichenes Korngrößenintervall gilt, kann auf Grund dieser Überlegungen nicht beurteilt werden. Auch KNÖLL [7] beobachtete bei Vielkristallen aus reinem Aluminium und Kupfer eine annähernde Korngrößenunabhängigkeit der Streckgrenzenverhältnisse bei verschiedenen Verformungstemperaturen. Das legt den Schluß nahe, daß bei den wichtigsten kubisch flächenzentrierten Metallen die Temperaturabhängigkeit der Streckgrenzenverhältnisse unabhängig von der Korngröße und somit auch an der Streckgrenze σ_S proportional σ_G unabhängig von der Korngröße erfüllt ist.

Die unreineren Nickelsorten zeigen gegenüber Ni 1 sowohl im Verlauf als auch im Betrag eine deutlich veränderte Temperaturabhängigkeit der Streckgrenzen. Besonders im Temperaturintervall von 90° bis 150 °K ist ein ausgeprägter Temperatureinfluß vorhanden. Der relativ starke Anstieg der Streckgrenze nach tiefen Temperaturen hin legt den Schluß nahe, daß ein Legierungseffekt vorliegt, wofür die kleinen Unterschiede in den Fremdatomkonzentrationen (vgl. Tabelle 1) verantwortlich zu machen sind. Für Temperaturen $T \leqq 160$ °K ist z. B. die Streckgrenze von Ni 3 etwa proportional zu $T^{-1/2}$, für $160° \leqq T \leqq 370$ °K etwa proportional zu $T^{-1/4}$. Die Temperaturabhängigkeit ist also sicher eine andere als man auf Grund der verschiedenen Ansätze von COTTRELL [25] ($\sigma_{St} \sim A - B\,T^{1/3}$) HAASEN [26] ($\sigma_{St} \sim T^{-3}$) bzw. FISHER [27] ($\sigma_{St} \sim T^{-1}$) für die elastische Wechselwirkung zwischen Versetzungen und interstitiellen Fremdatomen erwartet. Bemerkenswerter Weise lassen sich unabhängig von der Korngröße die Meßwerte jedoch recht gut durch Gleichungen der Form

$$\sqrt{\sigma} = a_i - b_i \sqrt{T}$$

beschreiben, wobei die a_i- und b_i-Werte unterhalb 160° bis 170 °K größer sind als bei höheren Temperaturen. Lediglich in Nähe der Sprungstelle (vgl. Fig. 5b) von Ni 2 treten geringe Abweichungen auf. Wir kommen an anderer Stelle darauf zurück.

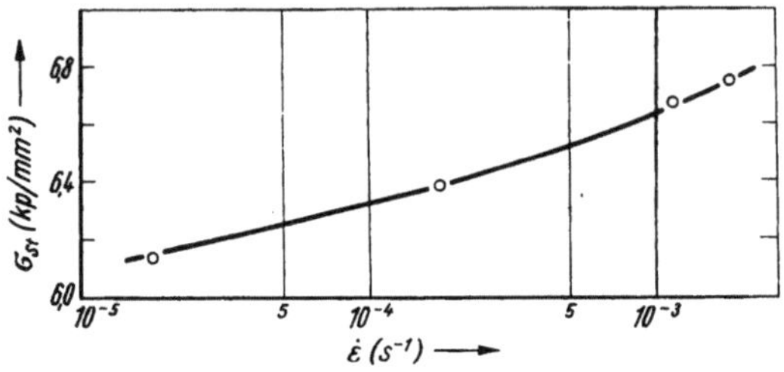

Fig. 10. Geschwindigkeitsabhängigkeit der Streckgrenze von Ni 2

Aus der relativ schwachen Temperaturabhängigkeit von σ_{St} für $T > 160$ °K darf nicht geschlossen werden, daß nur noch σ_G vorliegt. Die Geschwindigkeitsabhängigkeit von σ_{St}, die z. B. bei Ni2 mit $d = 95$ µm untersucht wurde, ist wie Fig. 10 zeigt bei 293 °K noch verhältnismäßig stark. Innerhalb der Geschwindigkeitsänderung von zwei Zehnerpotenzen ändert sich σ_{St} um etwa 10%, was nur auf einen noch vorhandenen σ_S-Anteil zurückgeführt werden kann.

Wie die kleinen Knicke in den für Ni2 gültigen Kurven (vgl. Fig. 5b) zeigen, führen offenbar kleine Unterschiede in den Beimengungen (vgl. Tabelle 1) zu relativ ausgeprägten Unterschieden im Temperaturverhalten zwischen 200° und 250 °K. Wie bereits angedeutet, kann die starke Temperaturabhängigkeit der Streckgrenze beim unreineren Nickel nicht allein der Wechselwirkung zwischen Versetzungen und Fremdatomen zugeschrieben werden. Temperaturwechselversuche [28] haben gezeigt, daß bei Ni2 Schneidprozesse eine relativ große Rolle spielen, so daß diese vermutlich mit verantwortlich sind, daß dort bei allen Korngrößen die auch bei Ni1 beobachteten Stufen auftreten.

Für Temperaturen $T < 160$ °K gilt (vgl. Fig. 9) für Ni2 und Ni3 unabhängig von der Korngröße $\frac{\sigma_{St}(T, d)}{\sigma_{St}(T_1, d)} = f(T)$. Nimmt man hier ein σ_S^{eff} an, das auch noch die Wechselwirkung der Versetzungen mit den Fremdatomen umfaßt, so ergibt sich ähnlich wie oben für Ni1 gezeigt σ_S^{eff} proportional σ_G.

5.3 *Temperaturabhängigkeit des Fließbereichs*

Mit abnehmender Verformungstemperatur und Korngröße nimmt die Länge des an die Streckgrenze anschließenden Fließbereichs zu (vgl. Fig. 6a und 6b). Bezieht man alle Werte für die Fließbereichslänge $\varepsilon_{Fl}(T, d)$ auf eine bestimmte Temperatur, z. B. $T_1 = 90$ °K, so ist das Verhältnis $\varepsilon_{Fl}(T, d)/\varepsilon_{Fl}(T_1, d)$ nahezu korngrößenunabhängig. Zwischen ε_{Fl}, der Verformungstemperatur T und der Streckgrenze σ_{St} ist bei Ni3 in guter Näherung (vgl. Fig. 11) die Beziehung

$$\ln \varepsilon_{Fl} = A + B/T\,\sigma_{St} \tag{7}$$

erfüllt. Dieser Zusammenhang ist bei Ni2 durch die Stufe in der Temperaturabhängigkeit der Streckgrenze gestört.

Es liegt nahe, das Auftreten des Fließbereichs mit der Ausbreitung einer Deformationsfront durch die Vielkristallprobe in Verbindung zu bringen. Nimmt man eine bestimmte Ausbreitungsgeschwindigkeit v_D über die Probe an, so ist der makroskopisch registrierte Fließbereich beendet, wenn sich das Deformationsband über die ganze Probenlänge erstreckt. Für die Zeit Δt_D, in der sich das De-

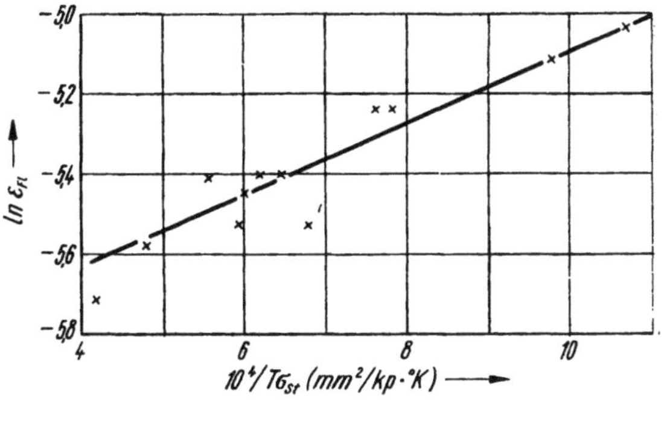

Fig. 11. ln ε_{Fl} als Funktion von $(T\,\sigma_{St})^{-1}$

formationsband über die ganze Probe der Länge l erstreckt, gilt

$$l/v_D = \Delta t_D = \frac{\varepsilon_{Fl}}{\dot{\varepsilon}} = \frac{\varepsilon_{Fl}}{v_W/l}, \tag{8}$$

wenn v_W die Wagengeschwindigkeit der Zugmaschine und $\varepsilon = v_W/l$ ist. Daraus ergibt sich für den Fließbereich

$$\varepsilon_{Fl} = v_W/v_D\,. \tag{9}$$

Bei kleiner Deformationsfrontgeschwindigkeit und konstanter Wagengeschwindigkeit der Zugmaschine ist also der Fließbereich länger als bei großen v_D-Werten. Setzt man (9) in (7) ein, so ergibt sich

$$v_D = C \exp\left(- B/T\, \sigma_{St}\right)\,. \tag{10}$$

Bei tiefen Temperaturen ist das Produkt $T\,\sigma_{St}$ kleiner als bei höheren, da die Temperaturabnahme gegen die Spannungszunahme überwiegt. Das bedeutet eine kleinere Geschwindigkeit der Bandfront bei tieferen Temperaturen. Entsprechend steigt ε_{Fl} an. Die kleinere Geschwindigkeit v_D bei tieferen Temperaturen kann auf verminderte Versetzungsgeschwindigkeiten in der Deformationsfront zurückgeführt werden [29]. Gleichung (10) ist identisch mit dem von FISHER und ROGERS [30] angegebenen Zusammenhang zwischen Lüdersbandgeschwindigkeit, Temperatur und Streckgrenze bei Eisen.

Durch Spannungs-Zeit- und Dehnungs-Zeit-Registrierungen gelang inzwischen der direkte Nachweis der Ausbreitung der Deformationsfronten bei den Nickelsorten Ni2 und Ni3. Außerdem konnte gezeigt werden, daß die kleinen Fließspannungsabnahmen beim sägezahnartigen Verlauf der Verfestigungskurve in Fig. 4 mit lokal begrenzten, inhomogenen Dehnungszunahmen verbunden sind.

5.4 Lineare Verfestigung

Die Verfestigungsanstiege in den Fig. 7a und 7b führen auf die in Tabelle 3 angegebenen Mittelwerte, die etwa um den Faktor 14 bis 20 größer sind als die Verfestigungsanstiege im Bereich II der Einkristallverfestigungskurve. HAASEN [2] und LEITZ [31, 32] beobachteten an Nickeleinkristallen (99,98% Ni) im Temperaturintervall 78° bis 300 °K einen Anstieg $\vartheta_{II} = 23$ kp/mm².

Der gegenüber den Einkristallen höhere Verfestigungsanstieg und die bei Ni2 und Ni3 beobachtete Korngrößenabhängigkeit von V_{II} läßt sich wie folgt verstehen: Bei Einkristallen nimmt bekanntlich die aktive Gleitlinienlänge L im Bereich II mit zunehmender Abgleitung a nach

$$L = \frac{\Lambda}{a - a^*} \tag{11}$$

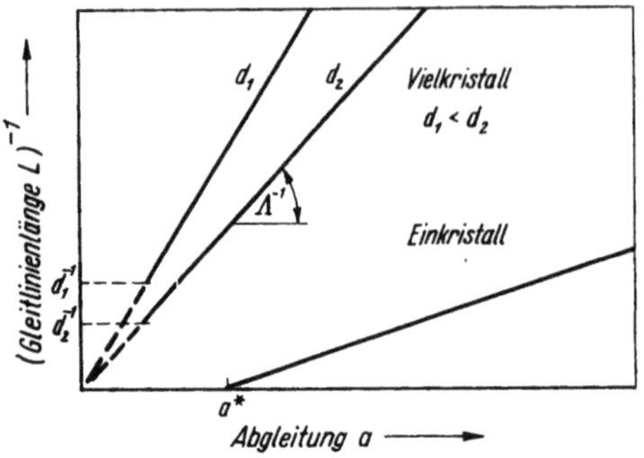

Fig. 12. Zusammenhang zwischen Gleitlinienlänge und Abgleitung bei Ein- und Vielkristallen, schematisch

ab [33], wobei Λ eine empirisch bestimmte Konstante ist. Beim Vielkristall wird die Gleitlinienlänge durch das engere Versetzungsnetzwerk beeinflußt und durch die Korngrenzen begrenzt, so daß immer $L_{\max} \leqq d$ gilt. Es ist zu erwarten, daß bei kleineren Körnern wegen der komplexeren Verformung die Gleitlinienlänge stärker mit der Verformung abnimmt als bei größeren. Geht man davon aus, daß (11) mit korngrößenabhängigem Λ und $a^* \approx 0$ auch für Vielkristalle gültig ist — das wird später bestätigt —, so erhält man das in Fig. 12 skizzierte Bild. Mit abnehmendem Korndurchmesser werden die Steigungen größer, also die Größe Λ kleiner. Sieht man für den Verfestigungsanstieg den Ansatz $V_{\mathrm{II}} \sim \Lambda^{-1/2}$ als erfüllt an, so erklärt sich der gegenüber den Einkristallen größere Verfestigungsanstieg und die Korngrößenabhängigkeit von V_{II} zwanglos als Folge stärkerer Hindernisbildung durch Mehrfachgleitung.

Einen experimentellen Beleg für diese Überlegungen bieten die quantitativen L-Bestimmungen von Zankl [11] an Nickelvielkristallen vergleichbarer Reinheit mit einer Korngröße von $d = 40\ \mu\mathrm{m}$ in Abhängigkeit vom Verformungsgrad. Rechnet man die zu den einzelnen Gleitlinienlängen gehörigen Dehnungen auf Abgleitungen mit dem Sachsschen Faktor $\overline{m} = 2{,}24$ um und trägt die L^{-1}-Werte gegen die Abgleitung a auf (vgl. Fig. 13), so erhält man wie bei Einkristallen einen linearen Zusammenhang, aber mit, wie oben vermutet, gegenüber den Einkristallwerten von Kronmüller [34] größerer Steigung. Für Nickelvielkristalle ergibt sich daraus ein Λ das etwa um den Faktor 4 kleiner ist als bei den Nickeleinkristallen.

Wir wollen nun einen quantitativen Vergleich vornehmen. Würden wir die Einkristallorientierung, an der die in Fig. 13 wiedergegebenen Meßwerte ermittelt wurden, der Abschätzung zugrundelegen, so würde nach obigem für den Anstieg im σ, ε-Diagramm gelten

$$V_{\mathrm{II}}^{\mathrm{EK}} = m_{\mathrm{EK}}^2\, \vartheta_{\mathrm{II}}^{\mathrm{EK}} \sim m_{\mathrm{EK}}^2\, \Lambda_{\mathrm{EK}}^{-1/2}\,, \tag{12}$$

wobei m_{EK} der Orientierungsfaktor ist. Für die Kristallite im Vielkristallverband postulieren wir entsprechend

$$V_{\mathrm{II}}^{\mathrm{VK}} = \overline{m}_{\mathrm{VK}}^2\, \vartheta_{\mathrm{II}}^{\mathrm{VK}} \sim \overline{m}_{\mathrm{VK}}^2\, \Lambda_{\mathrm{VK}}^{-1/2}\,, \tag{13}$$

wobei $\overline{m}_{\mathrm{VK}}$ der über alle Kristallitorientierungen gemittelte Orientierungsfaktor ist. Aus den Gleichungen (12) und (13) erhalten wir

$$V_{\mathrm{II}}^{\mathrm{VK}} = V_{\mathrm{II}}^{\mathrm{EK}} \frac{\overline{m}_{\mathrm{VK}}^2}{m_{\mathrm{EK}}^2} \sqrt{\frac{\Lambda_{\mathrm{EK}}}{\Lambda_{\mathrm{VK}}}} = \vartheta_{\mathrm{II}}^{\mathrm{EK}}\, \overline{m}_{\mathrm{VK}}^2 \sqrt{\frac{\Lambda_{\mathrm{EK}}}{\Lambda_{\mathrm{VK}}}}\,. \tag{14}$$

Aus Fig. 13 entnehmen wir $\Lambda_{\mathrm{EK}} = 6{,}5 \cdot 10^{-4}$ cm. Nicht unmittelbar erhalten wir Λ_{VK}, da wir die experimentellen L^{-1}-Werte der Vielkristallprobe über a, das mit Hilfe des Sachsschen Faktors umgerechnet wurde, aufgetragen haben.

Mit

$$a = \overline{m}_{\mathrm{VK}}\, \varepsilon$$

erhält man aus Fig. 13

$$\Lambda_{\mathrm{VK}} = \Lambda_{\mathrm{VK}}^* \cdot \frac{\overline{m}_{\mathrm{VK}}}{2{,}24}\,,$$

so daß

$$V_{\mathrm{II}}^{\mathrm{VK}} = \vartheta_{\mathrm{II}}^{\mathrm{EK}}\, \overline{m}_{\mathrm{VK}}^2 \sqrt{\frac{\Lambda_{\mathrm{EK}} \cdot 2{,}24}{\Lambda_{\mathrm{VK}}^* \cdot \overline{m}_{\mathrm{VK}}}}$$

wird.

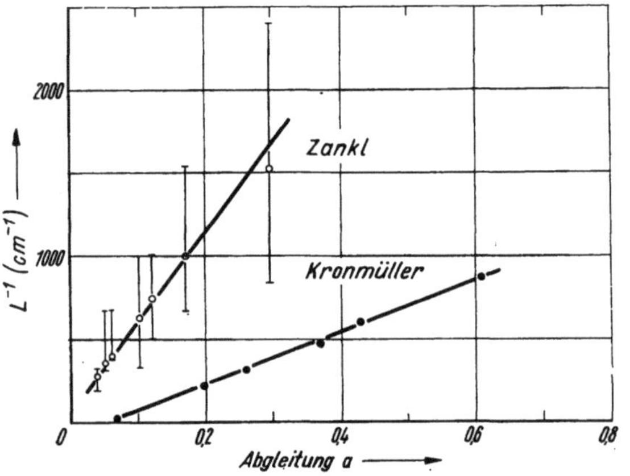

Fig. 13. Experimentell ermittelte L^{-1}-a-Diagramme. Die offenen Kreise sind die von Zankl ausgemittelten Meßwerte

Mit $\Lambda^{*}_{VK} = 1{,}86 \cdot 10^{-4}$ cm aus Fig. 13 und dem oben angegebenen Λ_{EK} ergibt sich

$$V_{II}^{VK} = 2{,}84 \sqrt{\bar{m}_{VK}^{3}}\; \vartheta_{II}^{EK} . \tag{15}$$

Setzen wir $\bar{m}_{VK} = 3{,}06$, also gleich dem Taylorfaktor, so wird

$$V_{II}^{VK}/\vartheta_{II}^{EK} = 15{,}2 ,$$

was in guter Übereinstimmung mit den experimentellen Feststellungen steht, nach denen $14 \leqq V_{II}^{VK}/\vartheta_{II}^{EK} \leqq 20$ ist.

Für den bei der reineren Nickelsorte beobachteten nahezu korngrößenunabhängigen Verfestigungsanstieg im Bereich II könnte eine mit zunehmender Korngröße schneller anwachsende Zahl von Subkörnern bei der Verfestigung verantwortlich sein. Inwieweit dafür auch Umorientierungen während der Deformation mit in Betracht kommen, wird zur Zeit untersucht.

Vergleicht man die bei verschiedenen Temperaturen und Korngrößen bestimmten Verfestigungsanstiege V_{II} von Nickel mit denen von Kupfer und Aluminium [7], so variieren alle Werte zwischen 95 kp/mm² (Al, 500 µm) und 440 kp/mm² (Ni3), also um den Faktor 4. Bezieht man die Verfestigungsanstiege auf den etwas temperaturabhängigen Schubmodel G der jeweiligen Metalle, so liegen die Werte V_{II}/G zwischen $3{,}72 \cdot 10^{-2}$ und $7{,}15 \cdot 10^{-2}$. Hier scheint eine etwa gleich große Schwankung zwischen den einzelnen V_{II}/G-Werten der Vielkristalle vorzuliegen wie bei den ϑ_{II}/G-Werten der Einkristalle [35]. Eine Gegenüberstellung der entsprechenden Zahlenwerte enthält Tabelle 4.

Tabelle 4

Bei Aluminium, Kupfer und Nickel beobachtete Höchst- und Kleinstwerte von ϑ_{II}/G bzw. V_{II}/G

	$(\vartheta_{II}/G(T)) \cdot 10^3$		$(V_{II}/G(T)) \cdot 10^3$	
	$T = 90$ °K	$T = 295$ °K	$T = 90$ °K	$T = 295$ °K
Al [4, 7]	3,35	2,46	61,2	37,2
Cu [4, 7]	3,15	3,09	71,5	53,0
Ni [31]	2,78	2,87	54,6	39,0

5.5 *Übergang zu Bereich III*

Die Auftragung von ln σ_{III} gegen T ergibt nach Fig. 8a und 8b einen linearen Zusammenhang. Auf Grund der beobachteten Korngrößenabhängigkeit von σ_{III} ist zu erwarten, daß auch die auf die Temperatur $T = 0$ °K extrapolierten Spannungen $\sigma_{III}(0)$ mit zunehmender Korngröße für die jeweilige Nickelsorte abnehmen. $\sigma_{III}(0)$ kann als die makroskopisch erforderliche Spannung angesehen werden, die ohne thermische Aktivierung notwendig ist, um Quergleitprozesse in den einzelnen Kristalliten hervorzurufen. Aus den Meßergebnissen erhält man mit dem Taylorschen Orientierungsfaktor für $\tau_{III}(0)$ die in Tabelle 3 angegebenen Zahlenwerte. Vergleichsweise findet Leitz [31] bei neueren Untersuchungen an Nickeleinkristallen $\tau_{III}(0) = 6{,}6$ kp/mm². Die Vielkristallwerte liegen in der Größenordnung der Einkristalldaten, was den Schluß auf Aktivierung von Quergleitung bei in beiden Fällen vergleichbaren äußeren Spannungen zuläßt. Auch die nach Gleichung (1) gefundenen und in Tabelle 3 aufgeführten Werte für B liegen über und unter dem Wert für Nickeleinkristalle. Leitz [31] berechnet aus dem ermittelten B-Wert von $3{,}11 \cdot 10^{-3}$ [°K^{-1}] die Stapelfehlerenergie bei Nickeleinkristallen zu 300 erg/cm². Eine formale Berechnung der Stapelfehlerenergie (nach Berner [4] sowie Seeger, Berner und Wolf [36]) führt bei den Vielkristallen zu Werten zwischen 160 und 330 erg/cm². Die beobachteten Unterschiede liegen außerhalb der Meßgenauigkeit. Ein eindeutiger Gang mit der Korngröße ist nicht zu beobachten. Die auftretenden unsystematischen Schwankungen der B-Werte zeigen zunächst, daß die τ_{III}-Methode zur Stapelfehlerenergiebestimmung nicht ohne weiteres für Vielkristalle übernommen werden kann.

Der deutschen Forschungsgemeinschaft danken wir für die Unterstützung dieser Untersuchungen, ebenso Herrn Dr. F. Assmus und der Vacuumschmelze AG, Hanau/Main, für das zur Verfügung gestellte Untersuchungsmaterial.

Literatur

[1] J. Diehl, Z. Metallk. **47**, 331 und 441 (1956).
[2] P. Haasen, Phil. Mag. **3**, 384 (1958).
[3] J. Diehl und R. Berner, Z. Metallk. **51**, 522 (1960).
[4] R. Berner, Z. Naturf. **15a**, 689 (1960).
[5] R. P. Carreker und W. R. Hibbard, Trans. AIME **209**, 1157 (1957).
[6] B. Jaoul, J. Mech. Phys. Solids **5**, 95 (1957).
[7] H. Knöll, Diplomarbeit, TH Stuttgart 1962.
[8] P. Feltham und J. D. Meakin, Phil. Mag. **2**, 105 und 1237 (1957).
[9] R. P. Carreker und W. R. Hibbard, Acta metall. **1**, 654 (1953).
[10] H. Christian, Diplomarbeit, TH Stuttgart 1960.
[11] G. Zankl, Z. Naturf. **18a**, 795 (1963).
[12] E. Macherauch, K. Kolb und H. Christian, Z. Naturf. **16a**, 1113 (1961).
[13] A. Portevin und F. Le Chatelier, C. R. Acad. Sci. (France) **176**, 507 (1913).
[14] A. H. Cottrell, Phil. Mag. **44**, 829 (1953).
[15] M. Boček, Czech. J. Phys. **10**, 841 (1960).
[16] W. A. Tiller, J. appl. Phys. **29**, 611 (1958).
[17] D. Kuhlmann-Wilsdorf und H. Wilsdorf, Acta metall. **1**, 394 (1953).
[18] J. S. Koehler, Phys. Rev. **85**, 480 (1952).
[19] A. N. Stroh, Phil. Mag. **46**, 968 (1955).
[20] J. Friedel, Les Dislocations, Gauthier Villars, Paris 1956.
[21] E. O. Hall, Proc. Phys. Soc. **B 64**, 747 (1951).
[22] N. J. Petch, J. Iron Steel Inst. **174**, 25 (1953).
[23] A. Seeger, Phil. Mag. **46**, 1194 (1955).

[24] A. SEEGER, Kristallplastizität, in Handbuch der Physik VII,2, Springer-Verlag, Berlin 1958.
[25] A. H. COTTRELL, Conference on high rates of strain, Institution of mechanical Engineers London 1957 (S. 448).
[26] P. HAASEN, Internal Stresses and Fatigue in Metals, Elsevier, Amsterdam 1959 (S. 205).
[27] J. C. FISHER, Trans. ASM **47**, 1423 (1955).
[28] E. MACHERAUCH und O. VÖHRINGER, in Vorbereitung.
[29] G. T. HAHN, Acta metall. **10**, 727 (1962).
[30] J. C. FISHER und H. C. ROGERS, Acta metall. **4**, 180 (1956).
[31] CH. LEITZ, Diplomarbeit, TH Stuttgart 1960.
[32] S. MADER, A. SEEGER und CH. LEITZ, J. appl. Phys. **34**, 3368 (1963).
[33] S. MADER, Z. Phys. **149**, 73 (1957).
[34] H. KRONMÜLLER, Z. Phys. **154**, 574 (1959).
[35] A. SEEGER, S. MADER und H. KRONMÜLLER, in Electron Microscopy and Strength of Crystals, Interscience 1963 (S. 665).
[36] A. SEEGER, R. BERNER und H. WOLF, Z. Phys. **34**, 3368 (1963).

(Received June 1, 1964)

phys. stat. sol. 6, 507 (1964)

Institut für Elektrowerkstoffe, Freiburg i. Br. (a), und
Physikalisches Institut der Universität Freiburg i. Br. (b)

An ESR Analysis of Thermally Activated Motional Effects in the ZnS–A-Centers

By

B. DISCHLER (a), A. RÄUBER (a), and J. SCHNEIDER (b)

The thermally activated motion of the unpaired spin in the ZnS–A-center is investigated over the temperature range 67 to 121 °K. The hopping rate is determined by an analysis of the motional effects in the ESR spectra.
The temperature dependence of the hopping rate is found to be described by an Arrhenius relation with an activation energy of 0.057 eV for the chlorine, and 0.058 eV for the bromine A-center. In addition, the components are given for the **g**-tensor of the bromine A-center.

Die thermisch aktivierte Bewegung des ungepaarten Spins innerhalb des ZnS–A-Zentrums wurde im Temperaturbereich 67 °K bis 121 °K untersucht. Die Sprungrate wurde durch Analyse der Veränderungen bestimmt, welche das Springen im ESR-Spektrum verursacht.
Die Temperaturabhängigkeit der Sprungrate kann durch eine Arrhenius-Beziehung beschrieben werden, wobei die Aktivierungsenergien 0,057 eV für das Chlor- und 0,058 eV für das Brom-A-Zentrum betragen. Ergänzend werden die **g**-Tensor-Komponenten für das Brom-A-Zentrum mitgeteilt.

An investigation of the chlorine-A-center in ZnS by electron spin resonance, ESR, was recently reported [1]. As a result of the ESR analysis, the following model for this center which is responsible for the 470 nm "self activated" emission of ZnS, was proposed: The chlorine A-center is formed by the association of a zinc vacancy with a monovalent chlorine impurity ion. The center is converted into the paramagnetic state by trapping a hole which is created during illumination of the crystal with ultraviolet light. At helium temperatures, the hole is permanently fixed on *one* of the three equivalent sulfur ions adjacent to the zinc vacancy. (The symmetry of this sulfur site, point group C_s, is reflected by the orthorhombic symmetry of the **g**-tensor). However, at temperatures above 77 °K, a rapid hopping of the unpaired spin among the three equivalent sulfur ions adjacent to the zinc vacancy was observed. Thus, at sufficiently high temperatures, the **g**-tensor of the A-center was found to approach axial symmetry.

The components of the **g**-tensor of the high temperature modification of the A-center can be constructed from those of the low temperature modification, where the unpaired spin is frozen in.[1])

---

[1]) It was kindly pointed out to us by Dr. K. NARITA, that equation (2) in [1] is not generally valid. The reason is, that averaging an anisotropic **g**-tensor around a given symmetry axis does not fully lift the original asymmetry. In the present case, $g_\perp$ should show a weak angular dependence with respect to a rotation around the symmetry axis. However, this effect is of the order 10^{-6} and therefore too small to be detected experimentally. The calculation of the g-factors of the high temperature modification of the Cl–A-center, given in [1], remains therefore valid.

In this paper we wish to report on an analysis of the kinetics of the thermally activated hopping motion of the unpaired spin. As a result, the activation energy required to promote the hole to an equivalent sulfur site was determined for the ZnS-Cl and ZnS-Br A-center in *cubic* ZnS-crystals from the temperature dependence of the ESR spectra.

In the following, the discussion of the hopping motion of the unpaired spin of the chlorine A-center will be restricted for simplicity to a special orientation of the crystal with respect to the static magnetic field H, i.e. $H \parallel [111]$. At helium temperatures, three different orientations of the A-centers can be distinguished in the ESR spectrum, having the statistical weight of 3:6:3, as shown in Fig. 1a.[2]) Due to the rapid hopping of the spin, only two different orientations exist at sufficiently high temperatures, with the statistical weights of 3:9 (see Fig. 1b).

To investigate the influence of the hopping motion on the ESR spectra quantitatively, we shall use the theory originally developed for an interpretation of

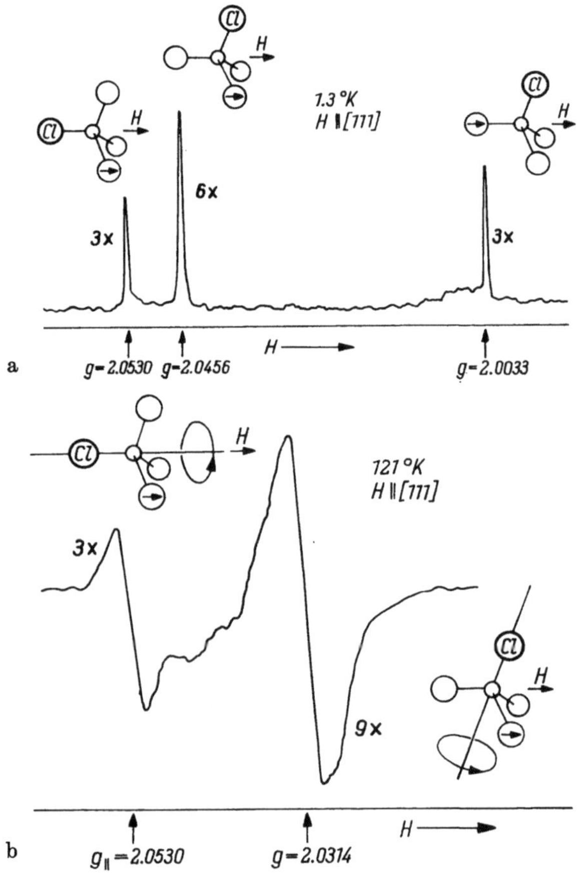

Fig. 1. ESR spectra of the chlorine ZnS-A-center for $H \parallel [111]$ at 9 GHz, taken at a) 1.3°K and b) 121 °K

[2]) This spectrum was taken by Dr. W. C. HOLTON during one of the authors (J.S.) stay at the Physics Research Laboratory of Texas Instruments.

motional effects in NMR spectra.[3]) Consider a spin, which is in a position A for a mean time τ_A, then makes a sudden rapid jump to a position B, where it rests for a mean time τ_B, and thereafter returns to A. This situation is encountered for the two differently oriented A-centers responsible for the two high field lines in the spectrum 1a. The Larmor frequencies for those two positions, which are different as a result of the anisotropic $\boldsymbol{g}$-tensor, will be denoted by ν_A and ν_B.

The frequency dependence of the absorption for such a system is given by the shape function [2, 3]

$$f(\nu) = K \frac{(c + \tau_A/T_2)\, P + Q\, R}{P^2 + R^2}, \tag{1}$$

where K is a normalizing constant and

$$P = 2\pi\tau_A \left[\frac{1}{4}(\Delta\nu)^2 - (\bar{\nu} - \nu)^2\right] + \left(c + \frac{\tau_A}{T_2}\right)\frac{1}{2\pi T_2},$$

$$Q = 2\pi\tau_A \left[\Delta\nu\left(\frac{1}{c} - \frac{1}{2}\right) + \bar{\nu} - \nu\right],$$

$$R = \Delta\nu\left(\frac{c}{2} - 1\right) + (\bar{\nu} - \nu)\left(c + \frac{2\tau_A}{T_2}\right),$$

with

$$c = 1 + \frac{\tau_A}{\tau_B}; \qquad \Delta\nu = \nu_A - \nu_B; \qquad \bar{\nu} = \frac{1}{2}(\nu_A + \nu_B).$$

The frequency of the microwave quantum absorbed is denoted by ν, and $1/T_2$ characterizes that part of the linewidth which is not directly influenced by the hopping motion. This quantity can be obtained from the low field line at $\boldsymbol{g} = 2.0530$, since this line originates from those centers, where the hopping occurs among three magnetically equivalent sulfur sites (see Fig. 1).[4])

The high field line in the spectrum 1b arises from those centers, where the spin can hop between one position A and two equivalent positions B; thus $\tau_B = 2\tau_A$.

By comparing the signal broadening, observed in the ESR spectra, with a diagram calculated from equation (1) (see Fig. 2), the temperature dependence of the hopping rate, $1/\tau_A$, was determined between 67 °K and 121 °K. It is in this range that the hopping motion can be studied conveniently at X-band frequencies.

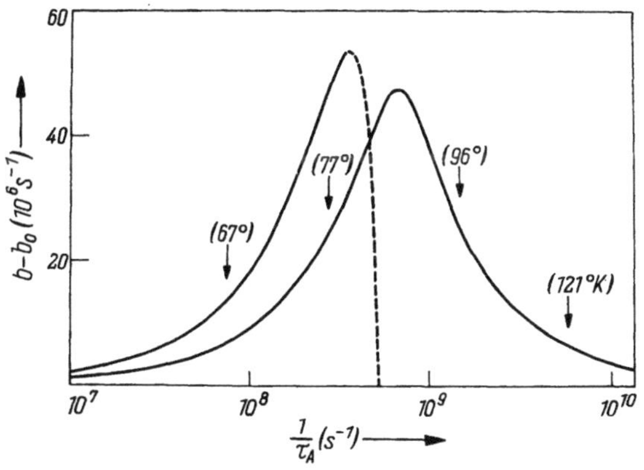

Fig. 2. Signal broadening, arising from the hopping motion calculated as a function of the hopping rate $1/\tau_A$ from equation (1) for $\Delta\nu = 1.85 \cdot 10^8\ s^{-1}$. This corresponds to the high field line of the Cl-A-center for $H \| [111]$. (Note that the first derivative is recorded.) The quantity b is the total peak to peak linewidth from which the natural linewidth b_0, i.e. that of the low field line, has been subtracted. In the left part of the diagram two lines exist with different broadening, which collapse into a single line as the hopping rate increases at higher temperatures

[3]) For a review see, for example, reference [2], chapter 10.

[4]) Only those jumps including a change in Larmor-frequency contribute to the motional effects in the spectrum (see [2]).

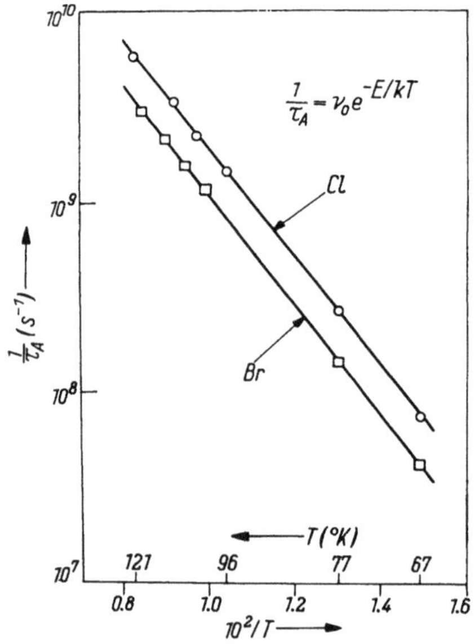

Fig. 3. Temperature dependence of the hopping rate in the chlorine and bromine A-center

The results are shown in Fig. 3, and it can be seen that the experimental points are well represented by an Arrhenius relation of the type

$$\frac{1}{\tau_A} = \nu_0 \cdot e^{-E/kT} \tag{2}$$

with

$$\nu_0 = (1.4 \pm 0.8) \times 10^{12}\ \mathrm{s}^{-1}$$

and

$$E = 0.057 \pm 0.005\ \mathrm{eV}$$

for the chlorine A-center in pure cubic ZnS-crystals.

Similar values have been observed for the isomorphic bromine A-center in ZnS[5]) where $\nu_0 = (0.9 \pm 0.6) \times 10^{12}\ \mathrm{s}^{-1}$ and $E = 0.058 \pm 0.005$ eV (see Fig. 3). The components of the $\boldsymbol{g}$-tensor of the bromine A-center in pure cubic crystals were determined as

$$g_{\parallel} = 2.0526 \pm 0.0003\ ,$$
$$g_{\perp} = 2.0304 \pm 0.0003$$

for high temperatures and

$$g_1 = 2.0025 \pm 0.0003\ ,$$
$$g_2 = 2.0535 \pm 0.0003\ ,$$
$$g_3 = 2.0572 \pm 0.0003$$

with a tilting angle of $2.5° \pm 0.5°$, when the hole is fixed at one S^{2-}-ion. The same notation is used as in [1].

Finally, it should be noted that the relaxation time was found to be longer than τ_A by two orders of magnitude. Apparently, the orientation of the spin and the phase of its Larmor precession are not seriously affected by the hopping process.

We wish to thank Professor R. MECKE and Professor K. DRANSFELD for the use of the spectrometer and for reading the manuscript.

References

[1] J. SCHNEIDER, W. C. HOLTON, T. L. ESTLE, and A. RÄUBER, Phys. Letters 5, 312 (1963).
[2] J. A. POPLE, W. G. SCHNEIDER, and H. J. BERNSTEIN, High-resolution Nuclear Magnetic Resonance, McGraw-Hill, New York 1959.
[3] H. S. GUTOWSKY and C. H. HOLM, J. chem. Phys. 25, 1228 (1956).

(Received June 1, 1964)

[5]) To avoid chemical contamination with chlorine, the bromine-activated crystals were prepared from elemental zinc and sulfur, purity grade 6N and 5N, by transport reaction using HBr as transporter.

phys. stat. sol. 6, 511 (1964)

IV. Physikalisches Institut der Humboldt-Universität zu Berlin und
Physikalisch-Technisches Institut der Deutschen Akademie der Wissenschaften zu Berlin,
Bereich elektrischer Durchschlag

Untersuchungen zur Polarisation der Exzitonen-Annihilationsstrahlung in CdS-Einkristallen

Von

O. GOEDE und E. GUTSCHE

Es wurde die spektrale Verteilung der Intensität der blauen Lumineszenzbande von CdS-Einkristallen bei Anregung mit ultraviolettem Licht für die beiden Polarisationsrichtungen $\mathfrak{E}$ senkrecht und $\mathfrak{E}$ parallel zur c-Achse zwischen 77 °K und 295 °K untersucht. Die Temperaturabhängigkeit der energetischen Lage der beiden Hauptmaxima und das Verhältnis χ der Maximalintensitäten für die beiden Polarisationsrichtungen wurden im angegebenen Temperaturbereich bestimmt. Ferner wurde die Winkelverteilung der Polarisation bei 90 °K gemessen. Es wird dargelegt, daß beide Banden der direkten strahlenden Annihilation von Eigen-Exzitonen zuzuschreiben sind. Auf der Basis dieser Vorstellung werden die Meßergebnisse im einzelnen diskutiert.

An investigation is made of the spectral distribution of the blue emission in CdS single crystals under excitation with ultraviolet light, for the directions of polarisation $\mathfrak{E} \perp$ (*c*-axis) and $\mathfrak{E} \parallel$ (*c*-axis), between 77 and 295 °K. The position of the two principal peaks, and the ratio (χ) of the peak intensities in the two directions of polarisation, are plotted as functions of temperature. The angular distribution of polarisation at 90 °K is also measured. It is shown that the two peaks can be attributed to the direct radiative annihilation of intrinsic excitons, and the results are discussed in detail on this basis.

1. Einleitung

Erstmalig wurden von KRÖGER [1] bei 90 °K und von KLICK [2] bei 4 °K Emissionslinien im blauen Spektralbereich bei der Photolumineszenz von CdS-Einkristallen beobachtet. Schon in älteren Deutungsversuchen [3] wurden diese Linien der strahlenden Annihilation von Exzitonen zugeschrieben. Erste Untersuchungen der Temperaturabhängigkeit der blauen Emissionslinien wurden von FURLONG und RAVILIOUS [4] durchgeführt. Die bei 4,2 °K beobachteten elf sehr schmalen Linien (Halbwertsbreite etwa 1 Å) verschoben sich mit wachsender Temperatur unter Zunahme der Halbwertsbreite zu langen Wellen, bei 300 °K konnte nur noch eine breite Bande bei 5077 Å festgestellt werden. Wegen der Abhängigkeit von Oberflächenbehandlungen der Kristalle nahmen diese Autoren an, daß die blauen Lumineszenzbanden durch adsorbierte Gasatome bedingt sind. Durch zahlreiche Untersuchungen anderer Autorengruppen in den folgenden Jahren [5, 6, 7, 8, 9] wurde jedoch der Zusammenhang der blauen Lumineszenzbanden von CdS-Einkristallen mit einer Exzitonen-Annihilation weitgehend gesichert. Insbesondere stellten GRILLOT et al. [5] erstmalig eine quasi-wasserstoffähnliche Serie von Linien bei Temperaturen unter 21 °K fest, wie sie für ein Mottsches Exziton zu erwarten wäre. GROSS et al. [6] konnten für zahlreiche Emissionslinien das resonanzartige Übereinstimmen mit Exzitonen-Absorptionslinien nachweisen. DIEMER et al. [7] schlossen aus der beobachteten Intensitätsabhängigkeit der blauen Lumineszenz von der UV-Anregungsintensität auf das

Vorliegen von Exzitonen-Annihilationsstrahlung, COLLINS [8] zog einen entsprechenden Schluß aus der festgestellten geringen Abklingzeit der Lumineszenz bei Impulsanregung mit 1 MeV-Elektronen. Sehr eingehende Untersuchungen bei 4 °K wurden auch von PEDROTTI und REYNOLDS [10] durchgeführt. Insbesondere versuchten diese Autoren eine Deutung der Polarisation des emittierten Lichtes in bezug auf die c-Achse der Kristalle durch Berücksichtigung der Valenzbandstruktur von CdS.

Einen wesentlichen Beitrag zum Nachweis der Exzitonennatur der blauen Lumineszenz von CdS-Einkristallen leisteten THOMAS und HOPFIELD durch eine Reihe experimenteller und theoretischer Arbeiten [11, 12, 13]. Sie konnten zunächst die bei 77 °K und 4,2 °K auftretenden Maxima im Reflexionsspektrum den Grund- bzw. angeregten Zuständen der Eigen-Exzitonen[1]) der drei Valenzbänder zuordnen [11], unter dieser Annahme die Polarisationseigenschaften deuten und Bindungsenergie und reduzierte Masse der Eigen-Exzitonen bestimmen. Durch Vergleich mit dem Reflexionsspektrum konnten auch im Lumineszenzspektrum die zu den beiden oberen Valenzbändern gehörenden Eigen-Exzitonen A und B infolge ihrer Annihilationsstrahlung nachgewiesen werden ([11], Fig. 3a und b). Die zahlreichen anderen bei Temperaturen unter 4 °K beobachteten Lumineszenzlinien wurden von den Autoren „gebundenen" Exzitonen bzw. Phonon-Satelliten zugeordnet [13].

Während die bisherigen Untersuchungen vorwiegend bei möglichst tiefen Temperaturen durchgeführt wurden, sind in einigen neueren Arbeiten [14, 15, 16, 31] die blauen Lumineszenzbanden auch bis zu 300 °K verfolgt worden. Die Autoren erhielten jedoch an verschiedenen Kristallen recht unterschiedliche Ergebnisse [15, 16], und die untersuchten Banden wurden, abgesehen von [31], nicht als Exzitonen-Annihilationsstrahlung interpretiert. Insbesondere wurden bisher noch keine Untersuchungen zur Polarisation dieser Lumineszenzbanden bei höheren Temperaturen durchgeführt.

Die Aufgabe der vorliegenden Arbeit sollte es sein, insbesondere die Polarisation der blauen Lumineszenzbanden an hochreinen CdS-Einkristallen bei Temperaturen zwischen 77 °K und 295 °K zu untersuchen und durch den Anschluß an entsprechende Messungen von THOMAS und HOPFIELD [11] bei tiefen Temperaturen die Exzitonennatur dieser Banden auch bei höheren Temperaturen nachzuweisen. Aus einer Analyse der Meßergebnisse können dann u. a. Aussagen über die Valenzbandaufspaltung in CdS als Funktion der Kristalltemperatur gewonnen werden.

2. Experimentelles

Die untersuchten (undotierten) CdS-Einkristalle wurden aus spektralreinem mikrokristallinen CdS-Pulver durch Sublimation in einem strömenden N_2-, H_2-, H_2S-Gasgemisch hergestellt. Sie hatten Plättchenform (etwa $3 \times 6 \times 0{,}05$ mm^3), die c-Achse lag in der Plättchenebene. Die Kristalle wurden spannungsfrei an einem Kristallhalter befestigt, der sich in einer Hochvakuumapparatur befand (Vakuum etwa $3 \cdot 10^{-6}$ Torr). Durch flüssigen Stickstoff bzw. durch einen regelbaren kalten N_2-Gasstrom wurden die Kristalle auf die jeweilige Meßtemperatur abgekühlt. Die Messungen bei 77 °K wurden an Kristallen durchgeführt, die sich auf einer Glimmerplatte in einer evakuierten abgeschmolzenen Quarzampulle befanden, die direkt in flüssigen Stickstoff getaucht wurde. Die Anregung der

[1]) „Eigen"-Exzitonen (intrinsic excitons) nach ELLIOT, Phys. Rev. **108**, 1384 (1957), keine Bindung an Störstellen.

Lumineszenz erfolgte durch Bestrahlung der Kristalle mit unpolarisiertem UV-Licht der Wellenlänge $\lambda = 365$ nm (Hg-Höchstdrucklampe HBO 500 und geeignete Kombination von Doppelinterferenz-, Farb- und Wärmeschutzfiltern). Die Anregungsintensität betrug 10^{16} Photonen/cm^2 s. Das von der bestrahlten Kristallfläche ausgehende Lumineszenzlicht wurde durch einen 3-Prismenspektrographen spektral zerlegt (Lineardispersion 11 Å/mm bei 4850 Å), zur Untersuchung der Polarisation diente ein Polarisationsfilter Bernotar (Durchlässigkeit bei gekreuzten Filtern im untersuchten Spektralbereich 0,3%). Die spektrale Verteilung der Intensität wurde automatisch registriert, indem ein Multiplier von einem Motor in der Plattenebene des Spektrographen bewegt und das Ausgangssignal des Multipliers nach Verstärkung von einem synchron laufenden Schreiber aufgezeichnet wurde.

Alle erhaltenen Meßergebnisse wurden auf gleiche spektrale Empfindlichkeit des Systems Multiplier-Spektrograph reduziert, die spektrale Verteilung dieser Empfindlichkeit wurde unter Benutzung einer geeichten Wolframbandlampe bestimmt. Ferner wurden die polarisierende Wirkung des Spektrographen als Funktion der Wellenlänge ermittelt und die Meßergebnisse dementsprechend korrigiert.

3. Meßergebnisse

Die im folgenden beschriebenen Meßergebnisse wurden an mehreren Kristallen verifiziert, die hier wiedergegebenen Meßkurven wurden alle an demselben Kristall gewonnen.

Fig. 1 zeigt die spektrale Verteilung der Intensität I der blauen Lumineszenz (Photonenzahl in willkürlichen Einheiten) für die beiden Polarisationsrichtungen $\mathfrak{E}$ senkrecht und $\mathfrak{E}$ parallel zur $\mathfrak{c}$-Achse des Kristalls für die Temperatur $T = 90$ °K. Für $\mathfrak{E} \parallel \mathfrak{c}$ ergibt sich ein Maximum bei $\lambda = 4858$ Å, für $\mathfrak{E} \perp \mathfrak{c}$ ein Hauptmaximum bei $\lambda = 4889$ Å. Das angedeutete langwellige Nebenmaximum ist bei höheren Temperaturen nicht mehr zu beobachten und wurde nicht weiter untersucht. Die Abhängigkeit der Lumineszenzintensität im Bandenmaximum $\lambda = 4889$ Å vom Winkel Θ zwischen der $\mathfrak{c}$-Achse des Kristalls und dem $\mathfrak{E}$-Vektor des Lumineszenz-

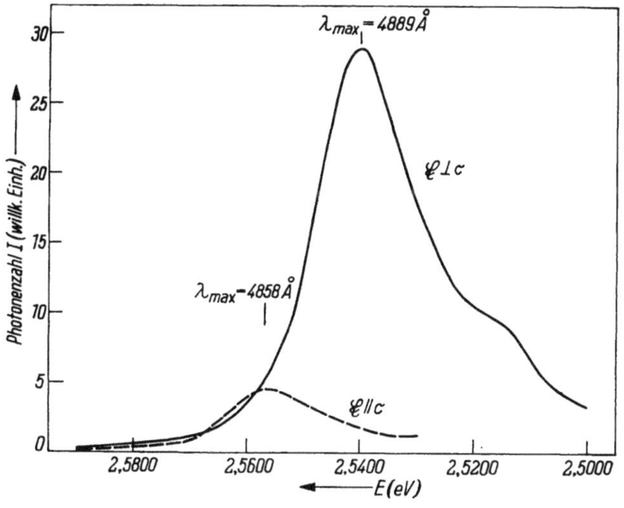

Fig. 1. Spektrale Verteilung der Intensität I der blauen Lumineszenz für die Polarisationsrichtungen $\mathfrak{E} \perp \mathfrak{c}$ und $\mathfrak{E} \parallel \mathfrak{c}$. Kristalltemperatur $T = 90$ °K

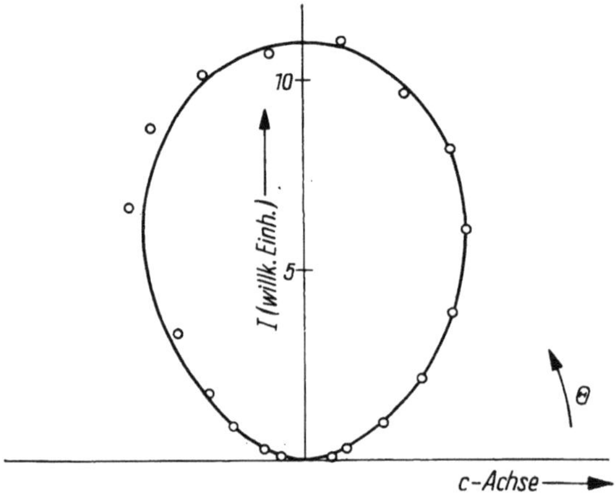

Fig. 2. Abhängigkeit der Intensität der Bande $\lambda = 4889$ Å bei 90° K vom Winkel Θ zwischen der c-Achse des Kristalls und dem $\mathfrak{E}$-Vektor des Lumineszenzlichts. Die ausgezogene Kurve stellt die Funktion $I(\Theta) = I_{\pi/2} \sin^2\Theta$ dar

lichtes zeigt Fig. 2. Die Meßkurve läßt sich gut durch die Funktion $I(\Theta) = I_{\pi/2} \sin^2\Theta$ darstellen.

Die Temperaturabhängigkeit der energetischen Lage der beiden Maxima für $\mathfrak{E} \perp \mathfrak{c}$ und $\mathfrak{E} \parallel \mathfrak{c}$ wurde im Bereich 77 °K bis 295 °K untersucht (Fig. 3). Für Temperaturen $T \gtrsim 120$ °K lassen sich die Meßkurven gut durch Geraden approximieren, die Anstiege ergeben sich zu

$$\frac{dE_{\max}^{\mathfrak{E}\perp\mathfrak{c}}}{dT} = -5{,}00 \cdot 10^{-4}\ \text{eV/Grad}\,, \qquad \frac{dE_{\max}^{\mathfrak{E}\parallel\mathfrak{c}}}{dT} = -4{,}78 \cdot 10^{-4}\ \text{eV/Grad}\,.$$

Bei Temperaturen unterhalb von etwa 120 °K ist die Temperaturverschiebung der Maxima wesentlich geringer ($dE/dT \approx -1{,}2 \cdot 10^{-4}$ eV/Grad).

Die Ergebnisse der Untersuchung der Temperaturabhängigkeit der energetischen Lage der Bandenmaxima (Fig. 3) stimmen relativ gut überein mit entsprechenden *ohne* Berücksichtigung der Polarisation gewonnenen Ergebnissen von Reynolds [14], Kulp [16] und Friedrich [15]. Diese Autoren finden für eine Bande, die offenbar der Bande für die Polarisation $\mathfrak{E} \perp \mathfrak{c}$ in dieser Arbeit entspricht, für Temperaturen $T > 120$ °K ebenfalls eine lineare Verschiebung des Maximums mit wachsender Temperatur (z. B. $dE/dT \approx -4{,}9 \cdot 10^{-4}$ eV/Grad [14]) und für tiefere Temperaturen auch eine wesentlich geringere Verschiebung.

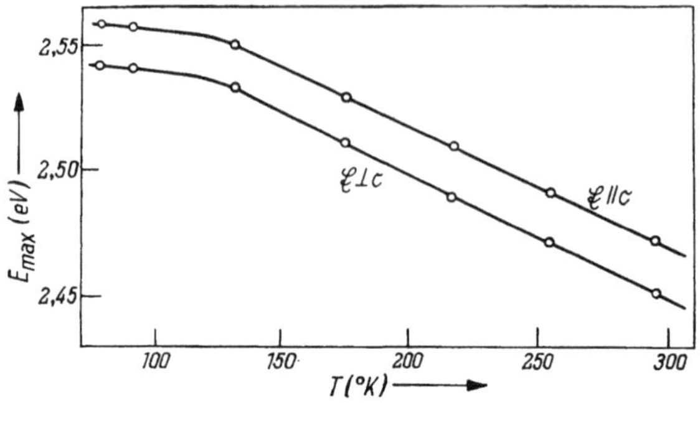

Fig. 3. Temperaturabhängigkeit der Lage der Maxima der blauen Lumineszenzbanden für $\mathfrak{E} \perp \mathfrak{c}$ und $\mathfrak{E} \parallel \mathfrak{c}$

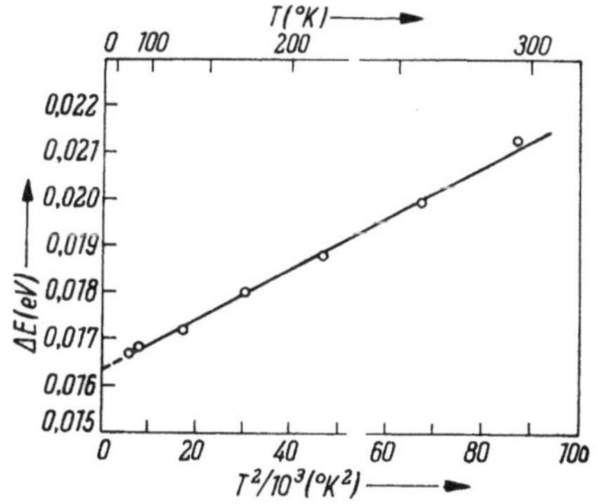

Fig. 4. Energetischer Abstand ΔE der Maxima der beiden Lumineszenzbanden für $\mathfrak{E} \perp c$ und $\mathfrak{E} \parallel c$ als Funktion der Kristalltemperatur T

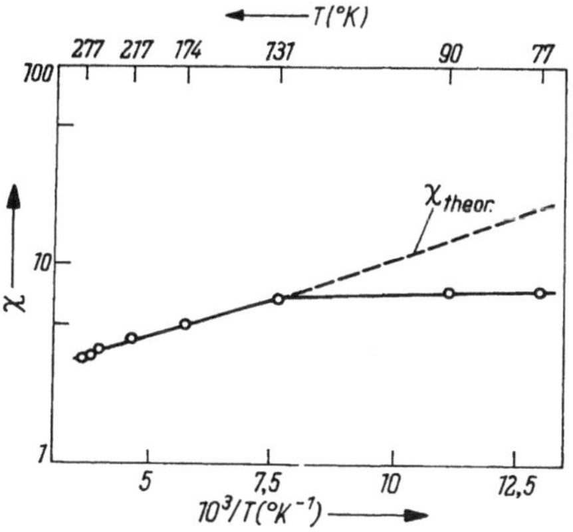

Fig. 5. Polarisationsverhältnis $\chi = \frac{I_{max}^{\mathfrak{E} \perp c}}{I_{max}^{\mathfrak{E} \parallel c}}$ als Funktion der Kristalltemperatur T Die gestrichelte Kurve stellt $\chi^{theor} = 1{,}6 \cdot \exp(\Delta E(T)/k\,T)$ dar

Der energetische Abstand ΔE der beiden Maxima steigt mit wachsender Temperatur an. Angenähert gilt im gesamten untersuchten Temperaturbereich $\Delta E = a + b\,T^2$ (vgl. Fig. 4) mit $a = 0{,}016_3$ eV und $b = 5{,}4 \cdot 10^{-8}$ eV/Grad².

Die Intensität I_{max} im Bandenmaximum nimmt für beide Banden mit wachsender Temperatur stark ab und ist bei 295 °K um mehr als eine Größenordnung geringer als bei 77 °K, während die Halbwertsbreite der Banden zunimmt (für die Bande $\mathfrak{E} \perp c$ von 0,02 eV bei 90 °K auf 0,07 eV bei 295 °K). In Fig. 5 ist das Verhältnis der Intensitäten im jeweiligen Bandenmaximum für beide Polarisationrichtungen $\chi = I_{max}^{\mathfrak{E} \perp c}/I_{max}^{\mathfrak{E} \parallel c}$ als Funktion der Kristalltemperatur dargestellt. Die gestrichelte Kurve gibt die Funktion $\chi^{theor} = 1{,}6 \cdot \exp(\Delta E(T)/k\,T)$ wieder (vergleiche (11)), wobei für $\Delta E(T)$ die gemessenen Werte nach Fig. 4 eingesetzt wurden. Für Temperaturen oberhalb von etwa 120°K werden die Meßwerte durch χ^{theor} gut approximiert. Dagegen weichen die experimentellen χ-Werte in dem Temperaturbereich $T \lesssim 120$ °K, in dem auch die Temperaturverschiebung der Bandenmaxima wesentlich geringer wird, von χ^{theor} ab und zeigen ein deutliches Sättigungsverhalten.

4. Diskussion

Die in dieser Arbeit untersuchten CdS-Einkristalle zeigten im Gegensatz zu den von Kulp [16] und Friedrich [15] untersuchten Kristallen eine einheitliche relativ einfache Struktur der blauen Lumineszenz, die für 77 °K gut mit den von Thomas und Hopfield [11] für diese Temperatur angegebenen Lumineszenzspektren übereinstimmt. Insbesondere stimmt die für 77 °K festgestellte energetische Lage der Bandenmaxima für die beiden Polarisationsrichtungen $E_{max}^{\mathfrak{E} \perp c} = 2{,}542$ eV und $E_{max}^{\mathfrak{E} \parallel c} = 2{,}558$ eV sehr gut überein mit den entsprechenden Werten von Thomas und Hopfield. Wie in Abschnitt 1 erwähnt, konnten diese Autoren durch Vergleich mit Reflexionsspektren von CdS-Einkristallen nachweisen, daß es sich bei den beobachteten Lumineszenzbanden um die Annihilationsstrahlung von Exzitonen handelt. Die in der vorliegenden Arbeit festgestellten Lumineszenzbanden können daher sowohl bei 77 °K als auch (wegen der stetigen Änderung aller Parameter) bei höheren Temperaturen der strahlenden Annihilation von Exzitonen zugeschrieben werden. Im folgenden sollen im Rahmen dieser Vorstellung die Meßergebnisse im einzelnen diskutiert werden.

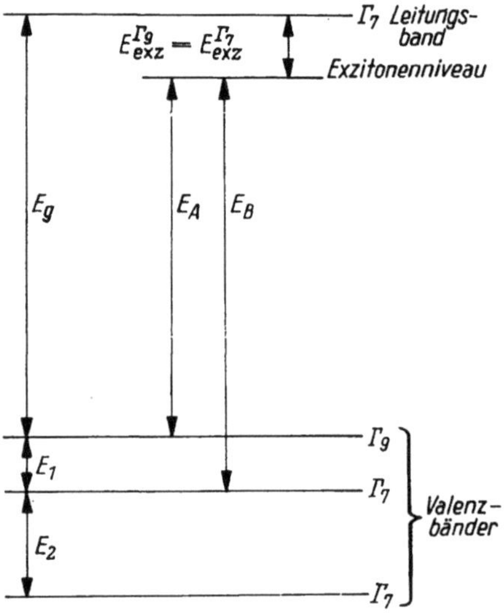

Fig. 6. Energieniveauschema für CdS für $\mathfrak{k} = 0$

Fig. 6 zeigt die bekannte Bandstruktur von CdS für den Wellenzahlvektor $\mathfrak{k} = 0$ unter Einschluß der Exzitonenniveaus. Die Bindungsenergien der zu den beiden oberen Valenzbändern gehörenden Exzitonen A und B sind mit $E_{exz}^{\Gamma_9}$ und $E_{exz}^{\Gamma_7}$ bezeichnet. Wie in [11] sei angenommen, daß

$$E_{exz}^{\Gamma_9} \approx E_{exz}^{\Gamma_7} \tag{1}$$

gilt.[2]) Für die Energie der bei der (direkten) Annihilation der Exzitonen A und B emittierten Photonen E_A und E_B gilt dann

$$E_A = E_g - E_{exz}^{\Gamma_9}, \qquad E_B = E_g - E_{exz}^{\Gamma_7} + E_1 . \tag{2}$$

Auf die Polarisation der emittierten Photonen in bezug auf die $\mathfrak{c}$-Achse der Kristalle kann aus gruppentheoretischen Betrachtungen geschlossen werden [12, 17]. Danach ergibt sich die Symmetrie des (Mottschen) Exzitons aus der Symmetrie des Leitungsbandes und des betreffenden Valenzbandes bei $\mathfrak{k} = 0$ und der Symmetrie der Elektron-Loch-Bahn: $\Gamma_{exz} = \Gamma_{L\text{-Band}} \times \Gamma_{V\text{-Band}} \times \Gamma_{Bahn}$. Da Γ_{Bahn} für den Exzitonen-Grundzustand in erster Näherung gleich Γ_1 ist, folgt daraus

$$\Gamma_{exz\,A} = \Gamma_5 + \Gamma_6 , \qquad \Gamma_{exz\,B} = \Gamma_1 + \Gamma_2 + \Gamma_5 . \tag{3}$$

Das bedeutet, daß die Photonen der (kleineren) Energie E_A senkrecht zur $\mathfrak{c}$-Achse polarisiert sind, während für E_B beide Polarisationsrichtungen $\mathfrak{E} \perp \mathfrak{c}$ und $\mathfrak{E} \parallel \mathfrak{c}$ erlaubt sind. Wenn man also identifiziert

$$E_A = E_{max}^{\mathfrak{E}\perp\mathfrak{c}} \quad \text{und} \quad E_B = E_{max}^{\mathfrak{E}\parallel\mathfrak{c}} , \tag{4}$$

wird damit auch die beobachtete Polarisation der Lumineszenzbanden ohne weiteres verständlich.[3]) Daß die emittierten Photonen E_A wirklich streng $\mathfrak{E} \perp \mathfrak{c}$ polarisiert sind, wie es die optischen Auswahlregeln fordern, zeigt auch Fig. 2. Die gemessene Abhängigkeit der Lumineszenzintensität vom Winkel Θ zwischen $\mathfrak{E}$ und $\mathfrak{c}$-Achse des Kristalls entspricht genau einem $\sin^2\Theta$-Gesetz, das in einem solchen Fall zu gelten hat.

---

[2]) Der Unterschied zwischen $E_{exz}^{\Gamma_9}$ und $E_{exz}^{\Gamma_7}$ ist durch die verschiedenen effektiven Massen der Löcher in den Valenzbändern bedingt. Da die Löchermassen für beide Valenzbänder merklich größer als die Elektronenmasse sind [28, 29, 30], liefert letztere den größten Beitrag zur reduzierten Masse des Exzitons, so daß (1) gerechtfertigt erscheint.

[3]) Für die Photonen E_B wird der jeweilige Anteil der beiden Polarisationsrichtungen $\mathfrak{E} \perp \mathfrak{c}$ und $\mathfrak{E} \parallel \mathfrak{c}$ durch das Verhältnis der Matrixelemente für die entsprechenden optischen Übergänge bestimmt. Aus dem Experiment ergibt sich, daß der Anteil $\mathfrak{E} \perp \mathfrak{c}$ offenbar so klein ist, daß ein entsprechendes Lumineszenzmaximum neben der breiten Hauptbande E_A nicht mehr aufgelöst werden kann.

Die Temperaturabhängigkeit der Exzitonenbindungsenergien ist sicher klein gegen die Temperaturverschiebung der Lumineszenzbanden, da erstere nur durch die relativ schwache Abhängigkeit der Dielektrizitätskonstanten und der effektiven Massen von der Temperatur bestimmt ist. Nach (2) läßt sich damit die Temperaturverschiebung der Lumineszenzbanden auf die Änderung des Abstandes zwischen dem Leitungsband und den Valenzbändern mit der Temperatur zurückführen.[4]) Das zeigt auch ein Vergleich mit Meßwerten für dE_g/dT (z. B. Fig. 4 und 5 in [21]).

Aus (1), (2) und (4) folgt, daß das experimentell bestimmte ΔE (Fig. 4) dem energetischen Abstand der beiden oberen Valenzbänder entspricht: $E_1 = E_{max}^{\mathfrak{E}\parallel c} - E_{max}^{\mathfrak{E}\perp c} = \Delta E$. Für T = 77 °K ergibt sich $E_1 = 0{,}0167$ eV in guter Übereinstimmung mit entsprechenden Werten aus Messungen der Absorptions- [18, 20, 21] und der Reflexionsspektren [11, 21, 23], der Anregungsspektren der grünen Lumineszenz [24] und der spektralen Verteilung der Photoleitung [22] von CdS-Einkristallen. Aus Fig. 4 ergibt sich, daß der Abstand der beiden oberen Valenzbänder mit wachsender Temperatur zunimmt.[5]) Diese beobachtete Zunahme kann ganz allgemein in zwei Anteile zerlegt werden:

$$\frac{dE_1}{dT} = \left(\frac{\partial E_1}{\partial T}\right)_V - \frac{\alpha}{\varkappa}\left(\frac{\partial E_1}{\partial p}\right)_T . \qquad (5)$$

Dabei ist $\varkappa = 1{,}66 \cdot 10^{-6}$ cm²/kp die kubische Kompressibilität von CdS [19] und $\alpha = (\alpha_{\parallel c} + 2\,\alpha_{\perp c}) = 1{,}7 \cdot 10^{-5}$ Grad^{-1} der thermische Ausdehnungskoeffizient für 300 °K [25]. Der zweite Term der rechten Seite der Gleichung (5) entspricht dem Anteil, der durch die mit einer Temperaturänderung verbundene Änderung der Gitterkonstanten bedingt ist (Dilatationsanteil), der erste ist der Anteil, der durch die Temperaturabhängigkeit der Wechselwirkung der Elektronen mit dem Gitter verursacht wird (Wechselwirkungsanteil) und auch bei konstantem Volumen auftritt. Das Verhältnis der beiden Anteile kann abgeschätzt werden, wenn man $\left(\frac{\partial E_1}{\partial p}\right)_T$ aus Messungen der Druckverschiebung der Absorptionskante entnimmt $\left(\left(\frac{\partial E_1}{\partial p}\right)_{T=300\,°K} = -0{,}6 \cdot 10^{-6}\ \text{eV/kp} \cdot \text{cm}^{-2}\,[19]\right)$ und dE_1/dT aus Fig. 4 für 300 °K bestimmt. Dann ergibt sich für den Wechselwirkungsanteil $\left(\frac{\partial E_1}{\partial T}\right)_V = (3{,}2 - 0{,}62) \times \times 10^{-5}$ eV/Grad, also nur das etwa vierfache des Dilatationsanteils. Im Gegensatz zur Situation bei der Temperaturabhängigkeit der Breite der verbotenen Zone (siehe dazu [19]) spielt also bei der Temperaturabhängigkeit der Valenzbandaufspaltung E_1 der Dilatationsanteil eine wesentliche Rolle.

Die Aufspaltung des Valenzbandes in drei Teilbänder bei $\mathfrak{k} = 0$ in Gittern mit Wurtzitstruktur wird durch die Kristallfeld- und die Spin–Bahn-Wechselwirkung bedingt. Für die Größe dieser Aufspaltungen E_1 und E_2 (siehe Fig. 6) wurden von verschiedenen Autoren [12, 26] unter Annahme eines vereinfachten Modells übereinstimmende Ausdrücke abgeleitet. In der Schreibweise

[4]) Zur Deutung der Temperaturabhängigkeit der Breite der verbotenen Zone im CdS vgl. z. B. [19].

[5]) Eine entsprechende Zunahme zeigen auch die E_1-Werte, die man aus Absorptionskurven von Dutton ([21], Fig. 2) entnehmen kann.

von Hopfield [12] gilt

$$E_1 = \frac{\delta + \Delta}{2} - \sqrt{\left(\frac{\delta + \Delta}{2}\right)^2 - \frac{2}{3}\,\Delta\delta}\,, \qquad E_2 = 2 \cdot \sqrt{\left(\frac{\delta + \Delta}{2}\right)^2 - \frac{2}{3}\,\Delta\delta}\,, \tag{6}$$

wobei δ die Spin–Bahn-Wechselwirkung und Δ die Kristallfeld-Wechselwirkung beschreiben. — Mit Hilfe von (6) und der gemessenen Temperaturabhängigkeit von E_1 soll im folgenden eine Aussage über die T-Abhängigkeit von Δ gewonnen werden.

Durch Auflösen von (6) nach δ und Δ können diese bisher nicht direkt experimentell bestimmbaren Parameter ermittelt werden. Mit den abgerundeten Werten $E_1 = 0{,}016$ eV und $E_2 = 0{,}057$ eV (nach [20]) ergibt sich für δ und Δ das Wertepaar 0,060 und 0,028 eV. Eine eindeutige Zuordnung nur auf Grund von (6) ist allerdings nicht möglich.[6]) Am wahrscheinlichsten dürfte jedoch die Zuordnung

$$\delta = 0{,}060 \text{ eV}\,, \qquad \Delta = 0{,}028 \text{ eV} \tag{7}$$

sein, wie sie schon von Pedrotti und Reynolds [20] und Sobolev [27] getroffen wurde. Damit würde dann die beobachtete Zunahme von E_2 um einen Faktor von maximal 7 bei CdS/CdSe-Mischkristallen mit wachsendem Se-Gehalt [20] der Spin–Bahn-Wechselwirkung zugeschrieben werden.[7]) Das ist aber plausibel, denn die Spin–Bahn-Wechselwirkung dürfte entscheidend durch das *Anion* bestimmt werden und bei Se wesentlich größer als bei S sein. Ein Vergleich zeigt ferner, daß dann δ bei *gleichem* Anion sich nur sehr wenig ändert [27, 30]. Außerdem kann nach [27] nur mit der Zuordnung (7) die bei einer Deformation in einer Ebene parallel zur $\mathfrak{c}$-Achse beobachtete unterschiedliche Verschiebung der beiden oberen Valenzbänder in bezug auf das Leitungsband gedeutet werden.

Aus (6) und (7) folgt

$$\left.\frac{dE_1}{dT}\right|_{T=77\,°\mathrm{K}} = 0{,}05 \cdot \frac{\partial\delta}{\partial T} + 0{,}43 \cdot \frac{\partial\Delta}{\partial T}\,. \tag{8}$$

Über eine eventuelle Temperaturabhängigkeit der Spin–Bahn-Wechselwirkung δ ist nichts bekannt, es dürfte aber wohl stets gelten

$$\left|\frac{\partial\delta}{\partial T}\right| \lesssim \left|\frac{\partial\Delta}{\partial T}\right|\,, \tag{9}$$

so daß der erste Term der rechten Seite von (8) vernachlässigt werden kann. Das bedeutet aber, daß die Temperaturabhängigkeit des Abstandes E_1 der beiden oberen Valenzbänder bei $\mathfrak{k} = 0$ im wesentlichen bedingt wird durch die Kristallfeld-Wechselwirkung Δ, die (nach (8) und (9)) mit wachsender Temperatur zunimmt,

$$\left.\frac{\partial\Delta}{\partial T}\right|_{T=77\,°\mathrm{K}} \approx 2 \cdot 10^{-5} \text{ eV/Grad}\,.$$

Im Rahmen der diskutierten Exzitonenvorstellung ist auch eine Deutung des gemessenen Verhältnisses χ der Intensitäten im Bandenmaximum für $\mathfrak{E} \perp \mathfrak{c}$ und $\mathfrak{E} \parallel \mathfrak{c}$ (Fig. 5) für $T > 120$ °K möglich, wenn man annimmt, daß das Verhältnis der Konzentrationen c_A und c_B der zu den beiden oberen Valenzbändern gehörenden Exzitonen A und B durch einen Boltzmann-Faktor bestimmt ist. Es

---

[6]) Beide Wurzeln der quadratischen Gleichung für δ und Δ erfüllen (6), man sieht sofort, daß (6) invariant gegen Vertauschung von δ mit Δ ist.

[7]) Für $\delta \gg \Delta$ wird E_2 nach (6) im wesentlichen durch δ bestimmt.

gilt dann

$$\frac{c_A}{c_B} = \exp\left(\frac{E_1}{k\,T}\right). \tag{10}$$

Berücksichtigt man, daß die Lumineszenzintensitäten in den Bandenmaxima $I_{max}^{\mathfrak{E}\perp c}$ und $I_{max}^{\mathfrak{E}\parallel c}$ durch die Produkte der Konzentrationen c_A und c_B mit den entsprechenden Wahrscheinlichkeiten für strahlende Annihilation gegeben sind, so erhält man mit (10)

$$\chi^{\text{theor}} = \gamma \cdot \exp\left(\frac{E_1}{kT}\right), \tag{11}$$

wobei γ das Verhältnis der Wahrscheinlichkeiten für strahlende Annihilation für die beiden Polarisationsrichtungen ist. Setzt man nach Tabelle 1 von [11] $\gamma = 25{,}6/15{,}7 \approx 1{,}6$, so erhält man mit den gemessenen E_1-Werten nach Fig. 4 die in Fig. 5 gestrichelt gezeichnete Kurve χ^{theor}, die für Temperaturen $T > 120$ °K die experimentellen χ-Werte gut approximiert.

Dieses Resultat läßt den Schluß zu, daß die Exzitonen sekundär aus den durch die anregende Strahlung primär gebildeten Elektron-Loch-Paaren entstehen, wobei die Konzentrationen c_A und c_B im wesentlichen durch die sich einstellende Boltzmannverteilung der Löcher über die Valenzbänder bestimmt sind.

Ein Grund für das Sättigungsverhalten von χ bei Temperaturen $T < 120$ °K und damit für das Abweichen von χ^{theor} ist bisher nicht bekannt. Bemerkenswert erscheint jedoch, daß die Abweichung von χ^{theor} bei einer Temperatur beginnt, bei der auch die Kurve der Temperaturverschiebung der Maxima der Lumineszenzbanden einen Knick hat.

Literatur

[1] F. A. Kröger, Physica **7**, 1 (1940).
[2] C. C. Klick, Phys. Rev. **89**, 274 (1952).
[3] F. A. Kröger und H. J. G. Meyer, Physica **20**, 1149 (1954).
[4] L. R. Furlong und F. Ravilious, Phys. Rev. **98**, 954 (1955).
[5] siehe z. B. E. Grillot und M. Bancie-Grillot, phys. stat. sol. **3**, 229 (1963).
[6] siehe z. B. E. Ф. Гросс и М. А. Якобсон, Изв. Акад. наук **25**, 531 (1961).
[7] G. Diemer und A. J. van der Houven van Oordt, Physica **24**, 707 (1958).
[8] R. J. Collins, J. appl. Phys. **30**, 1135 (1959).
[9] I. Broser und R. Broser-Warminsky, Z. Elektrochem. **61**, 209 (1956).
[10] L. S. Pedrotti und D. C. Reynolds, Phys. Rev. **120**, 1664 (1960).
[11] D. G. Thomas und J. J. Hopfield, Phys. Rev. **116**, 573 (1959).
[12] J. J. Hopfield, J. Phys. Chem. Solids **15**, 97 (1960).
[13] D. G. Thomas und J. J. Hopfield, Phys. Rev. **128**, 2135 (1962).
[14] D. C. Reynolds, Phys. Rev. **118**, 478 (1960).
[15] H. Friedrich, phys. stat. sol. **5**, K27 (1964).
[16] B. A. Kulp, R. M. Detweiler und W. A. Anders, Phys. Rev. **131**, 2036 (1963).
[17] M. Balkanski und J. des Cloizeaux, J. Phys. Radium **22**, 41 (1961).
[18] H. Radelt, Z. Naturf. **15a**, 269 (1960).
[19] E. Gutsche, phys. stat. sol. **1**, 716 (1961).
[20] F. L. Pedrotti und D. C. Reynolds, Phys. Rev. **127**, 1584 (1962).
[21] D. Dutton, Phys. Rev. **112**, 785 (1958).
[22] Y. S. Park und D. C. Reynolds, Phys. Rev. **132**, 2450 (1963).
[23] A. Lempicki, Phys. Rev. Letters **2**, 138 (1959).

[24] D. DUTTON, J. Phys. Chem. Solids **6**, 101 (1958).
[25] R. SEIWERT, Ann. Phys. **6**, 241 (1949).
[26] M. BALKANSKI und J. DES CLOIZEAUX, J. Phys. Radium **21**, 825 (1960).
[27] В. В.Соболев, Физ. тверд. тела **6**, 906 (1964).
[28] D. G. THOMAS und J. J. HOPFIELD, Phys. Rev. **122**, 35 (1961).
[29] E. GUTSCHE und H. LANGE, Proc. Internat. Conf. Phys. Semiconductors, Paris 1964, im Druck.
[30] M. Cardona, J. Phys. Chem. Solids **24**, 1543 (1963).
[31] C. E. Bleil und I. Broser, J. Phys. Chem. Solids **25**, 11 (1964).

(Received June 1, 1964)

phys. stat. sol. 6, 521 (1964)

Institute of Physics, Polish Academy of Sciences, Warsaw

Compatibility Relations and Selection Rules for the Dipole Electric Transitions in Crystals with Symmetry O_h^3

By

W. Gorzkowski

This paper is a continuation of [1]. Here the compatibility relations and selection rules for electric dipole transitions in crystals with the symmetry O_h^3 are given.

In der vorliegenden Arbeit, die eine Fortsetzung von [1] darstellt, werden Kompatibilitäts-Relationen und Auswahlregeln für elektrische Dipolübergänge in Kristallen der Symmetrie O_h^3 abgeleitet.

Compatibility Relations

Let P denote a point of high symmetry situated on a certain symmetry line L in the first Brillouin Zone, G_{k_P} — the wave vector group at the point P and G_{k_L} — the wave vector group on the line L. G_{k_L} is a subgroup of G_{k_P}. If we confine the group G_{k_P} to the elements of G_{k_L} then we can decompose the given representation of G_{k_P} as a sum of irreducible representations of G_{k_L}. From the physical point of view, this means that the energy band at point P corresponding to the given representation of G_{k_P} is, in general, split — when we pass from the point P to the general point on the line L — into several ones corresponding to the irreducible representations of G_{k_L}.

The decompositions of the representations connected with symmetry points into the irreducible ones connected with symmetry lines (the compatibility relations) are given in the Tables 1 to 4.

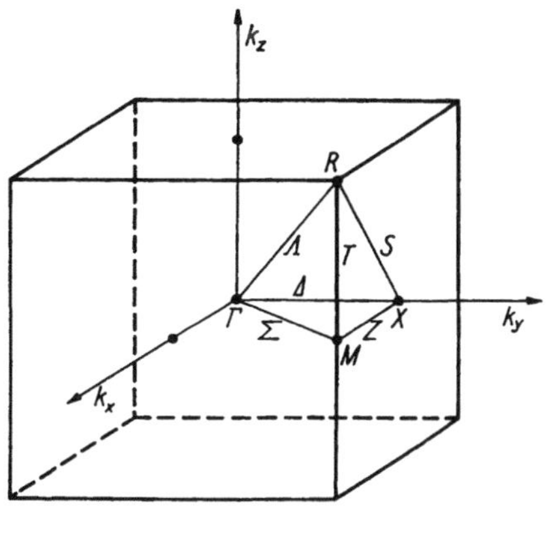

Fig. 1. The first Brillouin Zone for the crystals with symmetry O_h^3 (simple cubic lattice)

Table 1

Compatibility relations at the point $\Gamma(0, 0, 0) \frac{2\pi}{a}$

	Line Λ	Line Δ	Line Σ
Γ_1^+	Λ_1	Δ_1	Σ_1
Γ_1^-	Λ_2	Δ_4	Σ_3
Γ_2^+	Λ_2	Δ_2	Σ_2
Γ_2^-	Λ_1	Δ_3	Σ_4
Γ_3^+	Λ_3	$\Delta_1 + \Delta_2$	$\Sigma_1 + \Sigma_2$
Γ_3^-	Λ_3	$\Delta_3 + \Delta_4$	$\Sigma_3 + \Sigma_4$
Γ_4^+	$\Lambda_2 + \Lambda_3$	$\Delta_4 + \Delta_5$	$\Sigma_2 + \Sigma_3 + \Sigma_4$
Γ_4^-	$\Lambda_1 + \Lambda_3$	$\Delta_1 + \Delta_5$	$\Sigma_1 + \Sigma_2 + \Sigma_4$
Γ_5^+	$\Lambda_1 + \Lambda_3$	$\Delta_3 + \Delta_5$	$\Sigma_1 + \Sigma_3 + \Sigma_4$
Γ_5^-	$\Lambda_2 + \Lambda_3$	$\Delta_2 + \Delta_5$	$\Sigma_1 + \Sigma_2 + \Sigma_3$
Γ_6^+	Λ_4	Δ_6	Σ_5
Γ_6^-	Λ_4	Δ_6	Σ_5
Γ_7^+	Λ_4	Δ_7	Σ_5
Γ_7^-	Λ_4	Δ_7	Σ_5
Γ_8^+	$\Lambda_4 + \Lambda_5 + \Lambda_6$	$\Delta_6 + \Delta_7$	$2\,\Sigma_5$
Γ_8^-	$\Lambda_4 + \Lambda_5 + \Lambda_6$	$\Delta_6 + \Delta_7$	$2\,\Sigma_5$

Table 2

Compatibility relations at the point $R\left(\frac{1}{2}, \frac{1}{2}, \frac{1}{2}\right)\frac{2\pi}{a}$

	Line Λ	Line T	Line S
R_1	Λ_3	T_5	S_1
R_2	Λ_3	T_5	S_1
R_3	$\Lambda_1 + \Lambda_2$	T_5	S_1
R_4	$\Lambda_1 + \Lambda_2 + 2\,\Lambda_3$	$T_1 + T_2 + T_3 + T_4 + T_5$	$3\,S_1$
R_5	$\Lambda_4 + \Lambda_5 + \Lambda_6$	$T_6 + T_7$	$S_2 + S_3 + S_4 + S_5$
R_6	$\Lambda_4 + \Lambda_5 + \Lambda_6$	$T_6 + T_7$	$S_2 + S_3 + S_4 + S_5$
R_7	$2\,\Lambda_4$	$T_6 + T_7$	$S_2 + S_3 + S_4 + S_5$

Table 3

Compatibility relations at the point $X\left(0, \frac{1}{2}, 0\right)\frac{2\pi}{a}$

	Line Δ	Line Z	Line S
X_1	$\Delta_1 + \Delta_2$	$Z_1 + Z_4$	S_1
X_2	$\Delta_3 + \Delta_4$	$Z_2 + Z_3$	S_1
X_3	Δ_5	$Z_2 + Z_4$	S_1
X_4	Δ_5	$Z_1 + Z_3$	S_1
X_5	$\Delta_6 + \Delta_7$	$2\,Z_5$	$S_2 + S_3 + S_4 + S_5$

Table 4

Compatibility relations at the point $M\left(\frac{1}{2}, \frac{1}{2}, 0\right)\frac{2\pi}{a}$

	Line Z	Line Σ	Line T
M_1	Z_1	Σ_2	T_2
M_2	Z_4	Σ_2	T_1
M_3	Z_2	Σ_3	T_2
M_4	Z_3	Σ_3	T_1
M_5	Z_1	Σ_1	T_4
M_6	Z_4	Σ_1	T_3
M_7	Z_2	Σ_4	T_4
M_8	Z_3	Σ_4	T_3
M_9	$Z_1 + Z_4$	$\Sigma_1 + \Sigma_2$	T_5
M_{10}	$Z_2 + Z_3$	$\Sigma_3 + \Sigma_4$	T_5
M_{11}	Z_5	Σ_5	T_7
M_{12}	Z_5	Σ_5	T_6
M_{13}	Z_5	Σ_5	T_6
M_{14}	Z_5	Σ_5	T_7

Selection Rules

The probability amplitude of the dipole electric transitions per unit of time between the states $|\psi_i\rangle$ and $|\psi_f\rangle$ is proportional to the matrix element

$$M_{if} = \langle\psi_i\, |\boldsymbol{n}\cdot \mathrm{grad}|\, \psi_f\rangle\,,$$

where $\boldsymbol{n}$ is a vector parallel to the electric vector $\boldsymbol{E}$ of light [2]. The knowledge of representations, according to which $|\psi_i\rangle$, $|\psi_f\rangle$ and $\boldsymbol{n}\cdot\mathrm{grad}$ transform, allow to determine when M_{if} vanishes i.e. when the transition between the states $|\psi_i\rangle$ and $|\psi_f\rangle$ is forbidden.

The gradient transforms according to the vector representation D which is formed by 3-dimensional matrices α of the point operations. The representation D is, in general, reducible and we can decompose it into the sum of irreducible representations D_s.

Let $\boldsymbol{n}\cdot\mathrm{grad}$ transform according to some irreducible representation D_s included in D. In this case if $\boldsymbol{n}$ is perpendicular to the subspace S_s in which D_s acts, then $\boldsymbol{n}\cdot\mathrm{grad} = 0$ and then $M_{if} = 0$. This means that if $\boldsymbol{n}\cdot\mathrm{grad}$ transforms according to some representation D_s included in D, then the transitions with the vector $\boldsymbol{n}$ perpendicular to the subspace S_s are forbidden.

The integrand in M_{if} transforms according to the product representation $D^{i*}\times D_s\times D^f$, where D^{i*}, D^f, and D_s denote the representations according to which $|\psi_i\rangle$, $|\psi_f\rangle$, and $\boldsymbol{n}\cdot\mathrm{grad}$ respectively transform. If $D^{i*}\times D_s\times D^f$ does not contain the unit representation or equivalently if $D_s\times D^f$ does not contain the representation D^i, then $M_{if} = 0$ and then the transition between the states $|\psi_i\rangle$ and $|\psi_f\rangle$ is forbidden.

The selection rules obtained by the described method are given in the Tables 5 to 14.

Table 5

Selection rules at the point $\Gamma(0, 0, 0) \frac{2\pi}{a}$

Initial state	Final state
Γ_1^+	Γ_4^-
Γ_1^-	Γ_4^+
Γ_2^+	Γ_5^-
Γ_2^-	Γ_5^+
Γ_3^+	Γ_4^-, Γ_5^-
Γ_3^-	Γ_4^+, Γ_5^+
Γ_4^+	$\Gamma_1^-, \Gamma_3^-, \Gamma_4^-, \Gamma_5^-$
Γ_4^-	$\Gamma_1^+, \Gamma_3^+, \Gamma_4^+, \Gamma_5^+$
Γ_5^+	$\Gamma_2^-, \Gamma_3^-, \Gamma_4^-, \Gamma_5^-$
Γ_5^-	$\Gamma_2^+, \Gamma_3^+, \Gamma_4^+, \Gamma_5^+$
Γ_6^+	Γ_6^-, Γ_8^-
Γ_6^-	Γ_6^+, Γ_8^+
Γ_7^+	Γ_7^-, Γ_8^-
Γ_7^-	Γ_7^+, Γ_8^+
Γ_8^+	$\Gamma_6^-, \Gamma_7^-, \Gamma_8^-$
Γ_8^-	$\Gamma_6^+, \Gamma_7^+, \Gamma_8^+$

Table 6

Selection rules at the point $R\left(\frac{1}{2}, \frac{1}{2}, \frac{1}{2}\right) \frac{2\pi}{a}$

Initial state	Final state
R_1	R_4
R_2	R_4
R_3	R_4
R_4	R_1, R_2, R_3, R_4
R_5	R_5, R_6, R_7
R_6	R_5, R_6, R_7
R_7	R_5, R_6, R_7

Table 7

Selection rules at the point $X\left(0, \frac{1}{2}, 0\right)\frac{2\pi}{a}$

Initial state	Final state	
	α	β
X_1	X_3, X_4	X_1
X_2	X_3, X_4	X_2
X_3	X_1, X_2	X_4
X_4	X_1, X_2	X_3
X_5	X_5	X_5

α: $\boldsymbol{E} \perp (0, 1, 0)$; β: $\boldsymbol{E} \parallel (0, 1, 0)$

Table 8

Selection rules at the point $M\left(\frac{1}{2}, \frac{1}{2}, 0\right)\frac{2\pi}{a}$

Initial state	Final state	
	α	β
M_1	M_9	M_3
M_2	M_9	M_4
M_3	M_{10}	M_1
M_4	M_{10}	M_2
M_5	M_9	M_7
M_6	M_9	M_8
M_7	M_{10}	M_5
M_8	M_{10}	M_6
M_9	M_1, M_2, M_5, M_6	M_{10}
M_{10}	M_3, M_4, M_7, M_8	M_9
M_{11}	M_{13}, M_{14}	M_{14}
M_{12}	M_{13}, M_{14}	M_{13}
M_{13}	M_{11}, M_{12}	M_{12}
M_{14}	M_{11}, M_{12}	M_{11}

α: $\boldsymbol{E} \perp (0, 0, 1)$; β: $\boldsymbol{E} \parallel (0, 0, 1)$

Table 9

Selection rules on the line $\Delta(0, \eta, 0)\frac{2\pi}{a}$

Initial state	Final state	
	α	β
Δ_1	Δ_1	Δ_5
Δ_2	Δ_2	Δ_5
Δ_3	Δ_3	Δ_5
Δ_4	Δ_4	Δ_5
Δ_5	Δ_5	$\Delta_1, \Delta_2, \Delta_3, \Delta_4$
Δ_6	Δ_6	Δ_6, Δ_7
Δ_7	Δ_7	Δ_6, Δ_7

α: $\boldsymbol{E} \parallel (0, 1, 0)$; β: $\boldsymbol{E} \perp (0, 1, 0)$

Table 10

Selection rules on the line $Z\left(\xi, \frac{1}{2}, 0\right)\frac{2\pi}{a}$

Initial state	Final state		
	α	β	γ
Z_1	Z_1	Z_4	Z_2
Z_2	Z_2	Z_3	Z_1
Z_3	Z_3	Z_2	Z_4
Z_4	Z_4	Z_1	Z_3
Z_5	Z_5	Z_5	Z_5

α: $\boldsymbol{E} \parallel (1, 0, 0)$; β: $\boldsymbol{E} \parallel (0, 1, 0)$; γ: $\boldsymbol{E} \parallel (0, 0, 1)$

Table 11

Selection rules on the line $\Lambda(\xi, \xi, \xi)\frac{2\pi}{a}$

Initial state	Final state	
	α	β
Λ_1	Λ_1	Λ_3
Λ_2	Λ_2	Λ_3
Λ_3	Λ_3	$\Lambda_1, \Lambda_2, \Lambda_3$
Λ_4	Λ_4	$\Lambda_4, \Lambda_5, \Lambda_6$
Λ_5	Λ_5	Λ_4
Λ_6	Λ_6	Λ_4

α: $\boldsymbol{E} \parallel (1, 1, 1)$; β: $\boldsymbol{E} \perp (1, 1, 1)$

Table 12

Selection rules on the line $T\left(\frac{1}{2}, \frac{1}{2}, \zeta\right)\frac{2\pi}{a}$

Initial state	Final state	
	α	β
T_1	T_1	T_5
T_2	T_2	T_5
T_3	T_3	T_5
T_4	T_4	T_5
T_5	T_5	T_1, T_2, T_3, T_4
T_6	T_6	T_6, T_7
T_7	T_7	T_6, T_7

α: $\boldsymbol{E} \parallel (0, 0, 1)$; β: $\boldsymbol{E} \perp (0, 0, 1)$

Table 13

Selection rules on the line $S\left(\xi, \frac{1}{2}, \xi\right)\frac{2\pi}{a}$

Initial state	Final state		
	α	β	γ
S_1	S_1	S_1	S_1
S_2	S_3	S_2	S_5
S_3	S_2	S_3	S_4
S_4	S_5	S_4	S_3
S_5	S_4	S_5	S_2

α: $\boldsymbol{E} \parallel (0, 1, 0)$; β: $\boldsymbol{E} \parallel (1, 0, 1)$; γ: $\boldsymbol{E} \mid (1, 0, -1)$

Table 14

Selection rules on the line $\Sigma(\xi, \xi, 0)\frac{2\pi}{a}$

Initial state	Final state		
	α	β	γ
Σ_1	Σ_4	Σ_1	Σ_2
Σ_2	Σ_3	Σ_2	Σ_1
Σ_3	Σ_2	Σ_3	Σ_4
Σ_4	Σ_1	Σ_4	Σ_3
Σ_5	Σ_5	Σ_5	Σ_5

α: $\boldsymbol{E} \parallel (0, 0, 1)$; β: $\boldsymbol{E} \parallel (1, 1, 0)$; γ: $\boldsymbol{E} \parallel (1, -1, 0)$

Appendix

In the Table 15 are given the errors which have been found in the paper [1].

Table 15
Errors in the paper [1]

Table	Representation	Group element	Is printed	Must be
5	13 and 14	*38*	D_{14}	$-D_{14}$
12	5 and 6	*3*	1	-1
6	6	*5* (matrix element D_{44})	$-i\,\omega$	$i\,\omega$
1	—	*15*	D_{14}	D_{15}
1	—	*24*	$(d_2\|t)$, D_2	$(d_{24}\|t)$, D_{24}
3, 6	(below the table)		$\overline{12} = 5^2$	$12 = 5^2$
12	(in the title of table)		Δ	Λ

Acknowledgements

The author wishes to thank Prof. M. SUFFCZYŃSKI for valuable discussions and remarks. Also the author wishes to thank Mgr. K. OLBRYCHSKI for critical remarks.

References

[1] W. GORZKOWSKÍ, phys. stat. sol. **3**, 910 (1963).
[2] G. LUBARSKI, The group theory and its application in physics (in Russian), Moscow 1958.

(Received June 2, 1964)

phys. stat. sol. 6, 529 (1964)

Institut für angewandte Physik der Reinststoffe Dresden
der Deutschen Akademie der Wissenschaften zu Berlin

Eisenverteilung und Substruktur in zellular gewachsenen Aluminium-Einkristallen[1])

Von

J. Barthel, J. Kunze und R. Scharfenberg

Die Verteilung von Eisen und die Substruktur wird in durch konstitutionelle Unterkühlung hexagonal-zellular gewachsenen Aluminium-Einkristallen untersucht. In Verbindung mit einer Abschätzung der Zellgrabentiefe und der Eisenanreicherung in ihnen [8] wird nachgewiesen, daß die beobachteten eisenreichen Phasen eutektischen Ursprungs sind. Ihr alleiniges Auftreten an Zusammenstoßstellen dreier Zellen weist darauf hin, daß die Zellgrabensohlen an diesen Stellen durch fortgesetzte konstitutionelle Unterkühlung in das Kristallinnere abbiegen, so daß sich Schmelzröhren bilden. In der Nähe der Dekantierungsfläche entlang der gesamten Zellgrenze gefundene Ausscheidungen sind kein Hinweis dafür, daß das Eisen während des Wachstums zunächst an den Zellgrenzen stark angereichert wird. Sie werden auf das Erstarren der Restschmelze in den Zellgräben nach dem Dekantieren zurückgeführt. Die vorgeschlagene Zellenform erklärt ohne Widerspruch zu der Vorstellung von Tiller [4], warum die Zellen keine Substruktur darstellen. Die Substruktur ist weitmaschiger als die Zellstruktur. Die Subgrenzen folgen Zellgrenzen und beeinflussen die Eisenverteilung.

An investigation is made of the substructure and the distribution of iron in Al single crystals, hexagonal-cellular grown by constitutional supercooling. By estimating the depth of the cell ditchs, and the iron accumulation in them, it is shown that the observed iron-rich phases are of eutectic origin. The unique occurrence of these phases at points where three cells meet shows that at these points the bottoms of the cell ditchs turn, by continuous constitutional supercooling, into the inner part of the crystal, so that melting channels form. Segregations occur near decanting faces along the whole cell boundary. This does not indicate that during growth, iron-enrichment occurs initially at the cell boundaries. These segregations are believed to be due to the freezing of the residual melt in the cell ditches after decanting. The proposed cell form suggests, without disagreement with the ideas of Tiller, why the cells do not give rise to substructure. The substructure has a wider mesh than the cell structure. The subboundaries follow cell boundaries and influence the iron distribution.

1. Einleitung

Wächst ein Metallkristall aus seiner Schmelze, so kann die Erstarrungsfläche durch konstitutionelle Unterkühlung, deren Auftreten durch die Erstarrungsbedingungen und den Fremdstoffgehalt bestimmt wird, statt eben zellular oder dendritisch sein (z. B. [1—3]). Die Konzentration der Verunreinigungen sollte dann an den Zell- oder Dendritengrenzen je nach dem Verteilungskoeffizienten erhöht oder erniedrigt sein. Durch die Konzentrationsgradienten werden bei Unterschieden in den Atomradien des Grundmaterials und des Fremdstoffes an

---

[1]) Auszugsweise vorgetragen auf der Tagung *Festkörperphysik* in Warnemünde, 21. bis 24. Oktober 1963.

den Zellgrenzen Spannungen erzeugt, die zur Versetzungsentstehung und damit zu einer mit den Zellen zusammenfallenden Substruktur führen können [4].

Systematische Untersuchungen der Verteilung der Verunreinigungen und der Substruktur in hexagonal-zellular erstarrten Aluminiumkristallen [5] und der Substruktur in mit Cadmium verunreinigten Zinkkristallen [6] mit langgestreckten Zellen ergaben eine Anreicherung der Fremdelemente, deren Verteilungskoeffizient kleiner als eins war, an den Zellgrenzen wie auch eine mit den Zellen identische Substruktur nur in der Nähe der Dekantierungsfläche. In größerem Abstand wurde eine Konzentrationserhöhung nur noch an Zusammenstoßstellen dreier Zellen, den Tripelpunkten sowie den Grenzen einer weitmaschigeren Makromosaikstruktur festgestellt [5]. Auch die Substruktur änderte sich in Richtung weitmaschigerer Netze [5, 6].

Die Umwandlungen wurden auf eine Homogenisation der Verunreinigungen durch Diffusion im Festkörper, wodurch auch die Versetzungen und Substrukturen freigesetzt und beweglich werden, zurückgeführt.

Die autoradiographisch getroffene Beobachtung in zellular erstarrtem Aluminium, daß sich die Konzentration von Eisen (Verteilungskoeffizient 0,03) an den Tripelpunkten mit zunehmendem Abstand von der Dekantierungsfläche erhöht [7], widersprach dieser Deutung. Es wurde angenommen, daß das Eisen im Festkörper entlang der Zellgrenzen zu Ausscheidungen an Tripelpunkten diffundiert. Bei diesen Deutungen wurde vorausgesetzt, daß die in der Nähe der Dekantierungsfläche gefundene Anreicherung der Verunreinigungen an den Zellgrenzen der ursprünglichen Verteilung beim Erstarren entspricht und durch die größere Abkühlungsgeschwindigkeit nach dem Dekantieren erhalten geblieben ist.

Theoretische Abschätzungen von Kramer et al. [8] ergaben für Zinn sehr tiefe Zellgräben, so daß vermutet werden kann, daß die Zellgrabentiefe den Bereich, in dem Segregationen und Subgrenzen an Zellgrenzen auftreten, übersteigt. Beim Dekantieren verbleibt in den Gräben Restschmelze [9]. Sie erstarrt erst nachträglich unter geänderten Wachstumsbedingungen, die zu abweichenden Fremdstoffverteilungen und Substrukturen führen und somit die Änderungen in der Nähe der Dekantierungsfläche ebenfalls erklären können. Die Strukturen, Segregationen und Ausscheidungen an der Dekantierungsfläche könnten in diesem Falle nur bedingt zur Deutung der Entstehungsmechanismen der in größerer Entfernung vorliegenden herangezogen werden.

Eine erweiterte Untersuchung der Fremdstoffverteilung und der Substrukturen in der Übergangszone zellular erstarrter Kristalle sollte darüber Aufschluß geben, welchen Einfluß die Diffusion im Festkörper einerseits und die nachträgliche Erstarrung der Restschmelze in den Zellgräben andererseits auf die Umwandlung der Strukturen haben. Die Untersuchungen wurden an $\langle 100 \rangle$-orientierten, hexagonal-zellular erstarrten Aluminium-Einkristallen durchgeführt, die mit $1 \cdot 10^{-2}\%$ radioaktivem Eisen verunreinigt und unter definierten Wachstumsbedingungen, die es gestatteten, die Zellgrabentiefe und die Eisenanreicherung nach Kramer et al. abzuschätzen, gezüchtet waren.

2. Experimentelle Methoden

2.1 *Kristallherstellung und Probenvorbereitung*

Aus zonengeschmolzenem Aluminium mit einer Reinheit von 99,9995% hinsichtlich der Hauptverunreinigungen Fe, Cu, Si, Mg und Zn wurden im Vakuum Legierungen mit $1 \cdot 10^{-2}\%$ radioaktivem Eisen erschmolzen und zu Stäben vergossen. Ihre Homogenität wurde durch Strahlungsmessung mit einem Stabmeßgerät geprüft. Das Ziehen von ⟨100⟩-orientierten Einkristallen aus diesem Material vermittels angeschmolzener Keime und das Dekantieren zur Freilegung der Erstarrungsfläche erfolgte in einer bereits beschriebenen [7], der von ELBAUM und CHALMERS [10] analogen Apparatur. Die Wachstumsbedingungen waren so gewählt [3], daß eine hexagonal zellulare Erstarrungsfront entstand. Zur Untersuchung der Eisenverteilung und der Substruktur wurden von den Kristallen je ein ⟨100⟩-orientierter Längsschliff sowie Querschliffe von der Dekantierungsfläche beginnend, in 100, 250 μm Tiefe, weiterhin in je 250 μm, ab 2000 μm bis 5000 μm in je 500 μm Abstand sowie in größerer Entfernung von der Dekantierungsfläche hergestellt.

2.2 Methoden zur Untersuchung der Eisenverteilung

2.2.1 Autoradiographie

Die Eisenverteilung wurde autoradiographisch mit K 105 Abziehfilm (Filmfabrik Wolfen) untersucht. Reaktionen zwischen Film und Probe ließen sich durch Substrieren mit 0,75 μm dicken PVC-Folien vermeiden.

2.2.2 Elektrolytisches Polieren

Auf den mit einem Perchlorsäure-Alkohol-Gemisch (1:4) polierten Schliffflächen konnten bei sonst völlig ebener Oberfläche halbkugelförmige Löcher mit maximal 10 μm Durchmesser beobachtet werden. Sie waren auf den Längsschliffen wie die Schwärzungszentren der Autoradiographien in Reihen angeordnet (Fig. 1). Eine Zuordnung der Polierlöcher und der Schwärzungszentren gelang nur bei linienförmigen, schräg zu ihrer Achse angeschnittenen Eisenanreicherungen (Fig. 2). Der Vergleich des Polierbildes und der metallographischen Ätzung einer eutektischen Aluminium-Eisen-Legierung (1,8% Fe) ergab, daß die eisenreiche Phase beim Polieren herausgelöst wird. Auf Grund dieser Feststellungen wurden die Löcher in den zellular erstarrten Kristallen auf herauspolierte eisenreiche Phasen zurückgeführt und die Poliertechnik zur Bestimmung von Eisenausscheidungen herangezogen. Infolge ihres höheren Auflösungsvermögens ergänzte sie die Autoradiographie wesentlich.

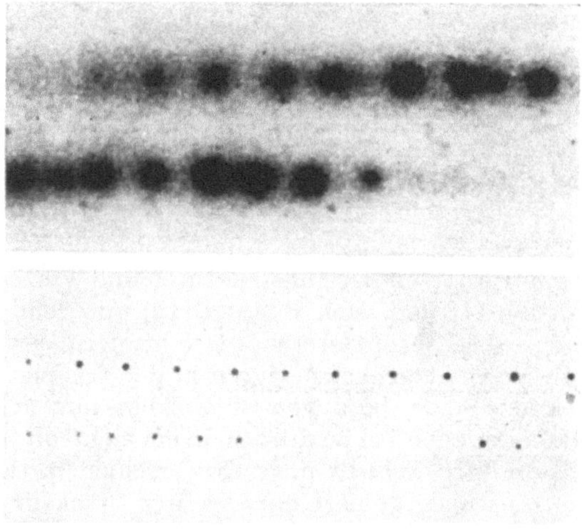

Fig. 1. Autoradiographie und Mikrophotographie polierter Längsschliffe. Die Punktreihen verlaufen in Erstarrungsrichtung. 90 ×

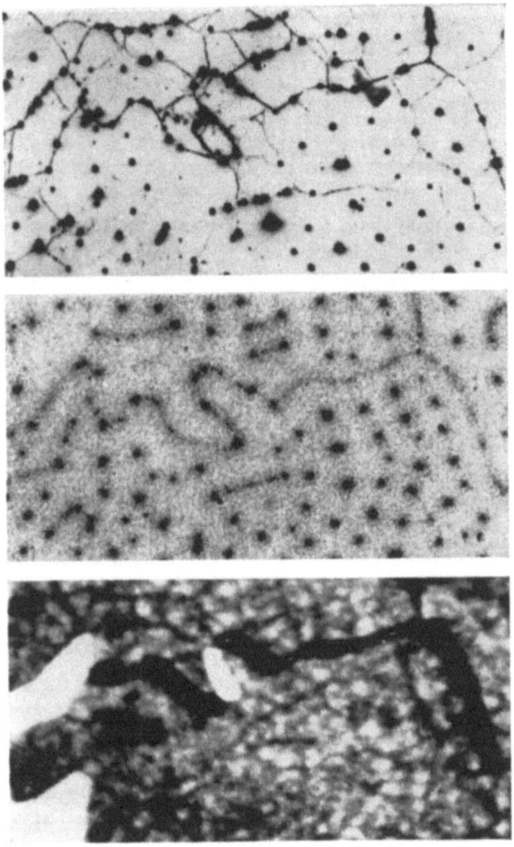

Fig. 2. Mikrophotographie, Autoradiographie und Berg-Barrett-Aufnahme eines Querschliffes in 1000 μm Abstand von der Dekantierungsfläche, leicht angeätzt. 35 ×

Biloni [5] berichtete von Löchern, die beim chemischen Ätzen an Tripelpunkten entstanden und führte sie auf eine große Versetzungsdichte zurück. Auf Polierflächen zellular erstarrter Aluminiumkristalle fanden Doherty und Davis [11] ebenfalls Löcher und brachten sie mit der Entstehung von Poren beim Wachstum in Verbindung. Da einerseits nicht angegeben ist, ob die Löcher bereits im Polierbild vorhanden waren und andererseits die Verunreinigungskonzentrationen in den Untersuchungsmaterialien unterschiedlich und nicht näher bekannt sind, ist ein unmittelbarer Vergleich der Erscheinungen und der Deutungen nicht möglich.

2.3 Methoden zur Untersuchung der Substruktur

2.3.1 Berg-Barrett-Methode

Mit einer von Wadewitz [12] modifizierten Berg-Barrett-Methode [13] konnten Substrukturen mit Verwinklungen zwischen einer Minute und einigen Grad abgebildet und bei Verwendung eines Drahtgitters die gegenseitige Verkippung der Subkörner vermessen werden. Um ungestörte Aufnahmen zu erhalten, mußte nach dem Schleifen der Probe eine etwa 0,6 mm dicke Schicht elektrolytisch abpoliert werden. Dabei wurde sorgfältig auf die Einhaltung einer ebenen Fläche geachtet, da sich Oberflächenunebenheiten ebenfalls abbildeten.

2.3.2 Chemisches Ätzen

Zum Anätzen der Substrukturen und Versetzungen wurden die Lösungen nach Wyon [14] und nach Barber [15] mit dem gleichen Ergebnis angewandt. Das Barbersche Ätzmittel erwies sich insofern als vorteilhafter, als es nicht zur Bildung von Makroätzgrübchen durch in der Lösung enthaltene Spurenverunreinigungen neigte. Beide Lösungen entwickeln nur dekorierte Versetzungen, so daß die Ätzbilder auch bei unvollständigem elektrolytischen Abpolieren der vom Schleifen beeinflußten Schicht ungestört erschienen. Das chemische Ätzbild stimmte mit der röntgenographisch entwickelten Struktur überein (Fig. 2).

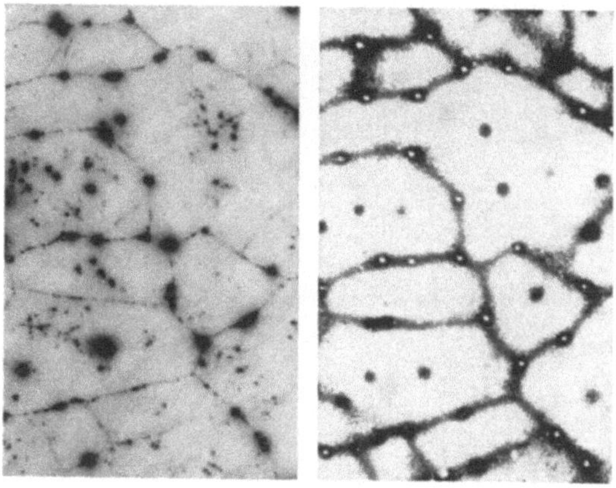

Fig. 3. Chemisches und thermisches Ätzbild desselben Querschliffes. Die Löcher der herauspolierten eisenreichen Phasen sind im thermischen Ätzbild von dunklen, grübchenfreien Säumen innerhalb der hellen, pitreichen Gebiete umgeben. Grübchenfreie Gebiete entstehen auch um dicht unterhalb der Oberfläche liegende, noch nicht herauspolierte Ausscheidungen. 128 ×

2.3.3 Thermisches Ätzen

Auch das thermische Ätzen nach Doherty und Davis [16] entwickelte grundsätzlich dieselbe Struktur (Fig. 3). Bei wiederholtem Ätzen und Zwischenpolieren ergab sich jedoch eine Veränderung der Substruktur in Richtung grobmaschiger Netzwerke. Eine analoge Veränderung der Substruktur von Aluminium beim Glühen konnte schon Wyon [17] beobachten. Die thermische Methode wurde deshalb nur in besonderen Fällen eingesetzt. Sie erwies sich analog der Berg-Barrett-Methode als sehr empfindlich gegenüber vom Schleifen zurückbleibenden Störungen.

3. Ergebnisse

Die Ergebnisse der Untersuchungen bezüglich der Eisenverteilung und der Substruktur stimmten für alle hexagonal-zellular erstarrten Kristalle qualitativ überein. Die quantitativen Angaben beziehen sich auf einen Aluminium-Einkristall, der mit einer Erstarrungsgeschwindigkeit von 14 cm/h bei einem Temperaturgradienten in der Schmelze von 6 °C/cm gezogen wurde.

3.1 Zellulares Wachstum und Eisenverteilung

3.1.1 Beobachtungen an der Oberfläche

Die hexagonalen Zellen auf der Dekantierungsfläche hatten einen Durchmesser von 160 μm, die Zellgräben waren 10 μm tief und wiesen besondere Vertiefungen an den Zusammenstoßstellen dreier Zellen, den Tripelpunkten, auf. Auf der Mantelfläche wurden als Zellgrenzen ausgeprägte Furchen parallel zur Wachstumsrichtung beobachtet (Fig. 4). Sie liefen in etwa 1 mm Abstand von der Dekantierungsfläche in flache Gräben, die eigentlichen *corrugations* [1] aus. Einige der ausgeprägten Furchen blieben jedoch auch in größerer Entfernung noch sichtbar. Der Vergleich mit Berg-Barrett-Aufnahmen ergab, daß in diesen Fällen die Zellgrenzen gleichzeitig Grenzen zwischen Subkörnern mit einer gegenseitigen Desorientierung von mehr als 0,1° waren.

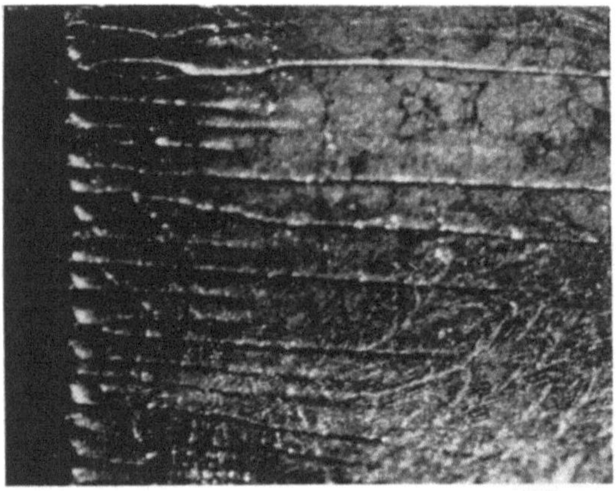

Fig. 4. Mikrophotographie der Mantelfläche eines zellular erstarrten Kristalls. Die Kante der Dekantierungsfläche ist links im Bilde sichtbar. 27×

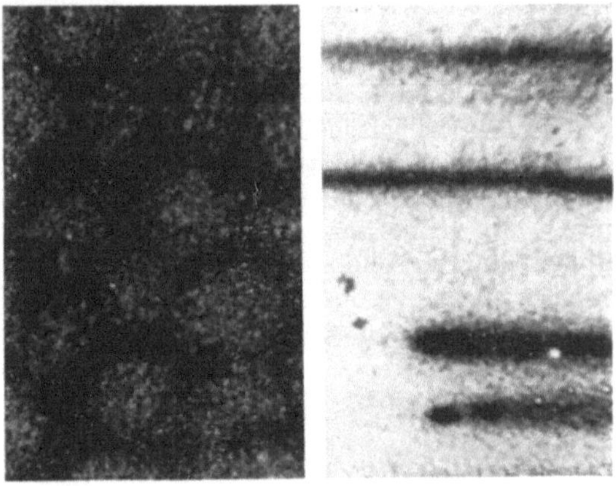

Fig. 5. Autoradiographien der Eisenverteilung auf einem Quer- und einem Längsschliff im Bereich zwischen 0 und 500 μm Abstand von der Dekantierungsfläche. 90×

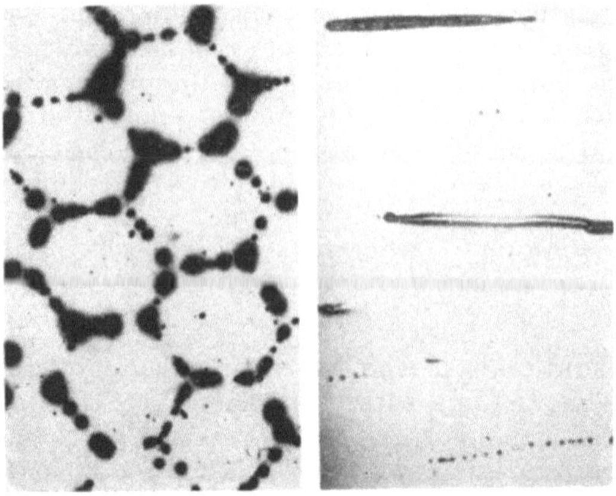

Fig. 6. Mikrophotographie polierter Quer- und Längsschliffe im Bereich zwischen 0 und 500 μm Abstand von der Dekantierungsfläche. 120×

3.1.2 Eisenverteilung

Die Autoradiographien der Querschliffe zeigten bis zu etwa 500 μm Abstand von der Dekantierungsfläche wie in [7] eine Eisenanreicherung an den Zellgrenzen. Die Schwärzung und damit die Eisenkonzentration an Tripelpunkten nahm mit dem Abstand von der Dekantierungsfläche zu. Auf den Längsschliffen wurde eine linienhafte Schwärzung in Erstarrungsrichtung sichtbar (Fig. 5). Die Mikrophotographien der polierten Quer- und Längsschliffe (Fig. 6) wiesen jedoch nach, daß die Eisenanreicherung in diesem Bereich aus diskreten Ausscheidungen eisenreicher Phasen in Form von in Erstarrungsrichtung angeordneten Röhren oder Punktreihen mit engem Punktabstand besteht. Die einzelnen Ausscheidungen konnten infolge ihres kleinen Abstandes autoradiographisch nicht getrennt werden.

Ab etwa 500 µm Abstand von der Dekantierungsfläche konnte eine erhöhte Schwärzung wiederum nur an Tripelpunkten und in Übereinstimmung mit Biloni [5] an den Zellgrenzen nachgewiesen werden, die einem Vergleich mit Ätzbildern und Berg-Barrett-Aufnahmen zufolge gleichzeitig Grenzen zwischen Subkörnern mit einer gegenseitigen Desorientierung von mehr als 0,1° waren (Fig. 2). Die Anzahl der so markierten Zellgrenzen nahm mit zunehmender Tiefe bei gleichbleibender Substruktur ab. Die Ausscheidungen verliefen röhrenförmig in Erstarrungsrichtung. Die Anzahl der mit Ausscheidungen versehenen Tripelpunkte in der Flächeneinheit stimmte mit der der Zellen überein. Da zu jeder hexagonalen Zelle zwei Tripelpunkte zu rechnen sind (Fig. 8), traten Ausscheidungen nur an jedem zweiten Tripelpunkt auf. Der Vergleich der von Querschliffen in verschiedenem Abstand von der Dekantierungsfläche hergestellten Autoradiographien ergab, daß die Ausscheidungen stets an denselben Tripelpunkten vorhanden waren.

Dieselben Verhältnisse wurden bei allen untersuchten Kristallen mit hexagonalzellularen Erstarrungsflächen unabhängig von den speziellen Wachstumsbedingungen und damit unterschiedlichen Zellendurchmessern gefunden.

Ab etwa 3500 µm Abstand von der Dekantierungsfläche waren auf den Autoradiographien neben jedem zweiten Tripelpunkt nur noch Zellgrenzen, die in unmittelbarer Nähe von Subgrenzen mit Verwinklungen um und über 1° lagen, durch Ausscheidungen entlang der Grenzen oder auch nur dadurch, daß die eisenreichen Phasen entlang dieser nahezu an jedem Tripelpunkt auftraten, ausgezeichnet (Fig. 7). Sie besaßen im allgemeinen einen ausscheidungsfreien Saum von der Breite eines Zellendurchmessers. In gleicher Tiefe nahmen die Ausscheidungen an Tripellinien ihre endgültige Anordnung in Form von in Erstarrungsrichtung verlaufenden Punktreihen mit einem Punktabstand von 90 µm an (Fig. 1).

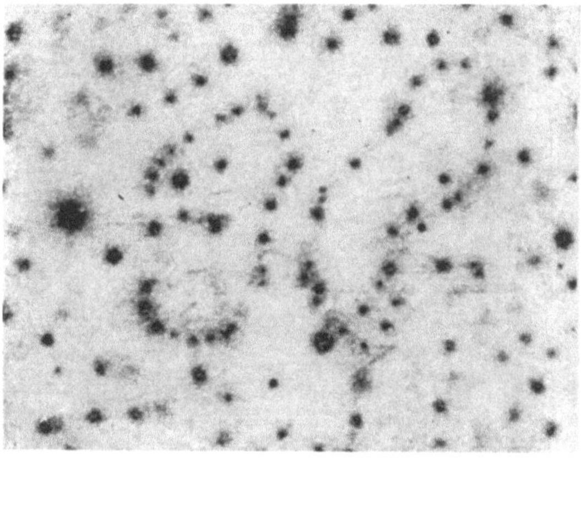

Fig. 7. Autoradiographie eines Querschliffes in 4000 µm Abstand von der Dekantierungsfläche. 35×

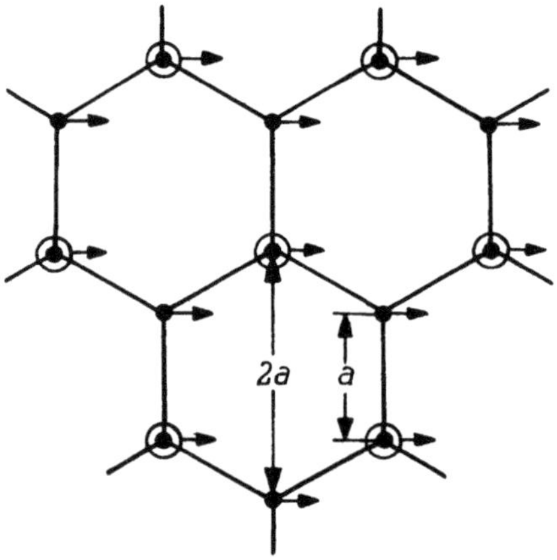

Fig. 8. Schematische Darstellung der hexagonalen Zellen. Die Pfeile zeigen die Zugehörigkeit der Tripelpunkte zu den einzelnen Zellen, die Kreise Tripelpunkte mit eutektischen Phasen an

Die Autoradiographien zeigten, daß die Schwärzung der Punkte innerhalb jeder Reihe in einer Richtung kontinuierlich abnahm. Dies wird erklärt durch ein leicht schräges Anschneiden der Punktreihe und die somit in einer Richtung zunehmende Tiefe der Ausscheidungen unterhalb des Schliffes, die zu einer erhöhten Strahlenabsorption führt. In der entgegengesetzten Richtung brach die Reihe der Schwärzungszentren plötzlich ab, da die Ausscheidungen beim Erreichen der Oberfläche herauspoliert wurden. Es blieb ein Streifen geringerer Schwärzungserhöhung bestehen, der darauf schließen ließ, daß die eisenreichen Phasen von einem Gebiet mit gegenüber dem Zellinneren erhöhter Eisenkonzentration umgeben waren. Diese Erscheinungen änderten sich auch mit zunehmendem Abstand von der Dekantierungsfläche nicht mehr.

3.2 Substruktur

3.2.1 Makromosaikstruktur

Die Strukturuntersuchungen ergaben eine in ⟨100⟩, der Erstarrungsrichtung, gewachsene Faserstruktur (Fig. 9). Die langgestreckten Subkörner waren sowohl um die Faser- als auch um dazu senkrechte Richtungen gegeneinander um Winkel verdreht, die im allgemeinen bei 0,1° lagen, jedoch auch vereinzelt Werte

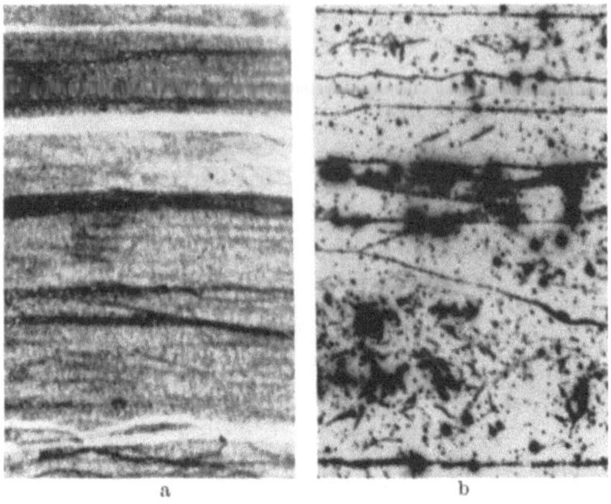

Fig. 9. a) Chemisch geätzter Längsschliff. 68 ×
b) Berg-Barrett-Aufnahme eines Längsschliffes in der Nähe der Dekantierungsfläche. 12 ×

Fig. 10. Mikrophotographie eines chemisch geätzten Querschliffes. 170 ×

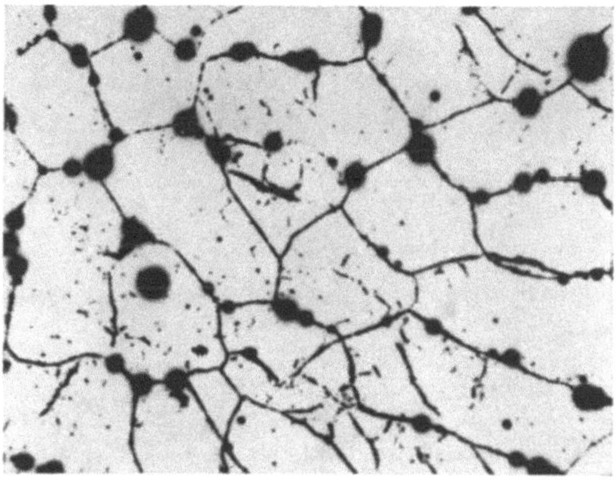

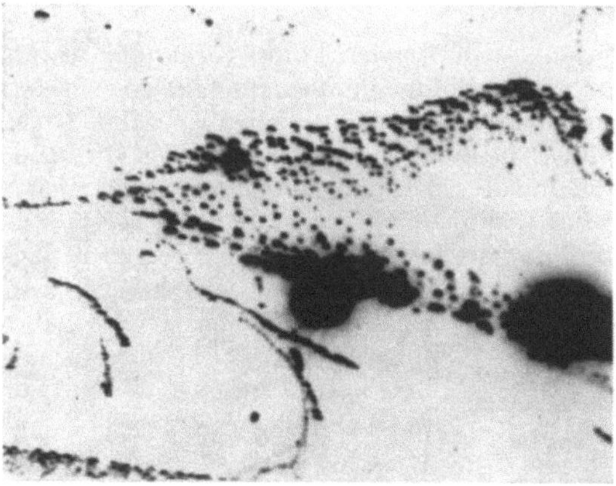

Fig. 11. Ausschnitt aus einem chemisch geätzten Längsschliff. 425 ×

bis zu 3° erreichten, mit statistisch verteiltem Drehsinn. Die Struktur entspricht damit der von CHALMERS et al. (z. B. [18]) für verschiedene Metalle beschriebenen *Striationstruktur*. Der Durchmesser der Subkörner schwankte zwischen dem der Zellen und etwa 1 mm. Sie umfaßten also in Übereinstimmung mit BILONI sowohl einzelne Zellen als ganze Zellbereiche. Aus der gestreckten Gestalt der Subgrenzen im Querschliff (Fig. 10) war zu entnehmen, daß sie nicht mit Zellgrenzen übereinstimmten. Im Längsschliff zeigten einige der Grenzen charakteristische Stufen, andere bestanden aus breiten Ätzgrubenbändern (Fig. 11), wobei diese besonders bei von ⟨100⟩ abweichenden Grenzen und an Verzweigungen auftraten.

3.2.2 *Mikrostruktur*

Innerhalb der *Striations* entwickelte das chemische Ätzen eine Mikrostruktur (Fig. 9 und 10) mit einer besonderen Häufung von Subgrenzen bzw. Versetzungen um die Eisenausscheidungen, die nicht mit Striationgrenzen zusammenfielen (Fig. 12). Diese Gebiete erhöhter Ätzgrubendichte wurden beim thermischen Ätzen neben den Makromosaikgrenzen durch grübchenfreie Bereiche sichtbar, und zwar auch dann, wenn die Ausscheidungen unterhalb der Probenoberfläche lagen, also nicht herauspoliert waren (Fig. 3). Sie wurden bereits von DOHERTY und DAVIS [11] auf zellular erstarrten Aluminiumkristallen gefunden und mit der Kondensation der Leerstellen an Poren in Verbindung gebracht. Ob in den in

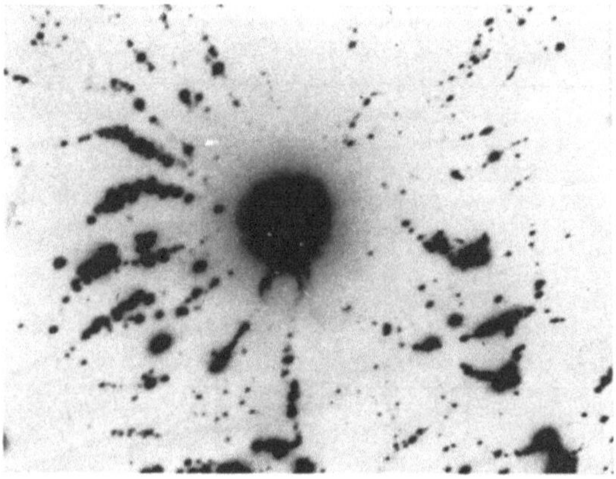

Fig. 12. Sternförmig angeordnete Ätzgrubenreihen um eine herauspolierte Ausscheidung auf einem chemisch geätzten Querschliff. 425 ×

der vorliegenden Untersuchung verwendeten Kristallen Poren vorhanden waren, konnte nicht geklärt werden. Die eindeutig gefundenen Zusammenhänge zwischen Polierlöchern und Eisenanreicherungen sowie zwischen den Ätzgrubenanhäufungen um die Ausscheidungen und den Gebieten, die keine thermisch erzeugten Grübchen aufwiesen, ließen jedoch den Schluß zu, daß die Spannungshöfe um die eisenreichen Phasen die Ursache für das Auftreten der grübchenfreien Gebiete darstellten. Für diese Auffassung sprach weiterhin, daß sie in einigen Fällen, besonders nach der mit dem thermischen Ätzen verbundenen Glühbehandlung, auf Berg-Barrett-Aufnahmen durch erhöhte Extinktion sichtbar wurden.

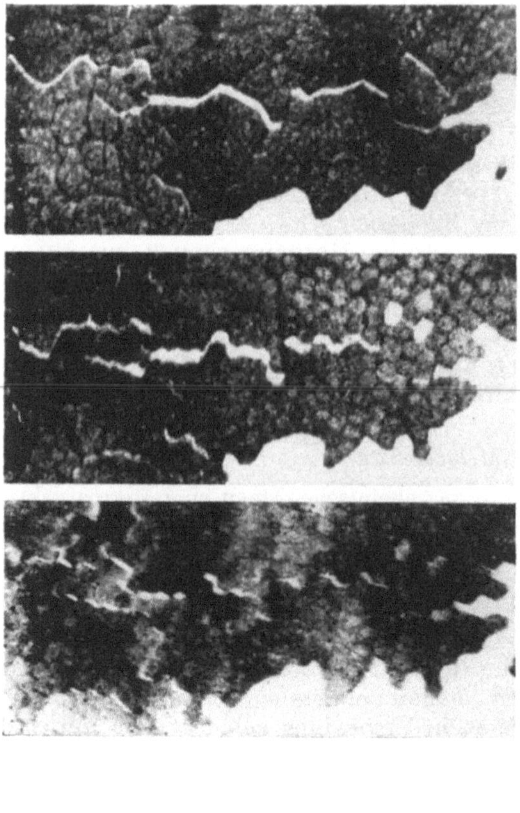

3.2.3 Änderungen in der Nähe der Dekantierungsfläche

Weder auf Längs- noch auf Querschliffen konnten mit Annäherung an die Dekantierungsfläche Änderungen der Substruktur hinsichtlich der Ausdehnung und der Verwinklung der Subkörner festgestellt werden. Die zunächst gestreckt verlaufenden Subgrenzen näherten sich lediglich den in den Berg-Barrett-Auf-

Fig. 13. Identische Ausschnitte aus Berg-Barrett-Aufnahmen von Querschliffen in 2000, 250 und 100 μm Abstand von der Dekantierungsfläche. Mit abnehmendem Abstand werden die Zellgrenzen durch Extinktionseffekte deutlich sichtbar. Die Subgrenzen zwischen Körnern mit einer kleinen gegenseitigen Verwinklung stimmen bereits in 250 μm Abstand, die Grenze im unteren Teil der Abbildungen, bei der das zweite Korn infolge einer Orientierungsabweichung von 2° in dem Ausschnitt nicht mehr sichtbar wird, erst in 100 μm Abstand mit den Zellgrenzen überein. 26 ×

nahmen durch Extinktionseffekte sichtbar werdenden Zellgrenzen (Fig. 13), wobei Grenzen zwischen stark desorientierten Subkörnern erst in unmittelbarer Nähe der Dekantierungsfläche mit Zellgrenzen zusammenfielen.

3.3 Das Verhalten von Korngrenzen

An zufällig entstandenen Bikristallen wurden in der Nähe der gestreckt erscheinenden Korngrenzen autoradiographisch kontinuierliche, gezackte Schwärzungslinien, die offenbar Zellgrenzen folgten, festgestellt. Die Korngrenzen stimmten bereits auf der Dekantierungsfläche nicht vollständig mit dem Verlauf von Zellgrenzen überein, sondern gingen unter Einnahme der kürzesten Länge durch Zellen hindurch.

4. Diskussion

4.1 Zellulares Wachstum und Eisenverteilung

4.1.1 Entstehung der Ausscheidungen

Die Abschätzung von Kramer et al. [8] ergibt für den vorliegenden Fall eine Tiefe der Zellgräben von 1 cm und eine Eisenanreicherung in der Schmelze auf 1,7 Gew.%. Die Berechnung ist sowohl hinsichtlich der theoretischen Ansätze als auch hinsichtlich der Materialkonstanten und Temperaturgradienten nur eine grobe Näherung. Sie zeigt jedoch, daß die Zellgrabentiefe den Bereich, in dem eine Eisenanreicherung an den Zellgrenzen beobachtet wurde, wesentlich übersteigt. Die Eisenanreicherungen weisen demnach nur darauf hin, daß in diesem Bereich vor dem Dekantieren in den Zellgräben eine an Eisen angereicherte Schmelze vorhanden war. Sie sind kein Hinweis dafür, daß das Eisen zunächst entlang der Zellgrenzen mit stark erhöhter Konzentration oder als eutektische Phase in den Kristall eingebaut wurde und erst im Verlaufe der mit fortschreitender Erstarrung erfolgenden Abkühlung des Kristalles zu Tripelpunkten diffundierte [7].

Die mit 1,7% berechnete Eisenanreicherung in den Sohlen der Zellgräben liegt in unmittelbarer Nähe der eutektischen Zusammensetzung (1,8% nach [19]). Es kann also im Gegensatz zu der in [7] vorgeschlagenen Deutung angenommen werden, daß die in größerer Tiefe beobachteten eisenreichen Phasen direkt durch eutektische Erstarrung während des normalen Wachstumsprozesses entstehen, wobei die sie umgebenden Bereiche mit einer gegenüber dem Zellinneren erhöhten Eisenkonzentration dem eutektischen Aluminiummischkristall mit 0,05 Gew.% Fe entsprechen.

4.1.2 Zellenform

Bei dieser Deutung bleiben zunächst die Fragen offen, warum die eutektischen Phasen

1. bis zu 500 μm Abstand von der Dekantierungsfläche entlang der Zellgrenzen und in größerer Entfernung nur an jedem zweiten Tripelpunkt auftreten,
2. an Subgrenzen jedoch auch in größerem Abstand gefunden werden,
3. in etwa 3500 μm Abstand von der Röhrenform in die von Punktreihen übergehen?

Die Annahme, daß die Schmelze beim Wachstum vor dem Dekantieren nur an Tripelpunkten die eutektische Konzentration erreicht, führt zwangsläufig zu dem

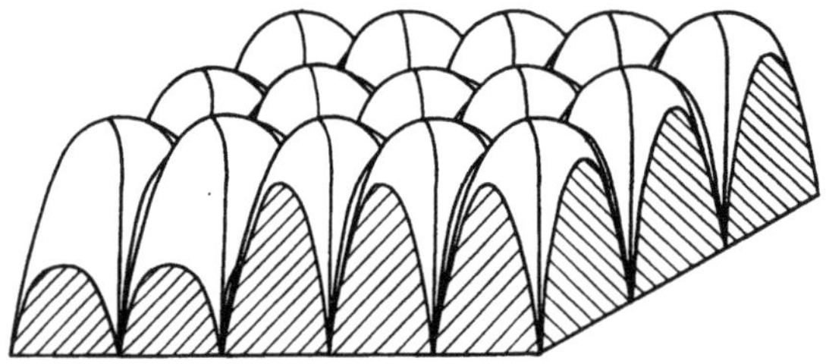

Fig. 14. Schematische Darstellung eines Teiles der zellularen Erstarrungsfläche. Die Schnittflächen sind schraffiert. Vereinfachend sind statt hexagonaler quadratische Zellen gezeichnet. Die zwei Zellenreihen links im Bild werden von einer Subgrenze begrenzt

Schluß, daß die Zellgrabensohle dort in die Tiefe abbiegt (Fig. 14), da die Liquidustemperatur konzentrationsabhängig ist. Die Ursache für die gebogene Sohle wäre wieder in der konstitutionellen Unterkühlung zu suchen, die zur zellularen Erstarrung führt, durch dieses Wachstum jedoch nicht aufgehoben wird [20] und sich in den Zellgräben fortsetzen kann. Das Eisen wird in dem gebogenen Teil der Zellgrenzen nur nach Maßgabe des Verteilungskoeffizienten in den Kristall eingebaut, es diffundiert in der Schmelze zu tiefen Tripelpunkten, so daß hier die eutektische Konzentration erreicht wird.

Die gebogenen Zellgrabensohlen werden durch dieselben Ursachen und Mechanismen wie die Zellen selbst als gebogene Erstarrungsfläche erzeugt. Infolgedessen sind auch dieselben Parameter für den Durchmesser der Bögen und damit den Abstand der ausgezeichneten Tripelpunkte wie für den Zellendurchmesser [1] verantwortlich. Daraus erklärt sich die Beobachtung, daß bei allen hexagonalzellularen Kristallen unabhängig von den Erstarrungsbedingungen jeder zweite Tripelpunkt eine eutektische Phase aufwies, der Abstand zweier mit eutektischen Phasen versehener Tripelpunkte also mit dem Zellendurchmesser übereinstimmte (Fig. 8). Die eisenreichen Phasen treten in Form von Punktreihen entsprechend dem Verhältnis von Mischkristall zur eisenreichen Phase auf.

Mit dieser Deutung lassen sich auch zwanglos die anderen Fragen klären, wenn man annimmt, daß die Sohle der Zellgräben im oberen Bogen etwa 500 μm und an den ausgezeichneten Tripellinien etwa 3500 μm hinter der Dekantierungsfläche liegt. Nach dem Dekantieren erstarrt infolge des erniedrigten Temperaturgradienten die Restschmelze von den Zellwänden ausgehend mit einer ständigen Erhöhung der Konzentration bis zur eutektischen Zusammensetzung und zur Bildung eisenreicher Phasen, die entsprechend der ursprünglichen Erstarrungsfront bis zu 500 μm an den Zellgrenzen und bis 3500 μm röhrenförmig an Tripellinien auftreten. Die Auswahl der Tripellinien wird durch die Verhältnisse vor dem Dekantieren vorbestimmt.

Die Abschätzung von Kramer et al. läßt sich auch auf das vorgeschlagene Modell anwenden. Das Verhältnis von Grabenfläche zur Zellfläche entspricht hier dem des Querschnitts der Schmelzsäulen an Tripelpunkten zur Zellfläche. Es ist klein gegenüber dem Verteilungskoeffizienten und dadurch ohne Bedeutung für die Rechnung. Diese ergibt ohne Berücksichtigung der Existenz eines Eutektikums eine Tiefe der Schmelzsäule an den ausgezeichneten Tripellinien von 10 cm und eine Eisenkonzentration von 9% in der Schmelze. Da das Eutektikum nicht überschritten werden kann, ist es möglich, aus dem Temperaturgradienten G_s im Kristall — der mit dem in der Schmelzsäule gleichgesetzt wird —, der Konzentrationsabhängigkeit der Liquidustemperatur m_L, der eutektischen Konzentration C_{Eu} und der Konzentration vor der Zellspitze C_0 den Abstand d von der Erstarrungsfläche zu berechnen, in dem die eutektische Temperatur erreicht wird, und d

gleich der Tiefe der Schmelzsäule zu setzen:

$$d = \frac{m_L (C_{Eu} - C_0)}{G_s}.$$

Mit $m_L = 2{,}8$ °C/%, $C_{Eu} = 1{,}8\%$ Fe, $C_0 = C_0\,(\text{solid})/k \approx 10^{-3}/0{,}03 = 3 \cdot 10^{-2}\%$ Fe und $G_s = 5$ °C/cm ergibt sich $d = 1$ cm. Damit haben die berechnete und die aus der Eisenverteilung bestimmte Tiefe der Schmelzröhren die gleiche Größenordnung. Die noch vorhandene Differenz kann darin begründet sein, daß der Temperaturgradient in der Nähe der Erstarrungsfläche größer als der in einem gewissen Abstand von ihr gemessene ist. Sie kann anderseits auch durch ein dem Dekantieren kurzzeitig nachfolgendes Erstarren unter nicht oder nur geringfügig geänderten Temperaturverhältnissen mit einer normalen Eisenverteilung vorgetäuscht werden.

Die auf der Mantelfläche in der Nähe der Dekantierungskante an Zellgrenzen gefundenen ausgeprägten Furchen sind entweder durch das hier erleichterte Dekantieren der Schmelze aus den Zellgräben oder durch den Volumenschwund beim Erstarren nach dem Dekantieren entstanden [8]. Ihre Länge von 1000 μm entspricht ihrer Mittelstellung zwischen Zellgräben und Tripellinien.

4.1.3 Eisenverteilung und Substrukturen

Der Verlauf der Subgrenzen in der Nähe der Dekantierungsfläche weist darauf hin, daß sie während des Wachstums mit Zellgrenzen übereinstimmen. In derartigen Zellgrenzen versucht die Grenzflächenenergie die Krümmung der Grabensohle zu reduzieren. Die Zellgräben werden also tiefer, so daß die in ihnen auch unterhalb von etwa 500 μm entlang der gesamten Zellgrenze beobachteten Ausscheidungen ebenfalls auf die Erstarrung der Restschmelze zurückgeführt werden können. Die Tiefe nimmt mit der Orientierungsabweichung der Subkörner und damit der Grenzflächenenergie zu. Da Korngrenzen in beliebigem Abstand von der Dekantierungsfläche homogen geschwärzt erscheinen, muß angenommen werden, daß die Zellgrabensohle in diesen Fällen eben ist.

Die größere Tiefe der mit Subgrenzen zusammenfallenden Zellgrenzen begünstigt die Diffusion zu und die Anreicherung in den Tripelpunkten dieser Grenzen zu Lasten der sie kreuzenden Zellgrenzen. Dies erklärt, daß speziell bei großen Orientierungsunterschieden alle Tripelpunkte derartiger Zellgrenzen eutektische Phasen aufweisen. Da im Mittel nur jeder zweite Tripelpunkt eutektisch erstarrt, ergibt sich ein ausscheidungsfreier Saum von der Breite eines Zellendurchmessers.

Die größere Tiefe der mit Subgrenzen übereinstimmenden Zellgrenzen äußert sich auch darin, daß die dazugehörigen ausgeprägten Furchen auf der Mantelfläche länger sind.

Die durchgeführten Untersuchungen lassen noch keinen Entscheid darüber zu, ob die vorgeschlagene Zellenform für das hexagonal-zellulare Wachstum charakteristisch ist. Diese Frage kann nur durch systematische Variation der Erstarrungsparameter und der Fremdstoffe hinsichtlich ihrer Schmelzdiagramme mit dem Grundmaterial geklärt werden. Die metallographisch von Biloni und Bolling [21] festgestellte Verteilung von Blei in zellular erstarrtem Zinn gestattet jedoch dieselbe Deutung.

Eigene, nicht systematisch durchgeführte Untersuchungen an Aluminium-Eisen (10^{-2}—10^{-3}%) zeigten, daß beim Übergang zu langgestreckten Zellen zunächst weiterhin relativ weit voneinander entfernte, diskrete eisenreiche Phasen auftraten, in der Nähe des Überganges zu ebenen Grenzflächen waren sie dagegen längs der gesamten Zellgrenze eng beieinander anzutreffen.

Obwohl die Änderungen der Eisenverteilung in der Nähe der Dekantierungsfläche mit Erstarrungsvorgängen begründet werden konnten, beeinflußt die Diffusion und die Ausscheidung im festen Zustand möglicherweise die endgültige Verteilung des zunächst echt gelösten Eisens längs der gesamten Zellgrenze. Mit den angewandten Untersuchungsmethoden war diese Beeinflussung nicht feststellbar.

4.2 *Substruktur*

Die Subgrenzen der Makromosaikstruktur sind nur insofern mit Zellgrenzen verbunden, als sie während der Kristallisation in der Tiefe der Zellgräben verlaufen. Sie trennen sich dann beim weiteren Wachstum des Kristalles durch Wanderung von der ehemaligen Lage der Zellgräben [22], richten sich aus und nehmen damit energetisch begünstigte Formen an. Nach dem Abschleudern der Schmelze wachsen die Subgrenzen zugleich mit den erstarrenden Zellgräben zur Oberfläche des Kristalls. Infolge der schnelleren Abkühlung des Kristalles oder auch infolge Behinderung durch Ausscheidungen entlang der Zellgräben wird die Wanderung erschwert oder unterdrückt, so daß Subgrenzen zwischen Körnern mit geringer Orientierungsdifferenz in diesem Bereich den Zellgrenzen noch folgen.

Weder im Inneren des Kristalles noch in unmittelbarer Nähe der Dekantierungsfläche stellten die Zellen Substrukturen dar. Die zellulare Erstarrung führt also trotz der unterschiedlichen Atomradien von Aluminium und Eisen in Übereinstimmung mit Doherty und Davis [11] nicht in jedem Falle zur Bildung von Substrukturen. Diese Beobachtung ist kein Widerspruch zu der Vorstellung von Tiller [4], da wesentliche Konzentrationsunterschiede nur an Tripelpunkten auftreten. Die Eisenverteilung und die daraus abgeleitete Vorstellung über die Zellenform erklärt also den Befund der Strukturuntersuchung, der andererseits als Bestätigung der Modellvorstellung angesehen werden kann.

Unterschiedliche thermische Ausdehnung der Tripelpunkten benachbarten Phasen sowie örtlich begrenzte Abweichungen in der Leerstellenkonzentration, infolge der Diffusion des Eisens in der Nähe der Ausscheidungen hervorgerufen, können während der Abkühlung des Kristalls zur Bildung und Vervielfachung von Versetzungen führen [23]. Sowohl die Konzentrationsgradienten im Mischkristallgebiet als auch die Ausscheidungen selbst sind also als Ursache der Anhäufung von Versetzungsätzgruben an Tripelpunkten anzusehen.

Schlußbemerkung

Die Untersuchung hat gezeigt, daß die Zellgrabentiefe die Stärke des Bereiches übersteigen kann, in dem sich die Strukturen und Fremdstoffverteilungen mit dem Abstand von der Dekantierungsfläche verändern. Diese Änderungen können deshalb nicht ohne weiteres als Spiegelbild von Umwandlungsvorgängen nach dem Erstarren betrachtet werden.

Weiterhin bestätigt sie das Ergebnis von Kramer et al. [8] sowie Biloni und Bolling [21], daß die Anreicherung der Verunreinigungen in den Zellgräben oder Röhren an Tripelpunkten zur eutektischen Erstarrung führen kann. Die eutektischen Phasen sind in unmittelbarer Nähe der Dekantierungsfläche längs der Zellgrenzen fein verteilt und von Versetzungsätzgrübchen oft schwer zu unterscheiden. Bei Strukturuntersuchungen mit chemischen Ätzmethoden [5, 6] ist deshalb zu prüfen, ob die Ätzgruben Versetzungen oder feinverteilten eutektischen Phasen zuzuordnen sind.

Wir danken Herrn Prof. Dr. E. Rexer für die Förderung der Arbeit durch Anregungen und Diskussionen. Die röntgenographischen Arbeiten wurden von Herrn Dipl.-Phys. H. Wadewitz und Herrn Dipl.-Phys. W. Reitzenstein durchgeführt, denen wir dafür und für wichtige Anregungen zu Dank verpflichtet sind.

Literatur

[1] C. Elbaum, Progress in Metal Physics 8, Pergamon Press 1959 (S. 203).
[2] W. C. Winegard, Met. Rev. **6**, 57 (1961).
[3] D. T. I. Hurle, Progress in Materials Science **10**, 81 (1962).
[4] W. A. Tiller, J. appl. Phys. **29**, 611 (1958).
[5] H. Biloni, Canad. J. Phys. **39**, 1501 (1961).
[6] V. V. Damiano und G. S. Tint, Acta metall. **9**, 177 (1961).
[7] J. Barthel und R. Scharfenberg, Ausscheidungsvorgänge in Legierungen, Akademie-Verlag Berlin 1964 (S. 222).
[8] J. J. Kramer, G. F. Bolling und W. A. Tiller, Trans. Amer. Inst. Min. Met. Engrs. **227**, 374 (1963).
[9] F. Weinberg, Trans. Amer. Inst. Min. Met. Engrs. **224**, 628 (1962).
[10] C. Elbaum und B. Chalmers, Canad. J. Phys. **33**, 196 (1955).
[11] P. E. Doherty und R. S. Davis, Trans. Amer. Inst. Min. Met. Engrs. **221**, 737 (1961).
[12] H. Wadewitz, Berichte zur 3. Tagung über Kristallwachstum, Moskau 1963 (im Druck).
[13] C. S. Barrett, Trans. Amer. Inst. Min. Met. Engrs. **161**, 15 (1945).
[14] G. Wyon, I. M. Marchin und P. Lacombe, Mem. Sci. Rev. metall. **56**, 549 (1959).
[15] D. J. Barber, Phil. Mag. **7**, 1925 (1962).
[16] P. E. Doherty und R. S. Davis, Acta metall. **7**, 118 (1959).
[17] G. Wyon, I. M. Marchin und P. Lacombe, C. R. Acad. Sci. **252**, 1467 (1961).
[18] P. E. Doherty und B. Chalmers, Trans. Amer. Inst. Min. Met. Engrs. **224**, 1124 (1962).
[19] M. Hansen, Constitution of Binary Alloys, McGraw-Hill Book Company Inc. 1958 (S. 90).
[20] W. A. Tiller, Canad. J. Phys. **37**, 1204 (1959).
[21] H. Biloni und G. F. Bolling, Trans. Amer. Inst. Min. Met. Engrs. **227**, 1351 (1963).
[22] M. D. Hunt und R. W. Smith, Canad. J. Phys. **40**, 1245 (1962).
[23] C. Elbaum, Proceedings of the Intern. Conf. on Crystal Lattice Defects, J. Phys. Soc. Japan **18**, 59 (Suppl. III).

(Received June 3, 1964)

phys. stat. sol. 6, 545 (1964)

Banaras Hindu University, Department of Physics

Equation of State of LiF

By

M. P. VERMA and B. DAYAL

The equation of state of LiF is studied by the method discussed in the authors' earlier paper on sodium chloride. The agreement between the calculated and the experimental values for thermal expansion becomes poor at high temperatures. Various possibilities for improving this agreement are discussed.

Die Zustandsgleichung von LiF wurde mit Hilfe einer Methode untersucht, die bereits in unserer früheren Arbeit am NaCl diskutiert wurde. Die Übereinstimmung zwischen der berechneten und der experimentell bestimmten thermischen Ausdehnung wird bei hohen Temperaturen schlecht. Es werden verschiedene Möglichkeiten diskutiert, um die Übereinstimmung zu verbessern.

1. Introduction

Amongst the alkali halides LiF is probably the most interesting substance from the point of view of lattice dynamics. Both the ions Li^+ and F^- have small mass and size and their polarisabilities have the smallest values amongst this class of compounds. The Debye characteristic temperature is very large resulting in a relatively large value for its zero point energy and zero point pressure. Different specimens of the salt show gross variations in its physical properties as reported by BALLARD et al. [1]. GOTT [2] has found as high as ten per cent difference in the lattice and macroscopic expansions even at a moderate temperature of about 200 °C.

The first calculations of the specific heats and thermal expansion of this substance were made by DAYAL [3]. He considered Raman's theory of crystal vibrations in which the vibration spectrum of a crystal consists of the frequencies corresponding to a division of the first Brillouin zone in eight equal parts. He made use of the Kellermann model [4] but included the contribution of the next nearest neighbours also to the non Coulomb repulsive interaction. The zero point energy was ignored and a good agreement between the theoretical and experimental specific heats and thermal expansion was obtained. KARO [5] used KELLERMANN's method in its entirety and found a very large difference between the theoretical and the experimental specific heat data.

Since the zero point energy in case of LiF is large it becomes necessary that it should be included in the theoretical calculations. The method of successive refinement used by us [6] enables this to be done in a convenient way. In the present paper we have calculated the equation of state of LiF by this method and the theoretical results have been compared with the experimental data. The method followed and the symbols used are the same as in the above paper which hereinafter will be referred to as I.

2. Results of Calculations and Discussion

As shown in I, the repulsive parameters A_0 and B_0 at the absolute zero are given by the following expression

$$\frac{\varepsilon^2}{4\,r_0^4}\left(\frac{2\,\alpha}{3} + B_0\right) + \frac{h}{4\,N\,r_0^3}\sum_i\sum_{\boldsymbol{k}}\left(\frac{\partial\nu(i,\boldsymbol{k})}{\partial g}\right)_0 = 0\,, \tag{1}$$

and

$$\frac{\varepsilon^2}{36\,r_0^4}[-\,8\,\alpha + 3\,A_0 - 6\,B_0] + \frac{h}{4\,N\,r_0^3}\sum_i\sum_{\boldsymbol{k}}\left(\frac{\partial^2\nu(i,\boldsymbol{k})}{\partial g^2}\right)_{00} = \frac{1}{\beta_0}\,, \tag{2}$$

where ε is the electronic charge, r the nearest neighbour distance, α the Madelung constant, $\nu(i, \boldsymbol{k})$ the frequency of the ith branch of the lattice vibrations with the wave vector $\boldsymbol{k}$, $g = \frac{v}{v_0} = \left(\frac{r}{r_0}\right)^3$, and β is the isothermal compressibility. The suffix $_0$ indicates the values of the quantities at 0 °K. The last terms in (1) and (2) are the contributions of the zero point energy. In I these were neglected in the first stage of approximations to get the preliminary values A_{01}, B_{01}, of the repulsive parameters. In the present case, the zero point energy is large. We, therefore, estimated the value of the last term in (1) by the Debye theory of specific heats as a first approximation. From this one readily obtains

$$\frac{h}{4\,N\,r_0^3}\sum_i\sum_{\boldsymbol{k}}\left(\frac{\partial\nu(i,\boldsymbol{k})}{\partial g}\right)_0 = \frac{9}{8}\,\frac{\gamma_0\,k\,\theta_{D_0}}{r_0^3}\,,$$

where γ_0 is the average Gruneisen parameter, k the Boltzmann constant, and θ_D the Debye characteristic temperature. The frequency dependent term in equation (2) could not be determined exactly and as a first approximation it was assumed that

$$\sum_i\sum_{\boldsymbol{k}}\left(\frac{\partial^2\nu(i,\boldsymbol{k})}{\partial g^2}\right)_{00} = \left|\sum_i\sum_{\boldsymbol{k}}\left(\frac{\partial\nu(i,\boldsymbol{k})}{\partial g}\right)_0\right|\,. \tag{3}$$

This was done because these were found to be of the same order in I. In this way we determined the first approximate values A_{01} and B_{01}. γ_0 and θ_{D_0} were taken to be 1.60 and 740 °K respectively as recorded in YATES and PANTER [7]. The values A_{01} and B_{01} were then used to calculate the frequency spectra as in I. In the present case, however, the first Brillouin zone was divided into one thousand equal parts and the dynamical matrices solved for the resulting 48 representative vectors on the TIFRAC electronic digital computer of the Tata Institute of Fundamental Research, Bombay, for nine volumes ($g = 1.00, 1.02, 1.04, \ldots, 1.16$).

It is interesting to point out here that $\sum_i\sum_{\boldsymbol{k}}\left(\frac{\partial\nu(i,\boldsymbol{k})}{\partial g}\right)_0$ averaged over the 1000 points of the Brillouin zone comes out to be almost the same as the average of the 125 points obtained in the coarser division. This shows that the coarser mesh of points adopted in I is capable of giving reliable results. This conclusion is also supported by the work of ARASE et al. [8] who find that the degree of fineness of the mesh does not have any significant effect on the frequency distribution.

The parameters A_0 and B_0 were calculated by the method of iteration described in I and their final values are

$$A_0 = 8.025\,, \qquad B_0 = -\,1.122\,.$$

Fig. 1. Thermal Expansion of LiF
——— Authors' theoretical curve; Experimental points are from: ○ — YATES and PANTER [7], △ — PATHAK et al. [9], ● — EUCKEN and DANNÖHL [10]

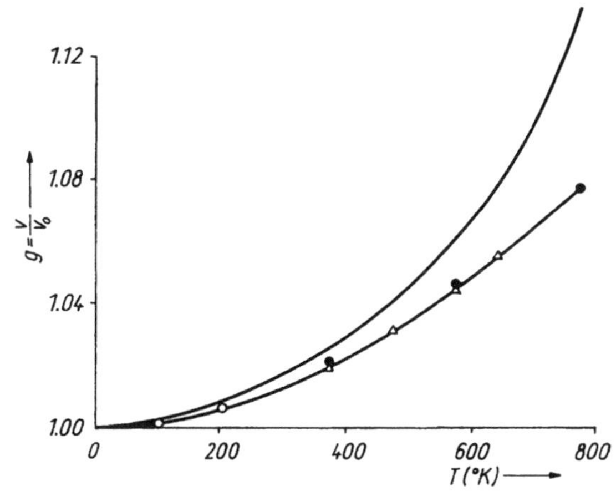

Isotherms were then plotted and the total expansion as read along the zero pressure axis is shown in Fig. 1. Experimental points have been taken from YATES and PANTER [7], PATHAK et al. [9], and EUCKEN and DANNÖHL [10]. Isothermal compressibilities β_T are plotted in Fig. 2 and compared with the values determined from the measurements of adiabatic compressibilities β_s by BRISCOE and SQUIRE [11]. Calculations of specific heats do not show any significant improvement over KARO's [5] results and so these are not being given here.

The agreement obviously is poor. The coefficient of expansion becomes infinite at about 800 °K which according to HERZFELD and MAYERS' theory [12] should be the melting point of the solid as against the experimental value of 1140 °K. This is not in accordance with our findings in the case of NaCl. A revision of calculations for NaCl revealed an arithmetical error and correction of the same brings down the melting point of this substance also to about 800 °K. The use of a slightly lower value of β_0 in the estimation of the parameter A is expected to lead to some improvement. In this case it is interesting to note that measurements of BERGMANN [13] give the room temperature value of the adiabatic compressibility of LiF about 5% lower than that of BRISCOE and SQUIRE [11] at the same temperature.

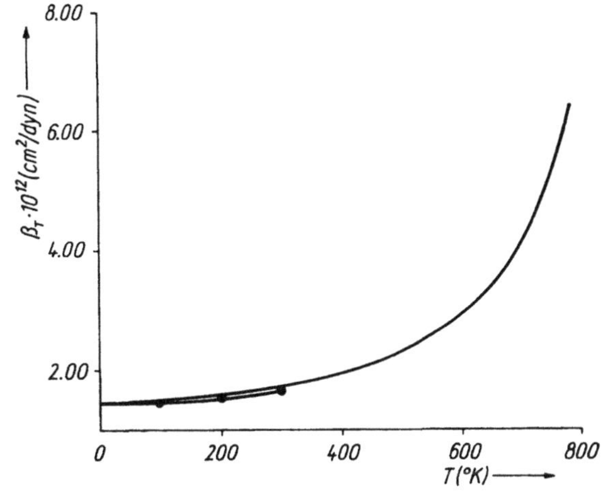

Fig. 2. Isothermal compressibility of LiF
——— Authors' theoretical curve; Experimental points are from: ● — BRISCOE and SQUIRE [11]

Since, however, the latter authors have taken the measurements at low temperatures going up to 4.2 °K their value has been selected for β_0 in our calculations.

The situation is, however, still not satisfactory because as discussed in I infinite expansion coefficient should actually appear at a much higher temperature than the melting point. It appears that under isothermal conditions the total pressure of the lattice falls rather slowly with distance towards the higher temperatures. TRIPATHI [14] has suggested that perhaps the constant ϱ appearing in the exponential repulsive potential

$$v(r) = b\,e^{-r/\varrho} \tag{4}$$

is a function of r. A suitable modification of (4) is the following

$$v(r) = b\,e^{-r^n/\varrho} \tag{5}$$

with three adjustable constants which is under examination.

Other ways by which one may expect to improve the agreements are (i) by including the effect of the next nearest neighbours also in the repulsive interaction, (ii) by introducing three body forces such as described by LUNDQVIST [15], and (iii) by considerations of anharmonicity. The first one does not seem to be very promising but the other two possibilities must be tried.

The authors are thankful to Dr. D. Y. PHADAKE and SHRI K. S. KANE of the Computer Section, T.I.F.R., Bombay, for making the facilities of their computer available for the present work.

References

[1] S. S. BALLARD, L. S. COMBES, and K. A. MCCARTHY, J. Opt. Soc. Amer. **41**, 772 (1951).
[2] A. GOTT, Ann. Phys. (Germany) **41**, 520 (1942).
[3] B. DAYAL, Proc. Indian Acad. Sci. **A 20**, 138 (1944).
[4] E. W. KELLERMANN, Phil. Trans. **A 238**, 513 (1940).
[5] A. M. KARO, J. chem. Phys. **31**, 1489 (1959).
[6] M. P. VERMA and B. DAYAL, phys. stat. sol. **3**, 901 (1963).
[7] B. YATES and C. H. PANTER, Proc. Phys. Soc. **80**, 373 (1962).
[8] E. M. ARASE, R. D. HATCHER, and J. NEUBERGER, J. chem. Phys. **39**, 353 (1963).
[9] P. D. PATHAK, N. V. PANDYA, and M. P. GHADIALI, Indian J. Phys. **46**, 293 (1963).
[10] A. EUCKEN and W. DANNÖHL, Z. Electrochem. **40**, 814 (1934).
[11] C. V. BRISCOE and C. F. SQUIRE, Phys. Rev. **106**, 1175 (1957).
[12] K. F. HERZFELD and M. G. MAYER, Phys. Rev. **46**, 995 (1934).
[13] L. BERGMANN, Der Ultraschall, S. Hirzel Verlag 1954.
[14] B. B. TRIPATHI, Indian J. pure appl. Phys. **1**, 278 (1963).
[15] S. O. LUNDQVIST, Ark. Fys. (Schweden) **9**, 435 (1955).

(Received June 4, 1964)

phys. stat. sol. 6, 549 (1964)

Institute of Physics, Polish Academy of Sciences, Warsaw

Determination of the Effective Mass in n-Type InSb by Means of Magneto-Plasma Reflection

By

L. ŚNIADOWER, J. RAUŁUSZKIEWICZ, and R. R. GAŁĄZKA

The refractive index for intraband transitions in degenerate semiconductors in the presence of a magnetic field is calculated for spherical and nonparabolical energy bands. The results are used for the description of plasma and magneto-plasma reflection, and for the determination of the effective mass of carriers in heavily doped InSb. Plasma and magneto-plasma reflections are observed in n-type InSb with carrier concentrations from 10^{18} to 10^{19} cm^{-3}. The values of the effective mass of electrons confirm Kane's band-structure up to concentrations exceeding 10^{19} cm^{-3}.

Der Brechungsindex n für Zwischenbandübergänge wird für entartete Halbleiter mit kugelförmigen und nicht-parabolischen Energiebändern bei Vorhandensein eines äußeren Magnetfeldes berechnet. Das Ergebnis wird zur Beschreibung von Plasma- und Magneto-Plasmareflektionen sowie zur Ermittlung von effektiven Massen in stark dotiertem InSb verwendet. Die Plasma- und Magneto-Plasmareflektionen sind in n-leitendem InSb mit Trägerkonzentrationen von 10^{18} bis 10^{19} cm^{-3} beobachtet worden. Die Werte für die effektiven Massen der Elektronen bestätigen die von KANE gefundene Struktur auch für Konzentrationen $> 10^{19}$ cm^{-3}.

1. Introduction

The following paper presents the analysis of the influence of the magnetic field on optical constants of semiconductors and reports some results concerning the determination of the effective mass in InSb.

The observations of the plasma edge in variety of semiconductors were first successfully applied for the measurements of effective masses by SPITZER and FAN [1]. Their results obtained for InSb were of particular interest for they enabled the determination of the variation of the effective mass with impurity concentration by observing the shift of the plasma edge for different values of electron concentration. This technique has been succesfuly used for other semiconductors as well. Its extention due to LAX and WRIGHT [2] for the case when a magnetic field is present has made the magnetoplasma effect a particulary valuable tool.

2. Theory

The reflectivity of the electromagnetic wave normal to the boundary between the vacuum and a conducting medium is expressed by

$$R = \frac{(n-1)^2 + k^2}{(n+1)^2 + k^2}\,, \tag{1}$$

where n is the refractive index and k the extinction coefficient. The parameters n and k are characteristic for a given medium and depend on the circular frequency of the electromagnetic wave ω and also on the external magnetic field H.

We will calculate the refractive index following the method described by Kołodziejczak [3], Kołodziejczak and Żukotyński [4, 5]. Although this approach is good for semiconductors with nonspherical and nonparabolic energy bands, we will limit for practical reasons our interest to the case of spherical energy bands and the final results will be given for strongly degenerate carriers.

For time dependent electric field we introduced a complex relaxation time [4]

$$\tau' = \frac{\tau}{1 + i\,\omega\,\tau}, \tag{2}$$

where τ is the ordinary transport relaxation time. By solving the Boltzmann equation the electrical conductivity tensor $\sigma_{\alpha\beta}$ can be found

$$\sigma'_{\alpha\beta} = q\,N \frac{\langle G_{\alpha\beta}\rangle}{\langle 1\rangle}, \tag{3}$$

where the prime denotes that the complex relaxation time should be introduced in place of τ. N is the carrier concentration and q is the charge of the carriers. The components of the tensor $G'_{\alpha\beta}$ for spherical but nonparabolic energy bands and for H in the z-direction are

$$\left.\begin{aligned}
G'_{11} = G'_{22} &= \frac{1}{2}\mu\left[\frac{1}{1+(\omega+\omega_c)^2\tau^2} + \frac{1}{1+(\omega-\omega_c)^2\tau^2} - \frac{i\,(\omega+\omega_c)\,\tau}{1+(\omega+\omega_c)^2\tau^2} - \right.\\
&\qquad \left. - \frac{i\,(\omega-\omega_c)\,\tau}{1+(\omega-\omega_c)^2\tau^2}\right],\\
G'_{12} = -\,G'_{21} &= \frac{1}{2}\mu\left[\frac{(\omega+\omega_c)\,\tau}{1+(\omega+\omega_c)^2\tau^2} - \frac{(\omega-\omega_c)\,\tau}{1+(\omega-\omega_c)^2\tau^2} + \frac{i}{1+(\omega+\omega_c)^2\tau^2} - \right.\\
&\qquad \left. - \frac{i}{1+(\omega-\omega_c)^2\tau^2}\right],\\
G'_{33} &= \mu\left[\frac{1}{1+\omega^2\tau^2} - \frac{i\,\omega\,\tau}{1+\omega^2\tau^2}\right].
\end{aligned}\right\} \tag{4}$$

$\mu = \frac{q\,\tau}{m^*}$ is the mobility and $\omega_c = \frac{q\,H}{m^*\,c}$ is the cyclotron frequency. $\frac{1}{m^*} = \frac{1}{\hbar^2 k}\cdot\frac{dE}{dk}$ is the momentum effective mass. The symbol $\langle A\rangle$ is defined as

$$\langle A\rangle \stackrel{\mathrm{df}}{=} \int_0^\infty \left(-\frac{df_0}{dE}\right) A(E)\,[\gamma(E)]^{3/2} dE\,, \tag{5}$$

where f_0 is the Fermi-Dirac distribution function. The function γ describes the nonparabolicity of the energy band $\gamma(E) = k^2$ (E is the energy and k is the modulus of the wave vector). For strong degeneracy we can put

$$-\frac{df_0}{dE} = \delta(E-\zeta) \tag{6}$$

and

$$\frac{\langle A\rangle}{\langle 1\rangle} = A(\zeta)\,, \tag{7}$$

where ζ denotes the Fermi level.

Assuming that the solution of the Maxwell equation in the form $\boldsymbol{E} = \boldsymbol{E}_0 e^{i\,\omega\,(t-\boldsymbol{N}\,\boldsymbol{r}/c)}$ one obtains

$$(\boldsymbol{N}\,\boldsymbol{N})\boldsymbol{E} - (\boldsymbol{N}\,\boldsymbol{E})\boldsymbol{N} + \frac{4\,\pi\,i}{\omega}\hat{\sigma}\,\boldsymbol{E} - \varepsilon\,\boldsymbol{E} = 0\,, \tag{8}$$

where $\boldsymbol{N}$ is the propagation vector and ε is the lattice dielectric constant.

We shall now discuss the equation (8) for Voigt's configuration, i. e. for $\boldsymbol{N}$ perpendicular to the direction of the external magnetic field $\boldsymbol{H}$. For $\boldsymbol{H} = (0, 0, H)$ and for polarization of light such that $\boldsymbol{E} \parallel \boldsymbol{H}$ we obtain

$$N_{\parallel}^2 = \varepsilon - \frac{4\pi i}{\omega} \sigma_{33}' \tag{9}$$

and for $\boldsymbol{E} \perp \boldsymbol{H}$

$$N_{\perp}^2 = \frac{\left(\frac{4\pi}{\omega} \sigma_{12}'\right)^2}{\frac{4\pi i}{\omega} \sigma_{11}' - \varepsilon} - \frac{4\pi i}{\omega} \sigma_{11}' + \varepsilon . \tag{10}$$

We can now determine the optical constants n and k making use of the relation

$$N_{\parallel,\perp} = n_{\parallel,\perp} - i\, k_{\parallel,\perp} . \tag{11}$$

From equation (11) it follows that

$$n_{\parallel,\perp}^2 - k_{\parallel,\perp}^2 = \mathrm{Re}\,(N_{\parallel,\perp}^2) , \qquad 2\, n_{\parallel,\perp}\, k_{\parallel,\perp} = -\,\mathrm{Im}\,(N_{\parallel,\perp}^2) . \tag{12}$$

Assuming $n_{\parallel,\perp}^2 \gg k_{\parallel,\perp}^2$, $\omega\tau \gg 1$ and making use of equations (3), (4), (7), (9), (10) and (12) we have

$$n_{\parallel}^2 = \varepsilon \left[1 - \frac{\omega_p^2}{\omega^2}\right], \tag{13}$$

$$n_{\perp}^2 = \varepsilon \left[1 - \frac{\omega_p^2\,(\omega^2 - \omega_p^2)}{\omega^2\,(\omega^2 - \omega_p^2 - \omega_c^2)}\right], \tag{14}$$

where $\omega_p^2 = \dfrac{4\pi q^2 N}{\varepsilon m^*}$ is the plasma frequency.

Of course for $H = 0$ we have $n_{\perp}^2(0) = n_{\parallel}^2$.

It is easily seen from equation (1) that the coefficient of reflectivity R attains its minimum value for $n = 1$, i. e. for

$$\omega_{min} = \omega_p \left(\frac{\varepsilon}{\varepsilon - 1}\right)^{1/2} \tag{15}$$

when $H = 0$ and

$$\omega_{\min 1,2} \simeq \omega_p \pm \frac{1}{2}\omega_c - \frac{1}{8}\frac{\omega_c^2}{\omega_p} \tag{16}$$

when $H \neq 0$. In the last case the minima are separated by the magnitude $\omega_c = \dfrac{q H}{m^* c}$.

Similar results can be deduced from Drude-Zener's theory, but their character is not so general. They do not include the distribution function of electrons, and there is no way for making account of degeneracy as well as the non-parabolicity of the energy band. Our considerations take into account these cases and the effective mass is taken at the Fermi level. It is possible to extend our results for the case of multicarrier model of semiconductor by appropriete summation in the conductivity tensor.

The experimental method of reflectivity measurements in the presence of a magnetic field is based on expressions (15) and (16). The effective mass m^* is measured directly from the splitting of the reflectivity minimum by the magnetic

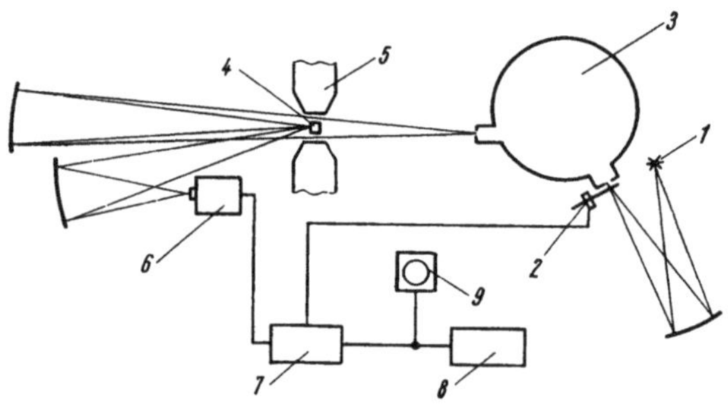

Fig. 1. The scheme of experimental arrangement for the reflectivity measurements in the presence of magnetic field

field (16) and the concentration of current carriers from the position of the reflectivity minimum for the vanishing magnetic field (15). For the later the knowledge of the lattice dielectric constant is required.

3. Experiment

The experimental arrangement used for reflectivity measurement in the presence of magnetic field is given schematically in Fig. 1. The radiation from the globar (1) was focused by a mirror on the entrance slit of the SPM-1 monochromator (3) equiped with prisms suitable for measurements up to 40 μm. Monochromatic and polarized radiation was focused on the Golay detector cell (6). The electric signal from the Golay cell was amplified by the amplifier (7) with phase detection. The beam was modulated by the chopper (2) with frequency 10 Hz. The resulting signal was observed on a scope (9) and registered by the digital voltmeter (8).

The reflectivity of a good aluminized mirror in reference to which the measurements were made was assumed to be 100%. The angle of reflection was about 5°. The radiation was polarized by the germanium polarizer consisting of a polished germanium plate and of alluminized mirrors. These mirrors assured the incidence of radiation on the germanium plate at the Brewster angle and the same direction of the entrance and exit beam.

The measurements were carried on n-type InSb samples with current carrier concentrations in the range of 10^{18} to 10^{19} cm^{-3}. Their dimensions were about $10 \times 3 \times 1$ mm. The surfaces were polished in the usual way as to obtain good optical surfaces.

4. Results

The results of the reflectivity measurements for two InSb samples in the absence of a magnetic field are presented in Fig. 2. It is seen that the reflectivity drops to a minimum at the frequency ω_{min} and then rises to high values for $\omega < \omega_{min}$. The influence of the magnetic field on the reflectivity in the case of sample B is shown in Fig. 3. The minimum of the reflectivity is distinctly split and the distance between the two new minima depends on the magnetic field according to formula (16).

Taking $\varepsilon = 15{,}4$ [6] the effective mass has been estimated making use of equation (16). The concentration of current carriers is now calculated from the position of the minimum of the reflectivity in the absence of magnetic field using the evaluated effective mass.

The effective mass can be also estimated from equation (15) if the current carrier concentration is known from an independent measurement of the Hall coeffi-

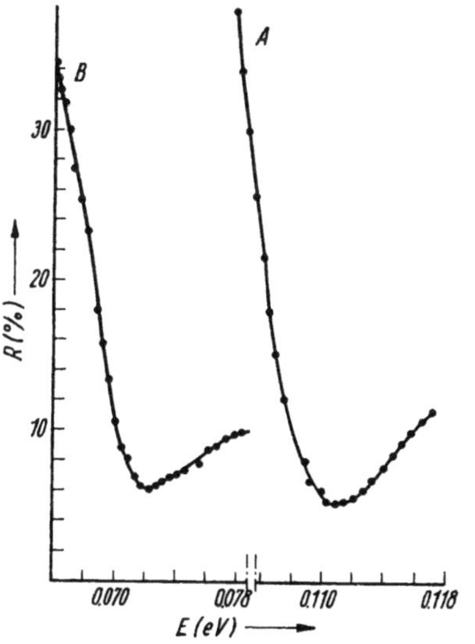

Fig. 2. The plasma reflection observed on two InSb samples

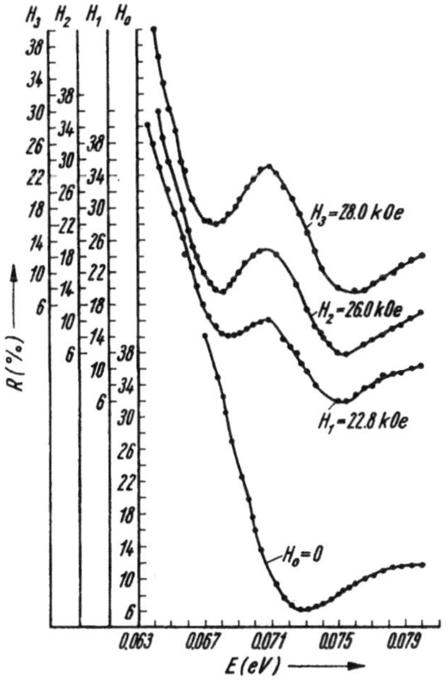

Fig. 3. The splitting of reflection minimum by different values of magnetic field. The vertical scales are adjusted to each curve respectively

cient. The Hall coefficient has been measured for all samples. In the case of sample A and F the splitting of the reflectivity curves by the magnetic field was not sufficient due to the fact that the available magnetic field was too weak. In these cases the effective mass was calculated from equation (15) using N obtained from Hall coefficient. The same was done in the case of sample B for comparison. The results are collected in Table 1.

Table 1

	Sample B	Sample A	Sample F
Position of the reflectivity minimum	0.0717 eV	0.1109 eV	0.1111 eV
$\hbar\,\omega_c$	0.0085 eV at 28 kOe	—	—
m^*/m_0 from ω_c	0.038 ± 0.005	—	—
N from magneto-plasma reflection	2.4×10^{18} cm^{-3}		
N from Hall coefficient	2.6×18^{18} cm^{-3}	8×10^{18} cm^{-3}	1.2×10^{19} cm^{-3}
m^*/m_0 from $\omega_{\min}$ and Hall coefficient	0.041 ± 0.006	0.055 ± 0.007	0.084 ± 0.007

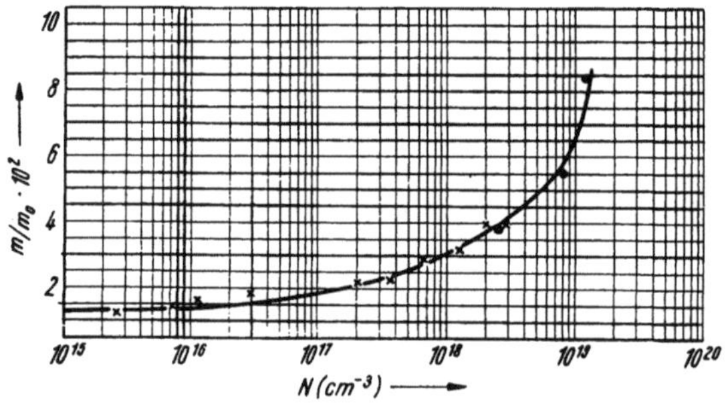

Fig. 4. The dependence of the effective mass in n-type InSb versus concentration of free carriers

5. Discussion

The dependence of the effective mass on current carrier concentration is well known in the case of InSb [1, 7, 8]. In Fig. 4 the theoretical dependence of effective mass versus current carrier concentration [8] as well as experimental points are shown. Our results are marked by full dots. The other experimental points shown on Fig. 4 are taken from the measurements of free carrier absorption and Faraday effect. Our results confirm the theoretical curve for higher current carrier concentrations. The highest point seems to fit quite well the theoretical curve for the concentrations exceeding 10^{19} cm^{-3}.

Acknowledgements

The authors are most grateful to Prof. Dr. L. SOSNOWSKI for the help and interest during the work. They are also greatly indebted to Dr. J. KOŁODZIEJCZAK for discussion concerning the theoretical problems.

References

[1] W. G. SPITZER and H. Y. FAN, Phys. Rev. **106**, 882 (1957).
[2] B. LAX and G. B. WRIGHT, Phys. Rev. Letters **4**, 16 (1960).
[3] J. KOŁODZIEJCZAK, Acta phys. Polon. **21**, 637 (1962).
[4] S. ŻUKOTYNSKI and J. KOŁODZIEJCZAK, phys. stat. sol. **3**, 990 (1963).
[5] J. KOŁODZIEJCZAK and S. ŻUKOTYŃSKI, phys. stat. sol. **5**, 145 (1964).
[6] T. S. MOSS and S. D. SMITH, Proc. Phys. Soc. **B 70**, 776 (1957).
[7] S. D. SMITH, T. S. MOSS, and K. W. TAYLOR, J. Phys. Chem. Solids **11**, 131 (1959).
[8] J. KOŁODZIEJCZAK, Acta phys. Polon. **20**, 289 (1961).

(Received June 5, 1964)

phys. stat. sol. 6, 555 (1964)

Physikalisch-Technisches Institut der Deutschen Akademie der Wissenschaften zu Berlin, Bereich Lumineszenzforschung, Liebenwalde bei Berlin

Untersuchungen zu unterschiedlichen Lumineszenzmechanismen in verschieden aktiviertem Zinksulfid[1])

Von

H. TREPTOW

An Cu- und Ag-aktiviertem Zinksulfid wurden die Form, Halbwertsbreite und Lage des Maximums der Emissionsbanden im Temperaturbereich von 100 bis 400 °K gemessen. Der Vergleich mit den Eigenschaften der Elektronenübergänge zwischen diskreten Termen und den Eigenschaften für Leitungsband-Akzeptorterm-Übergänge ergab, daß der langwelligen Emissionsbande bei Cu-Aktivierung ein Term-Term- und der Emissionsbande bei Ag-Aktivierung ein Band-Term-Übergang zuzuordnen ist. Die Erklärung der Eigenschaften der kurzwelligen Emissionsbande bei Cu-Aktivierung kann durch einen gemischten Band-Term- und Term-Term-Übergang möglich sein.

A study is made of the emission bands of ZnS activated with Cu or Ag. The form, half-width, and position of the maximum, of these bands, are given for temperatures between 100 and 400 °K. A comparison is made of the properties of electronic transitions between discrete levels and between the conduction band and an acceptor level. This shows that the long-wavelength emission band of Cu activated ZnS can be ascribed to the former type of transition but that the emission band of Ag activated ZnS is due to the latter type of transition. The properties of the short-wavelength emission band of Cu-activated ZnS can be explained by assuming both types of transition to be operative.

1. Einleitung

Die Grundzüge der heutigen Anschauungen über den Leuchtmechanismus in den Luminophoren vom Zinksulfidtyp konnten durch das Riehl-Schön-Modell [1, 2, 3] erklärt werden. Im ursprünglichen Modell wurde jeder Störstelle ein Term in der verbotenen Zone zugeordnet. Im Laufe der Untersuchungen der Lumineszenzerscheinungen stellte sich heraus, daß komplizierte Zusammenhänge oft nicht durch einzelne Terme in der verbotenen Zone, sondern durch direkte Wechselwirkung zwischen zwei oder mehreren diskreten Termen erklärt werden konnten [z. B. 4, 5]. Durch die Arbeiten von KRÖGER und Mitarbeitern [6, 7, 8] sowie PRENER und WILLIAMS [9, 10, 11] wurden Aktivator-Koaktivator-Paare als mögliche Lumineszenzzentren in den Sulfiden nahegelegt. Diese könnten Ursache dafür sein, daß Emissionsübergänge nicht mehr vom Leitungsband zum Akzeptorterm, sondern zwischen zwei diskreten Termen stattfinden. Weiterhin besteht die Möglichkeit, daß bereits der Einbau von einem einzelnen Aktivatorion mehrere Terme, die miteinander in Wechselwirkung stehen, in der verbotenen Zone hervorrufen kann. Ein solcher Fall wurde von BROSER und BROSER-WARMINSKY [12] sowie von SCHULZ [5] für Cu-aktiviertes ZnS zur Erklärung der Infrarotlumineszenz angenommen.

In der vorliegenden Arbeit soll gezeigt werden, daß durch Messung der Form von Emissionsbanden, ihrer Temperaturabhängigkeit und des Verhaltens bei Änderung der Konzentration der Elektronen im Leitungsband eine Entscheidung zwischen einem Band-Term und Term-Term-Emissionsübergang möglich sein kann.

---

[1]) Auszug aus der Dissertation, Humboldt-Universität, Berlin 1964.

Die Untersuchungen werden an den Emissionsbanden von Cu- und von Ag-aktiviertem Zinksulfid durchgeführt.

2. Experimentelle Ergebnisse

2.1 Eigenschaften der langwelligen Emissionsbande von Cu-aktiviertem Zinksulfid

Bei allen hier angegebenen Ergebnissen wurde die Lumineszenz mit den Quecksilberlinien bei 365 nm angeregt.

Die im folgenden beschriebenen Ergebnisse wurden an hexagonalen, Cu-aktivierten Zinksulfidkristallpulvern erhalten, die vorwiegend eine langwellige, grüne Emissionsbande aufweisen, die dem Kupfer als Aktivator zugeschrieben wird [13, 14, 15, 16]. Eine störende Beeinflussung durch die kurzwellige, dem Kupfer zuzuschreibende Bande [15] tritt selbst bei den tiefsten Meßtemperaturen nicht auf, obwohl mit abnehmender Temperatur das relative Intensitätsverhältnis beider Banden zugunsten der kurzwelligen Bande anwächst [17]. Zur Herstellung aller untersuchten Luminophore wurde chloridisches Schmelzmittel verwendet.

Die bei 104, 297 und 376 °K gemessenen Emissionsspektren sind in der Fig. 1 dargestellt. Dabei wurde als Ordinatenmaßstab nicht die relative Intensität, sondern der Quotient aus der relativen Intensität und der vierten Potenz der Photonenenergie ($I\,(h\,\nu)^{-4}$) benutzt. Bemerkenswert ist das Verhalten des Halbwertspunktes großer Photonenenergie. In dem kleinen Diagramm der Fig. 1 ist die Umgebung dieses Punktes mit vergrößertem Abszissenmaßstab nochmals

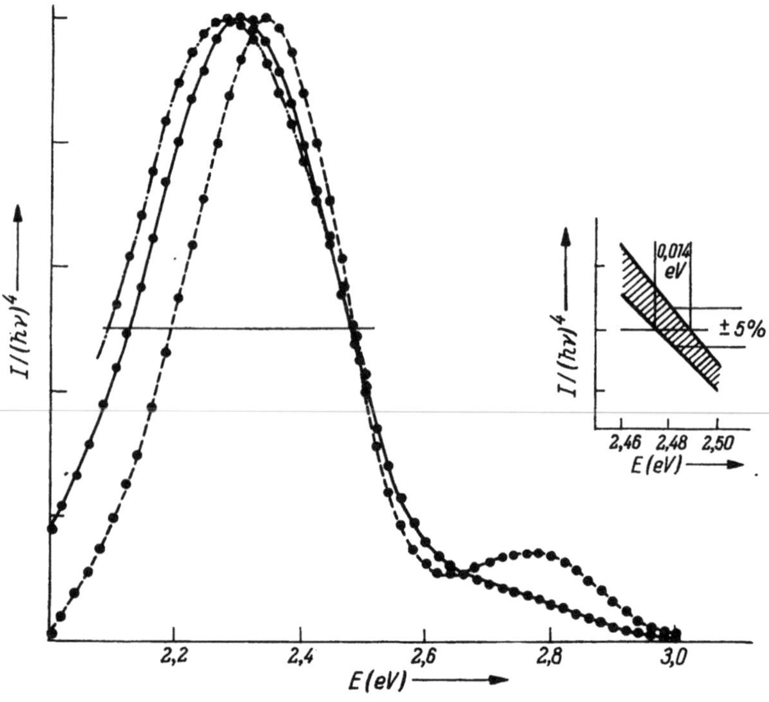

Fig. 1. Langwellige Emissionsbande von Cu-aktiviertem ZnS bei 104 °K (— — —), 297 °K (——) und 376 °K (— · — · —); im kleinen Diagramm ist die Umgebung des Halbwertspunktes großer Photonenenergie mit vergrößertem Abszissenmaßstab herausgezeichnet

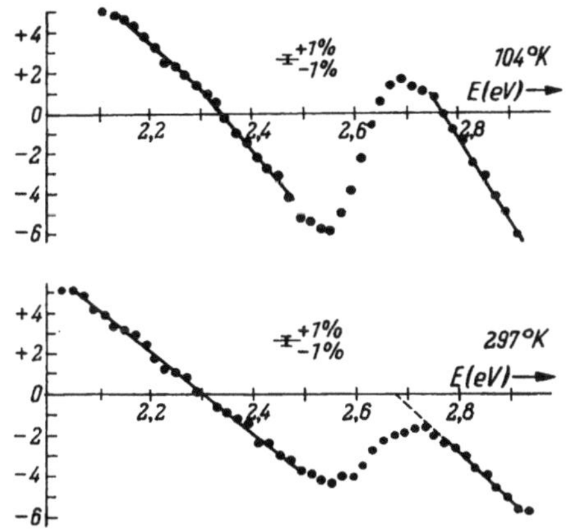

Fig. 2. Spektren von Cu-aktiviertem ZnS mit vorherrschender langwelliger Emissionsbande, umgerechnet in den Gaußkurvenmaßstab für 104 und 297 °K

dargestellt. Sämtliche für die Temperaturabhängigkeit gemessenen Kurven, von denen nur drei im großen Diagramm eingezeichnet sind, liegen innerhalb des schraffierten Bereiches. Da die Breite dieses Bereiches in halber Höhe nur 0,014 eV beträgt und dieser Wert nur wenig größer als die zur Messung benutzte spektrale Bandbreite ist, kann man vermuten, daß die Lage des Halbwertspunktes bei großer Photonenenergie temperaturunabhängig ist.

In der Fig. 2 sind die bei den Temperaturen 104 und 297 °K gemessenen Spektren umgerechnet in den Gaußkurvenmaßstab dargestellt.

Um zu untersuchen, ob die Bandenform sich durch eine Gaußkurve darstellen läßt, wurden die im $I \cdot (h\nu)^{-4} - h\nu$-Diagramm dargestellten Spektren logarithmiert und differenziert (Bildung des Differenzquotienten benachbarter Punkte). Diese Umrechnung, die als Umrechnung in den Gaußkurvenmaßstab bezeichnet wird, hat den Vorteil, daß sich für eine Gaußkurve eine Gerade ergibt, deren Schnittpunkt mit der Abszisse die Lage des Bandenmaximums (diese Eigenschaft gilt für jede beliebige Bandenform) und deren Richtungsfaktor die Halbwertsbreite H bestimmt. Jede Abweichung von einer Gaußkurvenform macht sich als gekrümmte Linie bemerkbar. Ein exponentieller Abfall würde einen zur Abszisse parallelen Verlauf ergeben. Diese Darstellung kann auch dazu verwendet werden, um für einen Teil einer Bande eine eventuell auftretende Darstellungsmöglichkeit durch eine Gaußfunktion nachzuweisen, und es können weiterhin genauere Aussagen über eine Überlagerung zweier Banden als aus der üblichen Darstellung gewonnen werden.

Um einen Eindruck von der Genauigkeit der Messungen und von der Güte der Aussage über die Kurvenform zu vermitteln, wurden in jedem Diagramm die umgerechneten Meßwerte und die Größe der Abweichung eingezeichnet, die eine um 1% fehlerhaft gemessene Intensität verursachen würde.

Beide Diagramme lassen die zwei dem Kupfer zuzuordnenden Banden erkennen. Die langwellige Hauptbande zeigt bei 104 °K einen leicht gekrümmten Verlauf, der sich bei allen Messungen bei Temperaturen unterhalb 200 °K feststellen läßt. Daraus ergibt sich bei tiefen Temperaturen eine asymmetrische Bandenform, deren Flanke bei geringer Photonenenergie breiter als die andere Flanke ist. Eine Beeinflussung durch die kurzwellige Cu-Bande kann nicht vorliegen, da dadurch nur die Flanke bei großer Photonenenergie verbreitert werden könnte. Im Bereich von 100 bis 190 °K beträgt das Verhältnis der Breite der Flanke großer zur Breite der Flanke geringer Photonenenergie in der Höhe des halben Maximalwerts 0,87 bis 0,93. Als Mittelwert ergibt sich 0,90.

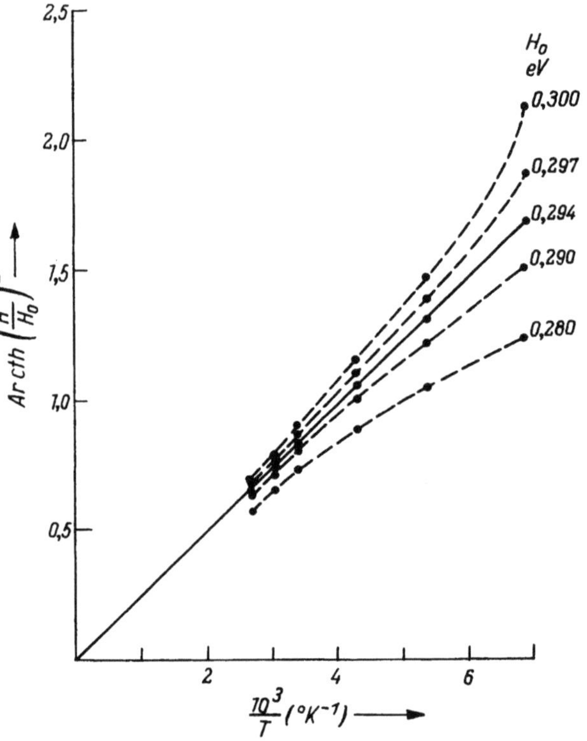

Fig. 3. Diagramm für die Temperaturabhängigkeit der Halbwertsbreite der langwelligen Cu-Emissionsbande nach Markham (siehe Text)

Bei allen gemessenen Temperaturen oberhalb 200 °K läßt sich mit der gleichen Genauigkeit, wie sie das Diagramm für 297 °K in Fig. 2 demonstriert, die Form einer Gaußkurve für die langwellige Emissionsbande des Cu-aktivierten Zinksulfids nachweisen.

Stellt man die gemessene Halbwertsbreite über der Wurzel aus der absoluten Temperatur dar — diese Darstellung wurde von Klick für die F-Zentren der Alkalihalogenide benutzt [18] —, so kann man vermuten, daß sich am absoluten Nullpunkt eine Halbwertsbreite zwischen 0,27 und 0,30 eV ergeben würde. Eine Messung der Halbwertsbreite bei 4 °K konnte nicht durchgeführt werden, und es liegen auch keine Angaben anderer Autoren vor.

Die bisher angegebenen Meßergebnisse lassen die Vermutung zu, daß eine Erklärung der Eigenschaften durch einen Emissionsübergang zwischen zwei diskreten Termen in Frage kommen kann. Die Berechnungen von Markham [19] und Pekar [20] ergaben für einen solchen Emissionsübergang bei hohen Temperaturen eine Gaußkurvenform, wenn der Faktor $(h\,\nu)^{-4}$ berücksichtigt wird, der für die relativ breiten Banden nicht zu vernachlässigen ist.

Eine wesentliche Voraussetzung für die Berechnung ist, daß der Energieaustausch zwischen den Zentren und der Umgebung über Phononen von nur einer effektiven Energie vor sich geht, d. h. über Phononen, deren Energie nur eine schmale, vernachlässigbare Verteilung um einen mittleren Wert besitzt. In diesem Falle müßte die Halbwertsbreite nach dem bekannten hyperbolischen Kotangensgesetz [19] von der Temperatur abhängen. Zur Überprüfung wurden deshalb in der

Fig. 4. Die berechnete Pekarkurve (ausgezogene Linie) im Vergleich zu den Meßwerten der langwelligen Cu-Emissionsbande bei 104 °K (eingezeichnete Punkte)

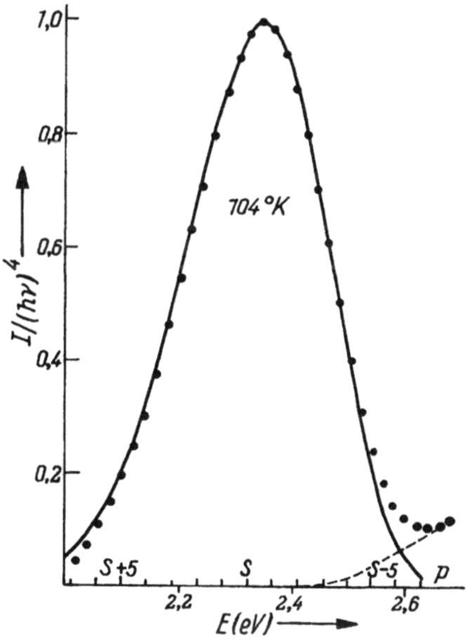

Fig. 3 nach einem Vorschlag von MARKHAM für verschiedene vorgegebene Werte der Halbwertsbreite H_0 am absoluten Nullpunkt $\text{Arcth}(H/H_0)^2$ über der reziproken absoluten Temperatur aufgetragen. Man entnimmt diesem Diagramm, daß die Meßpunkte für den vorgegebenen Wert $H_0 = 0{,}294$ eV alle auf einer Geraden liegen, die durch den Koordinatenursprung geht. Das bedeutet, daß für dieses H_0 alle Meßwerte dem hyperbolischen Kotangensgesetz genügen. Aus dem Anstieg der Geraden ergibt sich als Phononenenergie $\hbar\,\omega = 0{,}044$ eV.

Die theoretischen Berechnungen ergeben den folgenden Zusammenhang zwischen der Halbwertsbreite am absoluten Nullpunkt H_0, der Phononenenergie $\hbar\,\omega$ und dem Huang-Phys-Faktor S [19]:

$$H_0 = \hbar\,\omega\,\sqrt{a\,S}\,.$$

Der Faktor a, der auch als Smakula-Konstante bezeichnet wird, ist dimensionslos. Er hängt nur gering von der Form der Bande ab [19]. Aus dieser Gleichung läßt sich der Huang-Phys-Faktor zu $S = 8{,}0$ berechnen.

Die Phononenenergie und der Huang-Phys-Faktor gehen als bestimmende Größen in die asymmetrische Bandenform bei tiefen Temperaturen, die Pekarkurve, ein [19]. Einen Hinweis auf die Deutung der asymmetrischen Tieftemperaturform als Pekarkurve liefern bereits die oben angegebenen Verhältnisse der Breite der Flanken, da MARKHAM für die Pekarkurve ein Verhältnis von 0,90 angibt.

In der Fig. 4 ist die berechnete Pekarkurve als ausgezogene Kurve eingezeichnet. Die Meßpunkte des Emissionsspektrums bei 104 °K sind als einzelne Punkte eingetragen. Sieht man von der Abweichung bei großer Photonenenergie ab, die durch den Einfluß der kurzwelligen Bande hervorgerufen wird (vgl. Fig. 1), dann wird die asymmetrische Bandenform bei tiefen Temperaturen gut durch die berechnete Pekarkurve beschrieben. Im Gaußkurvenmaßstab (Fig. 5), der genauere

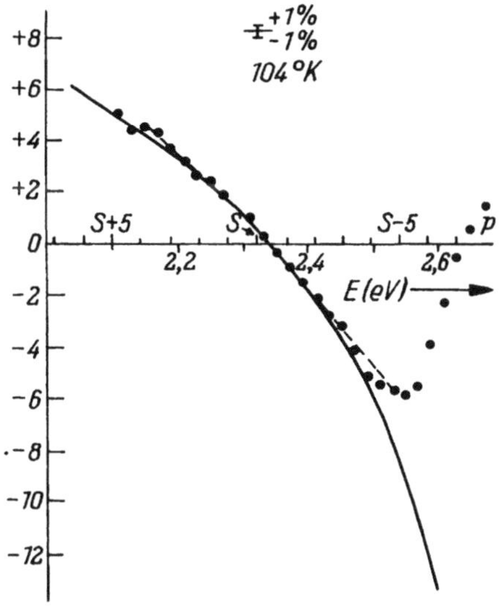

Fig. 5. Die berechnete Pekarkurve (ausgezogene Linie) im Vergleich zu den Meßwerten der langwelligen Cu-Emissionsbande bei 104 °K (eingezeichnete Punkte) im Gaußkurvenmaßstab

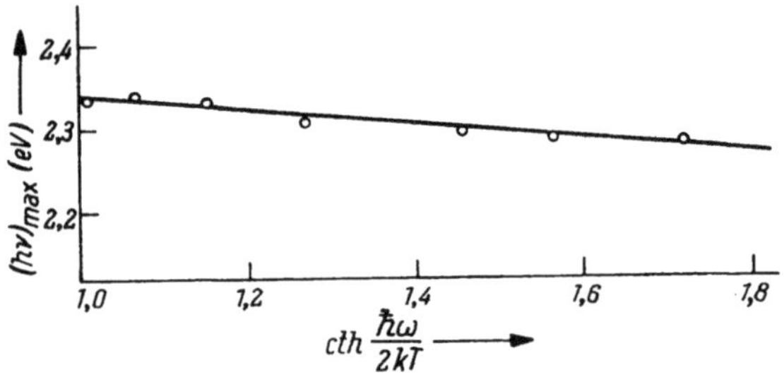

Fig. 6. Temperaturabhängigkeit der Lage des Maximums der langwelligen Emissionsbande von Cu-aktiviertem ZnS

Aussagen zuläßt, sind die Meßwerte (eingezeichnete Punkte) auch in sehr guter Übereinstimmung mit der berechneten Pekarkurve (ausgezogene Linie).

In Fig. 6 wurde die energetische Lage der Maxima $(h\,\nu)_{max}$ über cth $(\hbar\,\omega/2\,k\,T)$ aufgetragen. Die gute Darstellung der Meßwerte durch die eingezeichnete Gerade bestätigt die Gültigkeit der theoretisch berechneten Abhängigkeit [19].

Alle angegebenen Ergebnisse der langwelligen Emissionsbande von Cu-aktiviertem Zinksulfid sind unabhängig von der Anregungsintensität, d. h. unabhängig von der Konzentration der Leitungselektronen, wie experimentell bestätigt werden konnte.

Als Ergänzung zu den Eigenschaften der Emissionsbande wurde das Anregungsspektrum für die langwellige Emissionsbande des Cu-aktivierten Zinksulfids gemessen. Das Spektrum zeigt keine Andeutung einer vorgelagerten Bande. Dieses

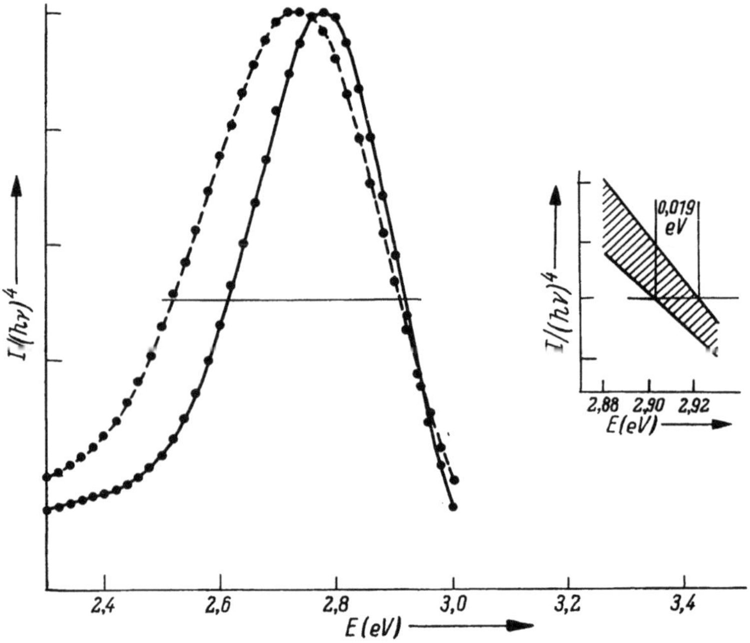

Fig. 7. Emissionsbande von Ag-aktiviertem ZnS bei 106 °K (——) und 296 °K (– – –); im kleinen Diagramm ist die Umgebung des Halbwertspunktes großer Photonenenergie mit vergrößertem Abszissenmaßstab herausgezeichnet

Ergebnis steht ebenso wie die ermittelte erforderliche Anregungsenergie von 2,79 eV in guter Übereinstimmung mit den Angaben anderer Autoren [5, 21, 22, 23].

2.2 *Eigenschaften der Emissionsbande von Ag-aktiviertem Zinksulfid*

An hexagonalen, Ag-aktivierten Zinksulfidkristallpulvern, deren Emissionsbande auf das Silber als Aktivator zurückgeführt wird [13, 14, 15, 16], wurden die folgenden Eigenschaften festgestellt.

In der Fig. 7 sind die bei 106 und 296 °K gemessenen Emissionsbanden dargestellt. In dem kleinen Diagramm der Fig. 7 ist die Umgebung des Halbwertspunktes bei großer Photonenenergie mit vergrößertem Abszissenmaßstab herausgezeichnet. In dem schraffierten Bereich liegen alle für die Temperaturabhängigkeit gemessenen Kurven. Wegen der geringen Breite dieses Bereiches von weniger als 0,02 eV kann man auch für die Emissionsbande des Ag-aktivierten Zinksulfids vermuten, daß die Lage des Halbwertspunktes bei großer Photonenenergie temperaturunabhängig ist.

Die in den Gaußkurvenmaßstab umgerechneten Banden für 106 und 296 °K sind in Fig. 8 dargestellt. Beide Diagramme zeigen eine asymmetrische Bandenform, die im gesamten untersuchten Bereich zwischen 100 und 400 °K festgestellt wurde. Bei jeder Meßtemperatur läßt sich die Flanke großer Photonenenergie gut durch eine Gaußfunktion beschreiben, während für die Flanke geringer Photonenenergie, die flacher abfällt, eine einfache Funktion nicht angegeben werden kann.

Beide Diagramme der Fig. 8, die den Kurvenverlauf bis 2,3 eV erkennen lassen, zeigen, daß zur Erklärung der Asymmetrie eine zusätzliche Emissionsbande bei geringer Photonenenergie nicht in Frage kommt, da kein erneuter Anstieg im Gaußkurvenmaßstab festzustellen ist (man vergleiche mit den Diagrammen der beiden sich beeinflussenden Cu-Banden, Fig. 2). Deshalb ist die Asymmetrie eine typische Eigenschaft der durch Ag-Aktivierung erhaltenen Emissionsbande. Die asymmetrische Bandenform kann auch keine Pekarkurve sein, da das Verhältnis der Breite beider Flanken in halber Höhe nicht 0,90 beträgt, sondern mit zunehmender Temperatur von 0,85 bei 106 °K bis 0,78 bei 375 °K abnimmt.

In Fig. 9 ist die gemessene Halbwertsbreite über der Meßtemperatur dargestellt. Die Halbwertsbreite zeigt einen Verlauf, der sich nicht durch einen hyperbolischen Kotangens darstellen läßt. Die Ursache für den bei 180 °K auftretenden ungleichmäßigen Verlauf kann aus den vorliegenden Untersuchungen nicht entnommen werden.

In Übereinstimmung mit den Beobachtungen anderer Autoren an Ag-akvitiertem Zinkkadmiumsulfid [24, 25] konnte für Ag-aktiviertes Zinksulfid keine Änderung der Lage und Form der Emissionsbande bei Änderung der Anregungsinten-

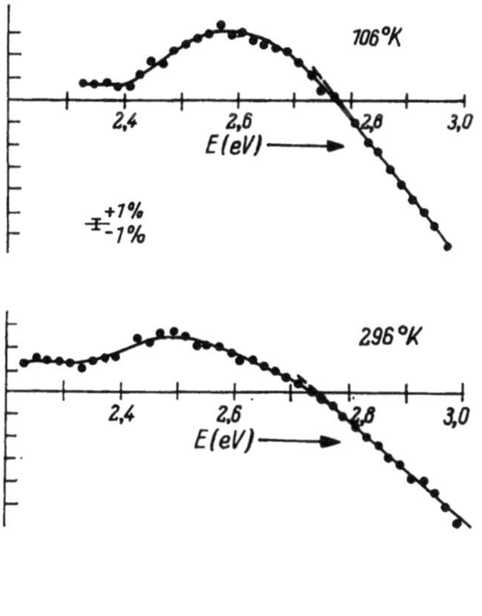

Fig. 8. Emissionsbande von Ag-aktiviertem ZnS für 106 und 296 °K im Gaußkurvenmaßstab

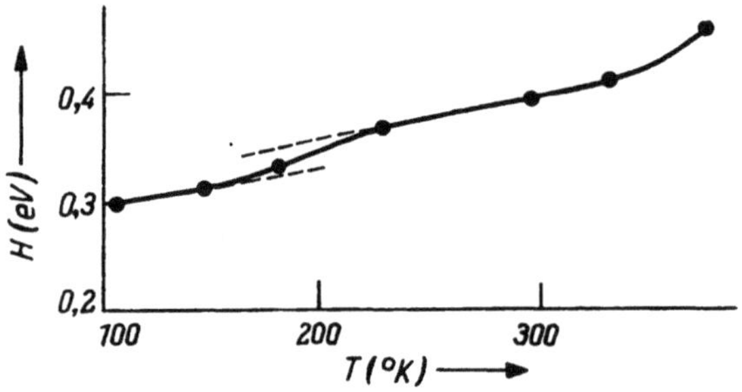

Fig. 9. Temperaturabhängigkeit der Halbwertsbreite der Emissionsbande von Ag-aktiviertem ZnS

sität festgestellt werden, d. h., die Emissionsbande ist bei gleichbleibender Verteilung der Elektronen im Leitungsband unabhängig von der Konzentration der Leitungselektronen.

2.3 Eigenschaften der kurzwelligen Emissionsbande von Cu-aktiviertem Zinksulfid

Untersucht wurden hexagonale, Cu-aktivierte Zinksulfidkristallpulver, die nur die kurzwellige, blaue Emissionsbande aufweisen, die ebenfalls dem Kupfer als Aktivator zugeschrieben wird [13, 14, 15, 16]. Alle verwendeten Luminophore wurden mit chloridischem Schmelzmittel hergestellt.

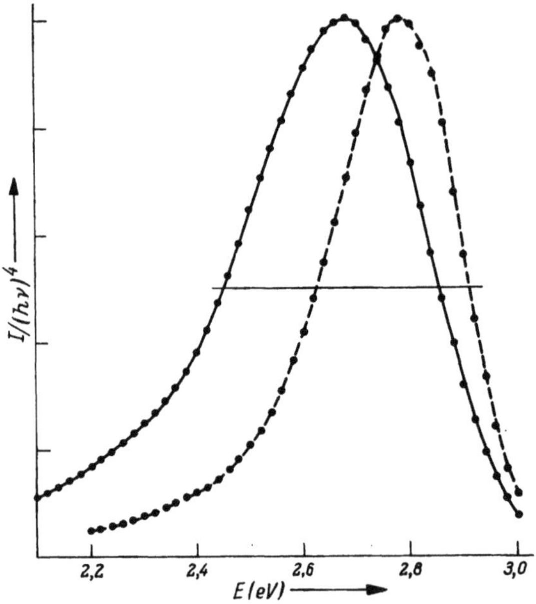

Fig. 10. Kurzwellige Emissionsbande von Cu-aktiviertem ZnS bei 104 °K (— — —) und 296 °K (——)

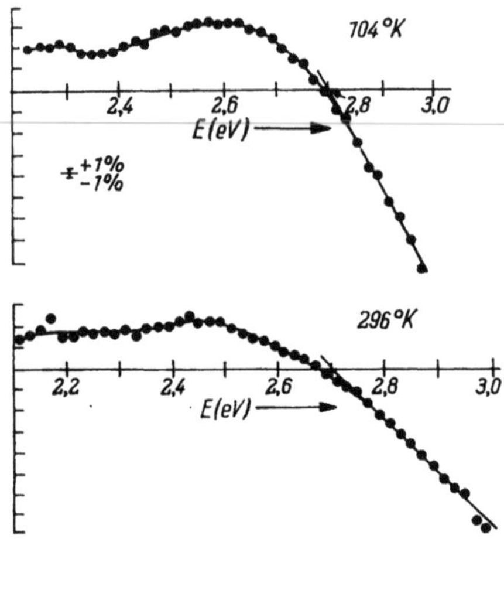

◀ Fig. 11. Kurzwellige Emissionsbande von Cu-aktiviertem ZnS bei 104 und 296 °K im Gaußkurvenmaßstab

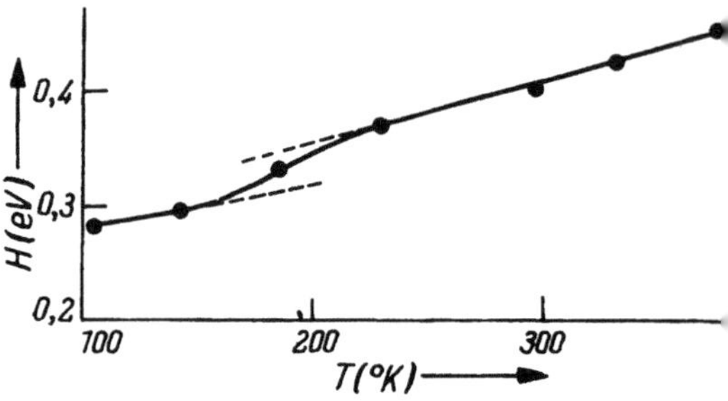

Fig. 12. Temperaturabhängigkeit der Halbwertsbreite der kurzwelligen Emissionsbande von Cu-aktiviertem ZnS

Aus der Fig. 10, die die Emissionsbande bei 104 und 296 °K zeigt, und der Fig. 11, in der die gleichen Banden im Gaußkurvenmaßstab dargestellt sind, können folgende Eigenschaften dieser Bande entnommen werden: Der Halbwertspunkt bei großer Photonenenergie ist im Gegensatz zu den vorher besprochenen Banden temperaturabhängig. Die Bande wird nicht durch die langwellige Bande bei Cu-Aktivierung beeinflußt (Fig. 11). Die Bandenform ist im gesamten untersuchten Temperaturbereich zwischen 100 und 400 °K asymmetrisch, mit flacher abfallender Flanke bei niedriger Photonenenergie. Die Flanke großer Photonenenergie läßt sich durch eine Gaußfunktion beschreiben, für die Flanke geringer Photonenenergie kann keine einfache Funktion angegeben werden. Die Halbwertsbreite in Abhängigkeit von der Meßtemperatur, die in Fig. 12 dargestellt ist, zeigt einen Verlauf, der bereits bei der Emissionsbande von Ag-aktiviertem Zinksulfid gefunden wurde.

Bis auf die Temperaturabhängigkeit des Halbwertspunktes bei großer Photonenenergie sind die bisher angegebenen Eigenschaften der kurzwelligen Cu-Bande qualitativ die gleichen wie die der Ag-Bande. Bei der kurzwelligen Cu-Bande ist jedoch die Lage und Form der Bande von der Anregungsintensität abhängig. Die gemessenen Werte sind in der folgenden Tabelle zusammengefaßt:

Relative Anregungsintensität	Lage des Maximums	Halbwertsbreite
1,00	2,68 eV	0,405 eV
0,32	2,66 eV	0,425 eV
0,08	2,64 eV	0,450 eV

3. Diskussion der Meßergebnisse

3.1 Langwellige Emissionsbande von Cu-aktiviertem Zinksulfid

Die festgestellten Eigenschaften der langwelligen Cu-Emissionsbande, d. h. die Gaußkurvenform für Temperaturen oberhalb 200 °K, die Form einer Pekarkurve für tiefere Temperaturen, die Gültigkeit des hyperbolischen Kotangensgesetzes für die Temperaturabhängigkeit der Halbwertsbreite und die Temperaturabhängigkeit der Lage des Emissionsmaximums, die mit dem Ergebnis der theoretischen Berechnungen von Markham [19] und Pekar [20] übereinstimmen, sprechen dafür, daß der Emissionsübergang innerhalb der verbotenen Zone zwischen zwei diskreten Termen abläuft. Die energetische Wechselwirkung des entsprechenden Zentrums mit seiner Umgebung erfolgt nur über eine Art von Phononen, deren Energie 0,044 eV beträgt. Weiterhin konnte ein Wert für die Halbwertsbreite H_0 am absoluten Nullpunkt erhalten werden, dessen experimentelle Bestätigung noch aussteht.

Das Ergebnis, daß die Lumineszenzübergänge nur von einem unter dem Leitungsband liegenden Voranlagerungsterm ausgehen, steht in Übereinstimmung mit einer Beobachtung von Riehl und Thoma [26]. Die Autoren fanden für die gleiche Emissionsbande durch Untersuchung von Glowkurven im Bereich unterhalb 77,4 °K und der reversiblen Umwandlung der beiden Cu-Banden ineinander einen Haftterm der Tiefe 0,07 eV, der nur für die langwellige Cu-Bande gefunden wurde. Bei der kurzwelligen Cu-Bande tritt ein solcher Term nicht auf.

Die durchgeführten Untersuchungen erlauben keine Aussage über die Ursache des Voranlagerungsterms, ob er dem Einbau des Aktivators selbst oder einer Assoziation zuzuschreiben ist.

Die Energie der Phononen, die nach erfolgtem Emissionsübergang an das Kristallgitter abgegeben werden, stimmt mit der Energie der longitudinalen optischen Gitterphononen des Wirtskristalls überein [27, 28, 29, 30]. Die gute Übereinstimmung bedeutet, daß für das Zentrum der langwelligen Bande des Cu-aktivierten Zinksulfids keine lokalen Gitterschwingungen auftreten. Deswegen kann das Zentrum für die Umgebung keine große Gitterverspannung mit sich bringen, wodurch die relative hohe, maximal erreichbare Zentrenkonzentration als auch die Stabilität dieser Zentren verständlich wird.

Unter Zugrundelegung der gemessenen Werte ist in Fig. 13 das Energiemodell des Zentrums für Zimmertemperatur maßstabsgetreu dargestellt. Im rechten Teil ist das Konfigurationskoordinatenmodell angegeben, das einen Emissionsübergang zwischen diskreten Termen am besten beschreibt. Für das Maximum der Emissionsbande beträgt die Energie, die über die Phononen nach erfolgtem Emissionsübergang abgegeben wird, $S \cdot \hbar\omega = 0{,}35$ eV. Ihr entspricht im links dargestellten Bändermodell die Franck-Condon-Verschiebung des Akzeptorterms. Aus der Photonenenergie des Emissionsmaximums, der erforderlichen Anregungsenergie ins Leitungsband und der Franck-Condon-Verschiebung läßt sich der Abstand des Voranlagerungsterms vom Leitungsband zu 0,1 eV bestimmen. Dieser Wert stimmt recht gut mit dem Ergebnis von Riehl und Thoma [26] überein. In dem Bändermodell sind entsprechend der erhaltenen Phononenenergie die verschiedenen Schwingungszustände des unbesetzten Akzeptorterms mit eingezeichnet. Die Gaußkurvenform der Emissionsbande bedeutet, daß die Zahl der Übergänge zu diesen Niveaus nach einer Gaußverteilung abfällt.

Für die experimentell gefundene Temperaturunabhängigkeit des Halbwertspunktes bei großer Photonenenergie liefern die bisherigen Berechnungen des Elektronenüberganges zwischen zwei diskreten Termen keine Erklärung. Mollwo [31] sowie Konitzer und Markham [32, 33] stellten die Temperaturunabhängigkeit des Halbwertspunktes großer Photonenenergie der Absorptionsbanden von F-Zentren in KCl und NaCl fest. Bemerkenswert ist, daß im Falle der langwelligen

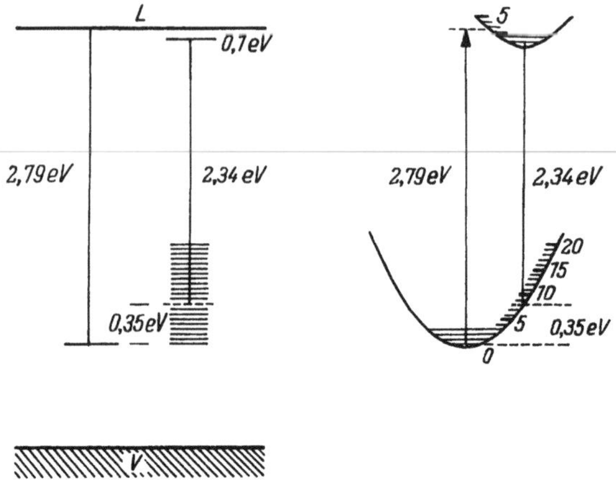

Fig. 13. Energiemodell des Zentrums der langwelligen Emissionsbande von Cu-aktiviertem ZnS (die angegebenen Werte gelten für Zimmertemperatur)

Emissionsbande von Cu-aktiviertem Zinksulfid und im Falle der Absorptionsbanden der Alkalihalogenide der Halbwertspunkt großer Photonenenergie temperaturunabhängig ist. Dieses Verhalten würde, wenn es am gleichen Material gefunden wird, der von PEKAR [20] gefundenen Spiegelsymmetrie zwischen Absorptions- und Emissionsbande widersprechen.

3.2 Emissionsbande von Ag-aktiviertem Zinksulfid

Alle Eigenschaften der Ag-Emissionsbande bis auf die Temperaturunabhängigkeit des Halbwertspunktes bei großer Photonenenergie lassen sich durch einen Elektronenübergang vom Leitungsband zum Akzeptorterm, der für einen Photohalbleiter typisch ist, erklären. Da für einen solchen Übergang bisher keine Berechnungen durchgeführt wurden, ist eine Diskussion nur anhand der qualitativen, schon vor längerer Zeit beschriebenen [34, 35, 2] charakteristischen Eigenschaften eines solchen Übergangs möglich. Die bei allen Temperaturen festgestellte Asymmetrie und die unterschiedliche Beeinflussung der beiden Flanken bei Temperaturänderung, die sich auch in dem variablen Verhältnis der Breite der Flanken der Bande bemerkbar macht, sind für den Band-Term-Übergang eines Photohalbleiters typisch.

Wegen der Abschätzung von MÖGLICH und ROMPE [36, 37], nach der sich jedes in das Leitungsband gehobene Elektron innerhalb einer Zeit der Elektronenverteilung angepaßt hat, die gegenüber der bis zur Rekombination vergehenden Zeit vernachlässigbar kurz ist, und dem experimentellen Befund von SCHÖN (siehe [2]), daß auch bei außerordentlich starker Anregung mit hoher Energie keine Beeinflussung der Form der Emissionsbanden beobachtet werden konnte, kann die Bandenform eines Band-Term-Überganges nur von der Verteilung der Elektronen im Band, nicht aber von der Konzentration der Leitungselektronen abhängen. Deshalb ist die Unabhängigkeit der Bandenform von der Anregungsintensität ein weiterer Hinweis für die Erklärung durch einen Band-Term-Übergang.

Da der Hauptanteil der Intensität des Halbwertspunktes bei großer Photonenenergie durch Übergänge vom Maximum der Elektronenverteilung im Band zum energetisch tieferen Halbwertspunkt der thermisch bedingten Verbreiterung des Akzeptorterms gegeben wird, bedeutet die experimentell gefundene Temperaturunabhängigkeit dieses Halbwertspunktes für einen Band-Term-Übergang eine direkte Verbindung der temperaturabhängigen Eigenschaften des Wirtskristalls mit denen des entsprechenden Zentrums. Die temperaturabhängigen Eigenschaften, die vom Wirtskristall abhängen, sind die Elektronenverteilung im Band und die Lage der unteren Kante des Leitungsbandes. Die temperaturabhängigen Eigenschaften, die dem Zentrum zuzuordnen sind, sind die mittlere Lage des Akzeptorterms und die Verbreiterung dieses Terms auf Grund thermischer Gitterschwingungen.

Erst nach der Berechnung des Problems des Band-Term-Überganges wird man eine Erklärung geben können, warum die asymmetrische Bandenform eine breitere Flanke geringer Photonenenergie besitzt, obwohl die Form der Elektronenverteilung im Leitungsband das Gegenteil vermuten läßt.

3.3 Kurzwellige Emissionsbande von Cu-aktiviertem Zinksulfid

Während die Form der langwelligen Emissionsbande von Cu-aktiviertem Zinksulfid und die Form der Emissionsbande von Ag-aktiviertem Zinksulfid unabhängig von der Konzentration der Elektronen im Leitungsband waren, gilt das für die kurzwellige Emissionsbande von Cu-aktiviertem Zinksulfid nicht. Die erstgenannten Banden können als Vertreter für reine Term-Term-Übergänge — d. h., es macht sich keine Störung durch Band-Term-Übergänge bemerkbar — bzw. reine Band-Term-Übergänge — d. h., es treten keine Störungen durch Term-Term-Übergänge auf — aufgefaßt werden. Nimmt man an, daß bei geeigneter energetischer Lage eines Terms in der Nähe der Leitungsbandkante Band-Term- und Term-Term-Übergänge in vergleichbarer Zahl möglich sind, dann ist eine direkte Abhängigkeit der Bandenform von der Konzentration der Leitungselektronen zu erwarten, da die Verteilung der Elektronen auf das Band und den oberen Term von der Zahl der vorhandenen Elektronen abhängig sein wird. Ein derartiger Übergang, d. h. ein gemischter Band-Term- und Term-Term-Übergang, der von vornherein nicht ausgeschlossen werden kann, könnte für die Erklärung der Eigenschaften der kurzwelligen Emissionsbande von Cu-aktiviertem Zinksulfid in Frage kommen. Eine genaue Entscheidung wird wahrscheinlich erst getroffen werden können, wenn eine Berechnung von Band-Term-Übergängen durchgeführt ist oder wenn wesentlich weiter gehende Kenntnisse der Eigenschaften dieser Emissionsbande, des Emissionsmechanismus und des Zentrenaufbaus vorhanden sind.

Für die Möglichkeit der Durchführung dieser Arbeit bin ich Herrn Prof. Dr. R. Rompe zu großem Dank verpflichtet. Herrn Dr.-Ing. H. Ortmann gebührt mein Dank für sein stetes und förderndes Interesse. Mein besonderer Dank gilt ferner den Herren Dr. E. Gutsche und Dr. W. Ludwig für wertvolle Diskussionen.

Literatur

[1] M. Schön, Phys. Z. **39**, 940 (1938).
[2] N. Riehl und M. Schön, Z. Phys. **114**, 682 (1939).
[3] M. Schön, J. Phys. Radium **17**, 689 (1956).
[4] S. Roberts und F. E. Williams, J. Opt. Soc. Amer. **40**, 516 (1950).
[5] H.-J. Schulz, phys. stat. sol. **3**, 485 (1963).
[6] F. A. Kröger, J. E. Hellingman und N. W. Smit, Physica **15**, 990 (1949).
[7] F. A. Kröger und J. Dikhof, Physica **16**, 297 (1950).
[8] F. A. Kröger und N. W. Smit, Physica **16**, 317 (1950).
[9] J. S. Prener und F. E. Williams, Phys. Rev. **101**, 1427 (1956).
[10] F. E. Williams, J. Opt. Soc. Amer. **47**, 869 (1957).
[11] F. E. Williams, J. Phys. Chem. Solids **12**, 265 (1960).
[12] J. Broser und R. Broser-Warminsky, J. Phys. Radium **17**, 791 (1956).
[13] F. A. Kröger, Some Aspects of the Luminescence of Solids, Amsterdam 1948.
[14] H. W. Leverenz, An Introduction to Luminescence of Solids, New York 1950.
[15] N. Riehl und H. Ortmann, Über den Aufbau der Zinksulfidluminophore, Monographie Nr. 72, Verlag Chemie, Weinheim/Bergstr. 1957.
[16] W. van Gool, Dissertation Amsterdam 1961, Philips Res. Rep. **3**, 1 (1961).
[17] M. Schön, Ann. Phys. (Germany) **3**, 333 (1948).
[18] C. C. Klick, Phys. Rev. **85**, 154 (1952).
[19] J. J. Markham, Rev. mod. Phys. **31**, 956 (1959).
[20] S. I. Pekar, Fortschr. Phys. **1**, 367 (1954).
[21] R. H. Bube, Phys. Rev. **80**, 655 (1950).
[22] A. Halperin und H. Arbell, Phys. Rev. **113**, 1216 (1959).
[23] A. Halperin und H. Arbell, J. chem. Phys. **34**, 879 (1961).
[24] N. R. Nail, F. Urbach und D. Pearlman, J. Opt. Soc. Amer. **39**, 690 (1949).
[25] H. Eder, Z. Phys. **166**, 328 (1962).

[26] N. Riehl und P. Thoma, Z. angew. Phys. **14**, 751 (1962).
[27] F. Abeles und J. P. Mathieu, Ann. Phys. (France) **3**, 5 (1958).
[28] A. Mitsuishi, H. Yishinaga und S. Fyjita, J. Phys. Soc. Japan **13**, 1235 (1958).
[29] F. A. Kröger und H. J. G. Meyer, Physica **20**, 1149 (1954).
[30] R. E. Halsted und B. Segall, Phys. Rev. Letters **10**, 392 (1963).
[31] E. Mollwo, Z. Phys. **85**, 56 (1933).
[32] J. D. Konitzer und J. J. Markham, J. chem. Phys. **32**, 843 (1960).
[33] J. J. Markham und J. D. Konitzer, J. chem. Phys. **34**, 1936 (1961).
[34] M. Schön, Z. techn. Phys. **19**, 361 (1938).
[35] K. Birus und M. Schön, Verh. Dtsch. Phys. Ges. **19**, 83 (1938).
[36] F. Möglich und R. Rompe, Phys. Z. **41**, 236 (1940).
[37] F. Möglich und R. Rompe, Z. Phys. **115**, 707 (1940).

(Received June 6, 1964)

See Short Note by P. KRATOCHVÍL and M. BOČEK

Fig. 2. The slip lines at the beginning of plastic deformation (magnification 3500 ×)

Fig. 3. Homogeneously covered surface with slip lines at the end of stage A (magnification 6000 ×)

See Short Note by P. Kratochvíl and M. Boček

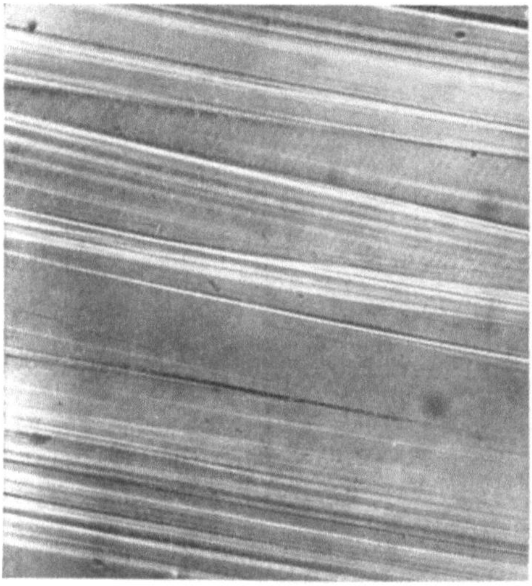

Fig. 5. The slip pattern at the beginning of stage B on the surface parallel to the slip direction (magnification 400×)

Fig. 6. Fanning of slip lines in stage B (magnification 400×)

Fig. 7. Slip lines due to the primary and secondary glide systems at the end of the stress-strain curve (magnification 3500×)

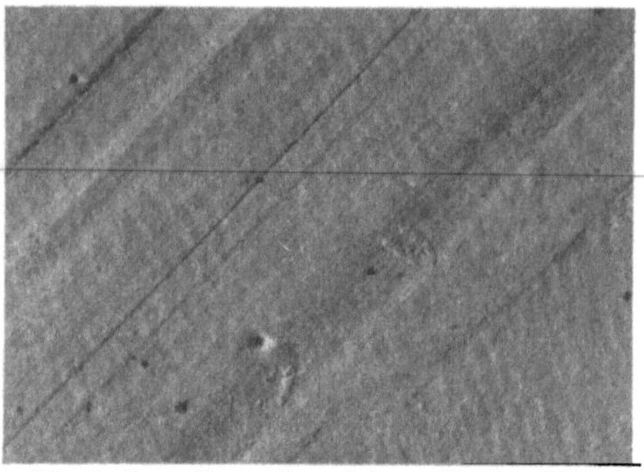

Fig. 8. Cross-slip in stage C (magnification 3500×)

Pre-printed Titles and Abstracts

Department of Solid State Physics, Faculty of Mathematics and Physics, Charles University, Prague

Slip Lines in Deformed Zinc Single Crystals

By

P. KRATOCHVÍL and M. BOČEK

Recently the work-hardening behaviour of bulk zinc crystals has received considerable attention (1 to 8). The observations of the surface of deformed crystals have been performed by SEEGER and TRÄUBLE (1), HAASEN and SIEMS (7) and LÜCKE et al. (8). In the present letter the surface slip patterns in zinc in different stages of the stress-strain curve have been studied to support some suggestions of previous papers (2 to 6).

Zinc single crystals (estimated purity 99.999 at.%) were grown from the melt in the manner described by BOČEK and KASKA (4). Crystals of the square cross section 0.5 x 0.5 cm were used, with an orientation giving the slip vector nearly parallel to a pair of surface planes. They were deformed at strain rates of approximately $10^{-4}s^{-1}$ in the Polanyi type tensile testing machine at room temperature. The method of subsequent deformations used by SEEGER and TRÄUBLE (1) has been employed.

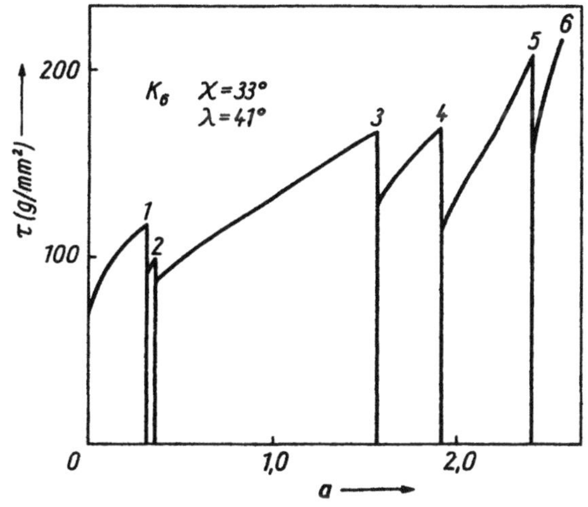

Fig. 1. An example of the resolved shear stress-strain curve (for crystal No. K_6). χ and λ are the angles between the direction of the applied force and the slip plane or the slip direction, respectively

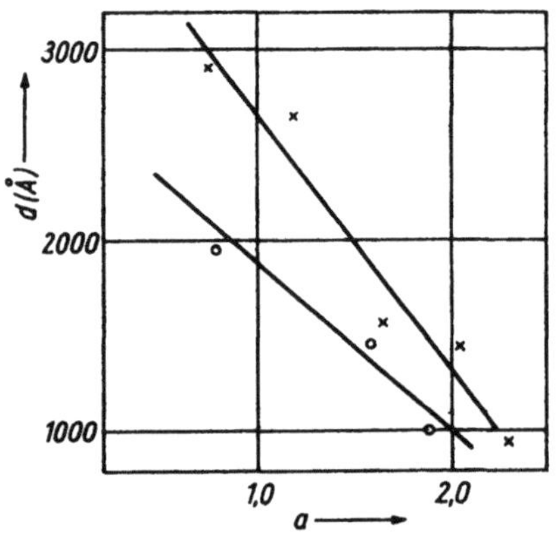

Fig. 4. The decrease of the distance d between the active slip planes with increasing strain in stage A

The erasing of slip patterns by chemical polishing and handling the sample during one step (at points 1 to 6 in Fig. 1) of the experiment did not exceed 10 min. The typical resolved shear stress-strain curve is illustrated in Fig. 1. Slip patterns of several crystals were studied, but only a few typical ones are discussed below.

At the beginning of deformation the distribution of slip lines is inhomogeneous (Fig. 2 - point 2). Slip proceeds not only along the existing slip lines, but is accompanied by appearing of new ones. At the end of the stage A of the stress-strain curve all the observed surface becomes covered with slip lines (Fig. 3, corresponds to point 3 in Fig. 1). The distance d between the neighbouring active planes decreases with the increasing strain in stage A (Fig. 4). In stage B undeformed regions appear between the slip bands (Fig. 5) and consequently the average distance d slightly increases. From the beginning of stage B the so-called fanning of slip lines is observed (Fig. 6) on the surface nearly parallel to the slip direction. The stage B is at its end characterized by appearing of non-basal glide system (Fig. 7) with the slip direction different from that of the basal glide. The plane $(11\bar{2}2)$ was identified (9) as the slip plane corresponding to the slip lines of the non-basal glide system in Fig. 7. There are two other phenomena observable in stage C together with the non-basal glide and fanning. These are 1) cross slip

(Fig. 8), 2) twinning and subsequent deformation of twinned regions.

SEEGER and TRÄUBLE (1) have observed the fanning of slip lines in all stages A, B, and C of the stress-strain curve. According to them the fanning of slip lines has been explained (1) by the thermally activated climb of edge dislocations, that enables activation of neighbouring slip planes and thus also spreading of the slip through all the material in stage A. In our case fanning does not appear in stage A. Fanning process might be connected only with the stages where glide obstacles appear and is one of possible processes to push the dislocations through the hardened crystal and to realise slip in stages B and C. The pile-ups of dislocations at the obstacles can spread out into neighbouring slip planes. The difference between the observations of SEEGER and TRÄUBLE and ours can be accounted for by different impurity contents. This view is supported by recent experiments on doped zinc performed in our laboratory (9). The decrease of work-hardening in stage C according to our experiments could be due to the activity of non-basal glide systems as well as to the cross slip. The observations are in agreement with the suggestion of possible activation of non-basal glide systems, when the suitable orientation of the sample has been reached (6). ŠVÁBOVÁ (5) deduced that the non-basal glide system in zinc to be activated is the $(11\bar{2}2)$ plane containing the $[11\bar{2}3]$ slip direction. Fig. 7 shows that the slip direction differs actually from the $[11\bar{2}0]$ direction of the basal glide system (note the different shadow character). The mentioned non-basal glide process can give rise to effective glide obstacles (6).

The authors wish to thank Mister P. OSIF for careful taking of replicas.

References

(1) A. SEEGER and H. TRÄUBLE, Z. Metallk. 51, 436 (1960).
(2) M. BOČEK, Czech. J. Phys. B10, 830 (1960).
(3) M. BOČEK, Czech. J. Phys. B10, 841 (1960).
(4) M. BOČEK and F. KASKA, phys. stat. sol. 4, 325 (1964).
(5) M. ŠVÁBOVÁ, phys. stat. sol. 3, 1819 (1963).

(6) M. BOČEK et al., phys. stat. sol. 4, 343 (1964).
(7) P. HAASEN and R. SIEMS, Z. Metallk. 46, 860 (1955).
(8) F. LÜCKE et al., Z. Metallk. 46, 860 and 792 (1955).
(9) R. KALINOVÁ, MSc-Thesis, Prague 1964.

(Received June 19, 1964)

Институт Кристаллографии Академии Наук СССР, москва

О влиянии гидростатического давления на фотопроводимость полупроводников

В.М. ФРИДКИН и Н.А. ТИХОМИРОВА

Влияние высокого гидростатического давления на фотопроводимость полупроводника (или диэлектрика) может быть обусловлено изменением времени жизни или подвижности неравновесных носителей. Согласно (1, 2) для монокристаллов серы и сернистого кадмия изменение подвижности электронов с давлением не вносит существенного вклада в зависимость фототока от давления. Ниже мы покажем, что на основе простой феноменологической модели РОУЗА (3) может быть получена зависимость времени жизни неравновесных носителей от давления.

Здесь мы рассмотрим частный случай, когда два уровня рекомбинации, электронный с энергией активации E_{r1}, и дырочный с энергией активации E_{r2}, расположены вблизи соответствующих демаркационных уровней с энергиями соответственно E_{dp} и E_{dn}. Согласно (3) выражения для времени жизни электронов τ_n и дырок τ_p в этом случае имеют вид:

$$\tau_n = \tau_{no} \exp\left(\frac{E_{dp} - E_{r1}}{kT}\right), \qquad (1)$$

$$\tau_p = \tau_{po} \exp\left(\frac{E_{dn} - E_{r2}}{kT}\right). \qquad (1')$$

Если концентрации электронных и дырочных уровней рекомбинации равны, то $E_{dn} = E_{fn}$ и $E_{dp} = E_{fp}$, где E_{fn} и E_{fp} энергии квази-уровней Ферми соответственно для электронов и дырок. Соотношения (1) и (1') должны быть дополнены выражениями, связавающими концентрации n и p свободных электронов и дырок с положением соответствующих квази-уровней Ферми:

$$n = N_c \exp\left(-\frac{E_{fn}}{kT}\right) = f\tau_n, \quad \frac{\partial \ln\tau_n}{\partial P} = -\frac{1}{kT}\,\frac{\partial E_{fn}}{\partial P}, \qquad (2)$$

$$p = N_v \exp\left(-\frac{E_{fp}}{kT}\right) = f\tau_p, \quad \frac{\partial \ln\tau_p}{\partial P} = -\frac{1}{kT}\,\frac{\partial E_{fp}}{\partial P}. \qquad (2')$$

Здесь f возбуждение, P давление, причем эффективные плотности состояний N_c и N_v по предположению не зависят от давления P. Изменение ширины запрещенной зоны E_g с давлением может привести к изменению расстояния между демаркационными уровнями (или квазиуровнями Фэрми) и уровнями рекомбинации и, следовательно, к изменению времени жизни носителей. При этом согласно (1) и (2), а также (1') и (2') должны быть всегда выполнены соотношения:

$$\frac{\partial E_{fn}}{\partial P} + \frac{\partial E_{dp}}{\partial P} = \frac{\partial E_{r1}}{\partial P}\ , \tag{3}$$

$$\frac{\partial E_{fp}}{\partial P} + \frac{\partial E_{dn}}{\partial P} = \frac{\partial E_{r2}}{\partial P}\ . \tag{3'}$$

Согласно (3) энергии квази-уровня Ферми и демаркационного уровня связаны следующими соотношениями:

$$E_{dn} = E_{fp} - kT\ \ln\left(\frac{S_p}{S_n}\right) - \frac{3}{2}\ kT\ \ln\left(\frac{m_p}{m_n}\right), \tag{4}$$

$$E_{dp} = E_{fn} + kT\ \ln\left(\frac{S_p}{S_n}\right) - \frac{3}{2}\ kT\ \ln\left(\frac{m_p}{m_n}\right), \tag{4'}$$

где сечения рекомбинации дырок s_p и электронов s_n , а также эффективные массы дырок m_p и электронов m_n по предположению не зависят от давления. Из (4) и (4') находим, что

$$\frac{\partial E_{fn}}{\partial P} = \frac{\partial E_{dp}}{\partial P} \quad \text{и} \quad \frac{\partial E_{fp}}{\partial P} = \frac{\partial E_{dn}}{\partial P}\ .$$

Подставляя эти соотношения в (3) и (3') и принимая, что

$$\frac{\partial E_{r1}}{\partial P} \approx \frac{\partial E_{r2}}{\partial P} \approx \frac{\partial E_g}{\partial P}\ ,$$

преобразуем (2) к следующему окончательному виду:

$$\frac{\partial \ln \sigma_n}{\partial P} = \frac{\partial \ln \tau_n}{\partial P} = -\frac{1}{2kT}\ \frac{\partial E_g}{\partial P}, \tag{5}$$

где σ_n фотопроводимость, обусловленная электронами. Аналогичное выражение может быть написано для дырок.

Согласно (5) уменьшение ширины запрещенной зоны с давлением приводит к росту времени жизни и, следовательно, увеличению фототока и наоборот. В таблице 1 мы приводим экспериментальные значения

$$\frac{\partial}{\partial P} \ln \frac{\sigma}{\sigma_0} \simeq \frac{\partial}{\partial P} \ln \frac{\tau}{\tau_0} ,$$

полученные для монокристаллов сернистого кадмия (2) и серы (1) для давлений до $\sim 10^4$ атм. Из данных таблицы видно, что соотношение (5) правильно передает не только знак, но и величину эффекта.

Обращает на себя вимание, что соотношение (5) было получено К.БЁРОМ, Е.ГУЧИ и Е.ЯНЕ на основе иной кинетической модили, учитывающей влияние уровней прилипания с энергией активации E_r на стационарное значение фототока (2). При этом, как и в нашем случае, авторы (2) существенно используют предположение

$$\frac{\partial E_r}{\partial P} \simeq \frac{\partial E_g}{\partial P} ,$$

которое в рамках выполненного рассмотрения остается необоснованным. Во всяком случае ясно, что из одних только измерений зависимости стационарной фотопроводимости от давления (без привлечения данных кинетики) нельзя сделать однозначных выводов в отношении детального механизма явления.

Таблица 1

кристалл	$\left(\frac{\partial E_g}{\partial P}\right)_T$	$-\frac{1}{2kT}\left(\frac{\partial E_g}{\partial P}\right)_T$	$\frac{\partial}{\partial P} \ln \frac{\sigma}{\sigma_0}$	Авторы
	(эb · атм$^{-1}$)	(атм$^{-1}$)	(атм$^{-1}$)	
S	$-7{,}2 \cdot 10^{-6}$	$+1{,}4 \cdot 10^{-4}$	$+2{,}2 \cdot 10^{-4}$	(1)
CdS	$+4 \cdot 10^{-6}$	$-0{,}8 \cdot 10^{-4}$	$-0{,}6 \cdot 10^{-4}$	(2)

Литература

(1) Н.А. ТИХОМИРОВА и В.М. ФРИДКИН, Физ. тверд. Тела 5, 2709 (1963).

(2) K.W. BÖER, E. GUTSCHE, and E. JAHNE, phys. stat. sol. 3, 922 (1963).

(3) Р. БЬЮБ, Фотопроводимость твердых тел, Издательство иностранной литературы, Москва 1962 (гл. 10, с. 372).

(Received June 15, 1964)

Institute of Physics, Czechoslovak Academy of Sciences, Prague

Anomalous Temperature Dependence of Magnetization of Hematite

By

J. KACZÉR and T.A. SHALNIKOVA

DZIALOSHINKY's paper (1) on the thermodynamical theory of weak ferromagnetism strongly increased interest in this subject in the last years. One of the classical materials exhibiting this property is hematite (α-Fe_2O_3). The temperature dependence of the magnetic properties of hematite single crystals was the subject of investigation in a number of recent papers (2 to 5). All these papers reported a phase transition from the antiferromagnetic to the weak ferromagnetic state. The transition temperature was found to be in the neighbourhood of -15 $^{\circ}$C, the width of the transition being in the temperature interval of a few degrees to a few tens of degrees. The saturation moment σ_0 was found by several authors to lie between 0.34 and 0.45 emu/g.

In connection with work on the magnetic properties of weak ferromagnets (6, 7) we investigated synthetic single crystals of hematite[1)]. The crystals, in the shape of thin platelets weighing about 0.5 mg, were measured on an automatic torsion balance (8) over a temperature range from -196 $^{\circ}$C to 20 $^{\circ}$C and for magnetic field strengths up to 6 kOe. The (111) direction was perpendicular to the suspension. Fig. 1 shows typical records of the torque L at three temperatures, i.e. below (a), in the region (b) and above the transition temperature (c), which, in agreement with the cited papers, was found in the neighbourhood of -13 $^{\circ}$C. Below the transition temperature curve (a) corresponds to the antiferromagnetic state with spins lying in the (111) direction which is state I in DZIALOSHINSKY's notation. In this region the paramagnetic susceptibility obtained by us $\chi_{\perp} - \chi_{\parallel} \sim 10^{-5}$ g^{-1} is in agree-

1) The crystals were prepared by M. VICHR of the Institute of Solid State Physics, as well as by Dr. Z. HAUPTMAN of the Institute of Physics, Czechoslovak Academy of Sciences, Prague.

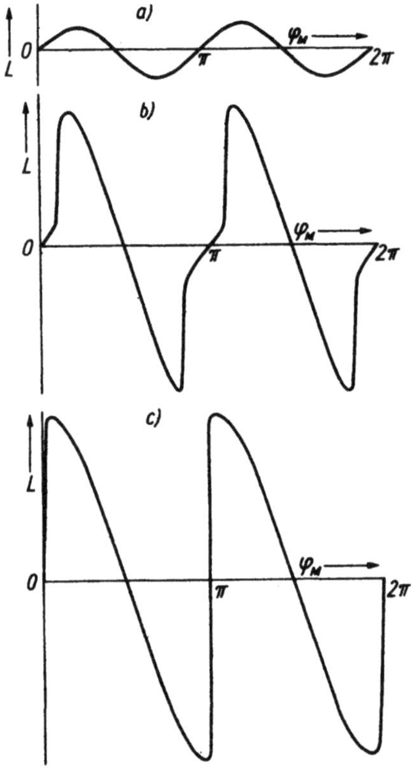

Fig. 1. Torque curves obtained at three temperatures in the neighbourhood of transition point: a) below, b) at and c) above transition

ment with earlier published data (3 to 5). Above the transition temperature the crystal exhibits weak ferromagnetism (curve c, in DZIALOSHINSKY's notation states II, III). For the saturation moment we obtained values ranging from 0.33 to 0.40 emu/g with an average (over 5 crystals) of 0.35 emu/g (± 5%).

Detailed measurement in the neighbourhood of the transition showed an anomalous dependence of the torque on the angle φ_M between the crystal axis (111) and the direction of the magnetic field. Fig. 1b shows this anomaly for a temperature of about −12 °C. The curves in this temperature region are composed of a paramagnetic portion, the amplitude of which is practically independent of temperature but is a quadratic function of the intensity of the applied field, and of a ferromagnetic portion, the amplitude of which as well as the angle φ_{MC} at which it suddenly appears, depends both on the temperature and the intensity of the applied field. Depending on the quality of the crystals, some variations in the shape of these curves were observed.

From these torque measurements we obtained a family of curves $\sigma(H,T)$, three of which, corresponding to the curves a,b,c in Fig. 1, are shown in Fig. 2. From the b-type curves it is evident that the transition from the antiferromagnetic

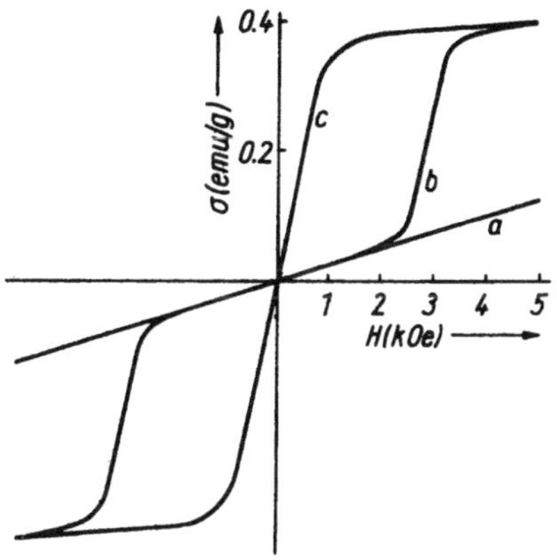

Fig. 2. Dependence of magnetization on external field and temperature for three different temperatures a,b,c corresponding to Fig. 1

state I to the weak ferromagnetic state (II, III) is field induced (12) by a critical field $H_c(T)$ and is to a certain extent analogous to field induced transitions from the antiferromagnetic to the ferromagnetic state (9 to 11). Extrapolation of $\sigma(H,T)$ to zero yields a linear dependence for $H_c(T)$ from which, if we put $H_c = 0$, a value for the transition temperature in zero field of $T_c = -11$ °C is obtained.

References

(1) I.E. DZIALOSHINSKY, Zh. eksper. teor. Fiz. 32, 1547 (1957).
(2) S.T. LIN, J. appl. Phys. S 32, 395 (1961).
(3) S.T. LIN, J. Phys. Soc. Japan 17 B-I, 226 (1962).
(4) A. TASAKI, K. SIRATORI, and S. IIDA, J. Phys. Soc. Japan 17 B-I, 232 (1962).
(5) A. TASAKI and S. IIDA, J. Phys. Soc. Japan 18, 1148 (1963).
(6) J. KACZÉR and N.M. KREINES, Zh. eksper. teor. Fiz. 41, 1691 (1961); Soviet Physics JETP 14, 1202 (1962).
(7) J. KACZÉR, Zh. eksper. teor. Fiz. 43, 2042 (1962); Soviet Physics JETP 16, 1443 (1963).
(8) J. KACZÉR, Czech. J. Phys. B 13, 386 (1963).
(9) S.S. SHALIT, Zh. eksper. teor. Fiz. 8, 518 (1938).
(10) H. BIZETTE, C. TERRIER, and B. TSAI, C.R. 243, 895 (1956).

(11) A.S. BOROVIK-ROMANOV and N.M. KREINES, Zh. eksper. teor. Fiz. 35, 1053 (1958).
(12) N.M. KREINES, Zh. eksper. teor. Fiz. 40, 762 (1961).

(Received June 25, 1964)

Laboratoire d'Électricité, Université Al.I. Cuza, Iassy

Sur la théorie de l'effet Procopiu

Par

M. SOROHAN

Le phénomène connu sous le nom effet Procopiu (1) a été découvert en 1930 (2). Il se manifeste par l'apparition, dans une boboine coaxiale avec un fil ferromagnétique, d'une f.e.m. de fréquence double par rapport à celle du courant alternatif (50 Hz) qui traverse le fil. Les caractéristiques de ce phénomène ont été étudiées expérimentalement par plusieurs auteurs (1 à 5), et par suit, il est utilisé à l'étude des propriétés ferromagnétiques des couches minces (6, 7, 8). En ce qui concerne la théorie de ce phénomène il y a quelques essais d'explications (2, 9, 10), mais seulement qualitatives et sans vérifications expérimentales directes.

Dans cette note nous présentons une relation établie entre l'intensité de l'effet Procopiu et les caractéristiques magnétiques du fil étudié (le champ coercitif H_c, le champ et l'induction maximum H_m et B_m, l'induction rémanente B_r).

Dans les conditions de l'effet Procopiu le fil ferromagnétique est soumis à un champ circulaire h_t sinusoïdal, et à un champ longitudinal H_l constant. Le fil, considéré homogène et isotrope, est aimanté d'après un cycle partiel d'aimantation dont les limites sont donnés par la valeur minimum H_l et la valeur maximum H_m du champ total $H = \sqrt{H_l^2 + h_t^2}$. Le champ total H prend une direction déterminée par l'angle α , mais l'induction magnétique B reste en retard avec l'angle δ , sa position étant déterminée par l'angle β (Fig. 1).

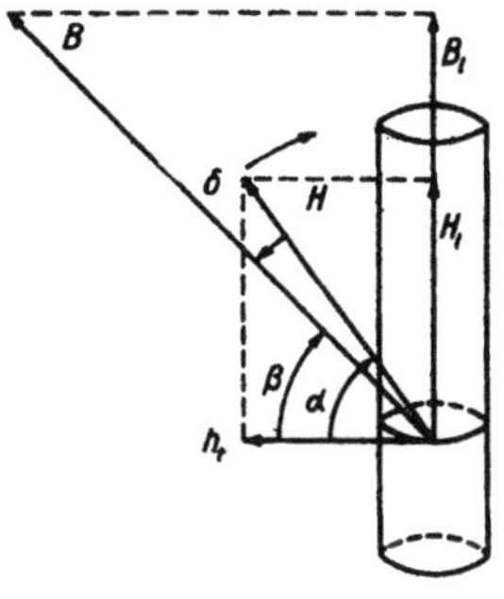

Fig. 1. Directions du champ et de l'induction par rapport à la direction du fil

La variation du flux qui produit la f.e.m. enregistré comme effet Procopiu peut être calculé si on connaît la variation de la composante longitudinale B_l de l'induction magnétique pendant une période du courant alternatif (50 Hz). Pour cela nous avons utilisé la relation (1)

$$I = I_s \left(1 - \frac{a_1}{H} - \frac{a_2}{H^2} \ldots\right) \tag{1}$$

donnée pour décrire l'approche de la saturation. Par une translation avec $2H_c$, et en supposant $B \simeq 4\pi I$ et $B_m \simeq B_s$, nous trouvons la relation (2)

$$B = B_m \left[1 - \frac{a_1}{H + 2H_c} - \frac{a_2}{(H + 2H_c)^2}\right], \tag{2}$$

qui décrit la branche décroissante du cycle d'hystérésis. Les constantes a_1 et a_2 sont déterminées par les conditions $B = B_r$ pour $H = 0$, et $B = B_m$ pour $H = H_m$. Donc nous trouvons

$$B = B_m - \frac{4H_c^2(B_m - B_r)}{H_m} \cdot \frac{H_m - \sqrt{H_l^2 + h_t^2}}{(\sqrt{H_l^2 + h_t^2} + 2H_c)^2} . \tag{3}$$

La composante longitudinale de l'induction a l'expression

$$B_l = B \cdot \sin(\alpha \mp \delta). \tag{4}$$

Pour vérifier expérimentalement la relation (4) nous avons utilisé un fil ayant les caractéristiques suivantes: H_c = 2,3 Oe, H_m = 10 Oe, B_r = 6500 G, B_m = 8250 G, H_l = 2,5 Oe. La variation de l'angle δ , en fonction du champ total H, a été déterminée expérimentalement. Pour cela nous avons aimanté le fil avec un champ tournant produit par la superposition d'un champ alternatif (50 Hz) longitudinal à un champ alternatif (50 Hz) circulaire (9). Ces deux champs sont égaux à la périphérie du fil et décalés avec 90°. A l'aide d'un oscillographe cathodique nous avons déterminé le décalage

Fig. 2. Variation de l'angle δ en fonction du champ total H

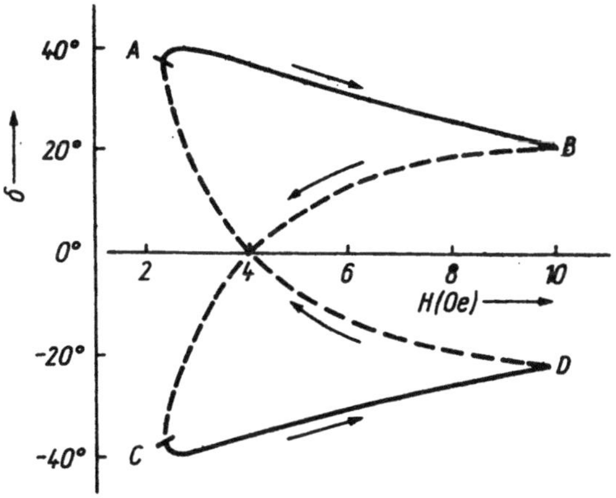

entre le champ longitudinal et la composante longitudinale de l'induction magnétique. Par suite, on détermine l'angle δ comme fonction de l'intensité du champ tournant (Fig. 2, AB et CD). Après le changement du sens de rotation du champ, l'angle δ a une autre variation. Des études effectuées sur des échantillons sous forme de disque, à l'aide d'un magnétomètre type Akulov, ont montré une variation représentée par la courbe BC et DA (Fig. 2). Nous avons considéré la même variation pour le fil à cause de leur caractéristiques magnétiques et de leur mode d'aimantation rapprochés.

En utilisant les caractéristiques du fil et les valeurs de l'angle δ nous avons calculé, à l'aide de la relation (4), la variation du flux en fonction du champ h_t. La Fig. 3 représente la variation du flux calculé et celle obtenue expérimentalement. La différence entre les données théoriques par rapport à celles expérimentales s'explique par les simplifications faites au commencement et par l'erreur de mesure sur l'angle δ . Bien que la relation (4) ne décrive pas parfaitement le phénomène, elle nous permet d'expliquer la variation de l'effet Procopiu avec différents facteurs.

Par exemple, l'apparition de ce phénomène est conditionnée de l'existence d'un champ total supérieur à une valeur bien déterminée approximativement le champ coercitif. Pour un cycle d'aimantation situé dans le domaine de Rayleigh nous avons utilisé, au lieu de la relation (1), la relation empirique de

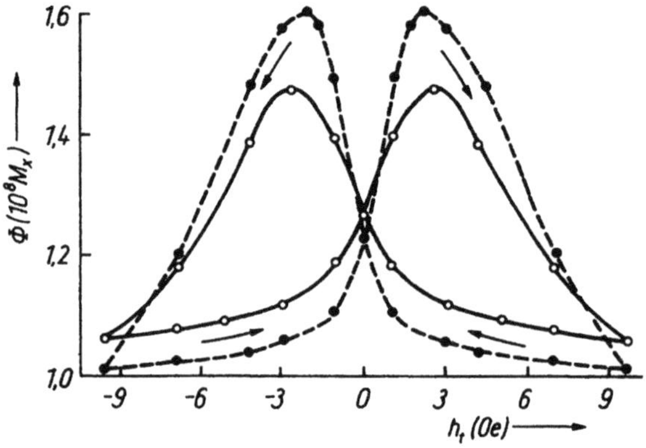

Fig. 3. Variation du flux en fonction du champ circulaire h_t
○ ○ ○ expérimentale
● ● ● théorique

Rayleigh (5) ou la relation théorique (6) établie par Procopiu (11).

$$B = aH + bH^2 , \tag{5}$$

$$B = a'H^3 - b'H^5 . \tag{6}$$

La variation du flux qui résulte dans ces conditions n'est pas semblable à celle indiquée dans la Fig. 3. Donc, l'effet Procopiu ne se manifeste pas, en concordance avec la constatation expérimentale mentionée plus haut. On peut aussi expliquer la saturation du phénomène et le passage par maximum avec la variation du champ longitudinal.

En abordant de ce point de vue l'étude de l'effet Procópiu, et en utilisant la grande sensibilité du phénomène à la variation de l'angle δ , nous pourrons établir d'importants résultats concernant l'anisotropie des fils et des couches minces déposées sur des fils nonferromagnétiques.

Bibliographie

(1) T. HOFBAUER et K.M. KOCH, Z. Phys. 130, 409 (1951).
(2) S. PROCOPIU, J. Phys. 1, 306 (1930).
(3) S. PROCOPIU et V. TUTOVAN, Stud. si cerc. Iassy 3-4, 143 (1951).
(4) F.A. KOCH, Z. Phys. 155, 475 (1959).
(5) S. PROCOPIU et I. VISCRIAN, Stud. şi cerc. Iassy XIV, 13, 285, 307 (1963).

(6) S. PROCOPIU, C.R. Acad. Sci. (France) 208, 1212 (1939).
(7) V. TUTOVAN et I. BURSUC, C.R. Acad.Sci. (France) 253, 405 (1961).
(8) I. BURSUC, A. OJOG et V. TUTOVAN, phys. stat. sol. 4, K61 (1964).
(9) M. SOROHAN, An.şt.Univ.Al.I.Cuza Iassy VIII, 253 (1962).
(10) G. VASILIU, Stud. si cerc. Iassy XI, 19 (1960).
(11) S. PROCOPIU, C.R. Acad.Sci. (France) 228, 475 (1949).

(Received July 1, 1964)

Institut für Experimentalphysik der
Technischen Universität Dresden

Sekundärelektronenemission und Reflexion von Ionen und Neutralteilchen an Kupfereinkristallen

Von
H. ZSCHEILE

Beobachtungen im Emissionselektronenmikroskop zeigen, daß die Sekundärelektronenausbeute einzelner Metallkristallite beim Beschuß mit schnellen Ionen von der Einschußrichtung abhängt. MAŠKOVA u.a. (1) haben an einem Einkristall die Abhängigkeit dieser Größe vom Einfallswinkel für eine feste Einfallsebene untersucht. FAGOT und FERT (2) bestimmten die Azimutabhängigkeit bei einigen Einfallswinkeln. Entsprechende Messungen über die Reflexion schneller Ionen gaben FLUIT u.a. (3) an.

Mitunter ist es vorteilhaft, im Emissionsmikroskop die Elektronen durch Beschuß mit schnellen Neutralteilchen auszulösen (4). Deshalb erschien es interessant, auch diese Auslöseart zu untersuchen. Es wurden die Sekundärelektronenemission und die Reflexion für Ionen und Neutralteilchen in Abhängigkeit vom Einfallswinkel ϑ und Azimut φ des Primärstrahles (Fig. 1) gemessen.

Die Ionen wurden aus einer Penningentladung extrahiert und in einem Wienfilter geringer Auflösung nach Massen getrennt. Durch Umladung in einer Gasstrecke konnten wahlweise auch Neutralteilchen erhalten werden. Als Objekt diente ein elektrolytisch polierter Kupfereinkristall, der auf einige hundert °C aufgeheizt war. Die Ionenstromdichte am Objekt war in der Größenordnung 10^{-5} A/cm^2, das Vakuum betrug etwa 10^{-6} Torr. Die

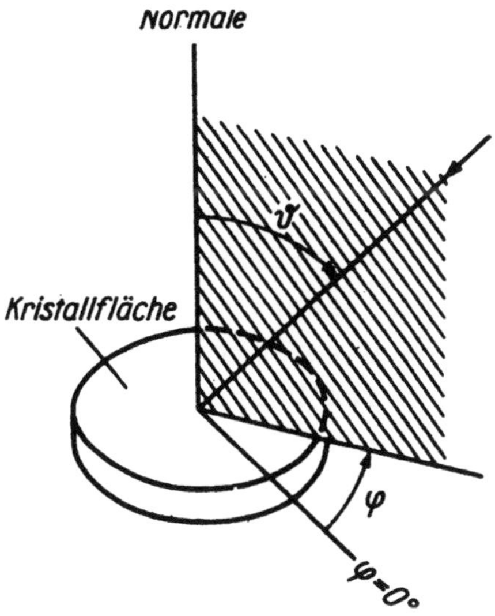

Fig. 1. Bedeutung der Winkel φ und ϑ

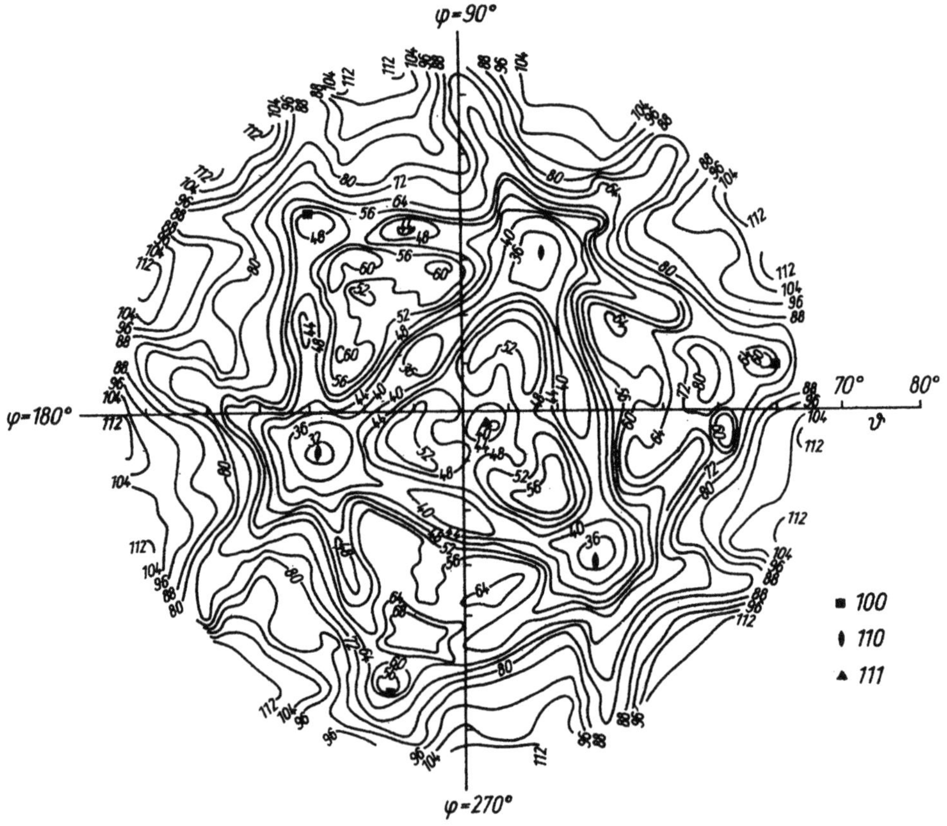

Fig. 2. Höhenliniendarstellung des Stromes $-I_-$ (in willkürlichen Einheiten) als Funktion der Einschußrichtung der Ionen

Sekundärströme wurden an einem Kollektor gemessen, der das Objekt umgibt. Zwischen ihm und dem Objekt lag eine Spannung von $\pm$ 30 V. Bei negativem Kollektor mißt man einen Strom I_+. Er besteht aus reflektierten Ionen und den von diesen oder von reflektierten Neutralteilchen am Kollektor ausgelösten Sekundärelektronen. Bei positivem Kollektor ergibt sich die Summe (Differenz der Beträge) I_- aus dem Sekundärelektronenstrom vom Objekt und dem Strom der schnellen reflektierten Ionen.

Fig. 2 zeigt die Richtungsabhängigkeit des Stromes I_- für eine ebene Kupferfläche, deren Normale ungefähr mit der

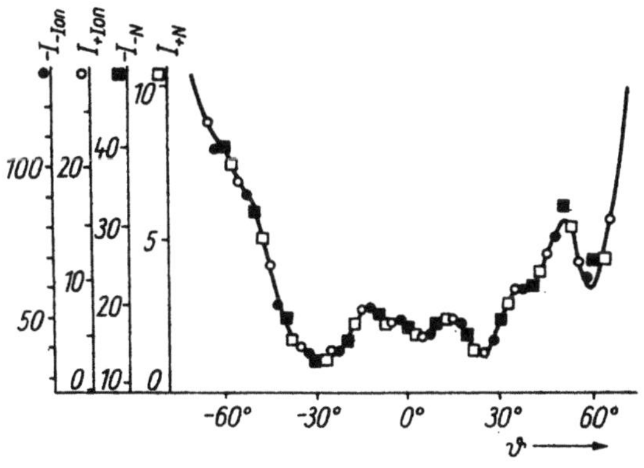

Fig. 3. I_- und I_+ für Ionen (Index Ion) und Neutralteilchen (Index N) in willkürlichen, jedoch gleichen Einheiten. Die primäre Teilchenstromstärke war für beide Vorzeichen der Kollektorspannung die gleiche; der Ionenstrom war etwa 2,8 mal größer als der Neutralteilchenstrom. Es ist nur ein Teil der Meßpunkte eingetragen

111-Richtung zusammenfällt. Die Einschußrichtungen sind in stereographischer Projektion dargestellt. Die eingetragenen Kristallrichtungen wurden aus einer Laueaufnahme bestimmt. Das Objekt wurde mit einem Gemisch von N_2^+- und wenig O_2^+-Ionen von 28 keV beschossen. Für niedrig indizierte Kristallrichtungen ergeben sich Minima des Stromes, in Übereinstimmung mit (2).

Einen Vergleich zwischen Ionen- und Neutralteilchenauslösung gibt Fig. 3. Das gleiche Objekt wie bei Fig. 2 wurde mit N_2^+ (und wenig O_2^+) bzw. N_2^0 (und wenig O_2^0) von 28 keV beschossen. Bei konstantem Azimut $\varphi = 8^0$ wurde der Einfallswinkel ϑ geändert. Beide Messungen von I_- lassen sich durch eine Kurve darstellen, wenn man verschiedene Ordinatenmaßstäbe verwendet, d.h., die Richtungsabhängigkeit ist für beide Fälle gleich. Auch die Werte I_+ lassen sich in diese Kurve einordnen, wenn man außer dem Maßstab auch den Ordinatennullpunkt ändert. Daraus folgt, daß sowohl der wahre Strom der Sekundärelektronen als auch derjenige der reflektierten Teilchen nach geeigneter Maßstabs- und Nullpunktswahl die gleiche Kurve ergibt. Offenbar sind die gleichen Ursachen für die Richtungsabhängigkeit der vier beobachteten Größen verantwortlich. Dagegen zeigt die Sekundäremission bei Elektronenbeschuß eine andere Winkelabhängigkeit (5). Wurden andere Ionenarten (Ar^+, H_1^+, H_2^+) der gleichen Energie verwendet, so änderten sich die relative Stärke und die Breite der Minima, ihre Lage blieb jedoch erhalten.

Die Einkristallpräparate und die Laueaufnahmen wurden im Institut für Röntgenkunde und Metallphysik der Technischen Universität Dresden angefertigt, wofür ich an dieser Stelle danken möchte.

Literatur

(1) E.S. MAŠKOVA, V.A. MOLČANOV und D.D. ODINCOV, Dokl. Akad. Nauk SSSR 151, 1074 (1963).
(2) B. FAGOT und C. FERT, C.R. 258, 1180 (1964).
(3) J.M. FLUIT, J. KISTEMAKER und C. SNOEK, Physica 30, 870 (1964).
(4) W. ENGEL und H. SCHLIEBE, 11. Tagung für Elektronenmikroskopie Zürich 1963.
(5) A.B. LAPONSKY und N.R. WHETTEN, Phys. Rev. 120, 801 (1960).

(Received July 3, 1964)

Max-Planck-Institut für Metallforschung, Stuttgart

Die mechanischen Eigenschaften von Aluminium-Zink- und Aluminium-Silber-Legierungen

Von

H. AUER, V. GEROLD, H. HABERKORN und W. ZÜRN

Es wird allgemein angenommen, daß die Erhöhung der mechanischen Festigkeit von Mischkristallen, die kugelförmige Guinier-Preston-Zonen enthalten, durch den erhöhten Widerstand hervorgerufen wird, den die Versetzungen beim Durchschneiden der Zonen erleiden (1). Für den erhöhten Widerstand werden verschiedene Ursachen diskutiert, wobei sich ergibt, daß die kritische Schubspannung τ_0 eine Funktion des Volumenanteils v_z der Zonen ist, jedoch völlig unabhängig von der Verteilung der Zonen, d.h. von ihrer Zahl und Größe sein sollte (1, 2). Nun ist jedoch bekannt, daß bei isothermer Auslagerung von Aluminium-Silber-Legierungen (3) bzw. von Aluminium-Zink-Legierungen (4, 5) die Härte wesentlich zunimmt. Da außerdem festgestellt worden ist, daß bei isothermer Auslagerung der Volumenanteil v_z der gebildeten Zonen konstant bleibt (abgesehen von einer kurzen Zeitspanne zu Beginn der Auslagerung), die Zahl und Größe der Zonen sich jedoch ändern (6, 7), kann man auch nach diesen Messungen eine Abhängigkeit der Härte von der Zonengröße vermuten.

Um diese Abhängigkeit von der Zonengröße direkt zu belegen, wurden sowohl an einer Aluminium-Silber-Legierung als auch an einer Aluminium-Zink-Legierung entsprechende Untersuchungen durchgeführt. Außerdem wurde mit der Messung der kritischen Schubspannung an Einkristallen verschiedener Aluminium-Zink-Legierungen begonnen.

In Fig. 1 sind die Resultate unserer bisherigen Untersuchungen zusammengestellt. Nahezu allen Messungen ist gemeinsam, daß die Härte und die kritische Schubspannung mit wachsendem Zonenradius zunächst stark zunehmen. Hat der Zonenradius einen Wert von etwa 30 Å erreicht, so tritt kaum noch eine Änderung auf. Eine Ausnahme bilden die Härtekurven der Aluminium-Silber-Legierung, die bei Temperaturen oberhalb 170 °C ausgelagert wurden. Die Silberkonzentration in den Zo-

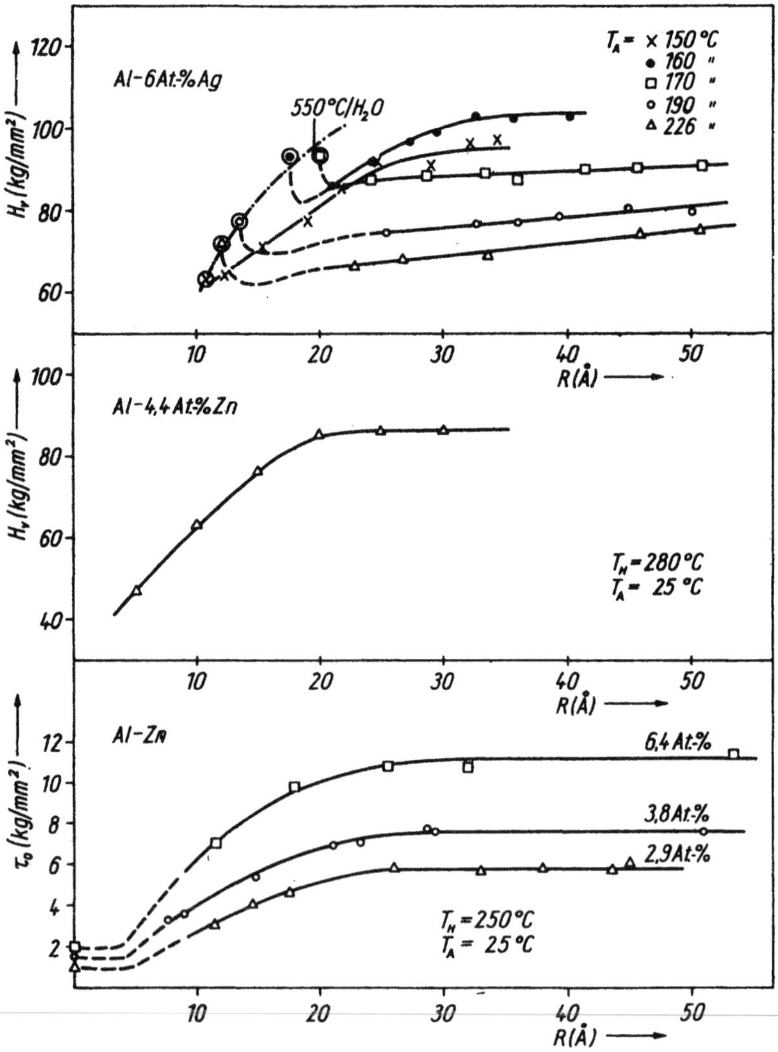

Fig. 1. Härte und kritische Schubspannung als Funktion des Radius der Guinier-Preston-Zonen

nen beträgt hier etwa 30 At.-%, während sie bei den niedrigeren Temperaturen auf etwa 55 At.-% ansteigt. Bei den untersuchten Aluminium-Zink-Legierungen beträgt die Zinkkonzentration in den Zonen etwa 80 At.-%.

Die kritische Schubspannung τ_o als Funktion der Teilchengröße R und dem Volumenanteil v_z der Zonen wird durch die folgende Formel gut wiedergegeben:

$$\tau_o = K\, v_z^{1/2}\, g(R). \qquad (1)$$

Dabei ist g(R) eine normierte Funktion, die für große Zonenradien den Wert eins erreicht. Die Größe v_z hängt von der Legierungskonzentration und von der Auslagerungstemperatur ab und läßt sich aus der metastabilen Mischungslücke berechnen.

Parallel dazu ausgeführte Strukturuntersuchungen zeigen, daß bei Aluminium-Silber-Legierungen bei Auslagerungstemperaturen unter 170 °C mit wachsendem Zonenradius eine Ordnung der Atome in den Zonen ausgebildet wird, die bei einem Zonenradius von etwa 30 Å ihren endgültigen Wert erhält (8). Bei den Aluminium-Zink-Legierungen ist keine geordnete Atomverteilung in den Zonen vorhanden. Dagegen spielen hier Verzerrungsfelder eine große Rolle. Jüngste Untersuchungen zeigen, daß diese Verzerrungsfelder bei einem Zonenradius von 30 Å eine wesentliche Änderung erfahren (9). Diese Strukturänderungen mögen die Abhängigkeit der kritischen Schubspannung vom Zonenradius beeinflussen, doch scheinen sie nicht den absoluten Betrag dieser Spannung zu erklären. Wir vermuten vielmehr, daß in den Zonen mit hohem Legierungsgehalt eine geringe Stapelfehlerenergie herrscht, die zu einer weiten Aufspaltung der Versetzungen in den Zonen führt.[1] Die hohe kritische Schubspannung ist dann darauf zurückzuführen, daß die aufgespaltenen Versetzungen beim Verlassen der Zonen wieder zusammengedrückt werden müssen. Auf Einzelheiten soll in einer späteren Arbeit näher eingegangen werden.

---

1) Die Idee der Versetzungsaufspaltung in G.P.-Zonen ist zuerst von A. KELLY im Zusammenhang mit Untersuchungen an Cu-Co-Legierungen geäußert worden (private Mitteilung).

Literatur

(1) A. KELLY und M.E. FINE, Acta metall. 5, 365 (1957).
(2) A. Kelly und R.B. NICHOLSON, Progr. in Mat. Science Vol. 10, Nr. 3, 1963 (S. 324).
(3) W. KÖSTER und F. BRAUMANN, Z. Metallk. 43, 193 (1952).
(4) R. GRAF, C.R. Acad. Sci. (France) 246, 1544 (1958).
(5) R.D. GARWOOD und A.L. DAVIES, J. Inst. Metals 88, 311 (1959/60).
(6) R. BAUR und V. GEROLD, Acta metall. 10, 637 (1962).
(7) V. GEROLD, phys. stat. sol. 1, 37 (1961).
(8) H. AUER, Diplomarbeit, Stuttgart 1962.
(9) W. MERZ, private Mitteilung.

(Received July 10, 1964)

IV. Physikalisches Institut der Humboldt-Universität zu Berlin

Über eine Änderung der Phononen-Emissionswahrscheinlichkeit bei der grünen Lumineszenz von CdS-Einkristallen

Von

O. GOEDE

Die grüne Lumineszenzbande, die bei UV-Anregung an undotierten CdS-Einkristallen bei hinreichend tiefen Temperaturen (unterhalb von etwa 150 °K) auftritt, besitzt bekanntlich eine charakteristische Struktur. Das Hauptmaximum liegt für eine Kristalltemperatur von 77 °K bei E_0 = 2,413 eV, zu langen Wellen hin folgen äquidistante Nebenmaxima, deren Abstand jeweils gleich der Energie $\hbar\omega_l$ = 0,038 eV eines longitudinalen optischen Phonons in CdS ist. Die einzelnen Maxima werden daher der Emission eines Photons bei gleichzeitiger Emission von 0,1,2... longitudinalen optischen Phononen zugeschrieben. Die relativen Höhen der einzelnen Maxima A_μ sollten nach HOPFIELD (1) einer Poisson-Verteilung folgen: $A_\mu \sim \bar{N}^\mu/\mu!$. Dabei ist μ = 0,1,2... die Zahl der im jeweiligen Maximum emittierten Phononen. $\bar{N}$ gibt die im Mittel pro Photon emittierte Anzahl von Phononen an und ist somit ein gewisses Maß für die Emissionswahrscheinlichkeit von longitudinalen optischen Phononen bei dem entsprechenden Elektronenübergang. Die Poisson-Verteilung der A_μ konnte bisher anhand von früher von KLICK (2) veröffentlichten Meßkurven der spektralen Verteilung der grünen Lumineszenz bei 4 °K verifiziert werden.

Im Rahmen des in (1) diskutierten vereinfachten Modells ist $\bar{N} = \sum_{\mathfrak{k}} |f_{\mathfrak{k}}|^2$, wobei die $f_{\mathfrak{k}}$ im wesentlichen durch die Fourierkoeffizienten der Ladungsdichte des im Lumineszenzzentrum getrappten Loches bestimmt werden, wenn die Lumineszenz durch die Rekombination freier Elektronen verursacht wird. Man sollte daher eigentlich erwarten, daß $\bar{N}$ bei fester Temperatur eine durch die Natur des Zentrums bestimmte für alle CdS-Kristalle charakteristische Konstante ist. Das ist jedoch nicht der Fall, wie Untersuchungen des Einflusses einer Vakuumtemperung der Kristalle auf die Struktur der grünen Lumineszenzbande ergaben, über die im folgenden berichtet werden soll.

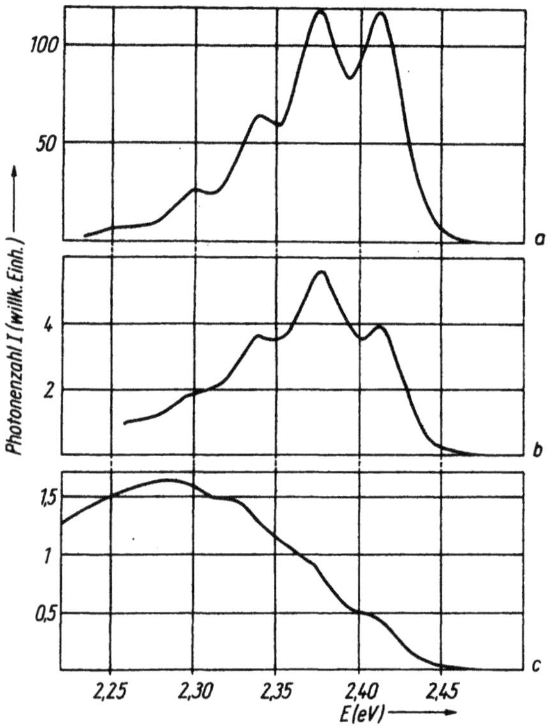

Fig. 1. Spektrale Verteilung der Intensität der grünen Lumineszenzbande für die Polarisationsrichtung $\mathfrak{E} \parallel \sigma$ (Kristalltemperatur T = 77 °K)
a) unbehandelter Kristall
b) nach einer kurzzeitigen Vakuumtemperung bei 400 °C
c) nach einer 15-minütigen Vakuumtemperung bei 400 °C

Fig. 1a zeigt die spektrale Verteilung der Lumineszenzintensität (korrigiert auf gleiche spektrale Empfindlichkeit des Empfängersystems) für einen unbehandelten Kristall bei 77 °K für die Polarisationsrichtung $\mathfrak{E}$ parallel zur c -Achse. Um bei der Bestimmung von $\bar{N}$ die bei dieser Temperatur vorhandene teilweise Überlappung der Teilbanden infolge der größeren Halbwertsbreite im Vergleich zu 4 °K zu berücksichtigen, wurde angenommen, daß sich die einzelnen Teilbanden in der Umgebung des jeweiligen Maximums durch Gaußkurven mit gleicher Halbwertsbreite approximieren lassen, was in guter Näherung erfüllt ist. Es zeigte sich, daß auch bei 77 °K die in dieser Weise bestimmten A_μ sehr gut einer Poisson-Verteilung folgen, wobei sich $\bar{N}$ = 0,98 ergibt.

In Fig. 1b ist die spektrale Verteilung der Lumineszenzintensität bei gleichen Anregungsbedingungen für denselben Kristall dargestellt, nachdem er sehr rasch im Hochvakuum auf etwa 400 °C erhitzt und dann sofort abgeschreckt wurde. Neben der Abnahme der Absolutintensität um fast eine Größenordnung stellt man eine charakteristische Änderung der relativen Höhe der Maxima A_μ fest. Sie folgen nach wie vor einer Poisson-Verteilung, jedoch ist jetzt das zweite Maximum das höchste[1], und es ergibt sich (nach entsprechender Korrektur) $\bar{N}$ = 1,5. Nicht geändert haben sich nach der Temperung die Lage des ersten Maximums E_0 und die Abstände der Maxima, die der Phononenenergie entsprechen. Eine noch größere Zunahme von $\bar{N}$ auf etwa 3 konnte durch eine länger (etwa 15 min) dauernde Vakuumtemperung erzielt werden (Fig. 1c).

Ein qualitativ gleicher Effekt der Zunahme von $\bar{N}$ bei gleichzeitiger Verringerung der Absolutintensität der grünen Lumineszenzbande als Folge einer Vakuumtemperung der Kristalle ergab sich auch für die Polarisationsrichtung $\mathfrak{E} \perp c$, die jeweils erhaltenen $\bar{N}$-Werte sind jedoch etwas kleiner als für $\mathfrak{E} \parallel c$. Die Lage der Maxima stimmt für beide Polarisationsrichtungen überein, die absolute Intensität ist für $\mathfrak{E} \perp c$ stets größer als für $\mathfrak{E} \parallel c$[2].

Der gleiche Effekt der Zunahme von $\bar{N}$ konnte auch durch Temperung der Kristalle im Schwefeldampf erzielt werden. Weitere Untersuchungen ergaben, daß $\bar{N}$ unabhängig von der Anregungsintensität (bei einer Variation über 3 Größenordnungen) ist.-Eine detaillierte Erklärung der beobachteten Zunahme der Phononen-Emissionswahrscheinlichkeit bei der grünen Lumineszenz von CdS als Folge einer Temperung und eine Deutung der anderen Effekte, insbesondere der Unterschiede von $\bar{N}$ für die beiden Polarisationsrichtungen, ist gegenwärtig noch nicht möglich. Es wird vermutet, daß die beobachtete Abnahme der absoluten Lumineszenzintensität nach der Temperung die Folge einer Zunahme des Anteils der strahlungslosen Übergänge in das Lumineszenzzentrum ist, wobei sich auch die Emissionswahrscheinlichkeit für longitudinale optische Phononen und damit $\bar{N}$ erhöht.

---

1) Über ein $\bar{N} > 1$ wurde auch in (3) berichtet.

2) Es zeigte sich, daß nach einer Vakuumtemperung das Verhältnis der Intensitäten $I^{\mathfrak{E} \perp c}/I^{\mathfrak{E} \parallel c}$ zunimmt, und zwar am stärksten für das Maximum mit $\mu = 0$.

Vor einer weiteren Diskussion sollen zusätzliche Messungen insbesondere auch bei tieferen Temperaturen durchgeführt werden.

Herrn Dr. GUTSCHE danke ich für wertvolle Diskussionen.

Literatur

(1) J.J. HOPFIELD, J. Phys. Chem. Solids 10, 110 (1959).
(2) C.C. KLICK, J. Opt. Soc. Amer. 41, 816 (1951).
(3) C.W. LITTEN und D.C. REYNOLDS, Phys. Rev. 125, 516 (1962).

(Received July 11, 1964)

Short Notes

Papers to be published in "physica status solidi"
Vol. 6, No. 3

Review Article

E.H. PUTLEY, Royal Radar Establishment, Malvern, Far Infra-Red Photoconductivity

Original Papers

T. PEČENÝ, Popov Research Institute of Radiocommunications, Prague, Proposal for Semiconductor Laser with Wide-Gap Emitters A known band structure is used to estimate the concentration of injected carriers necessary for inversion in the case of p-type GaAs. It is shown that a N-p-P structure with two wide-gap emitters N and P can improve the injection efficiency to such an extent that almost whole the current flowing through the laser results in recombination in the inverted region. A system is proposed which, besides employing N-p-P structure, uses semi-insulating GaAs for reduction of the junction area.

V. GALLINA and M. OMINI, FIAT Research Laboratories, Torino, Vacancy as a Phonon Field Perturbation(II) Using the general theory of Part I of this work, the lattice relaxation energy caused by a vacancy is given. Further, lattice centre displacements, produced by a vacancy pair, and the interaction energy between two vacancies have been evaluated. Some matrices required by our theory are tabulated; numerical results for argon at 80 ^{o}K are also given.

M. ROSENBERG, C. TANASOIU, and C. RUSU, Academia Republicii Populare Romine, Institutul de Fizica, Bucuresti, The Domain Structure of $Zn_{2-x}Co_xZ$ Single Crystals The domain structure of $Zn_{2-x}Co_xZ$ single crystals (with x = 0; 0.5; 2) was investigated. The observed patterns exhibit a gradually modification from a barium ferrite type structure (for Zn_2Z), passing through a cobalt type structure (for $Zn_{1.5}Co_{0.5}Z$), to a new characteristic structure of uniaxial single crystals with preferred plane of magnetization (for Co_2Z), consisting of a set of parallel domains separated by Bloch walls parallel to the hexagonal plane. All the observed domain structures are explained in terms of the domain theory and magnetic anisotropy dependence on composition.

CH. SCHWINK und G. GRIESHAMMER, II. Physikalisches Institut der Universität München, Untersuchungen des Fließbereichs neutronenbestrahlter Kupfereinkristalle (I) Elektronenmikroskopische Beobachtungen machen es sehr wahrscheinlich, daß bei vorsichtigem Anlassen die Zahl der Bestrahlungsfehler über 50 Å zunächst zunimmt und sich erst bei stärkerem Tempern langsam vermindert. Die Krümmung der Temperaturabhängigkeit der kritischen Schubspannung τ_0, die von 77 bis 510 °K gemessen wurde, im $\tau_0^{2/3} - T^{2/3}$ - Diagramm nichtangelassener Kristalle verschwindet mit dem Anlassen und kehrt sich dann sogar um. Es wird nahegelegt, letztere Krümmung unmittelbar mit der Dichteabnahme der Gleithindernisse in Zusammenhang zu bringen. Die Seegersche Theorie (1, 2) wird für diesen Fall modifiziert. Bei Raumtemperatur wächst τ_0 nichtangelassener Proben nicht proportional zu $\ln \dot{\varepsilon}$, erst mit zunehmendem Tempern nähert man sich einer Beziehung $\Delta\tau_0 \sim \ln \dot{\varepsilon}$. - Die Berechnung einer Aktivierungsenergie U_0 aus den Messungen ist nach den mitgeteilten Ergebnissen noch fragwürdiger als in älteren Arbeiten, lediglich die Zunahme von U_0 mit dem Anlassen scheint gesichert.

(1) A. SEEGER, Z. Naturf. 13a, 54 (1958).
(2) A. SEEGER, Proc. 2. UN Int. Conf. on Peaceful Uses of Atomic Energy 4, 250 (1958).

CH. SCHWINK und G. GRIESHAMMER, II. Physikalisches Institut der Universität München, Untersuchungen des Fließbereichs neutronenbestrahlter Kupfereinkristalle (II) Es wird eine kinematographische Methode beschrieben, mit deren Hilfe es gelingt, die Abgleitgeschwindigkeit $\dot{a}$ (zumindest eine untere Grenze) n-bestrahlter Cu-Kristalle zu messen. Da die Abgleitung inhomogen erfolgt, ergeben sich für ȧ viel größere Werte als bei homogener Gleitung. Bei einer Bestrahlung von rund $2x10^{18}$nvt (> 0,1 MeV) finden wir bei Zimmertemperatur ȧ = 1 bis 4 s^{-1}, verglichen mit ȧ ≈ $5x10^{-4}$ s^{-1} bei Annahme homogener Gleitung. Unser Verfahren erlaubt es uns weiter, den Zusammenhang zwischen Abgleitgeschwindigkeit ȧ und Dehngeschwindigkeit $\dot{\varepsilon}$ zu untersuchen. Wir finden eine lineare Beziehung zwischen diesen Größen, keine Proportionalität, wie bisher stets angenommen wurde. Dieses überraschende Ergebnis ermöglicht eine zwanglose Beschreibung der in Teil I mitgeteilten Messungen der Geschwindigkeitsabhängigkeit der kritischen Schubspannung. Alle mitgeteilten Messungen sind vorläufig nur bei Zimmertemperatur durchgeführt.

J.Z. DAMM and J. KOWALCZYK, Institute of Structural Research, Polish Academy of Sciences, Wrocław, The Influence of Plastic Deformation on Thermal Bleaching of Colour Centres in Potassium Chloride Crystals Irradiated by γ -Rays The isothermal bleaching of F-centres in the temperature range 106 to 185 °C was investigated in plastically deformed KCl crystals as well as in not deformed ones. In both cases the activation energy of the process was found to be equal to 1.30 ± 0.02 eV per F-centre. A bleaching mechanism based on diffusion of anion vacancies is proposed. The decrease of the bleaching rate in deformed materials is explained by the increase of the mean "micrograin" diameter.

H.G. GREWE und E. KAPPLER, Physikalisches Institut der Universität Münster, Strukturänderungen von vielkristallinem Kupfer nach extrem hoher plastischer Verformung Bei sehr großen plastischen Verformungen von Kupfer, wie sie durch Torsion zylindrischer Vollstäbe oder durch Walzen erreicht werden können, tritt eine Aufteilung der Kristallite in zahlreiche Bereiche verschiedener Orientierung auf. Die mittlere Größe der Bereiche wird bei hohen Verformungsgraden konstant. Sie ist umso kleiner, je tiefer die Verformungstemperatur ist. Hat man durch Tieftemperaturverformung ein feines Gefüge erzeugt, so erhält man nach anschließender Verformung bei Raumtemperatur ein gröberes Gefüge. Es wird angenommen, daß das bei hohen Verformungsgraden sich einstellende Gefüge das Ergebnis eines dynamischen Gleichgewichtes ist zwischen einem infolge zunehmender Verzerrung auftretenden Kornzerkleinerungsprozeß und einem dynamisch entweder durch Rekristallisation oder durch Subkornwachstum verursachten Kornwachstumsprozeß. Bei Raumtemperatur wurde statische Rekristallisation nachgewiesen. Die Gefügeuntersuchungen wurden elektronenmikroskopisch mit dem Nickelabdruckverfahren durchgeführt.

W. HEIMANN, Physikalisch-Technische Werkstätten, Wiesbaden-Dotzheim, Investigations of the Possibility of Reducing the Thermionic Sensitivity Occurring in Composite Photocathodes A study is made of the reduction of thermionic emissivity occurring in composite photocathodes of the Cs_2O-type. Proposals are given and experiments are reported to achieve this result by surface treatment of a Rb_2O-substrate in Cs-vapour. Reduction of at least two orders of magnitude of the thermionic emissivity could be reached, while still maintaining good sensitivity in the infra-red region. Furthermore interesting conclusions could be drawn from the experiments with regard to the mechanism of the emission processes.

H. HOFFMANN, II. Physikalisches Institut der Universität München, Wellenlänge und Kontrast der Magnetisierungsschwankungen einachsiger Permalloyfilme Der Einfluß der Schichtparameter einachsiger Permalloyfilme auf den Kontrast in der defokussierten elektronenmikroskopischen Abbildung wird theoretisch untersucht. Es zeigt sich, daß die Kristallitgröße und die lokale Anisotropiekonstante, die sich aus der Kristallenergie und Magnetostriktion des Einzelkristallits zusammensetzt, den Kontrast wesentlich beeinflussen. Die theoretische Analyse der Wellenlänge der Magnetisierungsschwankungen ergibt zwei verschiedene Wellenlängen, die sich im Ripple-Bild äußern. Die Wellenlänge λ_{SR} kurzer Reichweite gibt die Intensitätsschwankungen in der defokussierten Abbildung wieder, die Wellenlänge langer Reichweite λ_{LR} beschreibt die Vorzeichenänderung der Magnetisierungsschwankungen. Beide Wellenlängen zeigen je zwei unterschiedliche Bereiche der Abhängigkeit von der Kristallitgröße. In dem ersten Bereich ist λ konstant, im zweiten ist λ proportional zu D. Die experimentellen Ergebnisse von BALTZ und DOYLE (1) bestätigen die theoretischen Überlegungen.

(1) A. BALTZ und W. DOYLE, im Druck.

A.R.H. NIEDERMEYER, Institut für technische Elektronik der Technischen Hochschule München, Die Diffusionskoeffizienten von Antimon und Arsen in Germanium bei verschiedenen Konzentrationen und die Diffusion bestimmter Störstellenprofile durch Programmdiffusion Es wird beschrieben, wie die durch Diffusion unmittelbar an einer Germaniumoberfläche erreichten Dotierungskonzentrationen von Antimon und Arsen (Oberflächenkonzentrationen) ermittelt werden können. Bei dieser Bestimmung ergeben sich gleichzeitig die Diffusionskoeffizienten in Abhängigkeit von den Oberflächenkonzentrationen. Eine neue graphische Methode, die von einem mathematischen Verfahren abgeleitet ist, berücksichtigt die genannte Abhängigkeit und ermöglicht die Herstellung von verschiedenen Störstellenprofilen, auch von exponentiellen, durch Programmdiffusion (d.h. die Oberflächenkonzentration ist eine Funktion der Zeit). Die Übereinstimmung der theoretischen mit den experimentellen Werten ist recht gut.

R.W. DAVIDGE and P.L. PRATT, Metallurgy Department, Imperial College, London, Plastic Deformation and Work-Hardening in NaCl Single crystals of NaCl were compressed at room temperature. The stress-strain curve shows three well-defined stages similar to those found for fcc metals. The dislocation density, bulk density, and deformation characteristics were studied as functions of stress and strain. The recovery of the dislocation density and bulk density were compared with the recovery behaviour of flow stress and microhardness. During easy glide, slip occurs on a single system and hardening is interpreted in terms of the long-range stress between dislocations. In stage II, slip occurs on additional systems and it is suggested that the hardening is controlled by the forced intersection of dislocations on oblique systems through dislocation debris, in the form of dislocation dipoles, on the primary system.

M. BOČEK, Lehrstuhl für Festkörperphysik der mathematisch-physikalischen Fakultät der Karlsuniversität Prag und G. HÖTZSCH, Institut für Metallkunde und Materialprüfung der Bergakademie Freiberg, Die plastische Verformung von kadmiumlegierten Zinkeinkristallen (I) Es wird über das Verfestigungsverhalten kadmiumlegierter Zinkeinkristalle berichtet, die im dynamischen Zugversuch bei Temperaturen von 90 bis 573 °K verformt wurden. Gegenüber älteren Messungen an verunreinigten Zinkeinkristallen ergeben sich wesentliche Abweichungen im Einfluß des Fremdstoffgehaltes auf einige Kenngrößen der Verfestigungskurve. Es wird ferner gefunden, daß die Temperaturabhängigkeit der kritischen Schubspannung sowohl bei hohen als auch tiefen Temperaturen nicht mit den theoretischen Erwartungen für hexagonale Metalle übereinstimmt. Abschließend werden einige Ergebnisse über den Einfluß des Kadmiums auf die sprunghafte Verformung von Zinkeinkristallen mitgeteilt.

S. ARAJS and R.V. COLVIN, Edgar C. Bain Laboratory for Fundamental Research, United States Steel Corporation, Research Center, Monroeville, Electrical Resistivity of High Purity Iron from 300 to 1300 °K Electrical resistivity of high purity (99.99 %) iron has been studied as a function of temperature from 300 to 1300 °K. It has been found that the resistivity decreases by about 2 % going from α to γ iron. Such a large change has not been previously observed. The temperature dependence of the electrical resistivity of α and γ iron near the α-γ transition is not the same as has been reported in older investigations. The resistivity at the Curie temperature does not exhibit a sharp anomaly but changes gradually with temperature.

K. MEYER und E. GRAGERT, Institut für Physikalische Chemie der Deutschen Akademie der Wissenschaften zu Berlin, Deformations- und Rißstrukturen in NaCl-Kristallen bei der Stoßbearbeitung Es werden die bei der Stoßbearbeitung von (100)-NaCl-Oberflächen bei Temperaturen von -130 °C beobachteten Verformungsstrukturen in Abhängigkeit von der kinetischen Energie beschrieben. Bei geringen Stoßenergien bilden sich ausschließlich Schraubengleitbänder, denen sich bei höheren Energien Stufengleitbänder überlagern. Bei weiterer Steigerung der kinetischen Energie der Kugeln werden Risse sichtbar, die in (110)- und (100)-Ebenen verlaufen und sich nur selten in kristallografisch undefinierten Ebenen ausbreiten. Durch sukzessives Abtragen und Ätzen wird die räumliche Anordnung der durch den Stoß erzeugten Risse im Kristall bestimmt. Schematische Modellstrukturen verdeutlichen den komplizierten Rißverlauf. Die beobachteten Risse lassen sich z.T. durch Versetzungsreaktionen deuten. Es wird angenommen, daß die sich aus der Hertzschen Stoßtheorie (1) ergebenden maximalen Schubspannungen im Innern des bearbeiteten Kristalles Einfluß auf die Geometrie der beobachteten Rißstrukturen haben.

(1) H. HERTZ, Ges. Werke, Band 1, Leipzig 1895 (S. 135).

R. STRUMANE and R. DE BATIST, Solid State Physics Department, S.C.K. - C.E.N., Mol, Temperature Dependence of the Dislocation Line Charge in Sodium Chloride Single Crystals

The dislocation line charge has been shown to change sign twice between room temperature and 800 °C, in accordance with the theory of ESHELBY et al. (1). From the fact that at room temperature the specimens are in the so-called pseudo-intrinsic region and from the room temperature value of the line charge, derived from an earlier described model, the free energy of formation of cation vacancies is found to be 0.93 eV, in good agreement with the value determined by DAVIDGE (2).

(1) J.D. ESHELBY, C.W.A. NEWEY, P.L. PRATT, and A.B. LIDIARD, Phil. Mag. 3, 75 (1958).

(2) R.W. DAVIDGE, phys. stat. sol. 3, 1851 (1963).

V. MARINKOVIC and S. AMELINCKX, Solid State Physics Department, S.C.K. - C.E.N., Mol, Anti-Phase Domains and Dislocation Ribbons in Tealite ($PbSnS_2$) Evidence is presented for the presence of anti-phase boundaries in tealite ($PbSnS_2$). It is shown that the observed contrast effects are only consistent with one model of antiphase boundary. On intersecting anti-phase boundaries the dislocation ribbons change in width as a result of the difference in stacking fault energy. Such a behaviour is consistentwith the model deduced from the contrast effects.

R. GEMPERLE and M. ZELENÝ, Institute of Physics, Czechoslovak Academy of Sciences, Prague, Néel Walls in a Massive Uniaxial Ferromagnet in an External Field It was shown by calculating the energy of a Néel-type wall in an external field perpendicular to the easy axis and to the plane of the wall that such walls, hitherto known only in thin films, can exist also in a massive uniaxial ferromagnet. The theory put forward agrees with the results of experiments on magnetoplumbite.

J. BOK et C. GUTHMANN, Laboratoire de Physique de l'Ecole Normale Superieure, Paris, Effets des hauts champs électriques sur les phénomènes de transport dans InSb a basse température Nous avons étudié expérimentalement la conductivité électrique d'échantillons purs de type n d'InSb à 77 °K et à 20 °K en présence de hauts champs électriques. Nous avons mésuré l'effet Hall en régime d'impulsions. Pour interpréter les résultats nous avons utilisé l'approximation des "électrons chauds" dans le domaine de champ électrique compris entre la région ohmique et celle de l'avalanche. Les résultats sont en bon agrément avec la théorie de STRATTON (1) et montrent que la diffusion par les modes optiques polaires est le mécanisme principale d'échange d'énergie avec le réseau à 20 °K et 77 °K, et l'échange de quantité de mouvement à 77 °K. Ceci semble confirmer que la diffusion par le réseau est dominée par les collisions sur les modes optiques polaires dans InSb. A 20 °K, quand les chocs sur les impuretés ionisées gouvernent la diffusion à champ électrique nul, nous avons observé expérimentalement un maximum de mobilité en fonction du champ électrique pour un champ de l'ordre de 5 V/cm. Le fait que l'agrément entre théorie et expérience est moins bon à 20 °K pourrait enfin s'expliquer par l'influence de dislocations.

(1) R. STRATTON, Proc. Phys. Soc. A 242, 355 (1957).

J. GINTER, Institute of Experimental Physics, Warsaw University, On the Mixed Scattering by Optical Phonons and Ionised Impurities The temperature dependence of the Hall mobility for mixed scattering by optical phonons and ionised impurities has been calculated. It is assumed that below the Debye temperature a relaxation time exists which is given by $\tau_0 = C \varepsilon^0 (e^{\theta/T} - 1)$ for pure lattice scattering. The results are compared with own measurements for CdSe and Segall's experimental data for CdTe (1).

(1) B. SEGALL, M.R. LORENZ, and R.E. HALLSTED, Phys. Rev. 129, 2471 (1963).

K. FOSSHEIM and T. OLSEN, Fysisk Institutt, Universitetet i Oslo, Blindern, Magnetoacoustic Studies of the Fermi Surface of Aluminium Magnetoacoustic measurements in the frequency range 70 to 150 MHz have been done on super pure aluminium single crystals ($R_{293\,^oK}/R_{4.2\,^oK}$ up to 26 500). The directions of sound propagation were along the <100>, <110>, <111>, and <112> directions, and 16° off the <110> direction towards the nearest <100> direction, and a large number of oscillations were observed. The deduced Fermi surface dimensions for the second zone are in general agreement with Harrison's free-electron model. The data found with the direction of sound propagation along the three principal crystallographic directions agree with earlier investigations of KAMM and BOHM, and BEZUGLYI et al. Third zone dimensions in agreement with Ashcroft's calculations have been found. The conditions for magnetoacoustic oscillations are considered. Large angle intersection between the plane of the electron orbit and the contour at the point of interaction on the Fermi surface seem to be a fundamental requirement for pronounced oscillations. Orbits probably due to magnetic breakthrough at a field strength of 370 gauss are reported.

E. KAMIENIECKI, Institute of Physics, Polish Academy of Sciences, Warsaw, Hot Carriers in Microplasmas and their Radiation in Ge and Si The form of the distribution function for free carriers has been obtained by the use of Chambers' equation (1) under the assumption that the mean free path of the carriers is independent on energy. With this distribution function the spectrum of light emitted from reverse biased Ge and Si junctions has been calculated. The emitted light is believed to be due to bremsstrahlung of hot carriers in the

Coulomb field of charged centres. Good agreement is obtained with the experimental results of CHYNOWETH and GUMMEL (2), CHYNOWETH and MCKAY (3), and DAVIES and STORM (4) for Ge- and Si-junctions. The best fit is reached with field strengths of 10^5 - 10^6 V/cm (calculated for on assumed mean free path of 200 Å) and a lattice temperature in the microplasma region of approximately 350 to 700 °K.

(1) R.G. CHAMBERS, Proc. Roy. Soc. A 65, 458 (1952).
(2) A.G. CHYNOWETH and H.K. GUMMEL, J. Phys. Chem. Solids 16, 191 (1960).
(3) A.G. CHYNOWETH and K.G. McKAY, Phys. Rev. 102, 369 (1956).
(4) L.W. DAVIES and A.R. STORM, Phys. Rev. 121, 381 (1961).

J. MALINOWSKI and W. PLATIKANOWA, Institute of Physical Chemistry, Bulgarian Academy of Sciences, Decay of the Concentration of Photoexcited Holes in Silver Bromide The reaction between a silver deposit and the bromine formed by holes migrating up to the surface of a silver bromine crystal was used for checking the existence of mobile holes. The sensitivity of the method, greatly increased by the application of a process of development, allowed to follow up the concentration-decay curve of holes, generated by a light flash. For the different samples tested the effective life time, as determined from the initial slope of the decay curve, was found to be about 10^{-4} s. It was shown, however, that a small fraction of holes remains mobile up to about 10^{-2} s after their generation. This suggests that the process of concentration-decay is not likely to be described by a simple recombination mechanism.

W. RUSKE, Institut für Magnetische Werkstoffe der Deutschen Akademie der Wissenschaften zu Berlin, Jena, Über die Ummagnetisierung dünner polykristalliner Elektrolytnickelschichten bei spannungsinduzierter einachsiger Anisotropie (I) Der Einfluß von Korngrenzen auf die durch Dehnungsanisotropie erzeugte Wandverschiebungsfeldstärke wird untersucht. Die Theorie geht vom einfachen Beispiel des Bikristalls aus, dessen Einzelkristalle unterschiedliche Konstanten C uniaxialer Anisotropie und unterschiedliche Dicke D besitzen. C- und D-Inhomogenitäten an Korngrenzen sind eine Folge der Anisotropie der elastischen und magnetoelastischen Dehnungen bzw. eine Folge der Orientierungsabhängigkeit der Wachstumsgeschwindigkeit. Die am Bikristall erhaltenen Ergebnisse werden auf polykristalline Schichten erweitert. Es zeigt sich, daß die vorgeschlagene Theorie vor allem durch die Spannungs- oder Dehnungsabhängigkeit und durch die Schichtdickenabhängigkeit der Koerzitivfeldstärke definiert verformter Ni-Schichten geprüft werden kann.

R. GONTARZ, H. RATAJCZAK, Ferromagnetic Laboratory, Institute of Physics, Polish Academy of Sciences, Poznan, and P. SUDA, Institute of Physics, Czechoslovak Academy of Sciences, Prague, Magnetostriction of Thin Ni-Films Magnetostriction of thin Ni-films was investigated by the multibeam interference method. The magnetostriction constant λ_s was calculated from the changes in the mica substrate's deflection as it yielded to the stresses in the films. Its value does not differ considerably from that of bulk nickel. The deflections of the mica substrate due to the stresses stemming from the evaporation process were apparent in the interference pattern as a system of hyperbolas, which can be explained as originating in a saddle-like deflection of the substrate.

N. VAN HUONG, Institute of Physics, University of Hanoi and K. OLBRYCHSKI, Institute of Physics, Polish Academy of Sciences, Warsaw, Symmetrized Plane Waves for the Groups $D_{4h}^{17} - D_{4h}^{20}$ at Four Symmetry Points of the Brillouin Zone A method for the construction of symmetrized plane waves (SPWs) for space groups is described and tables of the SPWs for the groups $D_{4h}^{17} - D_{4h}^{20}$ at four points of high symmetry are given.

J. DZIESIATY, IV. Physikalisches Institut der Humboldt-Universität zu Berlin, Photoleitfähigkeitsuntersuchungen an CdS- und CdSe-Einkristallen mit einer kontaktfreien Mikrowellenmethode Unter Voraussetzungen, die für hochohmige Sulfidphosphore in einem weiten Leitfähigkeitsbereich gut erfüllt sind, hängen Absorptionskonstante und Reflexionsvermögen im Mikrowellenbereich linear von der Gleichstromleitfähigkeit ab, so daß Leitfähigkeitsänderungen kontaktfrei durch Mikrowellenmessungen ermittelt werden können. Es wurden Messungen der Photoleitfähigkeits-Beleuchtungscharakteristik, des Anregungsspektrums der Photoleitfähigkeit und der Relaxationszeiten an CdS- und CdSe-Einkristallen bei 9000 MHz im Temperaturbereich von -196 °C bis +150 °C durchgeführt. Dabei zeigte sich eine sehr gute Übereinstimmung mit Ergebnissen, die an den gleichen Proben nach Aufbringen "Ohmscher" Kontakte mit Gleichspannung gewonnen wurden.

J. DIEHL and F. HINZNER, Max-Planck-Institut für Metallforschung, Stuttgart, Slip Line Pattern and Work-Hardening of Neutron Irradiated Copper Single Crystals The development of the slip line pattern during deformation of neutron irradiated copper single crystals was investigated by electron microscopy at an intermediate neutron dose (5×10^{17} n/cm^2), at which according to the stress-strain curves the work-hardening behaviour is rather different from that at low or at high neutron doses. Using a selected area replica technique, specific areas of the crystal surface were observed up to seven times after successive deformation steps. The quite numerous stages and substages of the stress-strain curve are accompanied by differences in the arrangement of the slip traces as well as by changes of the characteristic properties of individual slip traces. Based on these results a tentative interpretation of work-hardening in the investigated dose range can be given.

R. v. JAN, Max-Planck-Institut für Metallforschung, Stuttgart, und Institut für theoretische und angewandte Physik der Technischen Hochschule Stuttgart, Defektverteilung in Verlagerungskaskaden (I) Die vorliegende Arbeit untersucht die räumliche Verteilung der in einer Verlagerungskaskade erzeugten Leerstellen und Zwischengitteratome. Ein von HOLMES und LEIBFRIED (1) angegebenes Verfahren zur Berechnung der Reichweite von Atomen läßt sich in einfacher Weise zu einer Kaskadentheorie erweitern. Durch zusätzliche Einführung eines Verlagerungsmodells gelingt es dann, Mittelwerte getrennt über die Verteilungen von Leerstellen und Zwischengitteratomen zu berechnen. Es zeigt sich, daß die Leerstellen auf ein kleineres Volumen verteilt sind als die Zwischengitteratome; diese

"Trennung der Defektarten" wird um so stärker, je härter das Wechselwirkungspotential der Atome ist, und kann (auch ohne Mitwirkung von fokussierenden Stoßketten) zur Ausbildung einer "Leerstellenzone" führen.

(1) D.K. HOLMES and G. LEIBFRIED, J. appl. Phys. 31, 1046 (1960).

M. WILKENS, Max-Planck-Institut für Metallforschung, Stuttgart, Streuung von Blochwellen schneller Elektronen in Kristallen mit Gitterbaufehlern Die Lösung der Schrödingergleichung für die Beugung schneller Elektronen in gestörten Gittern wird für den sogenannten N-Strahlfall durch N Blochfunktionen eines ungestörten Gitters beschrieben, deren Anregungsstärken ortsabhängig sind. Es werden Differentialgleichungen zwischen den Anregungsstärken aufgestellt, die eine von den Gitterstörungen induzierte Streuung zwischen den Blochwellen beschreiben. Diese Differentialgleichungen spezialisieren sich im symmetrischen Lauefall, N = 2 und "Säulennäherung" auf die in früheren Arbeiten (1, 2) auf anderem Wege abgeleiteten Differentialgleichungen. In Anwendung auf Fragen des elektronenmikroskopischen Kontrastes von Gitterfehlern wird ein Näherungsverfahren zur Lösung der Differentialgleichungen angegeben, bei dem (im Falle absorptionsfreier Kristalle) die Erhaltung der Teilchenzahl bis einschließlich quadratisch kleiner Glieder erfüllt ist. Vergleiche solcher Näherungslösungen mit numerisch integrierten Lösungen zeigen die Brauchbarkeit des Verfahrens. Die Verwandtschaft der vorliegenden Theorie der Beugung in gestörten Gittern mit einer kürzlich von KATO (3) veröffentlichten Theorie wird diskutiert.

(1) M. WILKENS, phys. stat. sol. 5, 175 (1964).
(2) A. HOWIE, Proc. Roy. Soc. A 271, 268 (1963).
(3) N. KATO, J. Phys. Soc. Japan 18, 1785 (1963) und 19, 67 (1964).

J. VAN LANDUYT, Solid State Physics Department, S.C.K. - C.E.N., Mol, An Electron Microscopic Investigation of Phenomena Associated with Solid Solution of Oxygen in Niobium A transformation process in the solid solution of oxygen in niobium is studied by means of electron microscopy and diffraction. Several stages of the process have been found by treating the thin foils in the electron microscope and in a furnace under reduced pressure of air. The first step consists of a nucleation phenomenon. Nuclei are formed which are composed of small parallel domains of an ordered phase. The projection of the domain boundaries on a {100} plane have the two perpendicular <110> directions. Lateral growth of the nuclei occurs until the whole region is covered with a cross-hatched pattern. Selected area diffraction shows the presence of many supplementary spots in a regular array between the niobium spots. The phenomenon is explained by ordering of the oxygen interstitials on a superlattice of the matrix, and shows striking similarity with the domains observed in niobium foils that were treated as bulk material. Larger lenticular-shaped domains have also been observed. These show internal boundaries of the twin type, which lie in {110} planes. A further stage is associated with the appearance of larger platelets, which are also oriented in crystallographic directions. Selected area diffraction over these area shows no significant difference from the previous stage. Irregular-shaped nuclei were observed in the same area, having a superstructure of the niobium lattice. The lattice spacing corresponds to twice the niobium spacing. During this study one of the suboxide phases, NbO_z , was observed as platelets growing in <110> directions on a {111} foil. This sub-oxide phase was epitaxially related to the niobium lattice. It is supposed that all the observed phenomena represent an intermediate stage in the transformation to the suboxide and oxide phases of niobium

H. PUFF, Physikalisch-Technisches Institut der Deutschen Akademie der Wissenschaften zu Berlin, <u>Photoströme in Halbleitern mit Barriere</u> Ausgehend von der sogenannten Ein-Gruppen-Näherung wird eine Theorie der von JUNGK und MENNIGER (1) angegebenen Methode für die Bestimmung der Diffusionslänge von Minoritätsträgern aus der Ortsabhängigkeit des p-n-Photoeffekts keilförmiger Proben entwickelt. Durch Berücksichtigung der Volumenanregung werden sowohl die Grenzen dieser Methode in Abhängigkeit von Schichtdicke und Oberflächenrekombinationsgeschwindigkeit bestimmt als auch das Gebiet, in dem sie zur Messung des Absorptionskoeffizienten führt.

(1) G. JUNGK und H. MENNIGER, phys. stat. sol. 5, 169 (1964).

Original Papers published in "Fizika Tverdogo Tela"
Vol. 6, No. 6

B.V. BUDYLIN and YU.V. KOZLOV, Bleaching of Neutron-Irradiated Alkali Halide Crystals During Isothermal Annealing Heat treatment of alkali iodide and bromide crystals, which were irradiated with neutron does above 10^{18} cm^{-2}, leads to an anomalous destruction of colour centres in regions near the surface and the formation and growth of macroscopic pores which give rise to intense scattering of visible light. It is found that the bleaching is most serious in the vicinity of the pores.

E.V. ZAROCHENTSEV and V.A. POPOV, Kharkov and Rostov-on-Don, Ground States of Biaxial Antiferromagnetics A theoretical study is made of the ground states of a biaxial antiferromagnetic dielectric, over a wide range of magnetic field strength.

V.B. FIKS, Leningrad, Drag of Ions by Electrons in a Crystalline Lattice A treatment is given of the drag of impurity ions by conduction electrons, for an arbitrary E(k) dependence. It is shown that, by a current of "holes" (in a metal), the ions are dragged in a direction opposite to that of electrons drag. Thus the direction of drag in a given metal depends upon the shape of the Fermi surface.

V.A. TYAGAI, Moscow, Nature of the Pulse Photoeffect at a CdS-Electrolyte Interface Measurements are made of the photo e.m.f. and the capacity of a CdS-electrolyte interface under pulse illumination. It is shown that the photoelectric effect is due to the separation of electron-hole pairs in the depletion layer. The lifetime of holes in low-resistivity CdS crystals is found to be 10^{-10} to 10^{-9} s. The measured capacity of the depletion layer coincides with the capacity deduced from photoelectric measurements. It is shown that for high-resistivity samples the photo e.m.f. is independent of resistivity.

G.M. GENKIN, Gorki, High-Frequency Conductivity due to Polarons A calculation is made of the high-frequency conductivity due to small-radius polarons in a semiconductor, at high temperatures at which conduction is governed by thermal jumps between lattice sites.

F.I. KREINGOLD and B.V. NOVIKOV, Leningrad, Origin of Changes in the Spectral Distribution of Photoconductivity of CdS Crystals Near the Absorption Edge A correlation is found between the changes in the photoconductivity spectrum which occur during cooling from 77 to 4.2 ^{o}K, and the changes which occur when the crystals are excited by modulated instead of unmodulated light. The influence of trapping processes and radiative recombination on these changes is studied by measurements of thermally stimulated conductivity and luminescent spectra, respectively.

L.L. BUISHVILI, Tbilisi, Acoustic Nuclear Magnetic Resonance in Antiferromagnetics A theoretical study is made of the ultrasonic excitation of waves due to moving domain walls in antiferromágnetics. The ultrasonic absorption coefficient is calculated for the case of acoustically induced nuclear magnetic resonance.

M.YA. GEN, E.A. VELICHENKOVA, I.V. EREMINA, and M.S. ZISKIN, Moscow, Preparation Conditions and Properties of Ag-Cu Alloys in a Highly Dispersive State Aerosol particles of Ag-Cu are prepared by simultaneous evaporation of Ag and Cu in a H or Ar atmosphere. It is found that in contrast to massive samples the lattice parameters are concentration-dependent also in the double-phase region. The changes of the lattice parameters are much larger than in massive samples and have their maximum values at 33 atomic percent Cu and 25 atomic percent Ag in particles of diameter 1.3×10^{-5} cm, and at 37.5 atomic percent Cu and 30 atomic percent Ag in particles of diameter 5×10^{-6} cm, and this indicates an extension of the miscibility range. The formation of highly dispersion alloys is discussed in general. Methods are proposed for the determination of their composition.

A.I. RAICHENKO, Kiev, Harmonic Vibrations of Atoms in an Imperfect Chain By the method of degenerate perturbations (1) expressions are derived for the frequency shift which occurs in the quasi-continuous vibrational spectrum of an atomic chain due to the presence of an impurity atom possessing different mass and coupling parameter. It is found that the twofold degeneracy is removed, and this involves shifts of both branches.

(1) I.M. LIFSHITS, Zh. eksper. teor. Fiz. 17, 1076 (1957).

YU.P. IRKHIN and SH.SH. ABELSKII, Sverdlovsk, Scattering by Inhomogeneous Spin-Distributions and Spontaneous Hall-Effect in Ferromagnetics A treatment is given of the spontaneous Hall-effect in iron-group ferromagnetics which is due to the scattering of conduction electrons by inhomogeneous spin-distributions ("magnetic scattering"), taking into account the spin-orbital interaction of the d-electrons, and the interaction between the orbital moments of the conduction electrons and the spins of the d-electrons. Making allowance for elastic and inelastic collisions the kinetic equation is derived in the Born approximation. It is found that the temperature dependence of the Hall coefficient (R_s) is determined by the temperature dependence of the "magnetic" contribution to resistivity. For $S = \frac{1}{2}$ this leads to $R_s(T) \sim M_s^2(0) - M_s^2(T)$ (M_s spontaneous magnetisation), as proposed in (1) on the basis of experimental data. The present results enable one to separate experimentally the contributions of phonons and spins to the resistivity of ferromagnetics.

(1) A.P. KOMAR and N.V. VOLKENSHTEIN, Dokl. Akad. Nauk SSSR 60, 785 (1948).

M.M. ZARIPOV and G.K. CHIRKIN, Kazan, EPR Spectrum of Cu^{2+} in NH_4Cl The EPR spectrum of Cu^{2+} in NH_4Cl is measured at 77 °K and about 4×10^{10} Hz. It is shown that the introduction of Cu leads to the formation of three magnetically non-equivalent centres for which the z-axis of the crystals field is parallel to an edge of the elementary cube. The compensation of the excess charge of Cu^{2+} is discussed in detail. Interpretations are given of the magnitudes of the g-factors and the hyperfine splitting constants.

E.V. ANDREEVA, N.V. KARLOV, A.A. MANENKOV, V.A. MILYAEV, and A.V. SHIRKOV, Moscow, Electron Paramagnetic Resonance of

Chromium Ions in Cadmium Tungstenate Measurements are made of the EPR spectrum of Cr^{3+} in $CdWO_4$ crystals grown by the Czochralski technique. The following values are obtained for the coefficients of the spin Hamiltonian:
$D = (42.9 \pm 0.05)$ GHz, $E = (2.35 \pm 0.02)$ GHz, $g_x = 1.97 \pm 0.01$, $g_y = 1.97 \pm 0.01$, $g_z = 1.98 \pm 0.01$. The spin-lattice relaxation time is found to be 3.6×10^{-4} s and 3.0×10^{-3} s at 4.2 and 1.6 °K, respectively. The results are compared with the EPR spectra of Cr^{3+} and Fe^{3+} in magnesium and zinc tungstenates.

N.A. VITOVSKII and T.V. MASHOVETS, Leningrad, Exact Determination of the Activation Energies of Impurities and Lattice Defects in Semiconductors A method is proposed for the determination of activation energies (ΔE) from non-linear parts of log (Hall coefficient) vs 1/T curves. It is shown that significant errors may occur if ΔE is, as usual, deduced from the slipe of these curves.

A.G. DEVYATKOV, SH.M. KOGAN, T.M. LIFSHITS, and A.YA. OLEINIKOV, Moscow, Conductivity of n-Type Indium Antimonide at Low Temperatures The behaviour of hot electrons in n-type InSb is studied by measurements of current-voltage characteristics. A method is used which enables the dependences of the conductivity (σ) and the coefficient $\beta = \frac{1}{\sigma} \cdot \frac{d\sigma}{d(E)}$ on lattice temperature and field strength (E) to be determined in a simple manner. The results indicate that the momenta of the electrons are mainly changed by collisions with charged impurities, whereas changes in their energy are mainly due to nonpolar and polar scattering by acoustic phonons.

B.I. SMIRNOV and YU.I. PATRIKEEV, Leningrad, Influence of the Deformation Conditions on the Yield Point and the Dislocation Density of LiF Crystals It is found that the dislocation density in a deformed LiF crystal is proportional to the applied stress, for all investigated velocities and temperatures of deformation. The limit of the linear part of the stress-strain curve (σ_1) and the yield point (σ_2) depend upon the deformation velocity ($\dot{\varepsilon}$) as $\dot{\varepsilon}^{r_1}$ and $\dot{\varepsilon}^{r_2}$, respectively. The results are related to the motion of dislocations.

V.P. SOLDATOV and V.I. STARTSEV, Kharkov, Elastic Twinning in Bismuth Crystals It is found that elastic-twin boundaries can be produced in Bi crystals by deformation at liquid nitrogen temperature. A discussion is given of the influence of slip on the twinning process.

V.G. SKOBOV, Leningrad, Quantum Theory of the Propagation of Electromagnetic Waves in Metals The high-frequency conductivity of a metal in a strong magnetic field is calculated for dominant scattering by neutral impurities, neglecting the effect of spatial dispersion. It is shown that the real part of the transverse conductivity (which causes an attenuation of magneto-hydrodynamic waves in metals containing equal numbers of electrons and "holes" (1)) is an oscillatory function of the magnetic field strength (Shubnikov - de Haas effect). Expressions are obtained for the height and shape of the maxima of this function. A study is also made of the oscillatory behaviour of the d.c. conductivity, taking into account the influence of scattering on the density of states.

(1) E.A. KANER and V.G. SKOBOV, Zh. eksper. teor. Fiz. 45, 610 (1963) and 46, No. 5 (1964).

V.M. GLAZOV and S.N. CHIZHEVSKAYA, Moscow, Susceptibility of Germanium, Silicon, and Compounds Possessing the Zincblende Structure near the Melting Point and in the Liquid State The magnetic susceptibility (χ) of (1) Ge, Si, InSb, GaAs, InAs, and (2) ZnTe, CaTe, Ga_2Te_3, In_2Te_3 is measured near the melting point and in the liquid state, by means of Faraday's method. It is found that all investigated substances are diamagnetic both in the solid and liquid state. During melting, χ decreases by 20 to 30 % and 9 to 10 %, for the substances (1) and (2), respectively. The results indicate that the substances (1) make a transition to a quasi-metallic state, whereas in the substances (2) the homopolar bonds are conserved.

I.A. ABROYAN and L.A. TSEKHNOVICHER, Leningrad, Additional Conductivity in Germanium due to Bombardment with Hydrogen Ions Measurements are made of the conductivity changes in Ge which are produced by bombardment with ions of atomic and molecular hydrogen possessing initial energies up to 7 keV. These changes are attributed to an excitation of bound electrons in Ge. The results are compared with electron bombardment data. It is found that the mean number of electrons excited by a hydrogen atom is by a factor 3 larger than the mean number excited by an electron which has the same initial energy.

E.E. VIOLIN and G.F. KHOLUYANOV, Leningrad, Electro- and Photoluminescence at Diffused p-n Junctions in Silicon Carbide Measurements are made of the spectral distributions of the electroluminescence (el) and photoluminescence (pl) at various temperatures in 6H and 4H SiC samples containing a p-n junction. These junctions are prepared by diffusion of B into n-type material. It is shown that the introduction of B leads to the appearance of new peaks in the el and pl.

S.A. SEREBROV and S.A. FRIDRIKHOV, Field-Induced Increase of the Secondary Electron Emission from Thin Sodium Chloride Layers The field-induced increase of the secondary electron emission from NaCl layers with thicknesses varying between 0.5 and 5 μm is measured at primary energies 7 to 10 keV and field strengths (in the layer) up to 10^6 V/cm. For porous layers estimates are made of the energy of the secondary electrons and their mean free path between two collisions with pore walls.

M.A. KRIVOGLAZ, Kiev, Theory of the Broadening of Zero-Phonon Lines in Mössbauer and Optical Spectra A study is made of the broadening of zero-phonon lines in Mössbauer and optical spectra due to lattice vibrations. Allowance is made for anharmonicity, and this leads to a decrease in the broadening due to local vibrations as obtained in (1) on the basis of the harmonic approximation. Expressions are given for the shape of the zero-phonon lines and the temperature dependence of their width. It is shown that, if anharmonicity is taken into account, the broadening may be connected with terms which are linear in the phonon operators.

(1) H.S. SNYDER and G.C. WICK, Phys. Rev. 120, 128 (1960).

I.P. AKIMCHENKO, V.S. VAVILOV, and A.F. PLOTNIKOV, Moscow, Properties of Radiation Defects Deduced from Photoconductivity Spectra of Electron-Irradiated Germanium The spectral distribution of photoconductivity in n- and p-type Ge irradiated with 1 MeV electrons is measured between 1.8 and 10 μm. It is shown that the level E_v + 0.42 eV is due to isolated interstitial atoms. Levels at E_v + 0.33 eV appear when Cu atoms are introduced into non-irradiated p-type samples (with residual impurity concentrations between 10^{11} and 10^{13} cm^{-3}). It is shown that levels of the same position are introduced by irradiation, due to the formation of Cu-vacancy complexes. The capture cross section of this centre for holes is found to be 5×10^{-19} cm^2 at 100 °K.

R.I. SHEKHMAMETEV and B.V. NOVIKOV, Leningrad, Excitation Spectra of Photoconductivity and Edge Emission in CdS Crystals The excitation spectra of the photoconductivity and edge emission in CdS crystals are measured at 77 °K, for modulated and unmodulated illumination. In the presence of modulation, a correlation is obtained between the two types of excitation spectra, and this is explained by an exciton mechanism.

I.A. BAEV and E.G. VALYASHKOV, Moscow, Flying-Spot Study of the Homogeneity of Semiconductor Crystals A method is developed which enables structural inhomogeneities in semiconductors to be detected by scanning the samples with a flying light-spot and visualising the spatial distribution of the resultant photo-e.m.f. on an oscilloscope. It is shown that, by means of this method, one can distinguish between different types of inhomogeneities (inhomogeneous impurity-concentrations, p-n junctions, grain boundaries), and, in the case of inhomogeneous impurity-concentrations, deduce the relative magnitude and direction of the concentration gradient. This method is applied to n-type Ge crystals grown by the Czochralski technique. The results are compared with the spatial distribution of conductivity which is measured directly after cutting the crystal into thin slices. It is found that the peaks in the oscillographs coincide with steps in the impurity distribution.

I.M. LIFSHITS and V.B. SHIKIN, Kharkov, Diffusion-Induced Viscous Flow in Porous Solids A discussion is given of the diffusion-induced viscous flow which occurs in porous polycrystalline solids under pressure. In the case of isotropic compression, simple formulae are obtained for the time dependence of density and viscosity after the application of pressure.

H. NEUMANN, E. KLOSE, and I.L. SOKOLSKAYA, Leningrad, Field-Emission Study of the Diffusion of Rhenium A study is made of the change of the shape of a monocrystalline Re point in field-emission microscope at elevated temperatures. The thermal activation energy of this change is found to be (1.5 ± 0.15) eV in the presence of an electric field, whereas two activation energies are obtained in the absence of an electric field: (1.5 ± 0.15) eV and (5.3 ± 0.5) eV. The latter value is attributed to volume diffusion and found to agree with theoretical results.

K.K. SVIDZINSKII, Moscow, Orientation of Centres and Polarisation of Luminescence in Crystals The angular distribution of Stokes' parameters is calculated for radiative transitions at an impurity centre. A general discussion is given of the effects of crystal-field symmetry of time-reversal symmetry on this distribution. The cases of dipole transitions in centres possessing trigonal and tetragonal symmetry are considered in detail. The results are applied to the two components of the R_1-line of ruby.

A.N. DREMIN and G.A. ADADUROV, Moscow, Behaviour of Glass under Dynamic Stress The mass velocities at the front of shock waves in glass are measured in a special arrangement at various applied pressures. The results are used to construct p-V diagrams for uniaxial compression.

E.S. ITSKEVICH, E. YA. ATABAEVA, and S.V. POPOVA, Moscow, Influence of Pressure on the Resistivity of Bismuth Selenide The resistivity (R) of n-type Bi_2Se_3 is measured at pressures up to 10.4 kBar and between 30 and 150 kBar. A steep decrease of R is found at about 100 kBar, and this is attributed to a transition to the metallic state.

S.A. DEMBROVSKII and A.A. VAIPOLIN, Leningrad, Properties of A_2Se_3 Crystals It is shown that the crystallisation rate of vitreous As_2Se_3 is inversely proportional to the viscosity. Single crystals prepared from the gas phase have the symmetry $P2_1/n$, with the lattice parameters a = (12.053 ± 0.001) Å, b = (9.890 ± 0.001) Å, c = (4.277 ± 0.001) Å, β = 90°28' ± 2', and are isomorphic to As_2Se_3. A comparison is made of the properties of As_2Se_3 in the vitreous and crystalline states.

YU.V. VOROBEV, Kiev, Non-Linear Photoconductivity in Cuprous Oxide The photoconductivity in Cu_2O is measured between 0.55 and 0.95 μm. The observed non-linear photocurrent vs light intensity curves are explained quantitatively by assuming that three types of recombination centres are present. It is found that the positions of the levels produced by two of these centres are equal, their capture cross-sections being 3×10^{-15} and 3×10^{-16} cm^2, respectively.

D.N. NASLEDOV, G.A. KALYUZHNAYA, and S.V. SLOBODCHIKOV, Leningrad, Electric and Photoelectric Properties of n-Type GaP The Hall coefficient, conductivity, and photoconductivity (σ_{ph}) in partially compensated n-type GaP are measured between 80 and 295 °K. Special investigations are made of the influence of additional unmodulated white light on σ_{ph}. The presence of levels 0.9 and 1.5 eV below the conduction band is deduced from the spectral distribution of σ_{ph}. Scattering by polar lattice-vibrations is found to be dominant.

I.L. SOKOLSKAYA and N.V. MILESHKINA, Leningrad, Field Emission and Migration of Germanium on Tungsten Surfaces The migration of Ge on a W surface is studied in a field-emission microscope. Two types of migration are observed:

1) At low temperatures (above room temperature) a migration of adsorbed atoms occurs, with an activation energy of 0.2 eV, and this leads to the formation of a monatomic Ge layer which has semiconducting properties.
2) Above 800 °K, a migration of chemi-sorbed atoms occurs, with activation energies of 1.24 and 1.7 eV. This process leads to the formation of incoherent layers possessing

poor emissional properties.

P.S. NESTERENKO, Rostov-on-Don, Formation Kinetics of the Photoelectret State in Cadmium Sulphide Crystals A study is made of the transition of CdS crystals to the photoelectret state due to near-infrared excitation, at room temperature.

E.I. GIVARGIZOV, Moscow, Crystallisation of Germanium and Silicon from Supercooled Drops, and the Formation of Defects in Epitaxial Layers Crystalline Ge and Si layers are grown by reduction of $GeCl_4$ and $SiCl_4$, respectively, in a H_2 stream. It is found that under different conditions of crystallisation, two different growth-mechanisms are operative ("high-temperature" and "low-temperature" mechanism), and this is related to earlier results on the electrical properties of epitaxial Ge and Si layers. It is shown that stacking faults are only formed by the "high-temperature" mechanism, and in the presence of an oxide layer on the substrate. This is explained on the basis of an interaction between supercooled Ge or Si drops with oxide "islands".

E.D. DEVYATKOVA, A.V. GOLUBKOV, E.K. KUDINOV, and I.A. SMIRNOV, Leningrad, Influence of the Spin-Phonon Interaction on the Thermal Conductivity of MnTe The thermal conductivity, thermal e.m.f., and resistivity of MnTe are measured in the vicinity of the Néel point (Θ_N). All these quantities show steep changes at Θ_N. The results are interpreted in terms of spin-phonon interaction.

E.V. BURSIAN and N.P. SMIRNOVA, Leningrad, Non-Linear Capacity of Thin Monocrystalline $BaTiO_3$ Layers Measurements are made of the permittivity (ε) of monocrystalline $BaTiO_3$ layers, extending earlier investigations (1) to smaller thicknesses. It is found that non-linear polarisation occurs at thicknesses (d) between 1 and 100 μm, the activation field for this polarisation being proportional to d^{-1}. At high field strengths the temperature dependence of ε agrees with that obtained in (1) for thicker crystals. The results confirm that the occurrence of non-linear polarisation is associated with surface layers.

(1) E.V. BURSIAN and N.P. SMIRNOVA, Fiz. tverd. Tela 4, 1675

(1962).

N.E. AINBINDER and V.S. GRECHISHKIN, Perm, Orientation of Single Crystals during Quadrupole Resonance A method is proposed which enables one to determine the directions of the principal axes of the crystal field tensor in the case of spin 5/2 assuming that the quadrupole interaction constant and asymmetry parameter are known from zero-field measurements. This method involves a comparison of coefficients in the secular equation with experimental resonance-frequencies.

N.I. KURDIANI, L.G. KHAVTASI, and N.V. BARAMIDZE, Tbilisi, Effect of Doping on the Reflection Spectrum of Indium Antimonide in the Fundamental Region The reflection coefficient of doped InSb samples is measured in the visible and near-infrared ranges. It is found that the introduction of impurities leads to a shift of the reflection peaks towards layer wavelengths, and this is explained by the presence of a "tail" of the density of states extending into the forbidden gap.

N.A. TSAL, M.V. PASHKOVSKII, and I.D. NABITOVICH, Lvov, Influence of Anion Impurities on the Coagulation of F-Centres in Potassium Chloride Crystals Optical and electron-microscopic studies are made of the coagulation of F-centres in KCl crystals containing O, S, OH^-, CO_3^{2-}, SO_4^{2-} ions or both divalent-cation and anionic impurities. It is shown that this coagulation is stimulated by the introduction of anionic impurities, and also strongly affected by the presence of Cl vacancies.

E.I. CHEGLOKOV, Tomsk, Symmetry of Hartree-Fock Functions in Crystals A study is made of the symmetry of the Hartree-Fock equations of a crystal, including and neglecting spin-orbital interaction. It is shown that, if the assumption of the double occupation of states is dropped, one may obtain solutions whose symmetry is different from that of the solutions of the Schrödinger equation for a periodic potential. Assuming that the orbitals localised on covalent bonds are equivalent a detailed discussion is given of the symmetry of the pertaining Hartree-Fock functions.

E.I. CHEGLOKOV, Tomsk, Possible Structure of the Valence Bands of Hexagonal Selenium and Tellurium Crystals On the basis of (1) a calculation is made of the structure of the valence bands in hexagonal Se and Te. The results enable experimental absorption-data (2, 3) to be explained by direct transitions between states in the valence bands, without including additional bands.

(1) preceding paper.
(2) R.S. CALDWELL and N.Y. FAN, Phys. Rev. 94, 1427 (1954) and 114, 664 (1959).
(3) H. GOBRECHT and A. TAUSEND, Z. Phys. 161, 205 (1961).

R.L. NEMCHENOK and K.E. PROKOFEVA, Leningrad, Photo- and Thermal Emission from the System Cu-Ba Investigations are made of the spectral distribution of the photo-emission, and the thermal emission at high temperatures, from Cu-Ba layers evaporated on a tungsten substrate. Special attention is given to the influence of evaporation conditions. It is found that the addition of Cu leads to a deterioration of the emissional properties.

S.G. RAUTIAN, Relaxational Terms in the Equation of the Density Matrix The effect of saturation in a monochromatic electromagnetic field is treated by introducing three relaxational terms into the equation of the density matrix.

M.D. MASHKOVICH, Moscow, Dielectric Properties of "Sital" Ceramics The influence of crystallisation temperature on the frequency dependence of ε and tg δ of "sital" ceramics is studied between 4×10^2 and 3.7×10^{10} Hz.

I.P. DZYUB, Kiev, Inelastic Incoherent Scattering of Slow Neutrons in Disordered Solid Solutions The incoherent single-phonon scattering of slow neutrons in disordered binary alloys is studied by the method of Green's functions. The differential scattering cross-section is obtained as a series in the concentration of one component. The contribution of local and "quasi-local" vibrations to the energy spectrum of scattered neutrons is calculated for the case of small concentrations, assuming that the two kinds of atoms differ only in mass. The initial stage of the formation of an impurity band in the vibrational spectrum is discussed in detail.

N.A. VITOVSKII, T.V. MASHOVETS, and S.M. RYVKIN, Leningrad, High-Temperature Annealing of Defect Produced by γ-Irradiation in n-Type Germanium It is shown that the annealing of γ-defects in Ge is governed by the recombination of donor- and acceptor-type defects ("bipolar" annealing). In addition, "monopolar" annealing may occur under various conditions. It is found that the decrease in the concentration of the defects is accompanied by a change in their energy spectrum.

Short Notes published in "Fizika Tverdogo Tela" Vol. 6, No. 6

L.N. OVANDER
Elementary Theory of the Emission of Raman Frequencies

I.A. GUSEV, A.N. MURIN, and P.P. SEREGIN
Diffusion of Cadmium in Indium Antimonide

V.L. VINETSKII, Kiev
Singlet and Triplet States and Paramagnetic Resonance of F_2-Centres

I.P. VOROMINA and S.A. SEMILETOV, Moscow
Structure and Electric Properties of Monocrystalline PbTe Layers

A.E. YUNOVICH, P.G. ELISEEV, I.A. NAKHODNOVA, A.B. ORMONT, L.A. OSADCHAYA, and V.M. STUCHEBNIKOV, Moscow
Radiative Recombination in GaAs p-n Junctions Prepared by Diffusion of Beryllium

I.L. DRICHKO and I.V. MOCHAN, Leningrad
Thermal E.M.F. in n-Type Indium Antimonide in the Presence of High Magnetic Fields

E.N. LOTKOVA, Moscow
Near-Infrared Absorption Bands in Irradiated Silicon Crystals

A.M. MESHKOV and G.S. ZHDANOV, Leningrad
Structure of n- and p-Type Di- and Triphenylmethane Layers

A.E. YUNOVICH, P.G. ELISEEV, A.B. ORMONT, L.A. OSADCHAYA, and V.M. STUCHEBNIKOV, Moscow
Spectra of the Coherent Radiation from GaAs p-n Junctions

G.A. KUROV, Moscow
Amorphous Germanium

S.N. ZHURKOV, V.A. ZAKREVSKII, and E.E. TOMASHEVSKII, Leningrad
Formation of Free Radicals during the Destruction and Deformation of Polymers Containing Sulphide Bonds

V.F. LYSOV, Leningrad
Connection of the Internal Photoeffect in CdS with Contact Phenomena

List of Short Notes

PRIERE d'adresser toute correspondance au

physica status solidi, Berlin C 2, Neue Schönhauser Str. 20.

Ce Journal est destiné à la publication des résultats de la recherche fondamentale dans le domaine entier de la Physique de l'Etat Solide.

Il paraîtra le Ier de chaque mois.

Il comportera:

1. Des exposés d'ensemble de mise au point sur les questions d'actualité de la Physique du Corps solide proposés par la Rédaction du Journal.
2. Des travaux originaux, n'ayant pas été ni ne devant pas être publiés ailleurs, contenant des résultats nouveaux, acceptés par la Rédaction du Journal (reçus au plus tard le 8 de chaque mois et paraissant dans le fascicule 50 jours après).
3. Des courtes communications de résultats de recherche, paraissant dans la partie du journal imprimée par procédé d'off-set sous la responsabilité personelle de l'auteur (reçues au plus tard le 12 de chaque mois et paraissant dans le fascicule 16 jours après).
4. Prétirage des titres et résumés des travaux devant paraître dans ce Journal ainsi que de ceux du journal soviétique „Физика Твердого Тела" (Fizika tverdogo Tela) (le derniers, traduits en anglais).

Recommandations aux auteurs

Les manuscrits seront acceptés et publiés en français, allemand, anglais et russe.
Les auteurs sont priés:

1. de présenter les manuscrits, si possible deux exemplaires par publication (excepté pour les courtes communications), écrits à la machine en recto seulement, avec double espacement. Les manuscrits des travaux originaux ne doivent pas dépasser, si possible, 20 pages et ceux des courtes communications 3 pages.
2. d'indiquer le nom et les initiales des prénoms des auteurs, ainsi que le lieu d'origine du travail.
3. de joindre un court résumé du travail d'au maximum 10 lignes, si possible aussi dans une deuxième des langues mentionnées plus haut.
4. de dessiner les croquis et figures sur des feuilles séparées et d'envoyer les photographies (seulement pour les exposés d'ensemble et les travaux originaux) sur papier blanc.
5. de présenter sur une feuille séparée les notes explicatives qui seront à joindre aux figures. Dans le cas de souhaits particuliers relatifs à la grandeur des figures, de le noter expressément sur chacune d'elles.
6. de joindre sur une feuille séparée les références bibliographiques dans l'ordre cité, rédigées dans l'ordre usuel (v. ce fascicule).
7. d'observer les recommandations de l'organisation „I.U.P.A.P." en ce qui concerne l'emploi des unités, des symboles etc.

Deux corrections d'épreuves seront adressées aux auteurs des exposés d'ensemble et des travaux originaux.
Les corrections de la partie imprimée par procédé d'off-set seront effectuées par la Rédaction.

Il sera délivré gratuitement 25 tirés à part pour les exposés d'ensemble et 75 tirés à part pour les traveaux originaux ainsi que pour les courtes communications. En outre, nous rétribuons les auteurs des exposés d'ensemble.

Указания авторам появляются на русском языке в следующем номере.

32913

Einsendungen und Zuschriften

bitten wir zu richten an:

physica status solidi, Berlin C 2, Neue Schönhauser Str. 20.

Die Zeitschrift dient der Publikation von Erkenntnissen der Grundlagenforschung des gesamten Gebietes der Festkörperphysik.

Sie erscheint am 1. eines jeden Monats.

Es werden aufgenommen:

1. Zusammenfassende Berichte über aktuelle Teilgebiete der Festkörperphysik nach Vereinbarung mit der Schriftleitung.
2. Originalarbeiten, die an anderer Stelle noch nicht veröffentlicht bzw. zur Veröffentlichung eingereicht sind, soweit sie neue Erkenntnisse enthalten, nach Rezension. Annahmeschluß am 8. jedes Monats (50 Tage vor Auslieferung des Heftes).
3. Kurze Mitteilungen von Forschungsergebnissen, die im Offsetteil der Zeitschrift in ausschließlicher Verantwortung des Autors erscheinen. Annahmeschluß am 12. jedes Monats (16 Tage vor Auslieferung des Heftes).
4. Vorabveröffentlichung der Titel und Abstracts von Arbeiten, die in dieser oder in der sowjetischen Zeitschrift „Физика Твердого Тела" (Fizika Tverdogo Tela) erscheinen (letztere in englischer Übersetzung).

Redaktionelle Hinweise

Manuskripte werden in englischer, deutscher, französischer oder russischer Sprache angenommen.

Die Autoren werden gebeten:

1. Manuskripte einseitig in Maschinenschrift mit mindestens zweizeiligem Abstand und (abgesehen von kurzen Mitteilungen) möglichst in 2 Exemplaren einzureichen. Der Manuskriptumfang sollte bei Originalarbeiten 20 Seiten, bei kurzen Mitteilungen 3 Seiten nicht überschreiten.
2. Namen und Vornameninitialen des Verfassers sowie den Ursprungsort der Arbeit anzugeben.
3. Eine kurze Zusammenfassung der Arbeit von höchstens 10 Zeilen, wenn möglich auch in einer zweiten der obengenannten Sprachen, beizufügen.
4. Zeichnungen auf besonderen Blättern und Photographien (möglichst nur für zusammenfassende Berichte und Originalarbeiten) auf weißem Papier beizufügen.
5. Abbildungsunterschriften auf einem gesonderten Blatt beizufügen und, falls bezüglich der Abbildungsgrößen besondere Wünsche bestehen, diese ausdrücklich auf den Abbildungen zu vermerken.
6. Literaturangaben in der Reihenfolge der Zitate in üblicher Weise (siehe Heft) ebenfalls auf einem gesonderten Blatt beizufügen.
7. Bei der Benutzung von Einheiten, Symbolen und dergleichen die von der IUPAP herausgegebenen Empfehlungen zu beachten.

Die Autoren erhalten von zusammenfassenden Berichten und Originalarbeiten zwei Fahnenabzüge übersandt.

Die Korrekturen des Offsetteiles werden in der Redaktion gelesen.

Von zusammenfassenden Berichten werden 25 Sonderdrucke, von Originalarbeiten und kurzen Mitteilungen 75 Sonderdrucke je Arbeit kostenlos geliefert. Zusammenfassende Berichte werden honoriert.

In the next issue, the above information will be published in English.

Bei Fragen zur Produktsicherheit wenden Sie sich bitte an:
If you have any questions regarding product safety,
please contact:

Walter de Gruyter GmbH
Genthiner Straße 13
10785 Berlin
productsafety@degruyterbrill.com